DATA STRUCTURES

UDIT AGRAWAL
M.Tech.
Associate Professor, CS & IT Deptt.
RBMI Group of Institutions, Bareilly

S.K. KATARIA & SONS®
Publisher of Engineering Books
4885/109, Prakash Mahal, Dr. Subhash Bhargav Lane,
Opposite Delhi Medical Association, Daryaganj, New Delhi–110002 (INDIA)
Phone: +91-11-43551243; Mobile: +91-9871775858, 7678425209
e-mail: katariabook@gmail.com, katariabooks@yahoo.com
Website: www.skkatariaandsons.com
(e-books, subscriptions and bulk buying also available on our website)

Published by:

S.K. Kataria & Sons®

4885/109, Prakash Mahal, Dr. Subhash Bhargav Lane,
Opposite Delhi Medical Association, Daryaganj, New Delhi-110002 (INDIA)
Phone: +91-11-43551243; Mobile: +91-9871775858, 7678425209
e-mail: katariabook@gmail.com, katariabooks@yahoo.com
Website: www.skkatariaandsons.com

First Edition: July, 2010; Third Edition: July, 2012; Reprint: 2026

Laser Typesetting by : Plus Computers, Delhi

Preface

The textbook of "DATA STRUCTURES" has been written according to the syllabus prescribed by PUNJAB TECHNICAL UNIVERSITY for B.Tech and MCA students.

Data structure is a centre module in the curriculum of almost every computer science program. This subject makes the students learn the art of analyzing algorithms as well as distinguishing between the specification of a data structure and it understands within an available programming language. The goal of this book is to help you understand different concepts of data structures. I am also attempt to present topics that discuss the theoretical basis of data structures as well as its applied aspects. The treatment of the text is methodical and systematic with a perfect balance in the presentation of theoretical and practical aspects. Each topic has been thoroughly covered in scope, content and also from the examination point of view. For each topic, several worked out examples, carefully selected to cover all aspects of the topic, are presented. An attempt has been made by me to write this book in a simple and lucid language with neat and self-explanatory diagrams, which could be easily understood by an average student.

The brief content of the book is as follows:

Chapter 1 gives the Dynamic Memory Management

Chapter 2 covers the Introduction to data structures

Chapter 3 deals with the introduction to Arrays and Matrices

Chapter 4 covers the concept of Linked list

Chapter 5-6 introduce the concept of Stack and Queues

Chapter 7 introduces recursion

Chapter 8 is devoted to the Trees

Chapter 9 introduces Binary search Tree

Chapter 10 covers AVL tree

Chapter 11 introduces B tree

Chapter 12 introduces Graphs

Chapter 13 covers different Sorting techniques

Chapter 14 introduces searching and hashing technique

The book is quite original in style and method and no pains have been spared to make it as compact, perfect and reliable as possible. Every attempt has been made to make the book a unique one. Some of the unique features of the book are as under:

- Each chapter contains an extensive list of exercises.
- Lucid, simple and conversational language.
- Systematic and sequential arrangement of different topics.
- **Exhaustive coverage :** Separate chapters for Trees, Binary Search trees, B-Trees and AVL Trees.
- Highly illustrative with over 250 diagrams.
- Emphasis on conceptual clarity.
- Concise and to-the-point description of all the topics.
- Eminently suitable for self-study.

In spite of our best efforts, some errors might have crept into the book. Report of any such error and all suggestions for improving the future editions of the book are welcome and will be gratefully acknowledged.

UDIT AGRAWAL

Acknolwedgements

The grace of GOD is above all. Then comes the blessings of the parents who always inspire me to take up a new task. I owe to my mother **Smt. Bina Devi Agarwal** and father **Dr. Nand Kishore Agarwal** for showing me the right path to fulfill the need of today. The desire to do something new always makes me restless and impatient. The present book is the result of that willingness of mine.

I am indebted to my elder sister **Mrs. Rachana Agarwal,** and younger brother **Er. Saurabh Agarwal,** for their benevolent support and proper guidance during the course of the preparation of this book.

I am grateful to my brother-in-laws **Mr. Anoop Agarwal** and **Mr. Himanshu Agarwal** for providing me valuable informations and suggestions to bring out this book in the present form.

I wish to express my deep sense of gratitude to **Mrs. Veena Mathur,** Executive Chairperson, RBMI Group of Institutions, **Dr. Navin P. Mathur,** Managing Director, RBMI Group. **Dr. Manish Sharma,** Director General, RBMI Group and **Dr. Neeraj Saxena,** Director, RBMI, Bareilly for encouraging and motivating me all the time to work efficiently and precisely.

I express my gratefulness to all my academic colleagues and friends for their valuable comments and suggestions.

The patience of my wife **Dr. Gunjan Agarwal** and the love of my two lovely kids **Unnati (Pari)** and **Parv** cannot be overlooked during the long and arduous task of producing such a book.

I would like to express my appreciation to the editorial and production staff of **S.K. Kataria & Sons** for the excellent production of the book.

Constructive criticism and meaningful suggestions are always welcome.

UDIT AGRAWAL

To

My Kids

Pari and Parv

Contents

PART—A

1. **Dynamic Memory Management:** Understanding pointers, usage of pointers, arithmetic on pointers, memory allocation, memory management functions and operators, debugging pointers - dangling pointers, memory leaks, etc
2. **Introduction:** Concept of data type, definition and brief description of various data structures, data structures versus data types, operations on data structures, algorithm complexity, Big O notation.
3. **Arrays:** Linear and multi-dimensional arrays and their representation, operations on arrays, sparse matrices and their storage.
4. **Linked List:** Linear linked list, operations on linear linked list, doubly linked list, operations on doubly linked list, application of linked lists.
5. **Stacks:** Sequential and linked representations, operations on stacks, application of stacks such as parenthesis checker, evaluation of postfix expressions, conversion from infix to postfix representation, implementing recursive functions.
6. **Queues:** Sequential representation of queue, linear queue, circular queue, operations on linear and circular queue, linked representation of a queue and operations on it, deque, priority queue, applications of queues.

Chapter 1

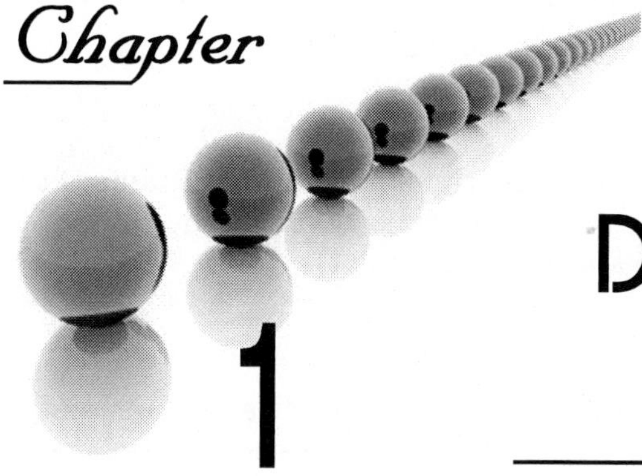

Dynamic Memory Management

1.1 INTRODUCTION

The memory of your computer can be imagined as a succession of memory cells, each one of the minimal size that computers manage (one byte). These single-byte memory cells are numbered in a consecutive way, so as, within any block of memory, every cell has the same number as the previous one plus one.

This way, each cell can be easily located in the memory because it has a unique address and all the memory cells follow a successive pattern. For example, if we are looking for cell 1776 we know that it is going to be right between cells 1775 and 1777, exactly one thousand cells after 776 and exactly one thousand cells before cell 2776.

Chapter Outline

- Introduction
- Reference Operator (&)
- Understanding Pointers
- Arithmetic on Pointers
- Arrays and Pointers
- Pointers and Structures
- Disadvantages of Pointers
- Memory Allocation
- Memory Allocation Process
- Library Functions for Dynamic Memory Allocation
- Dangling Pointers
- Memory Leak

1.2 REFERENCE OPERATOR (&)

As soon as we declare a variable, the amount of memory needed is assigned for it at a specific location in memory (its memory address). We generally do not actively decide the exact location of the variable within the panel of cells that we have imagined the memory to be—Fortunately, that is a task automatically performed by the operating system during runtime. However, in some cases we may be interested in knowing the address where our variable is being stored during runtime in order to operate with relative positions to it.

The address that locates a variable within memory is what we call a reference to that variable. This reference to a variable can be obtained by preceding the identifier of a variable with an ampersand sign (&), known as **reference operator**, and which can be literally translated as "**address of**". For example:

```
A = &ptr;
```

This would assign to A the address of variable ptr, since when preceding the name of the variable ptr with the reference operator (&) we are no longer talking about the content of the variable itself, but about its reference (i.e., its address in memory).

1.3 UNDERSTANDING POINTERS

A pointer works a little differently—it does not store a simple value directly. Instead, a pointer stores a **reference** to another value. The variable the pointer refers to is sometimes known as its "**pointee**".

Pointer variable

It is possible to store the address of a variable into another variable which is called a **pointer variable**.

Declaration

The declaration of a pointer variable takes the following form.

```
data-type * pointer-name
```

This tells the compiler three things:

1. The asterisk (*) tells that the variable is a pointer variable.
2. Pointer-name needs a memory location.
3. Pointer-name points to a variable of type data-type.

For example,

```
int * j ;
```

This declaration tells the compiler that j will be used to store the address of an integer value. In other words j points to an integer.

Remember that, addresses are always going to be whole number, therefore pointers always contain whole numbers.

The declaration float * x; does not mean that x is going to contain a floating-point value. What it means is, x is going to contain the address of a floating-point value.

Similarly, char * ch means that ch is going to contain the address of a char value.

A pointer is nothing but a variable that contains an address of a location in memory.

We can declare the variable ptr as a pointer to an integer as

```
int *ptr;
```

We can think of this as being a new type, which is **(int *)**, which means "pointer to an integer."

When it is declared, a pointer does not have a value, which means that it does not yet store an address in memory, and hence it does not point to an address in memory. Let's say we want to have the pointer that we defined point to the address of another integer. This is done with

```
int x;
int *ptr;
ptr = &x;
```

The unary operatory & returns the address of a particular variable. We can represent the above three lines of code graphically with

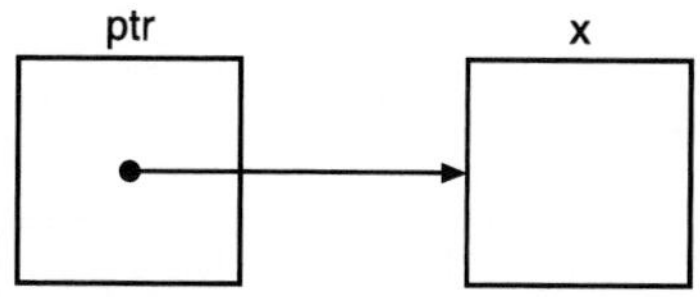

Since, we did not assign the value of x, it does not contain a value. We can assign the value of x in one of two ways. The first way is the most obvious way, which is simply

```
x = 5;
```

The pointer diagram now has a value at the location in which the variable is x is stored:

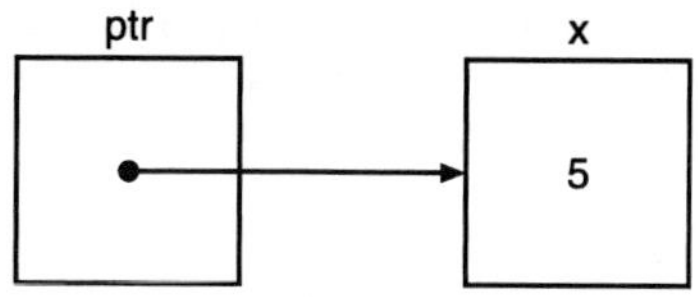

The other way we can assign the value of x is to use the pointer that points to its address. Since, ptr points to its address, we can assign a value to the address that this pointer points to with

```
*ptr = 5;
```

This is called **dereferencing the pointer** and it is identical to x = 5;. The dereferencing operator says to set the contents of location pointed to by the pointer to the specified value. Pointers themselves also take up memory, since they store a number that is an address to another location in memory. For the sake of argument, let's say that the pointer ptr is located at memory address 100 and that x is located at memory address 200, as depicted the following diagram, which results when you declare

```
int x, *ptr;
```

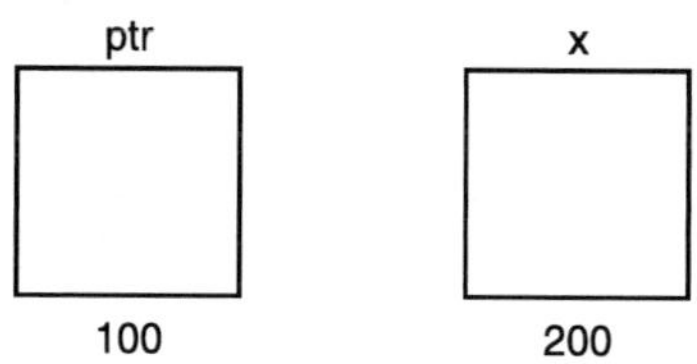

When you set the pointer to point to the address of x with

```
ptr=&x,
```

you are assigning a value of 200 to the pointer, so that the pointer diagram now looks like

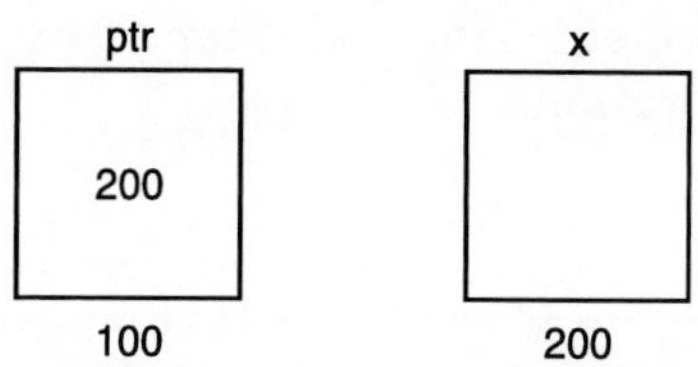

And when we set the value of x with either x = 5 or *ptr = 5, the pointer diagram looks like

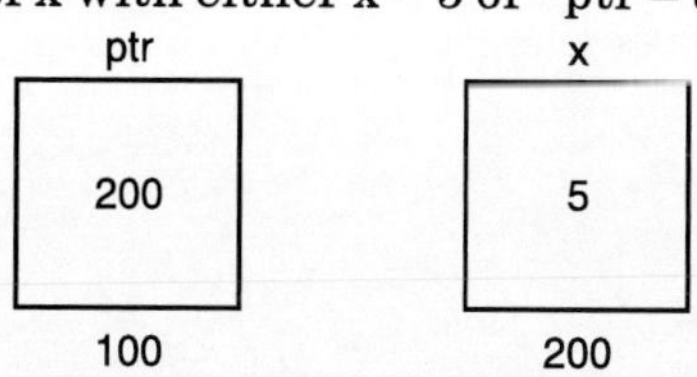

Since, the pointer simply stores the address of another variable, there is nothing to stop us from setting the value of the integer x to the pointer ptr with

```
x=ptr;
```

In this case the pointer diagram would look like

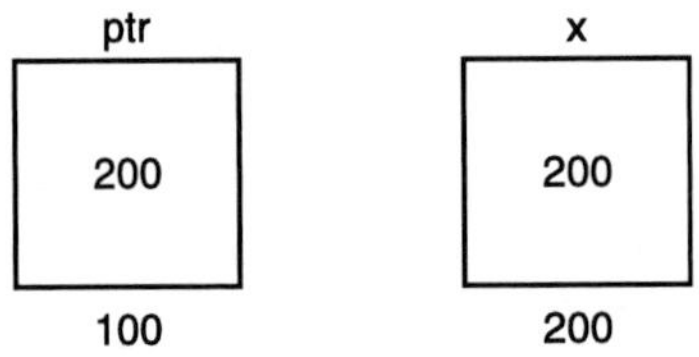

But it is very rare that you would ever want to do this. Since pointers in general only store addresses, it is much easier to understand pointers if we represent them in memory as blocks with a dot at their center, with an arrow that points to the particular variable that exists at the address of the pointer, as we did before, as in

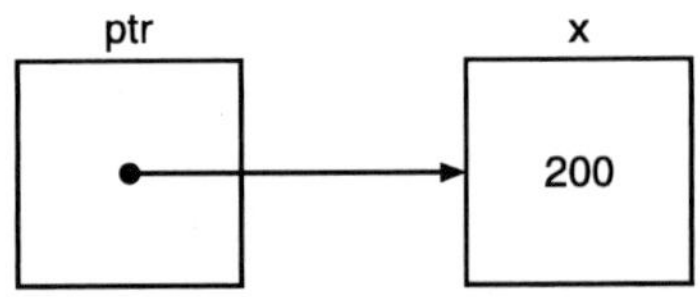

Since, x is not a pointer but it still contains the address of ptr, then its value is 200. Now consider the case when we declare more variables and pointers, such as

```
int x=10, y=5, *ptr1, *ptr2;
ptr1=&x;
ptr2=&y;
```

In this case the pointer diagram would look like

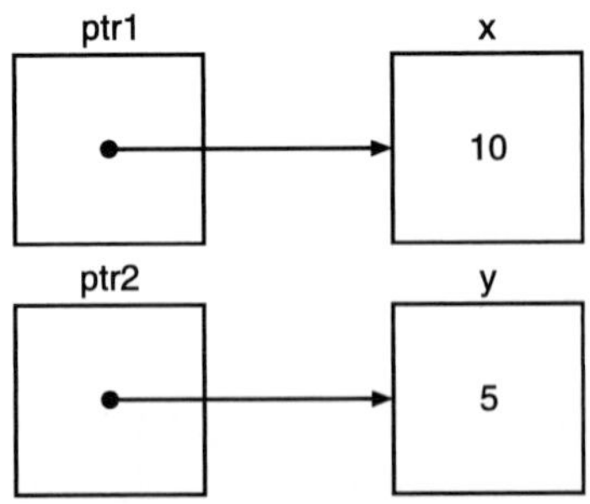

For this simple case, you can think of dereferencing a pointer as only changing the values at the addresses, while re-equating a pointer actually changes the pointers around. Consider the dereferencing operations

```
*ptr1=y;
*ptr2=x;
```

This first one changes the value of x to y, while the second one changes the value of y to the new value of x, which is the same as y, and hence does not change its value. The pointer diagram after this operation looks like

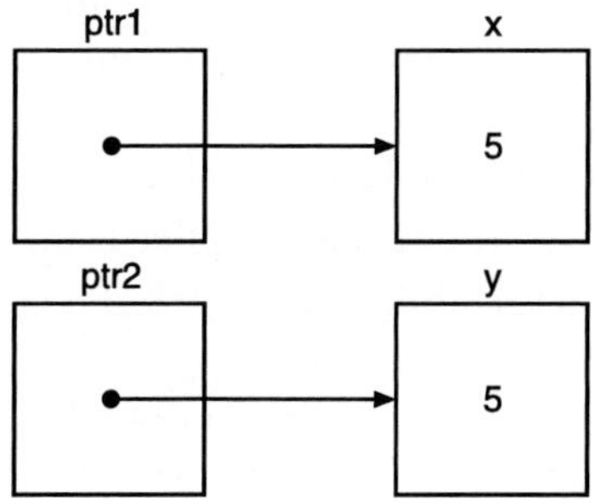

These operations are identical to

```
x=y;
y=x;
```

Now, consider the case in which we swap the pointers with

```
ptr1=&y;
ptr2=&x;
```

The pointer diagram now looks like

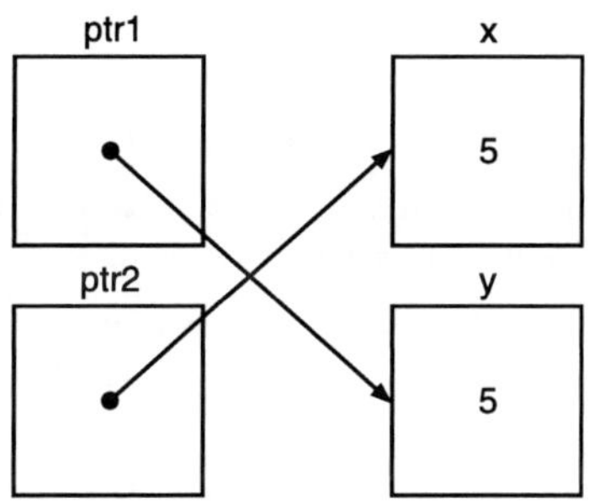

We can also set one pointer to the other,

```
ptr1=ptr2;
```

So, now the pointers both point to the memory location of x, as in

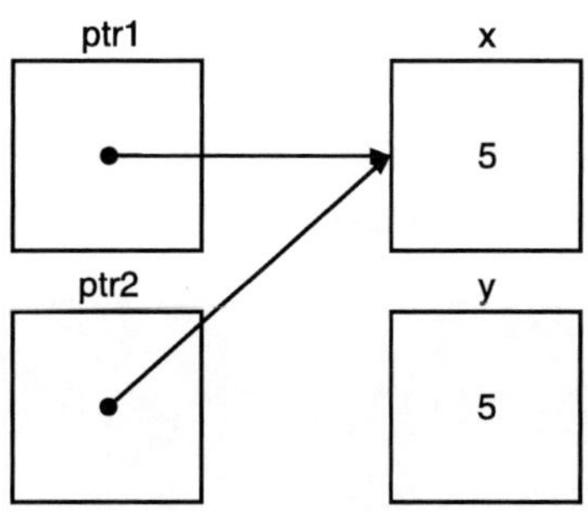

1.3.1 Why Have Pointers?

Pointers are an extremely powerful programming tool. They can make some things much easier, help improve your program's efficiency, and even allow you to handle unlimited amounts of data. For example, using pointers is one way to have a function modify a variable passed to it. It is also possible to use pointers to dynamically allocate memory, which means that you can write programs that can handle nearly unlimited amounts of data on the fly-- you don't need to know, when you write the program, how much memory you need.

1. Pointers allow you to implement sharing without copying, i.e., pass by reference v/s pass by copying. This allows a tremendous advantage when you are passing around big arrays as arguments to functions.
2. Pointers allow modifications by a function that is not the creator of the memory
3. Pointers allow us to use dynamic memory allocation.
4. Pointers obviously give us the ability to implement complex data structures like linked lists, trees, etc.
5. Pointers allow ease of programming, especially when dealing with strings. This is due to the fact that a pointer increment will move by the size of the pointee, i.e., easy coding to increment to the next memory location of an array, without worrying about how many bytes to move for each data type.
6. Pointers allow us to resize the data structure whenever needed. For example, if you have an array of size 10, it cannot be resized. But, an array created out of ***malloc*** and assigned to a pointer can be resized easily by creating a new memory area through malloc and copying the old contents over. This ability is very important in implementing sparse data structures also.
7. By using pointers, we can return more than one value from the function.

Pointers have many uses in C. They are:

- Pointers are more efficient method of accessing data.
- They provide efficient technique for manipulating data in an array and data table.
- Very compact and efficient code can be written using pointers and it results in reduction of the program length and complexity.
- Pointers can be used to return multiple values from a function using function arguments.
- A data manipulation in one function will reflect in other function by passing the reference (address).
- Pointer support dynamic memory allocation (Management).
- Pointer has efficient tool for manipulating dynamic data structures such as stacks, queues, linked list and trees.
- Using pointers, it is possible to access byte or word locations and also possible to access CPU registers.
- An experienced programmer can write very efficient programs using pointers.
- Pointers reduce length and complexity of programs.

Thus, the real power of C lies in the proper use of pointers.

1.4 ARITHMETIC ON POINTERS

We can perform a limited number of arithmetic operations on pointers. These operations are:

- Increment and decrement
- Addition and subtraction
- Comparison
- Assignment

The increment (++) operator increases the value of a pointer by the size of the data object the pointer refers to. For example, if the pointer refers to the second element in an array, the ++ makes the pointer refer to the third element in the array.

An expression like p++; will cause the pointer p to point to the next value of its type. For example, if p is an integer pointer with an initial value 1257, then after the operation p = p + 1; the value of p will be 1259. That is, when we increment a pointer, its value is increased by the 'length' of the data type that it points to. This length called the **scale factor**.

In general, a pointer to some type d-type (where d-type may be int, char, float, double or any user defined data type) when incremented by an integral value i will result is the following value:

```
(current address in pointer) + i * size of (d-type)
```

The decrement (--) operator decreases the value of a pointer by the size of the data object the pointer refers to. For example, if the pointer refers to the second element in an array, the -- makes the pointer refer to the first element in the array.

We can add an integer to a pointer but we cannot add a pointer to a pointer.

If the pointer p points to the first element in an array, the following expression causes the pointer to point to the third element in the same array:

```
p = p + 2;
```

If we have two pointers that point to the same array, we can subtract one pointer from the other. This operation yields the number of elements in the array that separate the two addresses that the pointers refer to.

We can compare two pointers with the following operators: ==, !=, <, >, <=, and >=.

Pointer comparisons are defined only when the pointers point to elements of the same array. Pointer comparisons using the == and != operators can be performed even when the pointers point to elements of different arrays.

You can assign to a pointer the address of a data object, the value of another compatible pointer or the NULL pointer.

Program

```
#include<stdio.h>
#include<conio.h>
void main()
{
        int num1,num2,add,mul,diff,div,*p,*q,*r,*s,*t,*u;
        clrscr();
        printf("Enter the value for num1 ");
        scanf("%d",&num1);
        printf("\nEnter the value for num2 ");
        scanf("%d",&num2);
        p=&num1;
        q=&num2;
        printf("\nADDITION:");
        printf("\n************\n");
        add=*p+*q;
        r=&add;
        printf("\nADDITION of %d and %d = %d\n\n",*p,*q,*r);
        printf("\nSUBSTRCTION");
        printf("\n************\n");
```

```
            diff=*p-*q;
            s=&diff;
            printf("\nSUBSTRCTION of %d and %d = %d\n\n",*p,*q,*s);
            printf("\nMULTIPLICATION");
            printf("\n************\n");
            mul=(*p)*(*q);
            t=&mul;
            printf("\n MULTIPLICATION of %d and %d = %d\n\n",*p,*q,*t);
            printf("\n DIVISION");
            printf("\n************\n");
            div=(*p)/(*q);
            u=&div;
            printf("\n DIVISION of %d and %d = %d\n\n",*p,*q,*u);
        getch();
    }
```

Rules of pointer operations

1. An integer value may be added or subtracted from a pointer variable.
2. A pointer variable cannot be multiplied by a constant.
3. Two pointer variables cannot be added.
4. When two pointers point to the same array, one pointer variable can be subtracted from another.
5. When two pointers point to the objects of the same data types, they can be compared using relational operators.

1.5 ARRAYS AND POINTERS

The concept of array is very much close to the one of pointer. As we know, the base address is the location of the first element (index 0) of the array. In fact, the identifier (name) of an array is equivalent to the address (pointer constant) of its first element. Array name is a pointer constant and its value cannot be changed. Since, the array name is a pointer constant to the first element, the address of the first element and array name both belongs to same memory location. Suppose, we declare an array 'a' as follows:

int a[5] = {5, 8, 2, 4, 7}

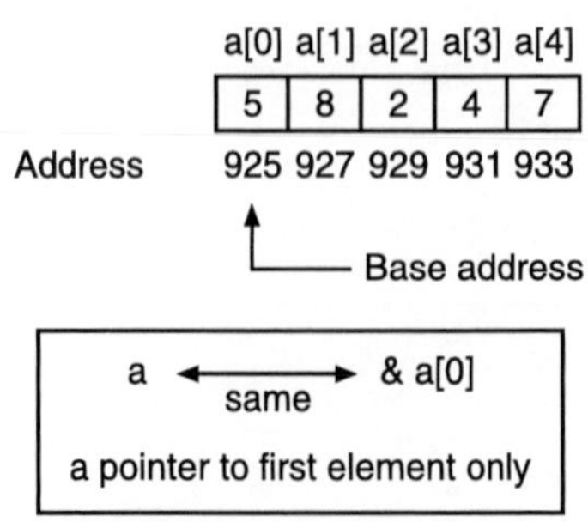

In this example, the value of a is 925, i.e., the location where a [0] is stored i.e,

```
a = & a[0] = 925
```

If we declare p as an integer pointer, then we can make the pointer p to point to the array 'a' by p = a;

The relationship between 'p' and 'a' is as follows.

```
p  =  &a  [0]  =  925

p+1  =  &  a[1]  =  927
p+2  =  &  a[2]  =  929
```

Thus the *(p+2) gives the value of a [2]. Hence, in a one dimensional array 'a' the expression ***(a + i) represents the element a[i]**

Similarly, an element in a two-dimensional array can be represented by pointer expression as ***(*(a+i)+j) represents the element a[i][j]**

1.6 POINTERS AND STRUCTURES

We know that the name of an array stands for the address of its 0th element. The same thing is true of the names of arrays of structure variables. Consider the following example:

```
Struct  abc
{
char  name  [20];
int  age;
float  sal;
}  emp  [5],  *ptr;
```

This statement declares 'emp' as an array of 5 elements, each of the type struct abc and ptr as a pointer to data objects of the type struct abc. The assignment

```
ptr  =  emp;
```

would assign the address of the 0th element of emp to ptr.

That is, the pointer ptr will now point to emp [0]. Its members can be accessed by

```
ptr  ->  name
ptr  ->  age
ptr  ->  sal
```

The symbol -> is called the **arrow operator** or member selection operator and is made up of a minus sign and a greater than sign.

Note that prt -> is simply another way of writing emp [0].

When the pointer ptr is incremented by 1, it is made to point to the next record i.e.,

```
emp  [1].  We  could  also  use  the  notation
     (*ptr).  name
```

to access the member 'name'.

1.7 DISADVANTAGES OF POINTERS

Pointer is most dangerous feature available in C when they are not used correctly. Pointers have some disadvantages and some of these are as follows:

1. No proper usage of pointer may cause crashing the system.
2. Pointers must be initialized otherwise pointer may contain garbage value results in crashing the system.

3. In wrong use of pointer, the compiler may not detect the error in most cases and therefore the program is likely to produce unexpected results.
4. Debugging becomes a difficult task.
5. Manipulation of address locations is not possible but it is possible to access and modify contents only.

1.8 MEMORY ALLOCATION

The high cost of RAM ensures that most embedded systems will continue to experience a shortage of memory. The software you use to implement these systems will use queues, linked lists, task-control blocks, messages, I/O buffers, and other structures that require memory only for a short time and that may return it to serve other functions. This is known as dynamic memory allocation. So, types of memory allocations in c:

1. Compile-time or Static allocation (using arrays)
2. Run-time or Dynamic memory allocation (using pointers)

Difference between Static & Dynamic memory allocation are:

1. In static memory allocation, memory is allocated to the variables at the start of the program while in dynamic memory allocation; memory is allocated at the runtime;
2. In static, memory allocated to the variable is fixed while in dynamic, memory allocated to the variable is dynamic.
3. Static memory allocation results in the inefficient use of memory while dynamic memory allocation makes the efficient use of memory.
4. In dynamic memory allocation, memory can be expanded & contracted during runtime as per requirement. This provides the flexibility to the programmer.

1.9 MEMORY ALLOCATION PROCESS

Before discussing these functions, let us look at the memory allocation process associated with a C program. Figure 1.1 shows a conceptual view of storage of a C program in memory. The program instructions and global and static variables are stored in a region known as permanent storage area and the local variables are stored in another area called stack.

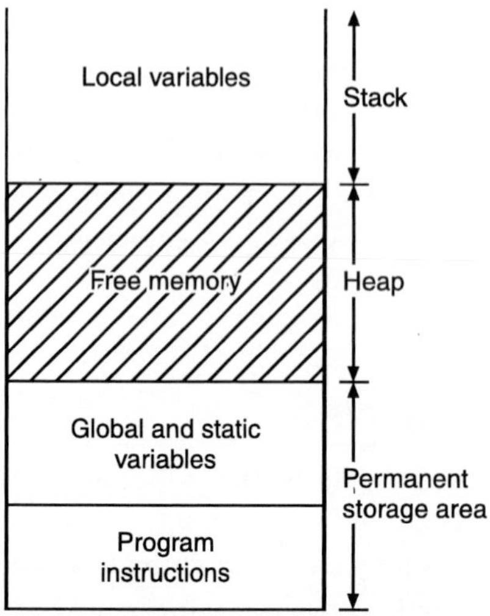

Fig. 1.1. Storage of a C program

The memory space that is located between these two regions is available for dynamic allocation during execution of the program. This free memory region is called the heap. The size of the heap keeps changing when program is executed due to creation and death of variables that are local to functions and blocks. Therefore, it is possible to encounter memory "overflow" during dynamic allocation process. In such situations, the memory allocation functions mentioned above return a NULL pointer (when they fail to locate enough memory requested).

Thus, the free memory space, called the heap, is available for dynamic memory allocation.

1.10 LIBRARY FUNCTIONS FOR DYNAMIC MEMORY ALLOCATION

There are many built-in standard library functions for efficient dynamic memory allocation/ de-allocation. The functions are: **malloc, calloc, realloc, free**.

malloc() meaning memory allocation, and free, well, that should be obvious. malloc() function returns a pointer to the memory block we requested.

Let's discuss the syntax of malloc(). It takes a single argument, a long integer representing the number of bytes we want to allocate from the heap. (Note: the heap is what we call all the memory) So, for example, to allocate memory for an array of characters for a 40 character string, we would do malloc(40);

To allocate memory though, we must have a pointer so we can know where the memory will be at when it gets allocated. Let's look at a small code example:

```
char *str = (char *)malloc(40); // allocate memory for a 40 character
string
```

This may look a little strange, because it uses a concept called "type casting". C has the ability to convert from one type of variable to another by using what are called type cast operators. The syntax is simple, put the variable type you want to convert to inside parentheses.

Because we want to convert the memory into a character pointer, which is specified by the type char *, we put char * in parentheses. Then C will give us a character pointer to the memory. This is very easy to do when a character pointer is involved. The important thing to note here is what malloc() returns to us. Because malloc() doesn't care how the memory we allocate will be used, malloc() just returns a pointer to a series of bytes. It does this by using a void pointer. Void pointers are just like character pointers, or integer pointers, or any other kind of pointer with one special difference: we do not know what the size of the void pointer is. There is no way to increment or decrement a void pointer. In fact, we can't do much of anything with a void pointer except acknowledge its existence until we convert (type cast) it to another variable type.

Characters are by definition 1 byte, and malloc always allocates in multiples of a single byte. Other types are more than 1 byte. Rather than remembering how many bytes an int variable takes on a system, we can use the sizeof operator. Let's see an example.

```
char *str = (char *)malloc(40 * sizeof(char));


// allocate memory for a 40 character string
```

The sizeof() operator will tell us the size of any variable or structure in our program. The one thing to keep in mind when using sizeof() is that the size of a pointer is how many bytes it takes to store the pointer, not the size of the memory block the pointer is pointing at. There is no need to know that information. In allocating memory like this, we know we have 40 characters instead of 40 bytes (even though they are technically the same). There is a difference between 40 integers and 40 bytes, and 40 long integers especially.

The **free()** function is basically the exact opposite of malloc(). So, instead of assigning a pointer the value returned, we give the function the pointer we got from malloc(). So, if we have the *str pointer we allocated above, to free that memory, we just need to call free with the pointer as an argument.

```
free(str);    // free the memory allocated by malloc()
```

You may be wondering how the *free()* function knows how much memory to free up, since we didn't tell it how much memory we allocated. Well, the short answer is: you don't need to worry about that. The system will take care of such minutia for us.

Calloc () function

It works similar to the malloc() function except that it requires two argument. Syntax for this function:

```
ptr = calloc (number of elements, size of each element);


Example: int *ptr;
ptr = (int*)calloc(10,2);
```

This functions is avallable in the header file <stdlib.h> or <alloc.h>. The memory allocated by the malloc() contains garbage values & the memory allocated by calloc() contains all zeros. Calloc is a memory allocation function that is normally used to request multiple blocks of storage each of the same size and then sets all bytes to zero.

Realloc() function

This function is used to resize the size of memory block, which is already allocated. This function is available in the header file <stdlib.h> or <alloc.h>. Note that before using the realloc() function, the memory block to be resized should be allocated using the following statement:

```
ptr = malloc(size);
Syntax of this function:
ptr = realloc (ptr, new_size);
```

Difference between malloc and calloc

Even though both malloc and calloc functions are used for allocating memory, there are some significant differences between these two functions.

Malloc	Calloc
1. The syntax of malloc is ptr = (data type*) malloc (number of elements * size of each element).	1. The syntax of calloc is ptr = (data type*) calloc (no. of elements, size of each element).
2. Allocates a contiguous single block of memory is allocated with specified size in parameter.	2. Allocates multiple blocks of memory and each block contains equal size.
3. Allocated memory locations are not initialized.	3. Initializes all bytes in the allocated block to zero.
4. Since there is no initialization then time efficiency is more than the calloc() function.	4. Calloc() is computationally more expensive because it set to zero but it is more convenient than malloc() function.

1.11 DANGLING POINTERS

Dangling pointers in computer programming are pointers that do not point to a valid object of the appropriate type. Dangling pointers arise when an object is deleted or deallocated, without modifying the value of the pointer, so that the pointer still points to the memory location of the deallocated memory.

We can have dangling pointers because of multiple reasons:

1. An un-initialized, non-static local pointer variable is a dangling pointer.

Auto variables are not initialized in C & C++. The value they contain is garbage. For example in the code

```
int * ptr;
```

ptr is a dangling pointer. Sometimes, such dangling pointers are also called **wild pointers**. Some languages (like java) give compile time error if you try to use a variable which is not explicitly initialized. Note that the below is not a dangling pointer because static variables are initialized to zero (NULL)

```
static int * ptr;    // Not a dangling pointer
```

2. Destroying the stack frame (the variable which the pointer points to goes out of scope)

```
int *ptr = NULL;
{
int temp = 20;
ptr = &temp
}// temp goes out of scope and ptr becomes dangling after this point.
```

3. Explicitly deleting the memory, which the pointer points to

```
int * ptr = new int;
delete ptr; // ptr is dangling pointer now.
ptr = NULL; // best practice to always assign NULL after deleting
a pointer.
Sometimes, this is done indirectly, as shown below,
int * ptr = new int;
int *new_ptr = ptr
delete ptr;
ptr = NULL;
```

In the above code ptr is not a dangling pointer but since new_ptr also points to the same memory location, new_ptr becomes a dangling pointer. (Because the memory location to which it points is deallocated).

4. Returning address of a local variable on stack.

```
int * function func(){
int abc;
return &abc; // returning address of a local variable
}
```

The stack-frame (activation record) of the function, on which abc resides will be deallocated when the function returns. So the address returned from the function does not point to the allocated memory.

1.12 MEMORY LEAK

A memory leak is a situation when there exist a memory, which is allocated to the program but cannot be accessed. Such memory develops on heap, when a user allocates the memory but change the pointer which holds the address to that location.

A memory leak in computer science is a particular type of unintentional memory consumption by a computer program where the program fails to release memory when no longer needed. That is memory leaks occur when memory allocated to a pointer is not deleted and the pointer goes out of scope. This allocated memory never gets de-allocated and remains in the heap until the system is restarted.

Example:

```
void abc()
{
    int* ptr = new int;
    *ptr = 10;
    ...
    ...
}
```

Since, ptr is allocated memory using 'new', but no 'delete' is called, hence memory allocated to ptr will leak.

Memory leaks are a common error in programming, especially when using languages that have no built-in automatic garbage collection, such as C and C++. Typically, a memory leak occurs because dynamically allocated memory has become unreachable.

Exercises

1. What is the function of a pointer variable?
2. What do you understand by a pointer to a pointer?
3. What does void main (void) mean?
4. What is a pointer? How is a pointer initialized?
5. What is dynamic memory allocation? What is the principal difference between the functions malloc and calloc?
6. What do you mean by Dangling pointers?
7. Define memory leak in detail?

□□□

Chapter

2 Introduction to Data Structures

Chapter Outline

- Introduction
- Data Structure: Definition
- Need of a Data Structure
- Types of Data Structures
- Data Types *vs.* Data Structures
- Data Structure Operations
- Overview of Various Data Structures
- What is an Algorithm?
- Desirable Attributes for Algorithms
- Algorithm Design Techniques
- Analysis of Algorithm
- Time-Space Trade-off
- Asymptotic Analysis
- Abstract Data Types (ADT)

2.1 INTRODUCTION

In computer science, a data structure is a particular way of storing and organising data in a computer so that it can be used efficiently. Different kinds of data structures are suited to different kinds of applications, and some are highly specialised to certain tasks. For example, B-trees are particularly well-suited for implementation of databases, while compiler implementations usually use hash tables to look up identifiers.

Data structures are used in almost every program or software system. Specific data structures are essential ingredients of many efficient algorithms, and make possible the management of huge amounts of data, such as large databases and internet indexing services. Some formal design methods and programming languages emphasise data structures, rather than algorithms, as the key organising factor in software design.

Basic Principle

Data structures are generally based on the ability of a computer to fetch and store data at any place in its memory, specified by an address — a bit string that can be itself stored in memory and manipulated by the program. Thus, the record and array data structures are based on computing the addresses of data items with arithmetic operations; while the linked data structures are based on storing addresses of data items within the structure itself. Many data structures use both principles.

Language Support

Assembly languages and some low-level languages such as BCPL generally lack support for data structures. Many high-level programming languages, on the other hand, have special syntax or other built-in support for certain data structures, such as vectors (one-dimensional arrays) in the C programming language, multi-dimensional arrays in Pascal, linked lists in Common Lisp, and hash tables in Perl. Many languages also provide basic facilities such as references and the definition record data types, that programmer can use to build arbitrarily complex structures. Most programming languages feature some sort of library mechanism that allows data structure implementations to be reused by different programs. Modern programming languages usually come with standard libraries that implement the most common data structures. Examples are the C++ Standard Template Library, the Java Collections Framework, and Microsoft's.NET Framework.

Modern languages also generally support modular programming, the separation between the interface of a library module and its implementation. Some provide opaque data types that allow clients to hide implementation details from client programs. Object-oriented programming languages, such as C++ and Java, use classes for this purpose.

2.2 DATA STRUCTURE: DEFINITION (UPTU 2003-04)

1. Structure means particular way of data organisation. So data structure is refers to the organisation of data in computer memory or the way in which data is efficiently stored, processed and retrieved is called data structure. Data structure simply means a structure that can be used to store a given collection of data in computer memory.
2. A data structure is an arrangement of data in a computer's memory (or sometimes on disk).
3. Data structure is the structural representation of logical relationships between elements of data.
4. A data structure is a group of data elements grouped together under one name. These data elements known as members can have different types and different lengths.
5. Data structure means how the data is organised in memory. There are different kinds of data structures. Some are used to store the data of same type and some are used to store different types of data. Data structure is a combination of two or more data type elements.
6. A data structure is defined as a way of representing data in computer memory it may into type to complex data involving many types, *e.g.*, structure.
7. A data structure is a method of representing data. It not only deals with raw data but also involves the relationship between data.
8. Data structures can be defined as triple {D.F.A.}, where D is Data F is a set of function and A is the set of axioms.
9. A data structure is a logical mode of a particular organisation of data. The choice of particular data structure depends on the following consideration:
   - It must able to represent the inherent relationship of the data in real world.
   - It must be simple enough so that it can process efficiently as when necessary.

2.3 NEED OF A DATA STRUCTURE

A data structure helps us to understand the relationship of one element with the other and organise it with in the memory. It is a mathematical or logical model of a particular organisation of data. The study of data structures includes:

- ***Logical description*** of the data structure
- ***Implementation*** of data structure
- ***Quantitative analysis*** of the data structure which includes determining the amount of memory needed to store the data structure and the time required to processing it.

Data structure is a representation of the logical relationship between data elements. In other words, a data structure is a way of organising all data items that considers not only the elements stored but also their relationship to each other.

Areas in which data structures are applied extensively are :

- Database Management System
- Compiler Design
- Network Analysis
- Numerical Analysis
- Artificial Intelligence
- Statistical Analysis Package
- Simulation
- Operating System
- Graphics

That is, in many applications, different data structures are used to do the operations on the data structures. In such a situation, there is a trade-off between memory utilisation and run time. That is, one data structure sacrifices memory compactness for speed; another utilises memory efficiently but results in a slow run time. So, each data structure has its own strengths and weaknesses. Table shows the characteristics of various data structures.

Characteristic of Various Data Structures

Data Structure	Advantages	Disadvantages
Array	Quick insertion, very fast access if index is known	Slow search, Slow deletion, Fixed size
Ordered Array	Quicker search than unsorted array	Slow insertion and deletion, Fixed size
Stack	Provides last-in first-out access	Slow access to other items
Queue	Provides first-in first-out access	Slow access to other items
Linked List	Quick insertion and quick deletion	Slow search
Binary Trees	Quick search, insertion and deletion if tree remains balance	Deletion algorithm is complex
Hash Table	Very fast access if key known, Fast insertion.	Slow deletion, access slow if key not known, inefficient memory usage
Heap	Fast insertion, deletion. Access to largest item	Slow access to other items
Graph	Models real world situation	Some algorithms are slow in access

EXAMPLE 2.1. What are the major data structures used in the following areas: RDBMS, Network data model and Hierarchical data model?

Solution: RDBMS - Array (*i.e.*, Array of structures)

Network data model - Graph

Hierarchical data model - Trees

2.4 TYPES OF DATA STRUCTURES

Depending upon the arrangement of data, data structures can be classified as arrays, records, link lists, stacks, queues, trees, graphs etc.

1. According to Nature of Size

(*i*) Static data structure.
(*ii*) Dynamic data structure.

Static data structure: A data structure is said to be static if we can store data upto a fix number *e.g.*, array.

Dynamic data structureless : As the name show a dynamic data structure is that data which allows the programmer to change its size during program execution to add or delete data. For example, link list, tree, graph etc.

2. According to Its Occurrence **(UPTU 2002-03)**

(*i*) Linear data structure.
(*ii*) Non-linear data structure.

Linear data structure: In linear data structure data is store in consecutive memory location or data is stored in a sequential form *i.e.*, every element in the structure has a unique predecessor and unique successor, *e.g.*, array, link-list, queue, stack etc.

Non-linear data structure: In non-linear data structure the data is stored in non-consecutive memory location or data is stored in inconsequential form *e.g.*, tree, graphs etc. A non-linear structure is mainly used to represent data containing a hierarchical relationship between elements. In this, elements do not form a sequence; there is no unique predecessor or unique successor.

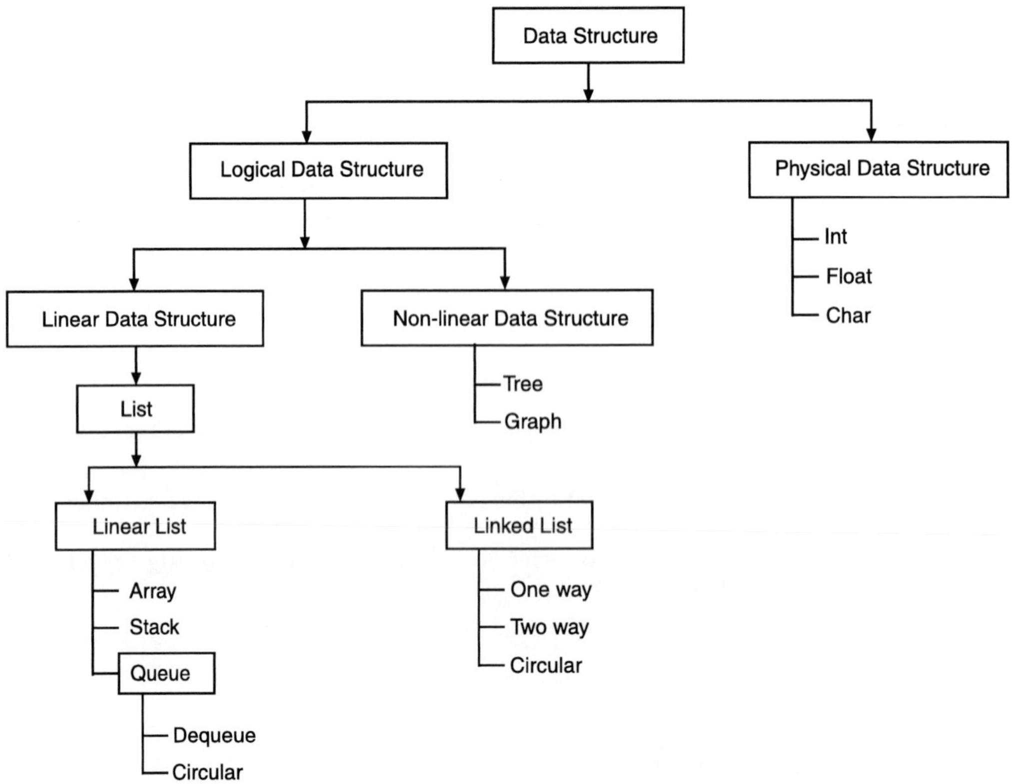

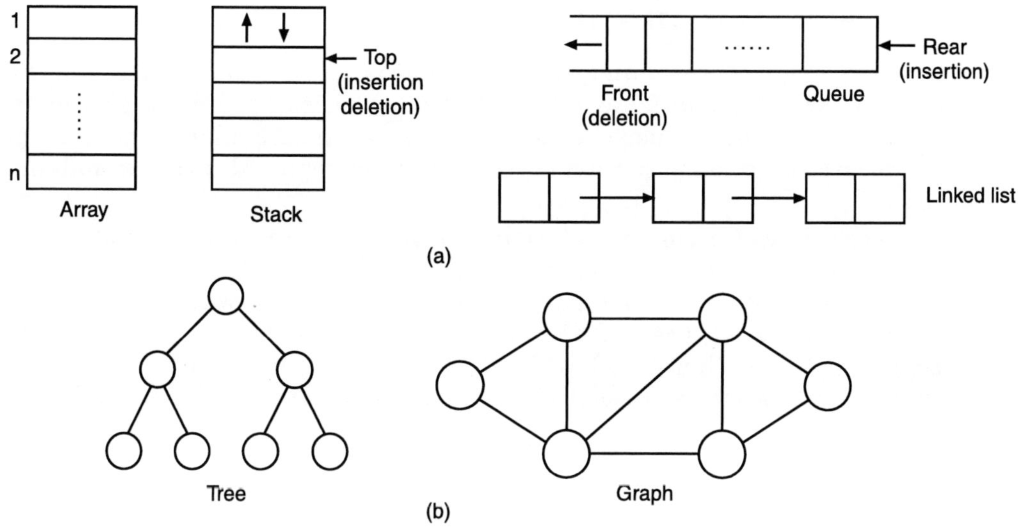

Fig. 2.1. (a) Linear data structures (b) Non-linear data structures

3. Primitive and Non-primitive Data Structures (UPTU 2003-04)

Primitive data structures are the basic data structures and are directly operated upon by the machine instructions, which is in a primitive level. These have different representations on different computers. They are integers, floating point numbers, characters, string constants, pointers etc.

Non-primitive data structures are the more complicated data structure emphasising on structuring of a group of homogeneous or heterogeneous data items. These are derived from the primitive data structures. Array, lists, files, linked lists, trees and graphs fall in this category.

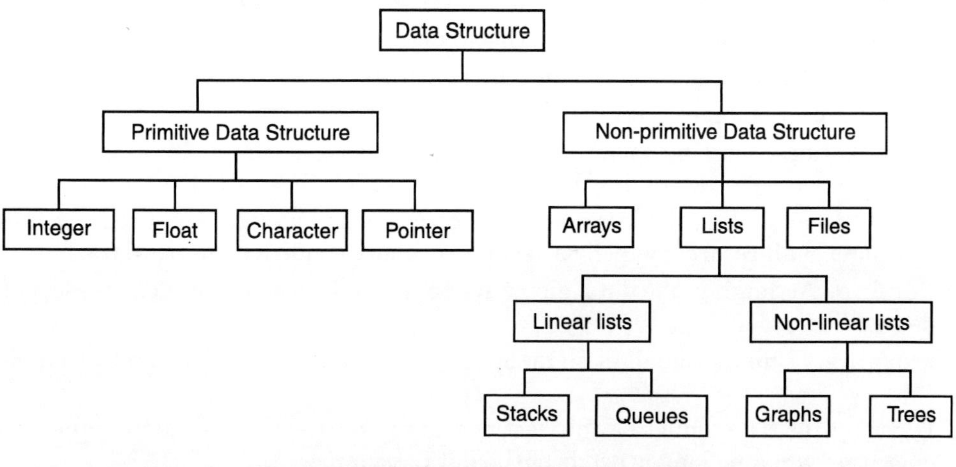

Fig. 2.2. Classifications of data structures

4. Homogeneous and Non-homogeneous Data Structures

In homogeneous data structures, the data elements are of same type like array. In non-homogeneous data structures, the data elements may not be of same type like structure in 'C'.

2.5 DATA TYPES *VS.* DATA STRUCTURES

In a programming language, the *data type* of a variable is the set of values that the variable may assume. For example, a variable of type *Boolean* can assume either the value true or the value false, but no other value. The basic data types vary from language-to-language. The rules for constructing composite data types out of basic ones also vary from language-to-language.

An *abstract data type* is a mathematical model, together with various operations defined on the model. As we have know, we shall design algorithms in terms of ADT's, but to implement an algorithm in a given programming language we must find some way of representing the ADT's in terms of the data types and operators supported by the programming language itself. To represent the mathematical model underlying an ADT we use *data structures,* which are collections of variables, possibly of several different data types, connected in various ways.

A data type is a term which refers to the kind of data. It is a well-defined collection of data with a well-defined set of operations on it.

A data *structure* is an actual implementation of a particular abstract data type, *i.e.*, this is an extension of the concept of the data type.

Data Type = Permitted Data Values + Operations

Data Structure = Organized Data + Allowed Operations

2.6 DATA STRUCTURE OPERATIONS

The data appearing in our data structure is processed by means of certain operations. Infact the particular data structure that one chooses for a given situation depends largely on the frequency with which specific operations are performed. The following data structure operations play a major role in processing of data:

1. **Creating.** This is the first operation to create a data structure. This is just declaration and initialisation of the data structure and reserved memory locations for data elements.
2. **Inserting.** Adding new records to the structure.
3. **Deleting.** Removing a record from the structure.
4. **Updating.** It changes data values of the data structure.
5. **Traversing.** Accessing each record exactly once so that certain items in the record may be processed. (This accessing or processing is sometimes called visiting the records.)
6. **Searching.** Finding the location of the record with a given—key value, or finding the locations of all records, which satisfy one or more conditions in the data.
7. **Sorting.** Arranging the data elements in some logical order, *i.e.*, in ascending and descending order.
8. **Merging.** Combine the data elements in two different sorted sets into a single sorted set.
9. **Destroying.** This must be the last operation of the data structure and apply this operation when no longer needs of the data structure.

2.7 OVERVIEW OF VARIOUS DATA STRUCTURES

(*a*) Arrays: The simplest type of data structure is a *linear array* and most often it is the only data structure that is provided in any programming language. An array can be defined as a collection of homogeneous elements, in the form of **index/value** pairs, stored in consecutive memory locations. An array always has a predefined size and the elements of an array are referenced by means of an index / subscript. Thus an array is a collection of variables of the same data type that share a common name. The general form of a one-dimensional array is

<type specifier> array name [size]

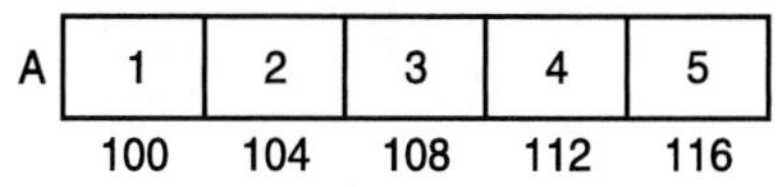

Memory Organisation of an Array

Advantages

- Searching is faster as elements are in continuous memory locations.
- Memory management is taken care of by the compiler itself.

Limitations

- Number of elements must be known in advance
- Inserting and Deletions are costlier as it involves shifting the rest of the elements

(*b*) Stack: *A stack is an ordered list in which items are inserted and removed at only one end called the TOP.* There are only 2 operations that are possible on a stack. They are the 'Push' and the 'Pop' operations. A Push operation inserts a value into the stack and the Pop operation retrieves the value from the stack and removes it from the stack as well. An example for a stack is a stack of plates arranged on a table. This means that the last item to be added is the first item to be removed. Hence a stack is also called as **Last-In-First-Out List or LIFO** list. A graphical representation of a stack is shown below:

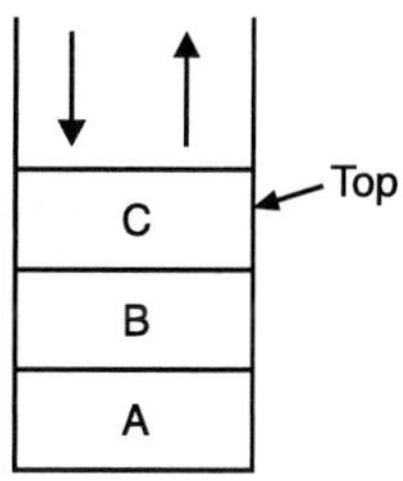

Fig. 2.3.

(*c*) Queue: A queue is an ordered list in which all insertions can take place at one end called the **rear** and all deletions take place at the other end called the front. The two operations that are possible in a queue are Insertion and Deletion. A real time example for a queue is people standing in a queue for billing on a shop. The first person in the queue will be the first person to get the service. Similarly, the first element inserted in the queue will be the first one that will be retrieved and hence a queue is also called as **First In First Out or FIFO** list.

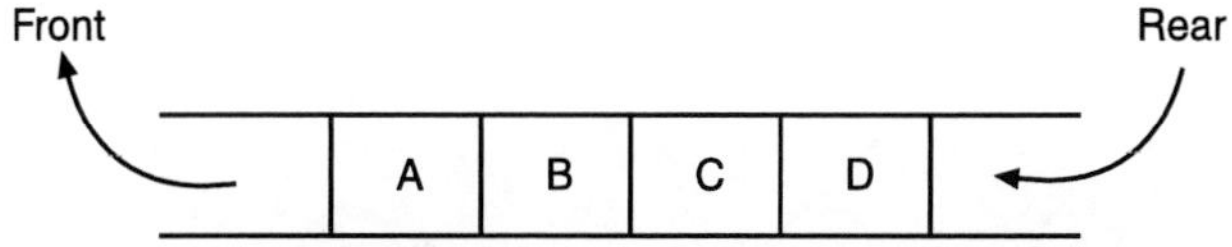

Fig. 2.4 Graphical Representation of a Queue.

The most common occurrence of a queue in computer applications is for scheduling of print jobs. For example, if several people are giving print requests, all the request get queued up in the printer and is processed on first come first serve basis.

(*d*) Linked Lists: In real time systems most often the number of elements will not be known in advance. The major drawback of an array is that the number of elements must be known in advance. Hence an alternative approach was required. This give rise to the concept called linked lists. A linear list is a linear collection of data elements called nodes, where the linear order is given by means of pointers. The key here is that every node will have two parts: first part contains the information/data and the second part contains the link/address of the next node in the list. Memory is allocated for every node when it is actually required and will be freed when not needed.

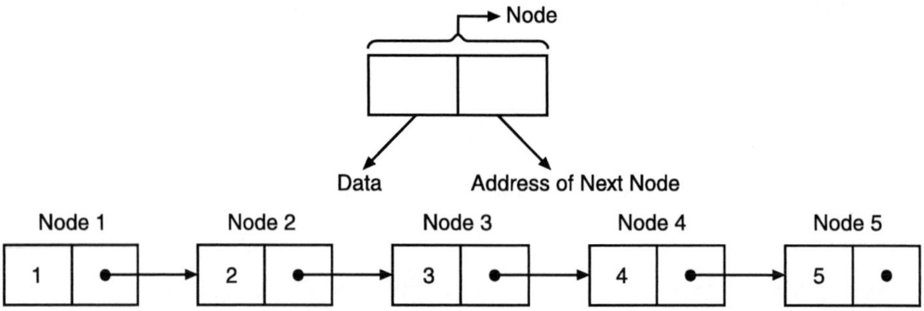

Fig. 2.5.

A linked list is used typically in computer applications for memory allocations. The system keeps tracks of all the memory available in the system with the help of a list. This area of memory is called as the memory pool. As and when the user requests for memory the system allocates memory to the user from this available pool. Once allocated these blocks will be deleted from the available list. Once the user frees the memory again the memory will be added to the list. In general linked lists are used in almost all places whenever a collection of elements are required and when the number of elements in the collection is not known in advance.

***(e)* Tree:** A tree is a non-linear data structure. This structure is mainly used to represent data containing a hierarchical relationship between elements like family tree, organisation chart etc.

A tree is a finite set of one or more nodes such that:

(*i*) There is a specially designated node called the **root.**

(*ii*) The remaining nodes are partitioned into $n >= 0$ disjoint sets T_1, ..., T_n where each of these sets is a tree. T_1,..., T_n are called *sub-trees* of the root.

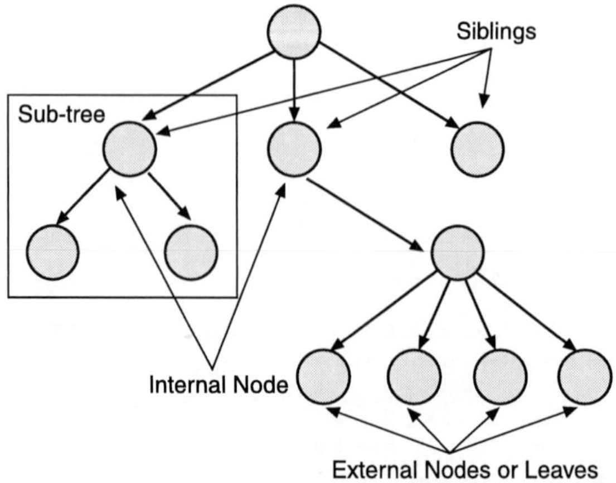

Fig. 2.6.

***(f)* Graphs:** Graph is a general tree with no parent child relationship. Graphs have many applications in computer science and other fields of science. In general, graphs represent a relatively less restrictive relationship between the data items. A graph G may be defined as a finite set V of vertices and a set E of edges (pair of connected vertices). The notation used is as follows: Graph G = (V, E). Consider the graph of Fig. 2.7.

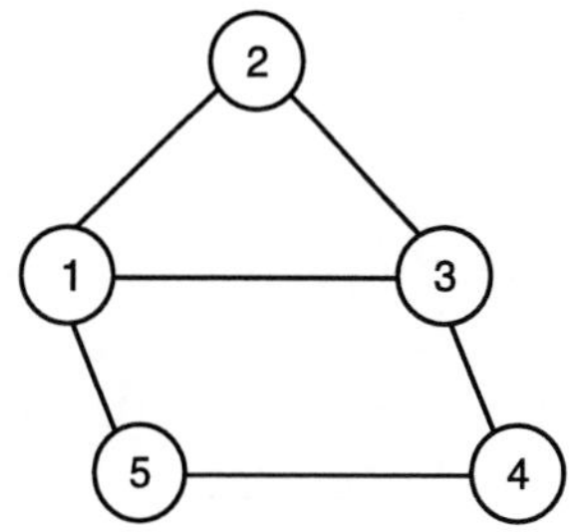

Fig. 2.7.

The set of vertices for the graph is V = {1, 2, 3, 4, 5}.

The set of edges for the graph is E = {(1,2), (1,5), (1,3), (5,4), (4,3), (2,3) }.

Graphs are used in a wide variety of applications. A graph can be used by airlines for maintaining the flight information such as the various flights, their routes and the distance between the places, etc. With the help of such a graph one will be easily able to find out whether there is any flight for a particular destination and if there are several routes to reach that destination which one is having the optimum distance or cost, etc.

2.8 WHAT IS AN ALGORITHM?

Once a data structure for a particular application is chosen, an algorithm must be developed that manipulates the related data items stored in it. Such an algorithm should have the following features.

1. It should be free of ambiguity.
2. It should be concise.
3. It should be efficient.

An *algorithm* is a precise plan for performing a sequence of actions to achieve the intended purpose. Each action is drawn from a well-understood selection of actions on data.

Definition: An *algorithm* is a step-by-step finite sequence of instructions, to solve a well-defined computational problem. That is, in practice to solve any complex real life problems; first we have to define the problems. Second step is to design the algorithm to solve that problem. In addition every algorithm must satisfy the following criteria:

(*i*) **Input:** There are zero or more quantities which are externally supplied;

(*ii*) **Output :** At least one quantity is produced;

(*iii*) **Definiteness :** Each instruction must be clear and unambiguous;

(*iv*) **Finiteness:** If we trace out the instructions of an algorithm, then for all cases the algorithm will terminate after a finite number of steps;

(*v*) **Effectiveness:** Every instruction must be sufficiently basic that it can in principle be carried out by a person using only pencil and paper. It is not enough that each operation be definite as in (*iii*), but it must also be feasible.

In formal computer science, one distinguishes between an algorithm, and a program. A program does not necessarily satisfy condition (*iv*). One important example of such a program for a computer is its operating system which never terminates (except for system crashes) but continues in a wait loop until more jobs are entered.

Therefore, *an algorithm can be defined as a sequence of definite and effective instructions, while terminates with the production of correct output from the given input.*

2.9 DESIRABLE ATTRIBUTES FOR ALGORITHMS

Although an algorithm may satisfy the criteria of the previous section by being precise, unambiguous, deterministic, and finite, it still may not be suitable for use as a computer solution to a problem. The basic desirable attributes that an algorithm should meet are explained below.

Generality: An algorithm should solve a class of problems, not just one problem. For example, an algorithm to calculate the average of four values is not as generally useful as one that calculates the average of an arbitrary number of values.

Good Structure: This attribute applies to the construction of the algorithm. A well-structured algorithm should be created using good building blocks that make it easy to:

- Explain,
- Understand,
- Test, and
- Modify it.

The blocks, from which the algorithm is constructed, should be interconnected in such a way that one of them can be easily replaced with a better version to improve the whole algorithm, without having to rebuild it.

Efficiency: An algorithm's speed of operation is often an important property, as is its size or compactness. Initially, these attributes are not important concerns. First, we must create well-structured algorithms that carry out the desired task under all conditions. Then, and only then, do we improve them with efficiency as an objective.

Ease of Use: This property describes the convenience and ease with which users can apply the algorithm to their data. Sometimes what makes an algorithm easy to understand for the user also makes it difficult to design for the designer (and vice-versa).

Elegance: This property is difficult to define, but it appears to be connected with the qualities of harmony, balance, economy of structure and contrast, whose contribution to beauty in the arts is prized. Also, as with the arts, we seem to be able to "know beauty when we see it". In algorithms, sometimes the term *beauty* is used for this attribute of an algorithm.

Other desirable properties, such as ***robustness*** (resistance to failure when presented with invalid data) and ***economy*** (cost-effectiveness), will only be touched upon in this chapter. Not because these properties are unimportant, but because the basic attributes of an algorithm should be understood first.

2.10 ALGORITHM DESIGN TECHNIQUES

There are two approaches for algorithm design; they are **top-down and bottom-up** algorithm design.

(*a*) Top-down Algorithm Design: The principle of top-down design states that a program should be divided into a main module and its related modules. Each module should also be divided into sub-modules according to software engineering and programming style. The division of modules processes until the module consists only of elementary processes that are basically understood and cannot be further sub-divided.

Top-down algorithm design is a technique for organising and coding programs in which a hierarchy of modules is used, and breaking the specification down into simpler and simpler pieces, each having a single entry and a single exit point, and in which control is passed downward through the structure without unconditional branches to higher levels of the structure. That is top-down programming tends to generate modules that are based on functionality, usually in the form of functions or procedures or methods.

In C, the idea of top-down design is done using functions. A C program is made of one or more functions, one and only one of which must be named *main*. The execution of the program always starts and ends with *main*, but it can call other functions to do special tasks.

***(b)* Bottom-up algorithm design:** Bottom-up algorithm design is the opposite of top-down design. It refers to a style of programming where an application is constructed starting with existing primitives of the programming language, and constructing gradually more and more complicated features, until the all of the application has been written. That is, starting the design with specific modules and builds them into more complex structures, ending at the top.

The bottom-up method is widely used for testing, because each of the lowest-level functions is written and tested first. This testing is done by special test functions that call the low-level functions, providing them with different parameters and examining the results for correctness. Once lowest-level functions have been tested and verified to be correct, the next level of functions may be tested. Since the lowest-level functions already have been tested, any detected errors are probably due to the higher-level functions. This process continues, moving up the levels, until finally the *main* function is tested.

For a given problem, there are many ways to design algorithms for it, *e.g.*, insertion sort is an *incremental approach,* Merge sort is a *divide and conquer approach.* The following is a list of several popular design approaches.

1. Incremental approach
2. Divide and Conquer approach
3. Greedy approach
4. Dynamic programming approach
5. Backtracking approach
6. Branch and bound approach
7. Randomised approach

Incremental Appraoch

— Insertion sort uses an incremental approach.

— Having sorted the sub-array [1 ... j-1], insert the single element $A[j]$ into its proper place, yielding the sorted sub-array A[1...j].

Divide and Conquer Approach

- Divide the original problem into a set of sub-problems
- Solve every sub-problem individually, recursively
- Combine the solutions of the sub-problems (top level) into a solution of the whole original problem.

Greedy Approach

Greedy algorithms seek to optimise a function by making choices (greedy criterion) which are the best locally but do not look at the global problem. The result is a good solution but not necessarily the best one. The greedy algorithm does not always guarantee the optimal solution however it generally produces solutions that are very close in value to the optimal.

Dynamic Approach

Dynamic programming is a technique for efficiently computing recurrences by storing partial results. It is a method of solving problems exhibiting the properties of overlapping sub-problems and optimal sub-structure that takes much less time than naive methods.

Branch-and-Bound

In a branch and bound algorithm a given sub-problem, which cannot be bounded, has to be divided into at least two new restricted sub-problems. Branch and bound algorithms are methods for global optimisation in non-convex problems. Branch and bound algorithms can be (and often are) slow, however, in the worst case they require effort that grows exponentially with problem size, but in some cases we are lucky, and the methods converge with much less effort.

Randomised Algorithms

A randomised algorithm is defined as an algorithm that is allowed to access a source of independent, unbiased random bits, and it is then allowed to use these random bits to influence its computation.

Backtracking Algorithms

Backtracking algorithms try each possibility until they find the right one. It is a depth-first search of the set of possible solutions. During the search, if an alternative doesn't work, the search backtracks to the choice point, the place which presented different alternatives, and tries the next alterative. When the alternatives are exhausted, the search returns to the previous choice point and try the next alternative there. It there are no more choice points, the search fails.

2.11 ANALYSIS OF ALGORITHM (UPTU 2009-10)

Why is it necessary to analyse an algorithm? After designing an algorithm, it has to be checked and its correctness needs to be predicted; this is done by analysing the algorithm. Thus, an algorithm analysis measures the *efficiency* of the algorithm. The ***efficiency of an algorithm*** can be checked by:

- Correctness of an algorithm
- Implementation of an algorithm
- Simplicity of an algorithm
- Execution time and memory requirements of an algorithm

The algorithm can be analysed by tracing all step-by-step instructions, reading the algorithm for logical correctness, and testing it on some data using mathematical techniques to prove it correct. Another type of analysis is to analyse the simplicity of the algorithm. That is, design the algorithm in a simple way so that it becomes easier to be implemented. However, the simplest and most straightforward way of solving a problem may not be sometimes the best one. Moreover there may be more than one algorithm to solve a problem.

2.11.1 Types of Analysis

1. Worst case running time
2. Average case running time
3. Best case running time

Worst Case Running Time (UPTU 2007)

- The behaviour of the algorithm with respect to the worst possible case of the input instance.
- The worst-case running time of an algorithm is an upper bound on the running time for any input. Knowing it gives us a guarantee that the item does not occur in data.
- There is no need to make an educated guess about the running time.

Average Case Running Time

- The expected behaviour when the input is randomly drawn from a given distribution.
- The average-case running time of an algorithm is an estimate of the running time for an "average" input.
- Computation of average-case running time entails "knowing all possible input sequences, the probability distribution of occurrence of these sequences, and the running times for the individual sequences".
- Often it is assumed that all inputs of a given size are equally likely.

Best Case Running Time

- The behaviour of the algorithm when input is in already in order. For example, in sorting, if elements are already sorted for a specific algorithm.
- The best case running time rarely occurs in practice comparatively with the first and second case.

The choice of a particular algorithm depends on following performance analysis and measurements:

1. Space complexity
2. Time complexity

2.11.2 Space Complexity

Analysis of space complexity of an algorithm or program is the amount of memory it needs to run to completion. Some of the *reasons* for studying space complexity are:

1. If the program is to run on multi-user system, it may be required to specify the amount of memory to be allocated to the program.
2. We may be interested to know in advance that whether sufficient memory is available to run the program.
3. There may be several possible solutions with different space requirements.
4. It can be used to estimate the size of the largest problem that a program can solve.

The space needed by a program consists of following components.

• ***Instruction Space:*** Space needed to store the executable version of the program and it is fixed.

• ***Data Space:*** Space needed to store all constants; variable values and has further two components:

(*a*) Space needed by constants and simple variables. This space is fixed.

(*b*) Space needed by fixed sized structural variables, such as arrays and structures.

(*c*) Dynamically allocated space. This space usually varies.

• ***Environment stack space:*** This space is needed to store the information to resume the suspended (partially completed) functions. Each time a function is invoked the following data is saved on the environment stack:

(*a*) **Return address:** For example, from where it has to resume after completion of the called function.

(*b*) Values of all lead variables and the values of formal parameters in the function being invoked.

The amount of space needed by recursive function is called the recursion stack space. For each recursive function, this space depends on the space needed by the local variables and the formal parameter. In addition, this space depends on the maximum depth of the recursion *i.e.,* maximum number of nested recursive calls.

2.11.3 Time Complexity (UPTU 2005, 07)

The time complexity of an algorithm or a program is the amount of time it needs to run to completion. The exact time will depend on the implementation of the algorithm, programming language, optimising the capabilities of the compiler used, the CPU speed, other hardware characteristics/specifications and so on. To measure the time complexity accurately, we have to count all sorts of operations performed in an algorithm. If we know the time for each one of the primitive operations performed in a given computer, we can easily compute the time taken by an algorithm to complete its execution. This time will vary from machine-to-machine. By analysing an algorithm, it is hard to come out with an exact time required. To find out exact time complexity, we need to know the exact instructions executed by the hardware and the time required for the instruction.

The time complexity also depends on the amount of data inputed to an algorithm. But we can calculate the order of magnitude for the time required. That is, our intention is to estimate the execution time of an algorithm irrespective of the computer machine on which it will be used.

> **DO YOU KNOW ?**
>
> The ***complexity of an algorithm*** is a function $f(n)$ which measures the time and/or space used by an algorithm in terms of the input size n.

Some of the reasons for studying time complexity are:

1. We may be interested to know in advance that whether an algorithm/ program will provide a satisfactory real time response.
2. There may be several possible solutions with different time requirements.

The time complexity of a program depends on all factors that the space complexity depends on.

2.12 TIME-SPACE TRADE-OFF

In computer science, a space-time or time-memory trade-off is a way of solving a problem or calculation in less time by using more storage space (or memory), or by solving a problem in very little space by spending a long time. So if our problem is taking a long time but not much memory, a space-time trade-off would let we use more memory and solve the problem more quickly or, if it could be solved very quickly but requires more memory than, we can try to spend more time solving the problem in the limited memory.

We may sometimes seek a trade-off between space and time complexity. For example, we may have to choose a data structure that requires a lot of storage in order to reduce the computation time. Therefore, the programmer must make a judicious choice from an informed point of view. The programmer must have some verifiable basis based on which a data structure or algorithm can be selected. Complexity analysis provides such a basis. The best algorithm (or program) to solve a given problem is one that requires less space in memory and takes less time to complete its execution. But in practice, it is not always possible to achieve both of these objectives. One algorithm may require more space but less time to complete its execution while the other algorithm requires less time space but takes more time to complete its execution. Thus, we may have to sacrifice one at the cost of the other. If the space is our constraint, then we have to choose a program that requires less space at the cost of more execution time. On the other hand, if time is our constraint such as in real time system, we have to choose a program that takes less time to complete its execution at the cost of more space.

> **DO YOU KNOW ?**
>
> The ***time-space tradeoff*** refers to a choice between algorithmic solutions of a data processing problem that allows one to decrease the running time of an algorithmic solution by increasing the space to store the data and vice versa.

Over 50 years of researches for algorithms related to different problem areas like decision-making, optimisation etc. and scientists have tried to concentrate on computational hardness of different solution methods. The study of *time-space trade-offs, i.e.*, formulae that relate the most fundamental complexity measures, time and space, was initiated by **Cobham,** who studied problems like recognising the set of palindromes on Turing machines. There are two main lines of motivation for such studies: one is the lower bound perspective, where restricting space allows you to prove general lower bounds for decision problems; the other line is the upper bound perspective where one attempts to find time efficient algorithms that are also space efficient (or vice-versa). Also, upper bounds are interesting for finding, in conjunction with lower bounds, the computational complexity of fundamental problems such as sorting. So, mainly algorithms are constrained under two resources *i.e.*, time and space. However, there are certain problems for which it is difficult or even impossible to find such a solution that is both time and space efficient. In such cases, the algorithms are written for specific resource configuration and also taking in consideration nature of input instance. Fundamentally, we measure either of the space and time complexities as a function of the size of the input instance. However, it is also likely to depend upon nature of the input. So, let us define the time and space complexity as below:

Time Complexity $T(n)$: An algorithm A for a problem P is said to have time complexity of $T(n)$ if the number of steps required to complete its run for an input of size n is always less than equal to $T(n)$.

Space Complexity $S(n)$: An algorithm A for a problem P is said to have space complexity of S(n) if the no. of bits required to complete its run for an input of size n is always less than equal to $S(n)$.

If we consider, both time and space requirements, general format is to use the quantity **(time * space)** for a given algorithm. $T(n)*S(n)$ is therefore quite handy to estimate the overall efficiency of an algorithm in general. The amount of space can easily be estimated for most of the algorithms directly and the measurement of run-time of algorithm mathematically or manually by testing can tell us the efficiency of the algorithm. From here we can establish, the range of this $T*S$ term, that we can afford. All algorithms that lie in this range are then acceptable to us.

2.13 ASYMPTOTIC ANALYSIS

We will learn about various techniques to bind the complexity function. In fact, our aim is not to count the exact number of steps of a program or the exact amount of time required for executing an algorithm. In theoretical analysis of algorithms, it is common to estimate their complexity in *asymptotic sense, i.e.*, to estimate the complexity function for reasonably large length of input '*n*'. Big *Oh* (O) notation, omega (Ω) notation and theta (θ) notation are used for this purpose.

2.13.1 Asymptotic Notation

The notations we use to describe the asymptotic running time of an algorithm are defined in terms of functions whose domains are the set of natural numbers $N = \{0, 1, 2 \ldots\}$. Such notations are convenient for describing the worst-case running-time function *T (n),* which is usually defined only on integer input sizes.

Why are asymptotic notations important?

1. They give a simple characterisation of an algorithm's efficiency.
2. They allow the comparison of the performances of various algorithms.

(*a*) Big-oh Notation

We have *O(f),* called *big-oh,* which represents the class of functions that grow no faster than *f*. This means that for all values of *n* greater than some threshold n_0, all of the functions in *O(f)* have values that are no greater than *f*. The class *O(f)* has *f* as an upper bound, so none of the functions in this class grow faster than *f*.

Formally this means that if $g(x) \in O(f)$, $g(n) = cf(n)$ for all $n = n_0$ (where *c* is a positive constant).

For a given function *g(n),* we denote by *O(g(n))* (pronounced "big-oh of *g* of *n*" or sometimes just "oh of *g* of *n*") the set of functions.
$O(g(n)) = \{f(n)$: there exist positive constants *c* and n_0 such that:

$$0 \le f(n) \le cg(n) \text{ for all } n \ge n_0\}.$$

We use O-notation to give an upper bound on a function, to within a constant factor. Figure shows the intuition behind O-notation. For all values *n* to the right of n_0, the value of the function *f (n)* is on or below *g(n).*

Thus, big-oh is the formal method of expressing the upper bound of an algorithm's running time. It is the measure of the longest amount of time it could possibly take for the algorithm to complete.

This is the class that will be of the greatest interest to us. Considering two algorithms, we will want to know **if the function categorising the behaviour of the first is in big oh of the second. If so, we know that the second algorithm does no better than the first in solving the problem.**

- **O (1):** Pronounced "order I" and denoting a function that runs in constant time.
- **O (n):** Pronounced "order N" and denoting a function that runs in linear time.
- **O (N²):** Pronounced "order N squared" and denoting a function that runs in quadratic time.
- **O (log N):** Pronounced "order log N and denoting a function that runs in logarithmic time.

DO YOU KNOW?

Asymptotic means a line that tends to converge to a curve, which may or may not-eventually touch the curve. It is a line that stays within bounds.

- **O (N log N):**Pronounced "order N log N" and denoting a function that runs in time proportional to the size of the problem and the logarithmic time.
- **O(N!):** Pronounced "order N factorial" and denoting a function that runs in factorial time.

Properties of the Big-oh Notation

1. Constant factors may be ignored such as $3n^2$ and $50n^2$ are both $O(n^2)$.
2. Higher powers of *n* grow faster than lower powers.
3. The growth rate of the sum of terms is the growth rate of its fastest growing term, *e.g.*, $5n^3+2n$ is $O(n^3)$.

4. If f is a polynomial of degree d, then *f* is $O(n^d)$.
5. Exponential functions grow faster than powers.
6. Logarithms grow more slowly than powers.
7. The product of upper bound of functions gives an upper bound for the product of the functions. For example, if *f* is $O(n^2)$ and g is $O(\log n)$, then *fg* is $O(n^2 \log n)$.

Limitation of Big-oh Notation

Big-Oh Notation has following two basic limitations:

1. It contains no effort to improve the programming methodology. Big-Oh Notation does not discuss the way and means to improve the efficiency of the program, but it helps to analyse and calculate the efficiency (by finding time complexity) of the program.
2. It does not exhibit the potential of the constants. For example, one algorithm is taking 1000 n^2 time to execute and the other n^3 time. The first algorithm is $O(n^2)$, which implies that it will take less time than the other algorithm which is $O(n^3)$. However in actual execution the second algorithm will be faster for $n < 1000$.

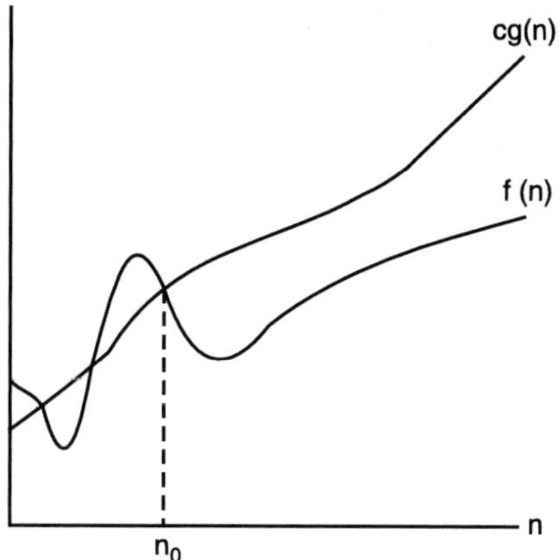

f(n) = O (g(n))

Fig. 2.8.

(*b*) Big-omega Notation

Just as O-notation provides an asymptotic *upper* bound on a function, Ω-notation provides an ***asymptotic lower bound.***

For a given function $g(n)$, we denote by $\Omega(g(n))$ (pronounced "big-omega of *g* of *n*" or sometimes just "omega of *g* of *n*")

$\Omega(g(n)) = \{ f(n)$: there exist positive constants c and n_0 such that:

$0 \le cg(n) \le f(n)$ for all $n \ge n_0\}$.

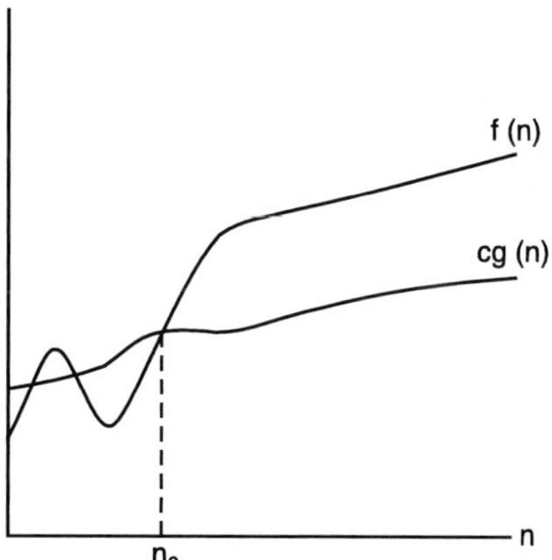

f(n) = Ω (g(n))

Fig. 2.9.

This is almost the same definition as Big-oh except that "$f(n) \ge g(n)$", this makes $g(n)$ a lower bound function instead of an upper bound function. It describes the best that can happen for a given data size.

Because we are interested in efficiency, $\Omega(f)$ will not be of much interest to us because $\Omega(n^2)$, for example, includes all functions that grow faster than n^2 including n^3 and 2^n.

(c) Theta Notation

The lower and upper bound for the function T is provided by the theta notation. For a given function $g(n)$.

We denote by $\theta(g(n))$ the set *of functions* as:

$\theta(g(n)) = \{ f(n)$: there exist positive constants c_1, c_2 and n_0 such that:
$0 \le c_1g(n) \le f(n) \le c_2g(n)$ for all $n \ge n_0\}$

A function $f(n)$ belongs to the set $\theta(g(n))$ if there exist positive constants c_1 and c_2 such that it can be "sandwiched" between $c_1g(n)$ and $c_2g(n)$, for sufficiently large n.

Formally, this class of functions is defined as the place where big-omega and big oh overlap, so $\boldsymbol{\theta(f) = \Omega(f) \cup O(f)}$.

When we consider algorithms, we will be interested in finding algorithms that might do better than the one we are considering. So, finding one that is in big theta (in other words, is of the same complexity) is not very interesting.

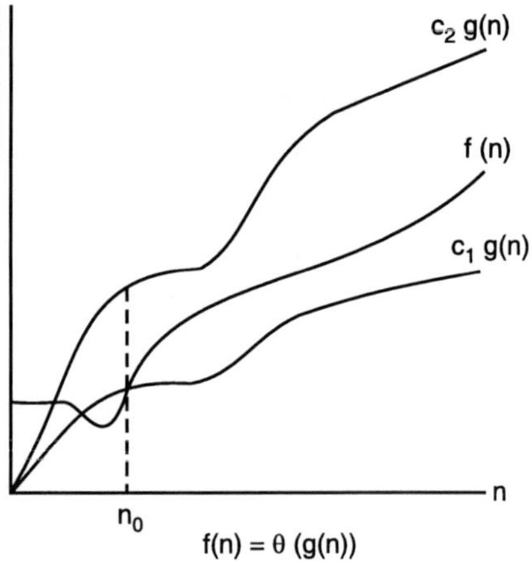

f(n) = θ (g(n))

Fig. 2.10.

Thus asymptotic notation gives us a way to express their relationship.

- If $f(n)$ is $O(g(n))$ this means f(n) grows no faster than g(n).
- If $f(n)$ is $\Omega(g(n))$ this means f(n) grows no slower than g(n).
- If $f(n)$ is $\theta(g(n))$ this means f(n) and g(n) grow at the same rate.

EXAMPLE 2.2. Suppose that it is known that running time of one algorithm is always N log N and the running time of another algorithm is always about N^3. What can you say about the relative performance of the algorithm?

Solution: The rate of growth of algorithms are shown in the given Fig. 2.11.

It can be seen that number of comparisons required in N logN is less than N^3. Thus, it can be seen that N logN is better than N^3. N^3 running time shows that the running time is of linear equation whereas N logN is the recursive equation and the running time of the recursive function depends upon divide and conquer technique. We know divide and conquer technique is always better than the linear equation. So, the performance of N logN is always better than the N^3.

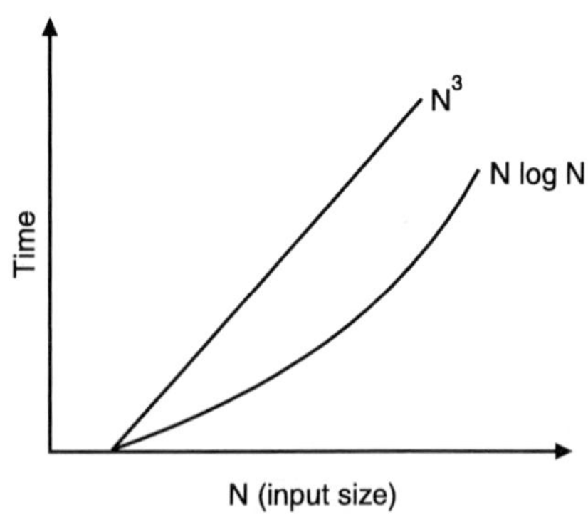

Fig. 2.11.

2.14 ABSTRACT DATA TYPES (ADT)

An Abstract Data Type (ADT) is defined as a mathematical model of the data objects that make up a data type as well as the functions that operate on these objects. We can think of an *Abstract Data Type* (ADT) as a mathematical model with a collection of operations defined on that model. Sets of integers, together with the operations of union, intersection, and set difference, form a simple example of an ADT. So, a useful tool for specifying the logical properties of a data type is abstract data type.

An Abstract Data Type (ADT) is the specification of the data type which specifies the logical and mathematical model of the data type.

An Abstract Data Type is the specification of logical and mathematical properties of a data type or data structure. A data type is a class of data objects together with a set of operations for creating and manipulating them. Data object represents space in the memory of actual processor, which can be recognised by variable name or identifier name by the programmer. Data object is also called a *container for data values.* **A data object is *elementary* if it contains a data value that is always manipulated as a unit. It is called a *data structure* if it is an aggregate of other data objects.** ADT acts as a useful guideline to implement a data type correctly. The specification of an ADT does not imply any implementation consideration. An *implementation* of an ADT is a translation, into statements of a programming language, of the declaration that defines a variable to be of that abstract data type, plus a procedure in that language for each operation of the ADT. An implementation chooses a *data structure* to represent the ADT; each data structure is built up from the basic data types of the underlying programming language using the available data structuring facilities.

The important step is the definition of ADT that involves mainly two parts:

1. Description of the way in which components are related to each other.
2. Statements of operations that can be performed on the data type.

For example, the **int** data type, available in the "C" programming language provides an implementation of the mathematical concept of an integer number. The int data type in "C" can be considered as an implementation of Abstract Data Type, ***INTEGER-ADT*.** **INTEGER-ADT** defines the set of numbers given by the union of the set {-l,-2,......,∞} and the set of whole numbers {0,1, 2,......+ ∞}. **INTEGER-ADT** also specifies the operations that can be performed on an integer number, for example, addition, subtraction, multiplication, etc. In the specification of the **INTEGER-ADT,** following issues must be considered:

(*a*) Range of numbers that will be represented.

(*b*) Format for storing the integer numbers.

This determines the representation of a signed integer value.

When we use abstract data types, our programs divide into two parts:

1. Application part
2. Implementation part

Application part describes the uses of ADT and implementation part implements the abstract data type.

Exercises

1. Define Data structure and also write down the difference between primitive data structure and non-primitive data structure.
2. Differentiate linear and non-linear data structures.
3. Name various data structures. Explain them briefly.
4. What are the operations to be performed on data structures?
5. What is an algorithm? Explain with example the time and space analysis of an algorithm.
6. Distinguish between a program and an algorithm.
7. Explain the different ways of analysing algorithm.
8. Define complexity of an algorithm. What is meant by time-space trade-off?
9. Distinguish between time and space complexity?
10. Define and explain Big-Oh notation.
11. What do you understand by best, worst and average case analysis of an algorithm?

Chapter 3

Introduction to Arrays and Matrices

3.1 INTRODUCTION

It is appropriate that we begin our study of data structures with the array. Arrays are most frequently used in programming. Mathematical problems like matrix, algebra and etc. can be easily handled by arrays. An array is a collection of homogeneous data elements described by a single name. Each element of an array is referenced by a subscripted variable or value, called *subscript*, or *index* enclosed in parenthesis. If an element of an array is referenced by single subscript, then the array is known as ***one-dimensional array*** or linear array and if two subscripts are required to reference an element, the array is known as ***two dimensional array*** and so on. Analogously the arrays whose elements are referenced by two or more subscripts are called ***multi-dimensional arrays***.

Arrays are a data type that is used to represent a large number of homogeneous values that is values that are all of the one data type. The data type could be of type char, in which case we have a string. The data type could just as easily be of type int, float or even another array.

Intuitively, an array is set of pairs, index and value. For each index which is defined, there is a value associated with that index.

An array can be defined as a collection of homogeneous elements, in the form of **index/value pairs**, stored in consecutive memory locations. An array always has a predefined size and the elements of an array are referenced by means of an *index/ subscript*.

Chapter Outline

- Introduction
- Declaration of Arrays
- One-Dimensional Array
- Representing One-Dimensional Array in Memory
- Operations on Arrays
- Limitations of Linear Arrays
- Multi-dimensional Arrays
- Two Dimensional Array
- Representation of Two Dimensional Array in Memory
- Address Calculation for 2D Array
- Traversing Two Dimensional Arrays
- Matrices as Two Dimensional Arrays
- General Multi-dimensional Array
- Ordered Arrays
- Sparse Matrices
- Representation of Sparse Matrix as an Array
- Character String Concept
- Classification of Strings
- C Strings
- Operations on Strings
- Applications of Array
- Passing an Array to a Function
- Vectors

An array can be defined as the collection of the sequential memory locations, which can be referred to by a single name alongwith a number known as the index, to access a particular field or data.

3.2 DECLARATION OF ARRAYS

Like any other variable arrays must be declared before they are used. The general form of declaration is :

```
type variable-name [size];
```

The type specifies the type of the elements that will be contained in the array, such as int, float or char and the size indicates the maximum of elements that can be stored inside the array. So, when we declare an array, we will have to assign a type as well as size. For example, when we want to store 10 integer values, then we can use the following declaration, int A[10];

By this declaration we are declaring A to be an array, which is supposed to contain in all 10 integer values. When the allocation is done for array A then in all 10 locations of size 2 bytes for each integer, *i.e.*, 20 bytes will be allocated and the starting address is stored in A. When we say A[0] we are referring to the first integer value in A.

Hence if we refer to the ith value in array we should write A[i-1]. *When we declare the array of SIZE elements, where, SIZE is given by the user, the index value changes from 0 to SIZE-1.*

Here it should be remembered that the size of the array is 'always a constant' and not a variable. This is because the fixed amount of memory will be allocated to the array before execution of the program. The method of allocating the memory is known as **'static allocation' of memory**.

3.3 ONE-DIMENSIONAL ARRAY

An array that can be represented by only one-dimension such as row or column and that holds finite number of same type of data items is called *one-dimensional* (*linear*) *array.*

One-dimensional array (or linear array) is a set of '*n*' finite numbers of homogeneous data elements such as :

1. The elements of the array are referenced respectively by an index set consisting of '*n*' consecutive numbers.
2. The elements of the array are stored respectively in successive memory locations.

'*n*' number of elements is called the length or size of an array. The elements of an array 'A' may be denoted in C language as:

A[0], A[1], A[2], ... A[*n*–1].

The number '*n*' in A[*n*] is called a subscript or an index and A[*n*] is called a subscriptial variable.

If '*n*' is 10, then the array elements A[0], A[1], ... A[9] are stored in sequential memory locations as follows :

A[0]	A[1]	A[2]	....	A[9]

In C, array can always be read or written through loop. To read a one-dimensional array, it requires one loop for reading and writing the array, for example :

For reading an array of '*n*' elements

```
for  (i=0;i<n;i++)
scanf("%d,&a[i];
```

For writing an array

```
for  (i=0;i<n;i++)
printf("%d,&a[i];
```

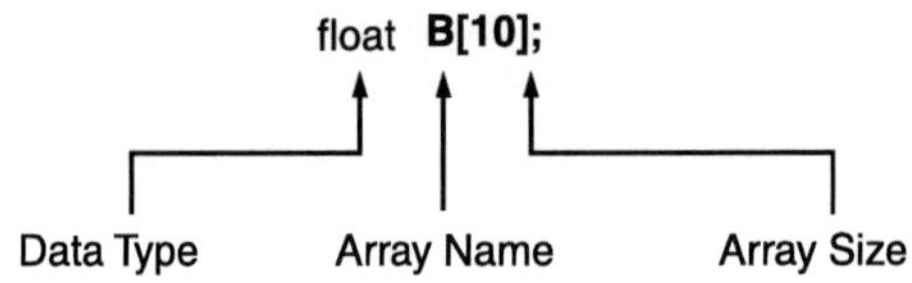

Fig. 3.1 Declaration of Array in C.

Program : /*Program to input element into Array and print all the elements */

```
#include <stdio.h>
#include <conio.h>
void main ( )
{
int a [100], i, n;
printf ("Input Range");
scanf ("%d", &n);
for (i = 0; i < n; ++i)
{
printf ("Input Element");
scanf ("%d", &a[i]);
}
printf ("Elements Are:\n");
for (i = 0; i < n; ++i)
printf ("%d\n", a[i]);
getch ( );
}
```

Program : /*Program to count the no. of positive and negative numbers */

```
#include <stdio.h>
void main ( )
{
int a [50], n,count_neg=0,count_pos=0,i;
printf ("Enter the size of the array\n");
scanf ("%d", &n);
printf ("Enter the elements of the array\n");
```

```
for (i = 0; i < n; i++)
scanf ("%d", &a[i]);
for (i = 0; i < n; i++)
{
if(a[i] < 0)
count_neg++;
else
count_pos++;
}
printf ("There are %d negative numbers in the array\n", count_neg);
printf ("There are %d positive numbers in the array\n", count_pos);
}
```

Program : /*Convert decimal number to binary equivalent*/

```
#include <stdio.h>
#include <conio.h>
void main ( )
{
int d, b[20], i, j;
printf ("Input Decimal No:");
scanf ("%d", &d);
i = 0;
while (d > 0)
{
b[i] = d % 2;
d = d/2;
i++;
}
printf ("Binary Equivalent:");
for (j = i-1; j > = 0; - -j)
printf ("%d", b[j]);
getch ( );
}
```

Program : /*Generate n number of terms of Fibonacci series. */

```
#include <stdio.h>
#include <conio.h>
void main ( )
{
int a [100], i, n;
printf ("Input total no. of terms");
scanf ("%d", &n);
a[0] = 0;
```

```
a[1] = 1;
for (i = 2; i < n; ++i)
a[i] = a[i - 1] + a[i - 2];
printf ("Terms of Fibonacci series :\n");
for (i = 0; i < n; ++i)
printf ("%d", a[i]);
getch ( );
}
```

3.4 REPRESENTING ONE-DIMENSIONAL ARRAY IN MEMORY

Let A be a one-dimensional array with *n* elements. As arrays are stored in consecutive memory location, the system need not keep track of the address of every element of A, but needs to keep track of the address of first element only, denoted by:

`Base (A)`

and called the **base address of A**. Using this address, the computer calculates the address of any element of A by the following formula.

LOC (A[k]) = base (A) + w(k-lower bound)	in general
LOC (A[k]) = base (A) + w*k	in C/C++

where w is the number of bytes per storage location for one element of the array.

EXAMPLE 3.1: Let the base address of the first element of the array is 400 and each element of the array occupies 2 bytes in the memory, then address of 4th element of a array A[10] will be given as

Solution: Address of 4th element = 400 + 2*4
= 400 + 8
= 408

EXAMPLE 3.2. Consider the linear array A(5 : 50), B(–5 : 10) and C(18)

(*a*) Find the number of elements in each array.

(*b*) Suppose Base (A) = 300 and w = 4 bytes for A. Find the address of A[15], A[40] and A[55].

Solution: (*a*) Number of elements in array is equal to the length of array. So

Length = UB – LB + 1

Thus Length (A) = 50 – 5 + 1 = 46
Length (B) = 10 – (–5) + 1 = 16
Length (C) = 18 – 1 + 1 = 18 (LB = 1)

(*b*) We know the formula

LOC(A[k]) = base (A) + w(k – LB)

∴ LOC(A[15]) = 300 + 4 (15 – 5) = 300 + 4 * 10 = 340
LOC(A[40]) = 300 + 4 (40 – 5) = 300 + 4 * 35 = 300 + 140 = 440

A[55] is not an element of A, since 55 exceeds UB = 50.

3.5 OPERATIONS ON ARRAYS

All the operations such as traversing, insertion, deletion, sorting, and merging can be performed on linear arrays.

3.5.1 Traversing

(UPTU 2010)

Traversing is the process of visiting each element of the array exactly once, *i.e.*, starting from first element upto the last element.

Algorithm :

1. I ← LB
2. For I = LB to UB
   PRINT A[I]
3. STOP

Program : /*Program to traverse an array*/.

```
#include <stdio.h>
#include <conio.h>
void main ( )
{
    int n, i, a[20];
    clrscr ( );
    printf ("Enter the length of array");
    scanf ("%d", &n);
    printf ("Enter the elements\n");
    for (i = 0; i < = n-1; i++)
        {
            scanf ("%d", &a[i]);
        }
    printf ("Traversing of the array \n");
    for (i = 0; i < = n-1; i++)
        {
            printf ("\n%d", a[i]);
        }
    getch ( );
}
```

3.5.2 Insertion in One-Dimensional Array

To insert the new element in the existing array following different positions of the array are used

1. At the end of the array.
2. At the beginning of the array.
3. At given position.
4. In the sorted array.

Before inserting the new element, we have to check the **overflow condition**. Suppose MAX is the total number of locations in the given array. If UB = MAX, then array is overflow and STOP.

(*i*) Insertion at the end of array

Algorithm

1. If UB = MAX then write array is OVERFLOW and STOP
2. READ DATA
3. UB ← UB + 1
   A[UB] ← DATA
4. STOP

Example :

Let we have an array

	0	1	2	3	4	
A[ ] =	7	5	2	8	9	

Suppose MAX = 10 and we want to insert element 15 at the end. Then

UB = UB + 1, *i.e.*, now UB = 4 + 1 = 5

and A[5] = 15 [Here UB = 4
LB = 0]

i.e., after insertion

0	1	2	3	4	5	
7	5	2	8	9	15	

(*ii*) Insertion at the beginning of the array

Algorithm

1. If UB = MAX then write "Array is overflow and STOP"
2. READ DATA
3. K ← UB
4. Repeat step (5) while K ≥ LB
5. A(k + 1) ← A(k)
   k ← k – 1
6. A(LB) ← DATA
7. STOP

Example :

Suppose the array is:

	0	1	2	3	4	
A[ ] =	7	5	2	8	9	

MAX = 10

Insert 1 at the beginning

Here UB = 4 and LB = 0

Now K = 4

4 ≥ 0 (True) then A[5] = A[4]

i.e., A[5] = 9 and k = 4 – 1 = 3

3 ≥ 0 (True) then A[4] = A[3]

i.e., A[4] = 8 and k = 3 – 1 = 2

2 ≥ 0 (True) then A[3] = A[2]

i.e., A[3] = 2 and k = 2 – 1 = 1

1 ≥ 0 (True) then A[2] = A[1]

i.e., A[2] = 5 and k = 1 – 1 = 0

0 ≥ 0 (True) then A[1] = A[0]

i.e., A[1] = 7

Now A[0] = 1

i.e., final array after insertion is

0	1	2	3	4	5	
1	7	5	2	8	9	

(*iii*) At the given location

Algorithm

1. If UB = MAX then write "array is overflow and STOP"
2. READ DATA
3. READ LOC as location at where insertion will be made.
4. K ← UB
5. Repeat step (6) while K ≥ LOC
6. A(k + 1) ← A(k)
   k ← k – 1
7. A(LOC) ← DATA
8. STOP

Example :

Suppose the array

	0	1	2	3	4	5	6	
A[ ] =	7	2	5	8	15	12	17	MAX = 10

We want to insert an element 10 at position 4

LOC = 4 and UB = 6

DATA = 10

k = 6

Repeat while k ≥ LOC, *i.e.* 6 ≥ 4 (True)

A[7] = A[6] *i.e.*, A[7] = 17

k = 6–1 = 5

$5 \geq 4$ (True) then A[6] = A[5] i.e., A[6] = 12

k = 5 – 1 = 4

$4 \geq 4$ (True) then A[5] = A[4] i.e., A[5] = 15

Now A[4] = 10

So, the final array after insertion is

0	1	2	3	4	5	6	7
7	2	5	8	10	15	12	17

(*iv*) In the sorted array

Algorithm

1. If UB = MAX then write "array is OVERFLOW and STOP"
2. READ DATA as element to be inserted
3. K ← LB
4. Repeat step (5) while A(k) < DATA
5. k ← k + 1
6. LOC ← k
   k ← UB
7. Repeat step (8) while k ≥ LOC
8. A(k + 1) ← A(k)
   k ← k – 1
9. A(LOC) ← DATA
10. STOP

Example :

Let we have an array

	0	1	2	3	4	5	6
A[ ] =	5	8	10	18	25	30	35

MAX = 10

Suppose we want to insert 15 in this sorted array at proper position.

Here k = 0 (LB = 0 UB = 6)

A(0) < 15 *i.e.,* 5 < 15 (True) then k = 0 + 1 = 1

A(1) < 15 *i.e.,* 8 < 15 (True) then k = 1 + 1 = 2

A(2) < 15 *i.e.,* 10 < 15 (True) then k = 2 + 1 = 3

A(3) $\not<$ 15 *i.e.,* 18 $\not<$ 15 (False)

So, now k = 3.

then LOC = 3 and k = 6 (Since LOC ← k and k ← UB)

Now k ≥ LOC, *i.e.,* 6 ≥ 3 (True)

A(7) = A(6) *i.e.,* A(7) = 35

and k = 6 – 1 = 5

Again 5 ≥ 3 (True)

A(6) = A(5) *i.e.,* A(6) = 30

and k = 5 – 1 = 4

Again $4 \geq 3$ (True)

A(5) = A(4) *i.e.,* A(5) = 25

and k = 4 – 1 = 3

Again $3 \geq 3$ (True)

A(4) = A(3) *i.e.,* A(4) = 18

and k = 3 – 1 = 2

Now $2 \geq 3$ (False)

So A(3) = 15

and the final array after insertion is

	0	1	2	3	4	5	6	7
A[ ] =	5	8	10	15	18	25	30	35

/*Insertion at beginning */

```
#include <stdio.h>
#include <conio.h>
#include <stdlib.h>
#define max 20
void main ( )
{
int a[max], i, j;
Clrscr ( );
printf ("\n Enter limit");
scanf ("%d", &j);
if (j>=max)
{
printf ("\n array size doesn't fulfill demand");
exit (0);
}
printf ("\n Enter Elements");
for (i = 0; i < j; i++)
scanf ("%d", &a[i]);
for (i = j; i > 0; i--)
a[i]=a[i - 1];
printf ("\n Enter new element");
scanf ("%d", &a[0]);
printf ("\n After Insert");
j++;
for (i = 0; i < j; i++)
printf ("%d", a[i]);
```

```
getch ( );
}
```

/*Insertion at last position */

```
#include <stdio.h>
#include <conio.h>
#include <stdlib.h>
#define max 20
void main ( )
{
int a[max], i, j, n;
printf ("\n Enter limit");
scanf ("%d", &j);
if (j>=max)
{
printf ("\n size of limit doesn't match");
exit (0);
}
printf ("\n Enter new element");
for (i = 0; i < j; i++)
{
scanf ("%d", &a[i]);
}
printf ("\n Enter Element");
scanf ("%d", &n);
a[j] = n;
printf ("\n Elements are :");
for (i = 0; i <= j; i++)
printf ("%d", a[i]);
getch ( );
}
```

/*Insertion at given position*/

```
#include <stdio.h>
#include <conio.h>
main ()
   {
      int i, j, k, n, pos;
      float a[50], item;
      printf("Size of the array");
      scanf("%d",&n);
      printf("\n Array elements \n");
      for (i=0; i<n; i++)
```

```
        scanf("%f", & a[i]);
    printf("\n Element to be inserted?");
    scanf("%f",&item);
    printf("\n Position of insertion ?");
    scanf("%d",&pos);
    /*Pushing down the elements*/
    n++;
    for (k=n; k>=pos; k--)
        a[k] = a[k - 1];
    a[--pos] = item;
    printf("\n Array after insertion; \n");
    for (i=0; i<n; i++)
        printf("%6.2f", a[i]);
    printf("\n");
}
```

3.5.3 Deletion

Like insertion, deletion is also possible at any position in the array, such as:

(a) From the end.

(b) From the beginning.

(c) If specific element is giving.

Note : Before deleting the element first we have to check the *underflow condition*, *i.e.*, if N = 0 (N = UB – LB + 1), then array is UNDERFLOW and STOP.

(*a*) From the end.

Algorithm

1. If N = 0 then array is UNDERFLOW & STOP.
2. A(UB) ← NULL
   UB ← UB – 1
3. STOP

(*b*) From the beginning.

Algorithm

1. If N = 0 then array is UNDERFLOW & STOP.
2. K ← LB
3. Repeat step (4) while K < UB
4. A(k) ← A(k + 1)
   K ← K + 1
5. A(UB) ← NULL
   UB ← UB – 1
6. STOP.

Example :

Suppose an array

A[ ] =

0	1	2	3	
2	8	10	15	

We want to delete element 2

Here k = 0 (∵ LB = 0 UB = 3)

K < UB *i.e.,* 0 < 3 (True)

A(0) = A(1) *i.e.,* A(0) = 8

and k = 0 + 1 = 1

1 < 3 (True) A(1) = A(2) *i.e.,* A(1) = 10

and k = 1 + 1 = 2.

2 < 3 (True) A(2) = A(3) *i.e.,* A(2) = 15

and k = 2 + 1 = 3.

$3 \not< 3$ (False)

Now A(3) = NULL

and UB = 3 – 1 = 2.

i.e., final array after deletion is

A[ ] =

0	1	2	
8	10	15	

Deletion of given element

Algorithm

1. If N = 0 then array is UNDERFLOW & STOP.
2. READ DATA as element to be deleted.
3. READ LOC as location at where deletion will be made.
4. K ← LB
5. Repeat step (6) while A(k) ≠ DATA
6. K ← K + 1
7. Repeat step (8) while K < UB
8. A(k) ← A(k + 1)
   K ← K + 1
9. A(UB) ← NULL
   UB ← UB – 1
10. STOP

/*Deletion at beginning */

```
#include <stdio.h>
#include <conio.h>
#include <stdlib.h>
#define max 20
void main ( )
```

```
{
int a[max], i, j;
printf ("\n Enter limit");
scanf ("%d", &j);
if (j>=max)
{
printf ("\n Array size doesn't exit");
exit (0);
}
printf ("\n Input element in an array");
for (i = 0; i <= j; i++)
{
scanf ("%d", &a[i]);
}
for (i = 0; i <= j - 2; i++)
{
a[i]=a[i + 1];
printf ("%d", a[i]);
}
j--;
getch ( );
}
```

/*Deletion at last */

```
#include <stdio.h>
#include <conio.h>
#include <stdlib.h>
#define max 20
void main ( )
{
int a[max], i, j ;
clrscr ( );
printf ("\n Enter limit");
scanf ("%d", &j);
if (j>=max)
{
printf ("\n size not exit");
exit (0);
}
printf ("\n Enter Element");
for (i = 0; i < j; i++)
{
scanf ("%d", &a[i]);
}
```

```
j--;
printf ("\n Array Element");
for (i = 0; i < j; i++)
{
printf ("\n%d", a[i]);
}
getch ( );
}
```

/*Deletion of specific element */

```
#include <stdio.h>
#include <conio.h>
#include <stdlib.h>
#define max 20
void main ( )
{
int a[max], k, i, j, n;
Clrscr ( );
printf ("\n Enter Limit:");
scanf ("%d", &j);
if (j>max)
{
   printf ("\n Array size doesn't exist.");
   exit (0);
}
printf ("\n Enter Elements in Array");
for (i = 0; i < j; i++)
{
   scanf ("%d", &a[i]);
}
printf ("\n Enter Element to delete");
scanf ("%d", &n);
for (i = 0; i < j; i++)
{
   if (n==a[i])
      {
         for (k=i; k<=j-2; k++)
            {
               a[k] = a[k + 1];
            }
         j--;
      }
```

```
    }
    printf ("\n Array Elements are :");
    for (i = 0; i < j; i++)
    {
       printf ("\n%d", a[i]);
    }
    getch ( );
}
```

EXAMPLE 3.3. Write an algorithm for deleting the kth element from a linear array.

Solution: Step (1) : ITEM ← A[k]

Step (2) : Repeat for $j = k$ to $n - 1$

Step (3) : A[j] ← A[j + 1]

Step (4) : $n \leftarrow n - 1$

Step (5) : Stop

3.5.4 Concatenation

In this, concatenate two different size arrays and create the third array, *i.e.*, combination of elements of both array.

Algorithm

1. K ← LBA
2. Repeat step (3) while K ≤ UBA
3. C(k) ← A(k)
   k ← k + 1
4. L ← LBB
5. Repeat (6) while L ≤ UBB
6. C(k) ← B(L)
   L ← L + 1
   K ← K + 1
7. STOP.

Example :

Let us consider array A as

1	2	3	4
5	2	8	3

and B[ ] is

1	2	3
7	9	12

LBA : lower bound of A

UBA : upper bound of A

LBB : lower bound of B

UBB : upper bound of B

Here K = 1

K ≤ UBA *i.e.*, 1 ≤ 4

Create new array C

and C[ ] =

5	2	8	3	

i.e., C[1] ← A[1]
C[2] ← A[2]
C[3] ← A[3]
C[4] ← A[4]

Now k = 5 and condition is false.

Now L = 1

L ≤ UBB *i.e.*, 1 ≤ 3 (True)

C[5] ← B[1]

and L = 1 + 1 = 2
K = 5 + 1 = 6

i.e.,

1	2	3	4	5	
5	2	8	3	7	

Now L = 2 and 2 ≤ 3 (True)

C[6] ← B[2]

and L = 2 + 1 = 3
K = 6 + 1 = 7

and C[ ] =

5	2	8	3	7	9	

L = 3 and 3 ≤ 3 (True)

C[7] ← B[3]

and L = 3 + 1 = 4
K = 7 + 1 = 8

and C[ ] =

5	2	8	3	7	9	12

Now L = 4 and condition is false.

So final array after concatenation is

5	2	8	3	7	9	12

/*Concatenation*/

```
#include <stdio.h>
#include <conio.h>
void main ( )
{
int a[5], b[5], c[10], i, j ;
clrscr ( );
printf ("\n Enter first array element");
```

```
for (i=0; i<=4; i++)
scanf ("%d", &a[i]);
printf ("\n Enter second array element");
for (i=0; i<=4; i++)
{
scanf ("%d", &b[i]);
}
for (i=0; i<=4; i++)
c[i] = a[i];
for (j=0; j<=4; j++)
c[i++] = b[j];
j = i;
printf ("\n After concatenation :");
for (i=0; i<j; i++)
   {
   printf ("%d", c[i]);
   }
getch ( );
}
```

3.5.5 Merging

This operation is use to merge two sorted array and create a third array in the sorted order *i.e.*, suppose two sorted arrays are:

	1	2	3	4	
A[ ] =	7	9	10	20	

	1	2	3	4
B[ ] =	1	5	8	15

After merging array C

C[ ] =	1	5	7	8	9	10	15	20

Algorithm

1. I ← LBA, J ← LBB, K ← 0
2. Repeat step (3) to (4) while I ≤ UBA and I ≤ UBB
3. If A(I) < B(J) then
   C(k) ← A(I)
   I ← I + 1
   else
   C(k) ← A(I)
   J ← J + 1
4. K ← K + 1

LBA = 1
UBA = 4

5. Repeat (6) while I ≤ UBA
6. C(k) ← A(I)
   I ← I + 1
   K ← K + 1
7. Repeat step (8) while J ≤ UBB
8. C(k) ← B(J)
   J ← J + 1
   K ← K + 1
9. STOP.

LBB = 1
UBB = 4

Example :

Suppose two arrays are:

	0	1	2	3
A[ ] =	5	8	15	18

	0	1	2
B[ ] =	2	9	12

Here I = 0
J = 0
K = 0

0 ≤ 3 and 0 ≤ 2 (True)

A[0] > B[0]

So C[0] ← B[0] *i.e.,* C[0] = 2

and J = 0 + 1 = 1

K = 0 + 1 = 1

Now 0 ≤ 3 and 1 ≤ 2 (True)

A[0] < B[1]

So C[1] ← A[0] *i.e.,* C[1] = 5

and I = 0 + 1 = 1

K = 1 + 1 = 2

Here, LBA = 0
UBA = 3
LBB = 0
UBB = 2

Now 1 ≤ 3 and 1 ≤ 2 (True)

A[1] < B[1]

So C[2] ← A[1] *i.e.,* C[2] = 8

and I = 1 + 1 = 2

K = 2 + 1 = 3

Now 2 ≤ 3 and 1 ≤ 2 (True)

A[2] > B[1]

So C[3] ← B[1] *i.e.,* C[3] = 9

and J = 1 + 1 = 2

K = 3 + 1 = 4

Now 2 ≤ 3 and 2 ≤ 2 (True)

A[2] > B[2]

$C[4] \leftarrow B[2]$ *i.e.,* $C[4] = 12$

and $J = 2 + 1 = 3$

$K = 4 + 1 = 5$

Now $2 \le 3$ and $3 \le 2$ (False)

Now $I = 2$, $J = 3$ and $K = 5$

$I \le UBA$ *i.e.,* $2 \le 3$ (True)

$C[5] \leftarrow A[2]$ *i.e.,* $C[5] = 15$

and $I = 2 + 1 = 3$

$K = 5 + 1 = 6$

Now $3 \le 3$ (True)

$C[6] \leftarrow A[3]$ i.e. $C[6] = 18$

and $I = 3 + 1 = 4$

$K = 6 + 1 = 7$

$4 \le 3$ (False)

and $3 \le 2$ (False)

So final array after merging is

2	5	8	9	12	15	18

3.5.6 Inversion

To invert the element of an array and stored inverted element in the same array without using another array.

Algorithm

1. $I \leftarrow LB$, $J \leftarrow UB$
2. Repeat step (3) to (4) while $I < J$
3. $TEMP \leftarrow A(I)$

   $A(I) \leftarrow A(J)$

   $A(J) \leftarrow TEMP$
4. $I \leftarrow I + 1$

   $J \leftarrow J - 1$
5. STOP

/*Inverse of an array*/

```
#include <stdio.h>
#include <conio.h>
void main ( )
{
int a[5], i, j, tmp ;
clrscr ( );
printf ("\n Enter Array Element");
for (i=0; i<=4; i++)
{
```

```
    scanf ("%d", &a[i]);
}
i = 0;
j = 4;
while (i < j)
{
    tmp = a[i];
    a[i] = a[j];
    a[j] = tmp;
    i++;
    j--;
}
printf ("\n Inverse of Array is :");
for (i=0; i<=4; i++)
{
    printf ("%d", a[i]);
}
getch ( );
}
```

3.5.7 Reverse the Order of Elements of an Array

```
#include <conio.h>
#include <stdio.h>
void main ( )
{
    int a[5], i;
    clrscr ( );
    printf ("\n enter the elements");
    for (i=0; i<=4; i++)
    scanf ("\n %d", &a[i]);
    printf ("\n the reverse order");
    for (i=4; i>=0; i--)
    printf ("\n %d", a[i]);
    getch ( );
}
```

3.5.8 Update the Element

```
#include <stdio.h>
#include <conio.h>
void main ( )
{
int a[5], i, j, no;
```

```
clrscr ( );
printf ("\n enter the element");
for (i=0; i<=4; i++)
scanf ("%d", &a[i]);
printf ("\n enter position");
scanf ("%d", &j);
if (j<0 && j>4)
printf ("\n invalid position");
for (i=0; i<=4; i++)
{
if (i==j)
{
printf ("\n enter new element");
scanf ("%d", &no);
a[j]=no;
}
}
printf ("\n no are");
for (i=0; i<=4; i++)
printf("\n%d", a[i]);
getch ( );
}
```

EXAMPLE 3.4. Write a program to find out the maximum and second maximum of an array of n elements.

Solution:

```
#include <stdio.h>
#include <conio.h>
#include <stdlib.h>
void main ( )
   {
   int n, a[20]], max = 0, Smax = 0, i, temp, loc;
   clrscr ( );
   printf ("\n Enter the size of array");
   scanf ("%d", &n);
   if (n > 20)
      {
         printf ("Size is not valid");
         getch ( );
         exit (0) ;
      }
   for (i=0; i<n; i++)
      {
```

```
            printf ("\n enter the elements %d :", i+1);
            scanf ("%d", &a[i]);
         }
    /* Find out the maximum no. */
    for (i=0; i<n; i++)
       {
          if (max < a[i])
          max = a[i];
          loc = i;
       }
    /* Interchange the maximum no. & last position */
    if (loc != n-1)
       {
          temp = a[loc];
          a[loc] = a[n-1];
          a[n-1] = temp;
       }
    for (i = 0; i < n - 2; i++)
          {
             if (Smax < a[i])
             Smax = a[i];
          }
       printf ("\n the maximum no. is %d", max);
       printf ("\n the second maximum no. is %d", Smax);
       getch ( );
    }
```

EXAMPLE 3.5. Write a program to interchange the odd & even position element of an array of n elements.

Solution:

```
    #include <stdio.h>
    #include <conio.h>
    #include <stdlib.h>
    void main ( )
       {
       int n, a[20], temp, i, j;
       clrscr ( );
       printf ("\n Enter the size");
       scanf ("%d", &n);
       if (n > 20)
          {
             printf ("\n Invalid size :");
             getch ( );
```

```
            exit (0) ;
         }
      for (i=0; i<n; i++)
         {
            printf ("\n Enter the numbers");
            scanf ("%d", &a[i]);
         }
   /* Print the values of different subscripts */
   for (i=0; i<n; i++)
         printf ("\na[%d]=%d", i, a[i]);
   for (j=0; j<n-1; j=j+2)
         {
         temp = a[j];
         a[j] = a[j+1];
         a[j+1] = temp;
         }
      printf ("\n after interchanging");
      printf ("\n");
      for (i=0; i<n; i++)
         printf ("\n a[%d] = %d", i, a[i]);
         getch ( );
      }
```

EXAMPLE 3.6. Write an efficient algorithm to find the kth element in a sequence of *n* elements.

(UPTU 2010)

Solution: Algorithm to find the kth element

1. For I = 0 to n
2. If I = k
   PRINT A[I]
3. STOP

```
for (i = 0: i < = n; i + +)
if (i = = k)
printf ("\n %d," a[i]);
```

3.6 LIMITATIONS OF LINEAR ARRAYS

1. Linear arrays are static structures, *i.e.*, memory used by them cannot be reduced or extended.
2. Previous knowledge of number of elements in the array is necessary.
3. Since the elements of these arrays are stored in consecutive locations, the insertions and deletions in these arrays are time consuming.

3.7 MULTI-DIMENSIONAL ARRAYS

An array can be of more than one dimension. There are no restrictions to the number of dimensions that we can have. But as the dimensions increase the memory requirements increase drastically which can result in shortage of memory. Hence a multi-dimensional array must be used with utmost care. For example, the following declaration

```
int a[50][50][50]
```

will result in occupying 50*50*50 = 1,25,000*4 = 5,00,000 bytes of memory.

3.8 TWO-DIMENSIONAL ARRAY

An array in which the elements need to be referred by two indices it is called a **two-dimensional array** or a **"matrix"** and if it is referred using more than two indices, it will be *Multi-dimensional Array*. The structure of the two-dimensional array is illustrated in the following figure : int A[10][10];

A_{00}	A_{01}	A_{02}						A_{08}	A_{09}
A_{10}	A_{11}								A_{19}
A_{20}									
A_{30}									
									A_{69}
A_{70}								A_{78}	A_{79}
A_{80}	A_{81}						A_{87}	A_{88}	A_{89}
A_{90}	A_{91}	A_{92}				A_{96}	A_{97}	A_{98}	A_{99}

Two dimensional array is an array that has two dimensions, such as **row** and **column**, *e.g.*, **int** B[6][8].

This defines a new array B containing six elements. Each of these elements is itself an array containing 8 integers. Fig. 3.2 illustrates such an array.

Array B

	1	2	3	4	5	6	7	8
1	0	10	12	13	19	20	18	23
2	56	51	62	73	79	70	80	63
3	......	......	......	......	......	......	......	......
4	......	......	......	......	......	75	......	......
5	......	......	......	......	......	......	......	......
6	20	31	32	33	39	40	48	33

Size = 6 × 8

Cell B[4] [6]

Fig. 3.2. Graphical representation of two dimensional array

Such an array is called a **two-dimensional array**. Total number of elements in a two dimensional array can be calculated by multiplication of the number of rows and the number of columns. If there are m rows and n columns, then the total number of elements is m × n, and m × n is called the **size of the array**. In mathematics, the two dimensional array is called a matrix and in business it is called table.

A two dimensional array can be expressed as follows :

A_{ij} or A[i, j] for $1 \le i \le m$ and $1 \le j \le n$ (where m and n are the number of rows and columns.

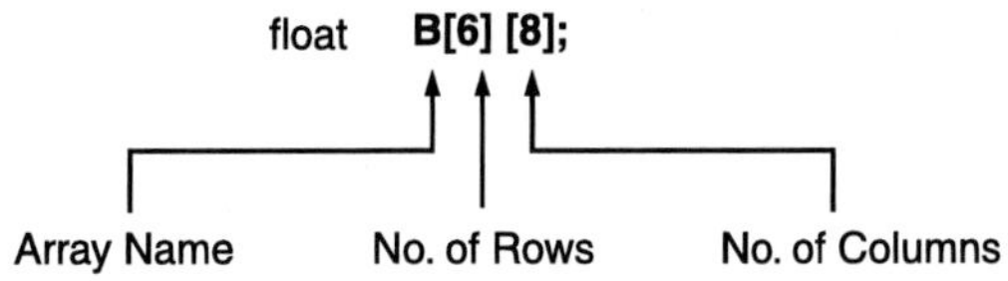

Fig. 3.3. Declaration of two dimensional array in C

A two dimensional array clearly illustrates the difference between a **logical** and a **physical** view of data. A two-dimensional array is a logical data structure that is useful in programming and problem solving. For example, such an array is useful in describing an object that is physically two dimensional, such as a map or a checkerboard. It is useful in organising a set of values that are dependent upon two inputs.

3.9 REPRESENTATION OF TWO-DIMENSIONAL ARRAY IN MEMORY (UPTU 2005)

However, although it is convenient for the programmer to think of the elements of such an array as being organised in a two dimensional table, the hardware of most computers has no such facilities. An array must be stored in the memory of a computer, and that memory is usually linear. By this mean that the memory of a computer is essentially a two dimensional array. So, to implement a two dimensional array, it is necessary to develop a method of ordering its elements in a linear fashion and of transforming a two dimensional reference to the linear representation.

The elements of a two-dimensional array are stored in computer's memory **row-by-row** or **column-by-column**. If the array is stored as *row-by-row*, it is called **row-major order**. If the array is stored as *column-by-column*, it is called **column-major order.**

Under *row-major representation*, the first row of the array occupies the first set of memory locations reserved for the array, the second row occupies the next set, and so forth. Under column-*major representation*, the first column of the array occupies the first set of memory locations reserved for the array, the second column occupies the next set, and so forth.

In *row-major order*, elements of a two dimensional array are ordered as :

$A_{11}, A_{12}, A_{13}, A_{14}, A_{15}, \mathbf{A_{16}}, A_{21}, A_{22}, A_{23}, A_{24}, A_{25}, \mathbf{A_{26}}, A_{31}, \ldots, \mathbf{A_{46}}, A_{51}, A_{52}, \ldots, \mathbf{A_{56}}$, and

In *column-major order*, elements are ordered as :

$A_{11}, A_{21}, A_{31}, A_{41}, \mathbf{A_{51}}, A_{12}, A_{22}, A_{32}, A_{42}, \mathbf{A_{52}}, A_{13}, \ldots, \mathbf{A_{55}}, A_{16}, A_{26}, \ldots, \mathbf{A_{56}}$.

Let us consider the following two-dimensional array :

$$\begin{bmatrix} a & b & c & d \\ e & f & g & h \\ i & j & k & l \end{bmatrix}$$

To make its equivalent **row-major representation**, we perform the following process :

Move the elements of the second row starting from the first element to the memory location adjacent to the last element of the first row. When this step is applied to all the rows except for the first row, you have a single row of elements. This is the row major representation.

By application of above mentioned process, we get {a, b, c, d, e, f, g, h, i, j, k, l}

Column-Major Representation

The second method of representing a two-dimensional array in memory is the column major representation. Under this representation, the first column of the array occupies the first set of the memory locations reserved for the array. The second column occupies the next set and so on. The schematic of a column major representation is shown in figure. Consider the following two-dimensional array :

$$\begin{bmatrix} a & b & c & d \\ e & f & g & h \\ i & j & k & l \end{bmatrix}$$

To make its equivalent **column-major representation**, we perform the following process :

Transpose the elements of the array. Then, the representation will be same as that of the row major representation. By application of above mentioned process, we get

{a, e, i, b, f, j, c, g, k, d, h, l}

EXAMPLE 3.7. Show how the matrix $\begin{bmatrix} 2 & 5 & 8 \\ 1 & 7 & 6 \\ 3 & 4 & 9 \end{bmatrix}$ **would appear in memory when stored in**

(*a*) Row-major order

(*b*) Column-major order

Solution: Given matrix can be represented by 2D-array with 3 rows and 3 columns

(*a*) Row-major order

1	2	3	4	5	6	7	8	9
2	5	8	1	7	6	3	4	9

row 1 row 2 row 3

(*b*) Column-major order

2	1	3	5	7	4	8	6	9

column 1 column 2 column 3

EXAMPLE 3.8. A 2D array with 2 rows and 3 columns stand in memory as

8 5 3 7 2 6

Show the matrix form of two-dimensional array.

Solution: Data may be stored in row major or column major. If the data is stored in row major order then the memory mapping and the matrix form would be as :

8 5 3 (row 1) 7 2 6 (row 2)

i.e. $\begin{bmatrix} 8 & 5 & 3 \\ 7 & 2 & 6 \end{bmatrix}$

If data stored in column major order then the memory mapping and the matrix form would be as

$$\underbrace{8 \quad 5}_{\text{column 1}} \quad \underbrace{3 \quad 7}_{\text{column 2}} \quad \underbrace{2 \quad 6}_{\text{column 3}} \qquad \text{i.e.} \begin{bmatrix} 8 & 3 & 2 \\ 5 & 7 & 6 \end{bmatrix}$$

3.10 ADDRESS CALCULATION FOR 2D ARRAY

Let us consider a two-dimensional array A of size m × n. Like linear array system keeps track of the address of first element only, *i.e.*, base address of the array (Base (A)). Using the base address, the computer computes the address of the element in the ith row and jth column *i.e.* LOC (A[i][j]).

Formulae :

(*a*) Column-Major Order

LOC(A[i][j]) = Base (A) + w[m(j-lower bound for column index)
+ (i-lower bound for row index)] in general

LOC(A[i][j]) = Base (A) + w[mj + i] in C/C++

(*b*) Row-Major Order

LOC(A[i][j]) = Base (A) + w[n(i-lower bound for row index)
+ (j-lower bound for column index)] (in general)

LOC(A[i][j]) = Base (A) + w[n × i + j] in C/C++

where w denotes the number of words per mercury location for the array A or the number of bytes per storage location for one element of the array.

EXAMPLE 3.9. Each element of an array Data[20][50] requires 4 bytes of storage. Base address of data is 2000. Determine the location of Data [10][10] when the array is stored as (*i*) Row-major (*ii*) Column major. **(UPTU 2003-04)**

Solution: Here Data[20][50] So, m = 20 and n = 50

w = 4 bytes

We want to find Data[10][10] So i = 10, j = 10

Suppose, first element of array is Data[1] [1]

(*i*) Row-Major

LOC [Data [10][10]] = Base [Data] + 4[50 (10 – 1) + (10 – 1)]
= 2000 + 4 [50 × 9 + 9] = 2000 + 4 × 459 = 2000 + 1836 = 3836.

(*ii*) Column-Major

LOC [Data [10][10]] = Base [Data] + 4[20 (10 – 1) + (10 – 1)]
= 2000 + 4 [20 × 9 + 9] = 2000 + 4 × 189 = 2000 + 756 = 2756.

EXAMPLE 3.10. Give an array A[6][6] whose first element is stored at location 100. Calculate the location A[2][5] if each element occupies 4 bytes and array is stored row-wise.

Solution: Suppose the first element of the given array is A[1][1] then in row-major method.

LOC [A[i][j]] = Base (A) + 4[6 (2 – 1) + (5 – 1)]
= 100 + 4 [6 × 1 + 4] = 100 + 4 × 10 = 140

EXAMPLE 3.11. Calculate the address of X[4, 3] in a 2D-array X(1..5, 1..4] stored in a row-major order in the main memory. Assume the base address to be 1000 and that each element requires 4 words of storage.

Solution: Here Base(X) = 1000

w = 4

n = 4. (No. of columns)

So, address of X[4, 3]=1000 + 4[4(4 – 1) + (3 – 1)]

= 1000 + 4[4 × 3 + 2] = 1000 + 4 × 14 = 1000 + 56 = 1056.

/* Program of two-dimensional array */

```
#include <stdio.h>
#include <conio.h>
void main ( )
   {
      int a [4][4], i, j;
      clrscr ( );
      printf ("Enter the elements");
      for (i=0; i<=3; i++)
         {
            for (j=0; j<=3; j++)
               {
               scanf ("%d", &a[i][j]);
               }
         }
      printf ("Elements of 2D array are :");
      for (i=0; i<=3; i++)
         {
            for (j=0; j<=3; j++)
               {
                  printf ("%d", a[i][j]);
               }
            printf ("\n");
         }
      getch ( );
   }
```

/* Program to find the sum of major diagonal elements of the 2D array */

```
#include <stdio.h>
#include <conio.h>
void main ( )
   {
      clrscr( );
      int a [5][5], i, j, sum, r, c;
      sum = 0;
```

```
        printf ("Enter No. of row and column of matrix is");
        scanf ("%d %d", &r, &c);
        printf ("Enter elements \n");
        for (i=0; i<r; i++)
           {
              for (j=0; j<c; j++)
                 {
                 scanf ("%d", &a[i][j]);
              if (i==j)
                 sum = sum + a[i][j];
           }
        }
        printf ("The sum of the diagonal elements = %d", sum);
     }
```

EXAMPLE 3.12. **Program to find sum of the elements above and below the diagonals elements.**

Solution:

```
  # include <stdio.h>
  # include <conio.h>
  void main ( )
     {
        int S1 = 0; S2 = 0;
        int i, j
        int a[5][5];
        clrscr ( );
     printf ("Enter 5 x 5" matrix);
     for (i = 0; i < = 4; i ++)
        {
           for (j = 0; j < = 4; j ++)
           {
              scanf ("%d", & a[i][j]);
           }
        }
     for (i = 0; i < = 3; i ++)
     {
        for (j = 0; j < = 3; j++)
        {
           if (i > j)
              S1 = S1 + a[i][j];
           else if (i < j)
              S2 = S2 + a[i][j];
        }
```

```
    }
    printf ("Sum of elements above the diagonal is %d," S1);
    printf ("Sum of elements below the diagonal is %d," S2);
    getch( );
}
```

3.11 TRAVERSING TWO-DIMENSIONAL ARRAYS

Traversing of a two-dimensional array can be classified into two categories depending upon the two type of representation :

1. Column-major order
2. Row-major order

Column-major order

Algorithm :

1. For J = 0 to Column-1
2. For I = 0 to Row-1
3. Print A[I][J]
4. Stop

Row-major order

Algorithm :

1. For I = 0 to Row-1
2. For J = 0 to Column-1
3. Print A[I][J]
4. Stop

C Program : /* Traversing a two-dimensional array */

```
#include <stdio.h>
#include <conio.h>
int i, j;
int A[10][10];
void traverse (int, int);
void create (int, int);
void create (int row, int col)
  {
    printf("\n Input value:\n");
    for (i=0; i<row; i++)
      {
        for (j=0; j<col; j++)
          {
            scanf("%d",&A[i][j]);
```

```
            }
        }
    }
void traverse (int row, int col)
    {
        printf ("\n Traversing in column major order : \n");
        for (j=0; j<col; j++)
            {
                for (i=0; i<row; i++)
                    {
                        printf ("%d", A[i][j]);
                    }
                    printf ("\n");
            }
        printf ("\n Traversing in row major order : \n");
        for (i=0; i<row; i++)
            {
                for (j=0; j<col; j++)
                    {
                        printf ("%d", A[i][j]);
                    }
                    printf ("\n");
            }
    }
void main ( )
    {
        clrscr( );
        int r, c;
        printf("\n Input the no. of rows");
        scanf("%d", &r);
        printf("\n Input the no. of columns");
        scanf("%d", &c);
        create (r, c);
        traverse (r, c);
        getch( );
    }
```

3.12 MATRICES AS TWO-DIMENSIONAL ARRAYS

We have seen that the two dimensional array elements are represented as double subscript values. This property makes them suitable to tackle matrices from the mathematical point of view. It is easy and convenient to translate arithmetic computations involving matrices into algorithms involving two-dimensional arrays.

A set of m × n numbers (real or complex) arranged in a array of m horizontal rows and n vertical columns is called a matrix of order m × n. A matrix is denoted by capital letters such as A, B, C, ... and its elements are denoted by corresponding lower letters suffixed with row index and column index such as a_{ij}, b_{ij}, c_{ij} respectively.

Algebra of Matrices

The operations most commonly performed an matrices are transpose, addition, subtraction and multiplication.

(*a*) Adding two matrices

Adding two similar matrices A and B means to generate a matrix C of same order, whose each element is the sum of corresponding elements of A and B.

Algorithm

1. For I = 0 to m – 1
2. For J = 0 to n – 1
   C[I, J] = A[I, J] + B[I, J]
3. STOP.

(*b*) Transpose of matrix **(UPTU 2009-10)**

Transpose of a matrix is defined as a matrix obtained by interchanging the [i, j] entries to [j, i] entries. That is, transpose of a matrix is obtained by making its columns as rows or *vice-versa*.

Algorithm

1. For i = 0 to m – 1
2. For j = i + 1 to n – 1
   temp = A[i, j]
   A[i, j] = A[j, i]
   A[j, i] = temp.
3. STOP.

(c) Multiplication of matrices

This is one of the most important algebraic operation which can be performed on the matrices. This operation requires a condition. The condition is that we have to apply multiplication as A × B when column of A must be equal to the rows of B, *i.e.*, suppose that A is m × n matrix and B is n × p matrix.

Algorithm : (Given two matrices A and B)

1. For i = 0 to m – 1
2. For j = 0 to n – 1
3. For k = 0 to p – 1
   C[i, j] = C[i, j] + A[i, k] * B[k, j]
4. STOP.

C Program to perform addition, subtraction, multiplication, transpose on matrices

```
#include <stdio.h>
#include <conio.h>
void create (int[3][3]);
```

```
void display (int[3][3])
void add(int[3][3], int[3][3], int[3][3]);
void sub(int[3][3], int[3][3], int[3][3]);
void mul(int[3][3], int[3][3], int[3][3]);
void transpose(int[3][3], int[3][3]);
void main ( )
{
int matrix 1[3][3], matrix 2[3][3], matrix 3[3][3];
clrscr ( );
printf ("Enter first matrix \n");
create (matrix1);
printf ("Enter second matrix \n");
create (matrix2);
printf ("First array");
display (matrix1);
printf ("Second array");
display (matrix2);
add(matrix1, matrix2, matrix3);/*Call to function add */
printf ("Addition");
display (matrix3);
sub(matrix1, matrix2, matrix3);/*Call to function sub */
printf ("Subtraction");
display (matrix3);
mul(matrix1, matrix2, matrix3);
printf ("Multiplication"); /* Call to function mul */
display (matrix3);
printf ("Transpose");
transpose(matrix1, matrix3);/*Call to function transpose */
display (matrix3);
getch ( );
}
/* Creates a matrix */
void create (int mat[3][3])
{
int i,j;
for(i=0; i<=2; i++)
{
for (j=0; j<=2; j++)
printf ("Enter the element");
scanf ("%d, &mat[i][j]);
}
}
```

```
}
/* Display matrix */
void display(int mat[3][3])
{
int i,j;
for(i=0; i<=2; i++)
{
for(j=0; j<=2; j++)
{
printf ("%d", mat[i][j]);
printf("\n");
}
}
}
/* Add matrices */
void add(int mat1[3][3], int mat2[3][3], int mat3[3][3])
{
int i,j;
for(i=0; i<=2; i++)
{
for(j=0; j<=2; j++)
{
mat3[i][j] = mat1[i][j] + mat2[i][j];
}
}
}
/* Subtract matrices */
void sub(int mat1[3][3], int mat2[3][3], int mat3[3][3])
{
int i,j;
for(i=0; i<=2; i++)
{
for(j=0; j<=2; j++)
{
mat3[i][j] = mat1[i][j] - mat2[i][j];
}
}
}
/* Multiply matrices */
void mul(int mat1[3][3], int mat2[3][3], int mat3[3][3])
{
int i,j,k;
```

```
for(k=0; k<=2; k++)
{
for(i=0; i<=2; i++)
{
mat3[k][i]=0;
for(j=0; j<=2; j++)
mat3[k][i]+=m1[k][j]  *m2[j][i];
}
}
}
/* Transpose of a matrix */
void transpose(int m1[3][3], int m2[3][3])
{
int i,j;
for(i=0; i<=2; i++)
{
for(j=0; j<=2; j++)
{
mat2[i][j]=m1[j][i];
}
}
}
```

Program : /*Sorting of whole matrix elements.*/

```
#include <stdio.h>
#include <conio.h>
void main ( )
   {
      int a[20][20], b[400], i, j, t, k, nr, nc;
      printf ("Input total number of rows and columns");
      scanf ("%d %d", &nr, &nc);
      k = 0;
      for (i=0; i<nr; ++i)
         for (j=0; j<nc; ++j)
            {
               printf ("Input elements");
               scanf ("%d", &a[i][j]);
               b[k++] = a[i][j];
            }
   /* Sorting Part */
   for (i=0; i<k-1; ++i)
         for (j=i+1, j<k; ++j)
```

```
            if (b[i] > b[j])
               {
                  t = b[i];
                  b[i] = b[j];
                  b[j] = t;
               }
      k=0;
      printf ("Sorted matrix :\n");
      for (i=0; i<nr; ++i)
         {
            for (j=0; j<nc; ++j)
               {
                  a[i][j] = b[k++];
                  printf ("%4d", a[i][j]);
               }
            printf ("\n");
         }
            getch ( );
      }
```

EXAMPLE 3.13. **Write a program to find the transpose of a matrix.**

Solution:

```
   #include <stdio.h>
   #include <conio.h>
   void main ( )
      {
         int A[3][3], B[3][3], i, j, m, n;
         clrscr ( );
         printf ("Enter the matrix A \n");
         for (i=0, i<=n-1; i++)
            for (j=0; j<=m-1; j++)
               {
                     scanf ("%d", &A[i][j]);
                     B[i][j] = A[i][j];
                  }
               }
            printf ("Transpose of matrix A is \n")
            for (i=0, i<=m-1; i++)
               {
                  for (j=0; j<=n-1; j++)
                     {
```

```
                    printf("%d",  B[i] [j]);
                }
                    printf  ("\n");
                }
            getch  ( );
        }
```

Special Matrices

A square matrix has the same number of rows and columns. Some forms of square matrices are as follows :

(1) Diagonal matrix

A matrix A is called a diagonal matrix if and only if

$$A[i][j] = 0 \quad \text{for} \quad i \neq j$$

e.g.,
$$\begin{bmatrix} 8 & 0 & 0 \\ 0 & 2 & 0 \\ 0 & 0 & 5 \end{bmatrix}$$

(2) Tridiagonal matrix

A matrix A is called a tridiagonal matrix if and only if

$$A[i][j] = 0 \quad \text{for} \quad |i\text{-}j|>1$$

e.g.,
$$\begin{bmatrix} 3 & 2 & 0 & 0 \\ 5 & 4 & 6 & 0 \\ 0 & 1 & 9 & 7 \\ 0 & 0 & 4 & 8 \end{bmatrix}$$

In n × n tridiagonal matrix A, the non-zero elements lie on one of the three diagonals.

(*a*) Main diagonal, for which i = j. It has n elements.

(*b*) Diagonal below main diagonal for which i = j + 1. It has n – 1 elements.

(*c*) Diagonal above main diagonal for which i = j – 1. It has n – 1 elements.

Therefore, an n × n tridiagonal matrix A has 3n – 2 non-zero elements on these three diagonals.

(3) Lower Triangular matrix

A matrix A is lower triangular matrix if and only if A[i, j] = 0 for i < j

e.g.,
$$\begin{bmatrix} 8 & 0 & 0 & 0 \\ 2 & 6 & 0 & 0 \\ 3 & 7 & 9 & 0 \\ 4 & 2 & 5 & 1 \end{bmatrix}$$

(4) Upper Triangular matrix

A matrix A is upper triangular matrix iff A[i, j] = 0 for i > j

e.g.,
$$\begin{bmatrix} 5 & 3 & 8 & 9 \\ 0 & 2 & 4 & 7 \\ 0 & 0 & 5 & 1 \\ 0 & 0 & 0 & 6 \end{bmatrix}$$

EXAMPLE 3.14. Write a program to find whether a matrix is upper triangular or not.

Solution:

```
# include <stdio.h>
 # include <conio.h>
void main ( )
   {
      int a[10] [10], i, j, n, flag;
      clrscr( );
      printf ("Enter the order of the matrix");
      scanf ("%d," & n);
      printf ("Enter elements:\n");
      for (i = 0; i < = n - 1; i ++)
      {
         for (j=0; j<=n-1; j++)
         {
            scanf ("%d," & a[i] [j]);
         }
      }
      flag = 1;
      for (i = 0; i < = n - 2; i ++)
      {
         for (j = i + 1; j < n; j ++)
         {
            if (a[i] [j]! = 0)
               flag = 0;
               break;
         }
      }
   if (flag)
      printf ("Upper triangle matrix");
   else
      printf ("Matrix is not a upper triangle matrix");
      getch ( );
   }
```

EXAMPLE 3.15. Suppose we want to place triangular (lower) matrix A in the memory.

(UPTU, 2011)

Solution:

$$A = \begin{pmatrix} a_{11} \\ a_{21} \to a_{22} \\ a_{31} \to a_{32} \to a_{33} \\ \dots\dots\dots\dots \\ \dots\dots\dots\dots \\ \dots\dots\dots\dots \\ a_{n1} \to a_{n2} \to a_{n3} \dots\dots\dots\dots \to a_{nn} \end{pmatrix}$$

clearly it would be wasteful to store those entries above the main diagonal of A. Since they are all 0's, hence, we store only the other entries of A in a linear array B as:

$B[1] = a_{11} \quad B[2] = a_{21} \quad B[3] = a_{22} \quad B[4] = a_{31} \dots$

Observe that B will contain only

$$1 + 2 + 3 + 4 + \dots + n = \frac{n}{2}(n+1) \text{ elements}$$

Now we will want the formula that gives us the integer L in terms of J and k where

$$B[L] = a_{Jk}$$

Observe that L represents the number of elements in the list upto and including a_{Jk}.

Now, $1 + 2 + 3 + \dots J - 1 = \dfrac{J(J-1)}{2}$ elements in the rows above a_{Jk} and there are k elements in row J upto and including a_{Jk}.

So,

$$\boxed{L = \frac{J(J-1)}{2} + k}$$

Let $J = 3 \quad k = 2$

then $$L = \frac{3(3-1)}{2} + 2 = 3 + 2 = 5$$

i.e., $$B[5] = a_{32}$$

EXAMPLE 3.16. Consider an n-square tridiagonal array A as follows:

Solution:

$$A = \begin{pmatrix} a_{11} \to a_{12} \\ a_{21} \to a_{22} \to a_{23} \\ a_{32} \to a_{33} \to a_{34} \\ \dots\dots\dots\dots \\ \dots\dots\dots\dots \\ a_{n,n-1} \to a_{nn} \end{pmatrix}$$

Note that A has n elements on the diagonal and $n - 1$ elements above and below the diagonal. Hence, A contains atmost $3n - 2$ non-zero elements. Suppose we want to store A in a linear array B as.

$$B[1] = a_{11} \quad B[2] = a_{12} \quad B[3] = a_{21} \quad B[4] = a_{22} \dots$$

Now we want the formula for L in terms of J and K such that $B[L] = a_{Jk}$

Note that there are $3(J - 2) + 2$ elements above a_{Jk} and $k - J + 1$ elements to the left of a_{Jk} so

$$L = 3[J - 2] + 2 + K - J + 1 + 1$$

$$L = 2J + K - 2$$

Let $J = 2 \quad K = 3$

$$L = 4 + 3 - 2 = 5$$

i.e., $B[5] = a_{23}$

3.13 GENERAL MULTI-DIMENSIONAL ARRAY

Most of the programming languages permit two or more than two-dimensional arrays. In multi-dimensional arrays, an element is referenced by more than one subscripts. In general terms, a multi-dimensional array can be represented as :

Storage_class data-type array-name [exp 1][exp 2] ... [exp n]

where srorage_class refers to the storage class of the array. The storage-class is optional and the default value for storage class is automatic for arrays that are defined inside of a function and external for arrays defined outside of a function. Data_type is its data type, array-name is the array name and exp1, exp2, and expn are positive integer expressions that indicate the number of array elements associated with each subscript.

A three-dimensional array with m matrices, each having r rows and C columns will contain m*r*c elements and the array is declared as:

```
int A[m][r][c]
```

In three-dimensional array, we think of data being arranged in array of matrices.

EXAMPLE 3.17. Write a program to input a matrix, determine if it is symmetrical matrix, *i.e.*, (A = A') **UPTU, 2005, 2011**

Solution:

```
#include <stdio.h>
#include <conio.h>
void main ( )
   {
      int a[10][10], n, i, j, flag;
      printf ("Enter the order of the matrix;");
      scanf ("%d", &n);
      printf ("Enter matrix : \n");
      for (i=0, i<=n-1; i++)
         {
            for (j=0; j<=n-1; j++)
               {
                  scanf ("%d", &a[i][j]);
               }
         }
```

```
    flag = 1;
    for (i=0; i<=n-1; i++)
      {
        for (j=i+1; j<=n-1; j++)
          {
            if (a[i][j] - a[j][i] ! = 0)
              flag = 0;
          }
      }
    if (flag)
      {
        printf ("Given matrix is symmetric matrix");
        }
    else
      {
        printf ("Not Symmetric matrix");
      }
    getch ( );
  }
```

EXAMPLE 3.18. Write a C program to read and write array and copy the contents of array A in array B in reverse order then find sum of array A and array B and store it in array C.

Solution:

```
  #include <stdio.h>
  #include <conio.h>
  #define max 50
  void main ( )
    {
      int A[max], B[max], C[max], i, j, n;
      printf ("Enter the value of n");
      scanf ("%d", &n);
      for (i=0, i<=n-1; i++)
        scanf ("%d", &A[i]);
      j = n - 1;
      for (i=0;, i<=n-1; i++)
        {
          B[j] = A[i];
          j = j - 1;
        }
      for (i=0; i<n-1; i++)
        C[i] = A[i] + B[i];
      for (i=0; i<n-1; i++)
```

```
        printf ("The array C is \n \t %d", C[i]);
    getch ( );
}
```

An element of three-dimensional array is specified by three subscripts. The first subscript specifies a plane number, the second subscript a row number and the third a column number. Such an array is useful when a value is determined by three inputs. For example, an array of temperatures might be indexed by latitude, longitude and altitude.

C also allows arrays with more than three-dimensions. For example, a six-dimensional array may be declared by:

```
int A[7][15][3][5][8][2];
```

Referencing an element of this array would require six subscripts.

Suppose that A is an n-dimensional array declared by:

```
int A[r1][r2] ... [rn];
```

and stored in *row-major order*. Each element of A is assumed to occupy w storage locations and Base(A) is defined as the address of the first element of the array, *i.e.*,

```
A[0][0] ... [0]
```

Thus the address of $A[i_1]\,[i_2]\,...\,[i_n]$ may be written as:

$$\text{Base (A)} + w * [in + rn * (i(n-1) + r(n-1) * (... + r3 * (i_2 + r_2 * i_1) ...))]$$

3.14 ORDERED ARRAYS

An array in which the data elements are stored in ascending or descending order is called an *ordered array*. In ascending order, smallest value at index 0, and each successive index holding the value larger than the previous index.

Insertion and Deletion in Ordered Arrays

When we insert an element into ordered array, the correct location must be found for the insertion just above a smaller value and just below a larger one. So, all the larger values must shift downward towards the upper bound to make a space for the inserted element.

Deletion in the ordered arrays required searching such as binary search. Binary search works for ordered arrays.

Advantage and Disadvantages of Ordered Arrays

The biggest advantage of ordered array is quick searching using a binary search. Ordered arrays are therefore useful in the applications where searches are frequent, but insertions and deletions are not. Ordered arrays might be useful for an application of workers personnel information system, where hiring new workers and laying off existing ones would probably be infrequent compared with accessing an existing worker's information.

The disadvantage of ordered array is that it slows down insertion because all the data elements with higher value must be moved towards the upper bound.

3.15 SPARSE MATRICES

Sparse : many elements are zero
dense : few elements are zero

There are special type of matrices in which most of the elements are "0s" (zeros). That is, if a lot of elements from a matrix have a value 0 then the matrix is known as a *sparse matrix.*

There is no precise definition of when a matrix is sparse and when it is not, but it is a concept, which we can all recognise intuitively. If the matrix is sparse, we must consider an alternate way of representing it rather than the normal row-major or column-major arrangement. This is because a sparse matrix is a matrix containing very few non-zero elements. If the user stores the entire matrix including zero elements then there is a wastage of storage space.

Consider a 100 × 100 matrix have 5 non-zero elements per row then the non-zero percentage of the matrix can be calculated as :

$$\left[\left\{\frac{\dfrac{5 \times \text{non-zero elements}}{\text{row}}}{100 \times 100 \text{ elements}}\right\} \times 100 \text{ rows}\right] \times 100\%$$

$$= 0.05\% \text{ non-zero elements.}$$

Thus, to represent this 100 × 100 matrix in memory only 0.05% of the memory is required.

So, significant storage and computational savings can be realised by using sparse storage and solution techniques.

There are two ways of representing sparse matrices.

(a) Array representation

(b) Linked-list representation

There are many types of sparse matrices which can be dealt in the mathematics and engineering fields, the most common of which are the triangular matrices (upper or lower). One such triangular matrix (lower triangular) is shown as :

$$\begin{bmatrix} 4 & 0 & 0 & 0 & 0 & 0 \\ 3 & 2 & 0 & 0 & 0 & 0 \\ 2 & 8 & 5 & 0 & 0 & 0 \\ 7 & 5 & 9 & 10 & 0 & 0 \\ 10 & 6 & 12 & 7 & 11 & 0 \\ 18 & 3 & 8 & 5 & 3 & 1 \end{bmatrix}$$

Figure shows a sparse matrix of dimension 5 × 5

	0	1	2	3	4
0	0	0	0	5	0
1	0	4	0	0	7
2	0	3	0	0	9
3	1	0	0	2	0
4	0	0	8	0	0

3.16 REPRESENTATION OF SPARSE MATRIX AS AN ARRAY

In the array representation of a sparse matrix, only the non-zero elements are stored so that storage space can be reduced. Each non-zero element in the sparse matrix is represented as (Row, Column, Value).

For this, a two-dimensional array containing 3 columns can be used. The first column is for storing the row numbers, the second column is for storing the column numbers and the third column represents the value corresponding to the non-zero element at (row, column) in the first two columns. For example, consider the following sparse matrix.

$$\begin{bmatrix} 2 & 0 & 0 & 0 \\ 0 & 1 & 0 & 0 \\ 0 & 4 & 3 & 0 \\ 0 & 0 & 6 & 0 \end{bmatrix}$$

The above matrix can be represented as

Row	Column	Value
0	0	2
1	1	1
2	1	4
2	2	3
3	2	6

The structure declaration for array is as:

```
#defined MAX 50;
struct triplet {
   int row;
   int column;
   int element;
} spmatrix[MAX];
```

The maximum number of non-zero elements in the sparse matrix is defined by constant MAX.

The array representation will use. [2*(n + 1)*size of (int) + n * size of (T)] bytes of memory where n is the number of non-zero bytes and T is the data type of elements.

The major *limitation* of array representation of a sparse matrix is that we need to know the number of non-zero elements in the matrix.

/* Array implementation of sparse matrix */ **(UPTU 2002)**

```
#include <stdio.h>
#include <conio.h>
#define MAX 50
Struct triplet {
         int row;
         int col;
         int element;
      } spmatrix [MAX];
void main ( )
      {
         int i, j, n, row, col;
         int k = 0;
         printf ("\n Enter matrix dimension :");
```

```
        scanf ("%d %d", &row, &col);
        printf ("\n Enter no. of non-zero elements in sparse
        matrix");
        scanf ("%d", &n);
        for (i=0, i<n; i++)
        {
           printf("\n Enter element row, column and its value");
           scanf ("%d %d %d", &spmatrix[i].row, &spmatrix[i].col,
           &spmatrix[i].element);
        }
     printf ("\n sparse matrix is :");
     for (i=0, i<row; i++)
        {
           printf ("\n");
           for (j=0, j<col; j++)
              {
                 if ((spmatrix [k].row==(i+1)&&(spmatrix[k].col==j+1)
                    {
                       printf ("\t%d", spmatrix [k].element);
                          k++;
                    }
                 else
                       printf ("\t0");
                    }
              }
        }
```

Linked Representation of a Sparse Matrix

Representing a sparse matrix as an array of 3-tuples suffers from one important limitation. When carry out addition or multiplication, we need to know the number of non-zero terms in each of the sparse matrices. As a result, it is not possible to predict the size of the resultant matrix before hand. So instead of an array, we can represent the sparse matrix in the form of a linked list.

In the linked list representation a separate list is maintained for each column as well as each row of the matrix, *i.e.*, if the matrix is of size 4 × 4, then there would be 4 lists for 4 columns and 4 lists for 4 rows.

A node in a list stores the information about the non-zero element of the sparse matrix. The head node for a column list stores the column number, a pointer to the node which comes first in the column and a pointer to the next column head node.

A head node for a row list stores, a pointer to the node, which comes first in the row list, and a pointer to the next row head node.

A node on the other hand stores the row number, column number and the value of the non-zero element of the sparse matrix. It also stores a pointer to the node that is immediately to the right of the node in the row list as well as a pointer to the node that is immediately below the node in the column list.

For example : A sparse matrix (5 × 6)

$$\begin{bmatrix} 0 & 0 & 0 & 8 & 0 & 0 \\ 0 & 0 & 1 & 0 & 0 & 7 \\ 0 & 3 & 0 & 0 & 0 & 0 \\ 0 & 0 & 2 & 0 & 0 & 0 \\ 0 & 0 & 0 & 0 & 0 & 5 \end{bmatrix}$$

The linked list representation of the above sparse matrix is:

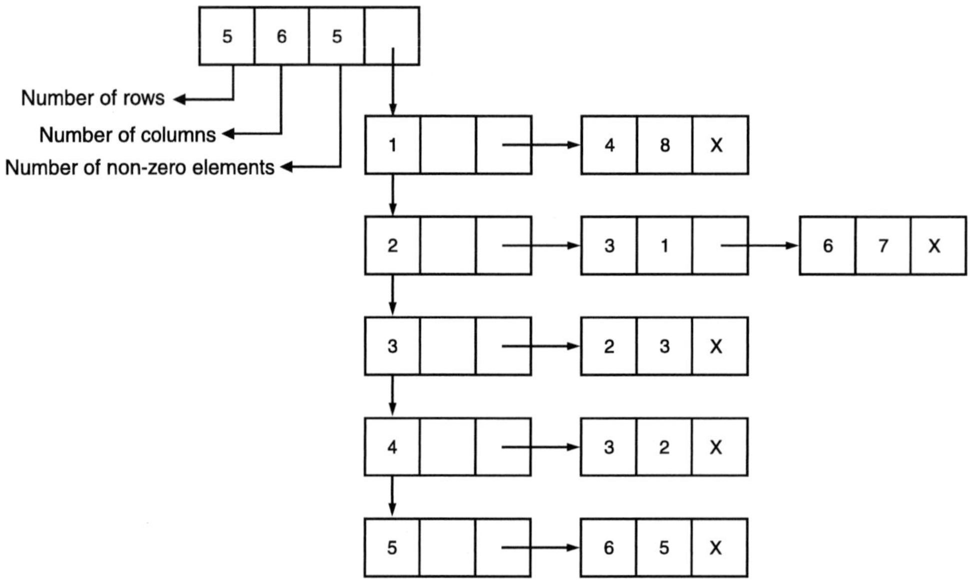

Fig. 3.4.

Different Forms of a Sparse Matrix

A sparse matrix, which is also a square matrix, can fall in the following different categories.

(a) Diagonal Matrix : When the non-zero elements are stored on the leading diagonal of the matrix.

$$\begin{bmatrix} 2 & 0 & 0 & 0 \\ 0 & 3 & 0 & 0 \\ 0 & 0 & 4 & 0 \\ 0 & 0 & 0 & 5 \end{bmatrix}$$

(b) Tridiagonal Matrix : In this non-zero elements are placed on the diagonal or placed below or above the leading diagonal.

$$\begin{bmatrix} 0 & 2 & 0 & 0 \\ 0 & 0 & 5 & 0 \\ 0 & 0 & 0 & 7 \\ 0 & 0 & 0 & 0 \end{bmatrix}$$

(c) Lower Triangular Matrix : Where the non-zero elements are placed below the leading diagonal.

$$\begin{bmatrix} 0 & 0 & 0 & 0 \\ 2 & 0 & 0 & 0 \\ 9 & 3 & 0 & 0 \\ 6 & 0 & 12 & 0 \end{bmatrix}$$

(d) Upper Triangular Matrix : In this, the non-zero elements are placed above the leading diagonal.

$$\begin{bmatrix} 0 & 2 & 0 & 1 \\ 0 & 0 & 5 & 0 \\ 0 & 0 & 0 & 6 \\ 0 & 0 & 0 & 0 \end{bmatrix}$$

EXAMPLE 3.19. Show through appropriate data structure representation of the following 4 × 4 sparse matrix. **(UPTU, 2005)**

$$\begin{bmatrix} 0 & 0 & 11 & 0 \\ 12 & 0 & 0 & 0 \\ 0 & -4 & 0 & 0 \\ 0 & 0 & 0 & -25 \end{bmatrix}$$

Solution: The given 4 × 4 sparse matrix can be represented in array representation a

Row	Column	Value
0	2	11
1	0	12
2	1	– 4
3	3	– 25

and the linked representation is

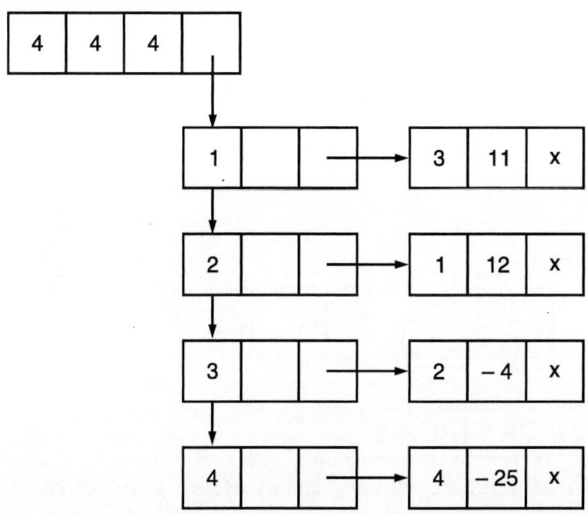

Fig. 3.5.

EXAMPLE 3.20. Let *A* be *N* × *N* sparse matrix array. Write algorithms for the following:

(UPTU, 2005)

(*i*) Find the number NUM of non-zero element in *A*.

(*ii*) Find the product PROD of the diagonal elements.

$(a_{11}, a_{22}, \ldots a_{nn})$

Solution: (*i*) This algorithm find the number of non-zero elements in matrix *A*.

Step 1. Set $k = 0$

Step 2. Repeat for $I = 0$ to $N - 1$

Step 3. Repeat for $J = 0$ to $N - 1$

Step 4. IF A[I, J] $\neq 0$

Step 5. then $k = k + 1$

Step 6. PRINT "Number of non-zero elements", *k*

Step 7. STOP

(*ii*) This algorithm find the product of diagonal elements.

Step 1. PROD = 1

Step 2. Repeat For $I = 0$ to $N - 1$

Step 3. Repeat for $J = 0$ to $N - 1$

Step 4. IF $(I \neq J)$

Step 5. then PROD = PROD * *A* [*I*, *J*]

Step 6. PRINT 'PROD'

Step 7. STOP

3.17 CHARACTER STRING CONCEPT

Just as group of numbers can be stored in an array, a group of characters can also be stored in an array, such an array of character is called as string. In general, a string is a sequence of characters that is treated as a single entity. Any group of characters (except double quote sign) defined between double quotation makes is a string constant.

The end of the string is marked with a special character *null character*, which is the character escape sequence \0.

For example, the word "STRING" is a string of 6 characters. This is stored in the main memory as :

0	1	2	3	4	5	6
S	T	R	I	N	G	\0

3.18 CLASSIFICATION OF STRINGS

The classification of strings are represented as shown in the figure.

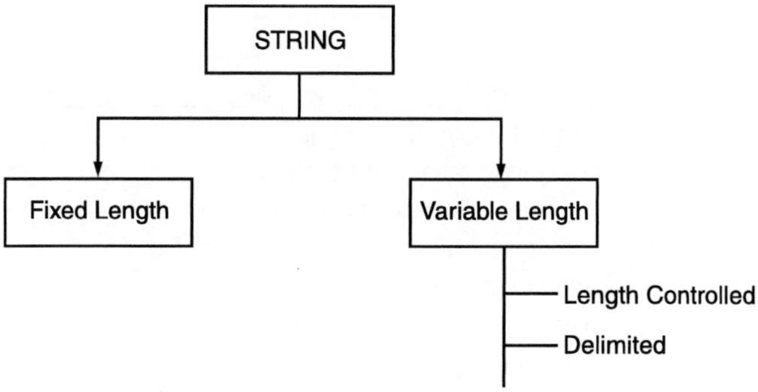

Fig. 3.6.

Figure 3.6 shows a fixed-length, length-controlled and delimited strings are represented in memory.

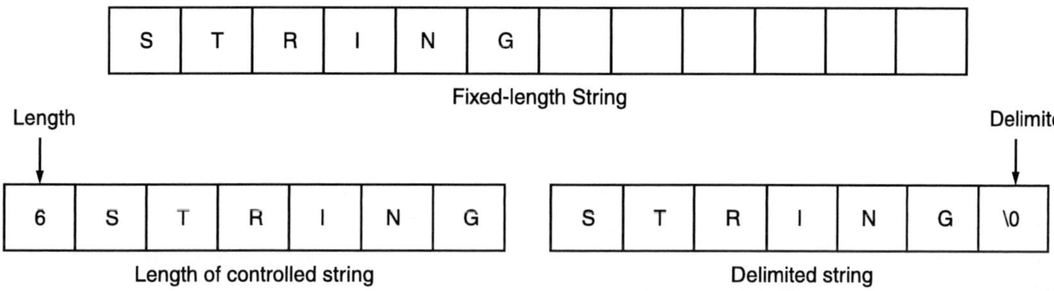

3.19 C STRINGS (UPTU 2005, 07)

A C string is a variable-length sequence of characters that is delimited at the end by the null character (\0). In general, string characters are selected only from the printable ASCII character set. In C, any ASCII characters other than the null character, *i.e.*, delimiter, from being used in a string. A string in computer memory can be represented using an array or a linked list. In array representation, a string of length *n* can be represented by a one-dimensional array with n elements where the ith element of the array stored i^{th} character of the string. All string library functions accepts strings represented as array of characters.

In link list representation each node contains a character and a pointer to the next character in the string. The pointer or link field of the last node of the linked list set to NULL to indicate the end of string.

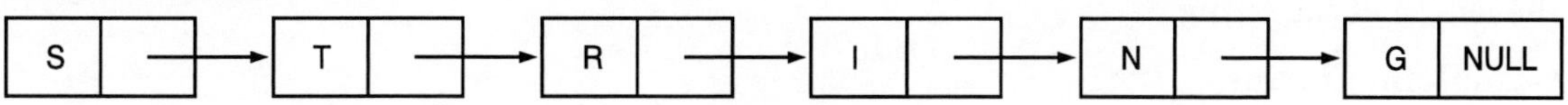

Fig. 3.7.

Important Facts about Strings

* Array elements are stored in contiguous memory locations.
* Last element is always a null character.
* Null character points to the end of a string.
* The string with zero character is called empty string or null string.
* The total number of characters in a string is called string length.

Common standard library functions of C language are:

S.No.	Function	Use	Header File
1	strlen	finds length of a string	string.h
2	strcpy	to copy one string to other	string.h
3	strcmp	to compare two strings	string.h
4	strcat	appends one string at end of other	string.h
5	strupr	converts string to upper case	string.h
6	strlwr	converts string to lower case	string.h
7	strrev	reverse a string	string.h

3.20 OPERATIONS ON STRINGS

Some basic operations that can be performed on strings are as follows :

Operation	Description
Length	Finds the length of the string.
Copy	Copies the contents of one string to another string.
Concatenate	Appends one string at the end of another string.
Compare	Compares two strings to find whether they are identical or not. If they are not identical which comes first in the dictionary order.
Reverse	Reverses the string.
Left Substring	Extracts substring of specified number of characters from the left, *i.e.*, from the beginning of the string.
Right Substring	Extracts substring of specified characters from the right, *i.e.*, from the end of the string.
Mid Substring	Extracts substring of specified number of characters starting from a given position, *i.e.*, from the middle of the string.
Uppercase	Converts lowercase alphabets of string to uppercase.
Lowercase	Converts uppercase alphabets of string to lowercase.

(*i*) Length of String

Length of a string is the number of characters in the string. All spaces in a string are part of the string. However, the null character ('\0') is not part of the string. To find the length of a string we use an integer variable as a counter. The string is scanned from the beginning and the counter is incremented till we find the null character and the final value of the counter gives the length of the string.

C implementation

```
#include <stdio.h>
#include <conio.h>
#define MAXSIZE 100
void main ( )
{
   int len=0;
   char *p,str [MAXSIZE];
   clrscr ( );
```

```
    printf ("enter the string");
    scanf ("%[^\n]", str);
    p=str;
    while(*p!='\0')
       { p=p+1;
       len = len+1;
       }
       printf ("THE LENGTH OF THE STRING IS %d, len+1");
       getch ( );
       }
```

(ii) Comparing Two Strings

To compare two strings, we keep on compare the corresponding characters of the both strings till these corresponding characters are same and none of the string is exhausted. If both the strings are exhausted simultaneously, then it means that the strings are identical else strings are different.

C implementation

```
#include <stdio.h>
#include <conio.h>
#include <stdlib.h>
main ( )
{
   char *cl, *c2, str1[10], str2[10];
   clrscr ( );
   printf ("enter the string no 1");
   scanf ("%s", strl);
   printf ("enter the string no 2");
   scanf ("%s", str2);
   cl=strl;
   c2=str2;
   while(*c1==*c2 &&*c1!='\0' &&*c2!='\0')
      {
         cl=cl+1;
         c2=c2+1;
         }
      if (*c1=='\0' && *c2=='\0')
            printf ("strings are equal\n");
      else
            printf ("strings are not equal\n");
      getch ( );
}
```

(*iii*) Copying a String

To copy a string to another string, the character of the source string are copied one-by-one to the destination string.

/*C Program to copy two strings */

```
#include <stdio.h>
#include <conio.h>
#include <stdlib.h>
void main ( )
   {
      int i=0;
      char *cl, str1[10], str2[10];
      clrscr ( );
      printf ("Enter the string to be copied");
      scanf ("%s", str1);
      c1=str1;
      while(*c1!='\0')
      {
         str2[i]=*c1;
         c1=c1+1
         i=i+1;
      }
         str2[i]='\0';
         printf ("print the string copied %s", str2);
         getch ( );
   }
```

(*iv*) Concatenate two Strings

/*C Program to concatenate a string at the end of a given string*/

```
#include <stdio.h>
#include <conio.h>
#include <stdlib.h>
main ( )
   {
      char *c1,*c2, str1[10], str2[10], str3[10];
      int i=0;
      printf ("Enter the string no 1");
      scanf ("%s", str1);
      printf ("Enter the string no 2");
      scanf ("%s", str2);
      c1=str1;
      c2=str2;
      while(*c1!='\0')
```

```
        {
           str3[i]=*c1;
           c1=c1+1;
           i=i+1;
        }
           str3[i]=' ';
        while(*c2!='\0')
           {
              i=i+1;
              str3[i]=*c2;
              c2=c2+1;
           }
              str3[i+1]='\0';
        printf ("the new string is %s", str3);
        getch ( );
     }
```

(v) Reversing a String

To reverse a string, we use two pointers. One pointer points to the beginning of the string, *i.e.*, first letter of the string. The second pointer is taken so that it can be advanced to the last character of the string. Then we swap the characters pointed by these pointer till the left pointers is less than the right pointer. In each iteration, we increment the left pointer and decrement the right pointer.

```
void strReverse (char *str1)
{
     char *ptr1, *ptr2, temp;
     ptr1 = str1;
     ptr2 = str1;
     while (*ptr2 !='\0')
        ptr2++;
     ptr2--;
     while (ptr1 < ptr2)
     {
        temp = *ptr1;
        *ptr1 = *ptr2;
        *ptr2 = temp;
        ptr1++;
        ptr2--;
     }
  }
```

(vi) Converting Alphabets of String to Uppercase

As we know that ASCII code of uppercase alphabets is from 65 to 91 whereas ASCII code of lowercase alphabets is from 97 to 123. Therefore, to convert a string to uppercase, the string is scanned for each character and we check whether the ASCII code of that character is in the range 97-123. If it is in this range, we subtract 32 from it, which gives corresponding uppercase alphabet.

```
void touppercase (char *str1)
{
        char *ptr1;
        ptr1 = str1;
        while (*ptr1)
        {
            if ((*ptr1>='a') && (*ptr1<='z'))
            *ptr1-=32;
            ptr1++;
        }
}
```

(*vii*) Converting Alphabets of String to Lowercase

To convert a string to lowercase, the string is scanned for each character in the string, we check whether the ASCII code of that character is in the range of 65-91, *i.e.*, it is uppercase alphabet. If it is in this range, we add 32 to it, which gives corresponding lowercase alphabets.

```
void tolowercase (char *str1)
{
        char *ptr1;
        ptr1 = str1;
        while (*ptr1)
        {
            if ( (*ptr1>='A') && (*ptr1<='Z')
                *ptr1+=32;
            ptr1++;
        }
}
```

Note that, to read a string from the user any of the following methods can be used.

1. Declare a string variable, then read the string using scanf( )
   e.g., char name[5];
   scanf ("%s", name);
2. Declare a string variable, then read the string using gets( )
   e.g., char name[25];
   gets (name);

There is a difference between the above two methods. In the case of the first method, the user cannot give the string separated by a white space. Suppose, if the user specifies more than one word for the string, only the first word will be read and assigned to the string variable; whereas in the case of the second method the user can specify any number of words for the string and the entire string will be read and assigned to the string variable.

EXAMPLE 3.21. **Write a program to sort the strings :**

Solution:

```
#include <stdio.h>
#include <conio.h>
#include <string.h>
void main ( )
   {
      int ch, i, j, n;
      char names[25][30], t[30];
      clrscr ( );
      printf ("Enter number of strings");
      scanf ("%d", &n);
      printf ("Enter the strings : \n");
      for (i=0; i<=n; i++)
         gets(names[i]);
      for (i=0; i<=n; i++)
         printf ("\n%s", names[i]);
      printf ("Enter your choice \n");
      printf ("1. Ascending order....\n");
      printf ("\t\t 2. Descending order....\n");
      scanf ("%d", &ch);
      switch (ch)
         {
            case 1 :
               for (i=0; i<=n; i++)
                  {
                     for (j=i+1; j<=n; j++)
                        {
                           if(strcmp(names[i],names[j])>0)
                              {
                                 strcpy (t,names[i]);
                                 strcpy(names[i],names[j]);
                                 strcpy(names[j],t);
                              }
                        }
                  }
               printf ("sorted in Ascending order \n");
               for (i=0, i<=n; i++)
                  puts (names[i]);
               break;
            case 2 :
```

```
            for (i=0; i<=n; i++)
               {
                  for (j=i+1; j<=n; j++)
                     {
                        if (strcmp (names[i],names[j])<0)
                           {
                              strcpy (t,names[i]);
                              strcpy (names[i],names[j]);
                              strcpy (names[j],t);
                           }
                     }
               }
            printf ("Sorted in Descending order \n");
            for (i=0; i<=n; i++)
               puts (names[i]);
            break;
         }
            getch ( );
         }
```

Example 3.22. **Write a C program to check whether the string is palindrome or not.**

Solution:

```
#include <stdio.h>
#include <conio.h>
#include <string.h>
void main ( )
   {
      clrscr ( );
      char str[20], flag;
      int i, j;
      printf ("\n Enter the string");
      gets (str);
      flag = 'y';
      printf ("\n the string is");
      for (i=0; str[i] !='\0'; i++)
         printf ("%c",str[i]);
      for (j=0, i=i-1; i>j; j++, i--)
            {
               if (str[i]!=str[j])
                  {
                     flag = 'n';
```

```
                    break ;
                }
            }
        if (flag =='n')
            printf ("The string is not palindrome");
        else
            printf ("String is palindrome");
        getch ( );
    }
```

3.21 APPLICATIONS OF ARRAY

Polynomials like $6x^3 + 2x^2 + 5x + 2$ can be maintained using an array. Arithmetic operations like addition and multiplication of polynomials are common and often we need to write a program involving these operations. Once we build an array to represent a polynomial, we can use such an array to perform common polynomial operations like addition and multiplication.

3.22 PASSING AN ARRAY TO A FUNCTION

As we know, the name of the array is a pointer to the first element, so we can send the array name to a function for further operations. When we are passing the array name, we are not prefixing the address operator. Remember, the array name itself is a pointer constant, so the name is already contains the starting address of the first element in the array. The calling function looks like

xyz (array name);

In the called function, programmer can declare the array in one of two ways. First, it can be used as traditional array notation.

i.e., int xyz (int array name [ ])

Second, we can also declare the array as a pointer in the function header, *i.e.*,

int xyz (int *array name)

Note that, if we want pass multi-dimensional array, we should use the array syntax in the header declaration and definition. The compiler needs to know that the size of the dimensions for computing offset for pointer arithmetic. Thus, to receive a two-dimensional and three-dimensional array, we need the following declarations in the function header.

int xyz (int array name [ ] [10]) // two-dimensional array

int xyz (int array name [ ] [10][10]) // three-dimensional array

3.23 VECTORS

Some programming languages provide us a facility of vector, which generalises the concept of array. A vector is a one-dimensional ordered collection of numbers. Normally, a number of contiguous memory locations are sequentially allocated to the vector. A vector size is fixed and therefore, requires a fixed number of memory locations. A vector can be *column vector*, which represents a 'n' by 1 ordered collections or a *row vector* which represents a 1 by 'n' ordered collections.

Column Vector appears as:

$$A = \begin{bmatrix} A_1 \\ A_2 \\ A_3 \\ \vdots \\ A_n \end{bmatrix}$$

and a *row vector* appears as

$$A = (A_1, A_2, \ldots A_n)$$

When a vector contains real numbers this is called as **real vector** when they contain complex numbers, they are called **complex vectors**.

Exercises

1. What is an array ? How is it represented in memory ?
2. Give an example to show the usefulness of an array.
3. What are the uses of an array ? What is an ordered array ?
4. Why we need an array ?
5. Write short notes on :
   (*a*) Row-major implementation
   (*b*) Column-major implementation
6. For a single dimensional array if base address is 2000. Find 6th index elements address. If a data stored in this array needs only 3 byte ?
7. How memory is allocated to arrays ? Explain the access mechanism for the array elements.
8. What do you understand by multi-dimensional array ? How they are represented in memory ?
9. How address calculation is done in the array ? Derive the formula for two-dimensional, three dimensional and N dimensional array.
10. What is a sparse matrix ? How is it represented in the memory ?
11. What is an ordered list ? Write a C function to insert an element in ordered stored in an array.

Chapter

4 Linked List

4.1 INTRODUCTION

An array is a very useful data structure provided in programming languages. However, it has some limitations such as:

1. Memory storage space is wasted, as the memory remains allocated to the array throughout the program execution even few nodes are stored. Because additional nodes are still reserved and their storage can't be used for any purpose.
2. The size of array can't be changed after its declaration *i.e.*, its size has to be known at compilation time.
3. The data in the array are separated in computer memory by same distance, which means that inserting an item inside the array requires shifting other data in this array.

These limitations can be overcome by using linked list data structure. Linked list is the most commonly used data structure used to store similar type of data in memory. The elements of a linked list are not stored in adjacent locations as in arrays. Thus, a Linked List has been defined to consist of an ordered set of elements which may increase or decreased as and when required.

In array, once memory space is allocated it cannot be extended. That is why this data structure is called ***static data structure.*** In linked-list, memory space allocated for the elements of the list can be extended at any time. That is why; this linked-list data structure is called **dynamic**

Chapter Outline

data structure. In the case of static data structure (array), each element will have only the data field, whereas in the dynamic data structure, each element will have two types of fields, namely data fields and pointer fields. Data fields have the actual data. The pointer field has the address of the next element. Figure 4.1 shows the structure of an element in a linked list.

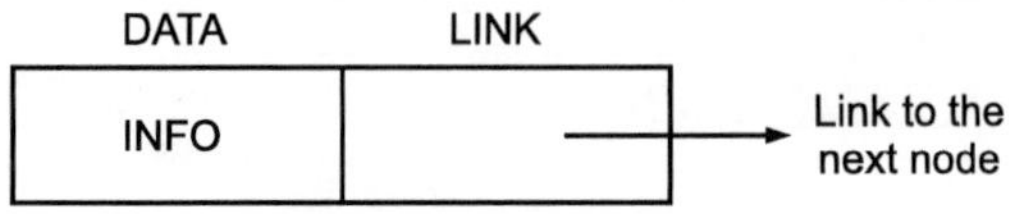

Fig. 4.1.

4.2 DEFINITION

A linked list is an ordered collection of finite homogeneous data elements called nodes where the linear order is maintained by means of links or pointers. That is, each node is divided into two parts : the first part contains the information of the element and the second part, called the link field or the next pointer field, contains the address of the next node in the list.

The number of pointers is maintained depending on the requirements and usage. Based on this, linked list are classified into three categories :

1. Linear linked list
2. Doubly linked list
3. Circular linked list
4. Header linked list
5. Circular doubly linked list

Linear Linked List : In linear linked list, each node is divided in two parts : First part contains the information of the element. And the second part contains the address of the next node in the list. That is, each node has a single pointer to the next node, and in the last node's case a null pointer representing that there are no more nodes in the linked list.

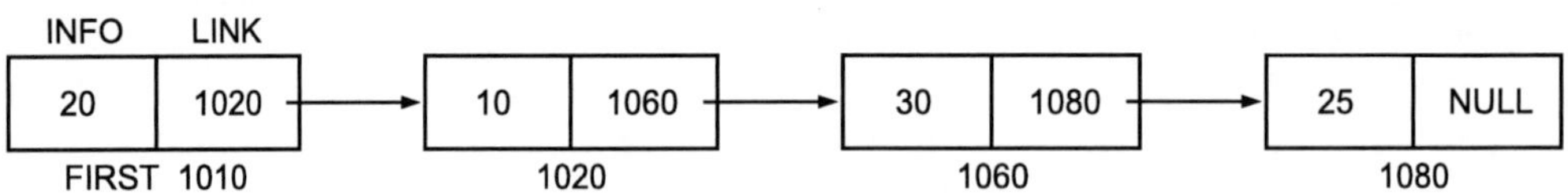

Fig. 4.2. Logical representation of singly linked list

Circular Linked List : It is just like a linear linked list in which the second part of the last node contains the address of the first node in the list *i.e.,* the second field of the last node does not point to the null pointer rather it points back to the beginning of the linked list.

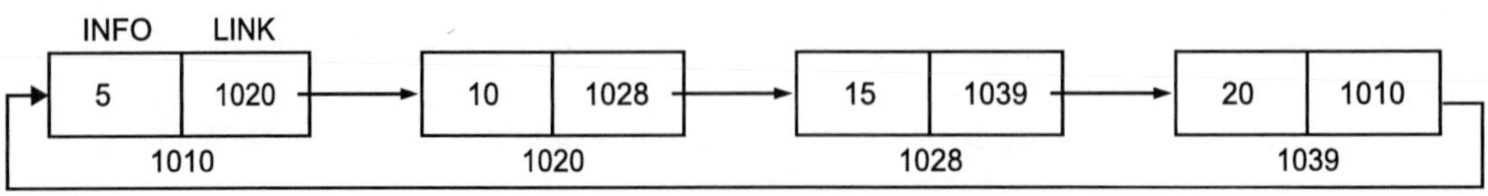

Fig. 4.3.

Doubly Linked List : Doubly linked lists are also called two-way lists. In the case of doubly linked list, two link fields are maintained instead of one as in singly linked list which help in accessing both the successor and predecessor nodes. A structure of the doubly linked list is shown in figure.

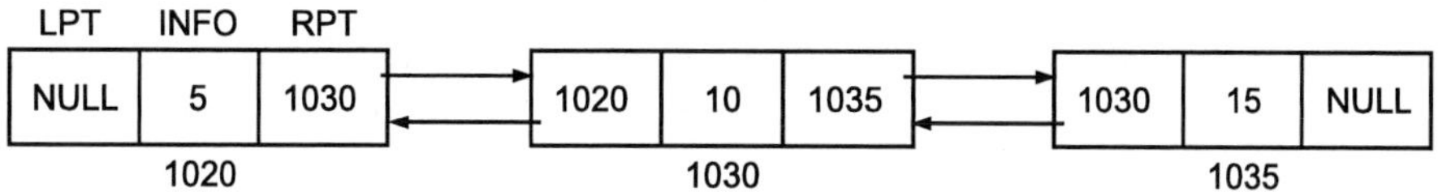

Fig. 4.4.

Header Linked List : In header linked list always contains a single node, called *header node,* at the beginning of the linked list. This header node contains vital information about the linked list such as the number of nodes in the lists, whether the list is sorted or not.

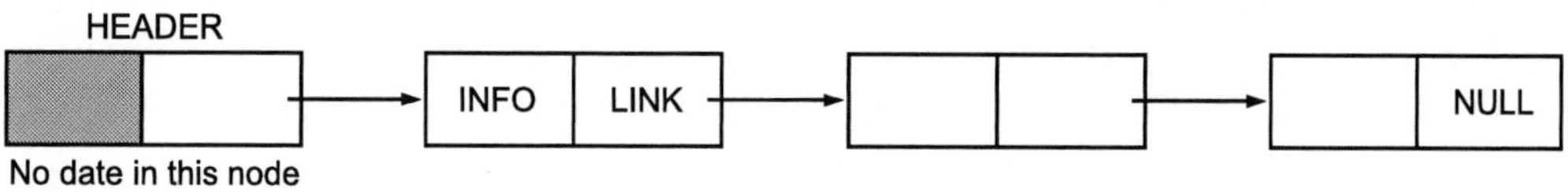

Fig. 4.5.

Circular Doubly Linked List : A circular doubly linked list is one which has both successor and predecessor pointer in circular manner. The aim of considering circular doubly linked list is to simplify the insertion and deletion operations performed on doubly linked list.

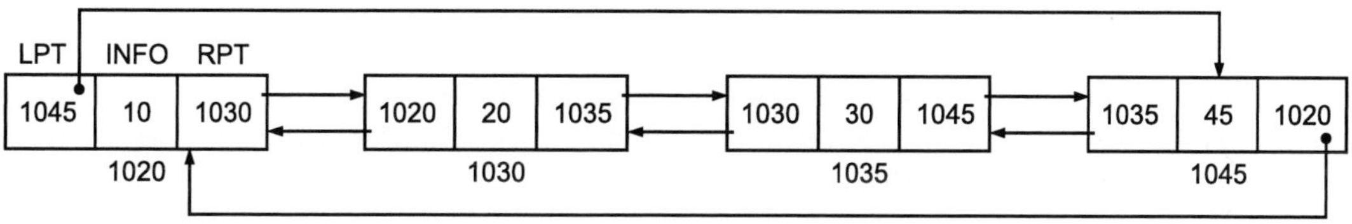

Fig. 4.6. Circular doubly linked list

In circular doubly linked list, the LPT of first node contains the address of the last node and the RPT of the last node contains the address of the first node.

Doubly-linked vs. Singly-linked

Double-linked lists require more space per node (unless one uses xor-linking), and their elementary operations are more expensive; but they are often easier to manipulate because they allow sequential access to the list in both directions. In particular, one can insert or delete a node in a constant number of operations given only that node's address. To do the same in a singly-linked list, one must have the previous node's address. Some algorithms require access in both directions. On the other hand, they do not allow tail-sharing, and cannot be used as persistent data structures.

Circularly-linked vs. Linearly-linked

A circularly linked list may be a natural option to represent arrays that are naturally circular, *e.g.,* for the corners of a polygon, for a pool of buffers that are used and released in FIFO order, or for a set of processes that should be time-shared in round-robin order. In these applications, a pointer to any node serves as a handle to the whole list.

With a circular list, a pointer to the last node gives easy access also to the first node, by following one link. Thus, in applications that require access to both ends of the list (*e.g.,* in the implementation of a queue), a circular structure allows one to handle the structure by a single pointer, instead of two.

4.3 ADVANTAGES AND DISADVANTAGES OF USING LINKED LIST

Advantages

Linked list has number of advantages, some of the very important advantages are :

1. **Linked lists are dynamic data structures :** They can grow or shrink during the execution of a program.
2. The size is not fixed.
3. Data can store non-continuous memory blocks.
4. **Insertion and deletion of nodes are easier and efficient.** Unlike array a linked list provides flexibility in inserting a node at any specified position and a node can be deleted from any position in the linked list.
5. Many more complex applications can be easily carried out with linked lists.

Disadvantages

Some disadvantages of linked list dynamic allocation method are :

1. **More memory :** In the linked list, there is an special field called **link** field which holds address of the next node, so linked list requires extra space.
2. Accessing to arbitary data item is the little bit cumbersome and also time consuming.

In general, a sequential allocation method is used to represent the data, it may result an inefficient or wastage of memory. If we use a linked allocation method, we can efficiently utilise the memory whenever an application demands, we can allocate or deallocate memory dynamically using an pointer declaration.

4.4 SINGLY LINKED LIST

In a linear or singly linked list or one-way list, each node is divided in two parts.

- First part contains the information of the element *i.e.,* INFO.
- Second part contains the link field, which contains the address of the next node in the list.

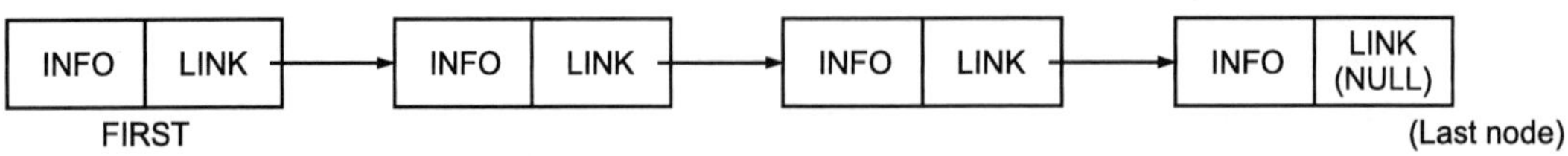

Fig. 4.7.

The linked list is accessed from an external pointer variable say FIRST, which points the first node in the list. The last node of list contains a special value in the LINK field known as NULL and indicates the end of the list.

The list with no nodes is called **empty list** or the **null list** and is denoted by the null pointer in the list pointer variable FIRST. That is:

FIRST = NULL

How can a linked list be created ? A linked list of student having its information as integer data type can be declared as:

```
Struct node
   {
      int info ;
      struct node * link ;
```

```
        } ;
    struct  node    *  first  ;
```

Note that link is declared as a pointer to the structure node itself. This means that **link** points to a location in memory which is of the struct node. This is because **link** contains the address of the next node in the linked list. This type of structure is called **self-referential structures.**

Now apply the dynamic allocation features to implement linked lists. In dynamic memory allocation we use a pointer variable and its size is allocated during the execution of the program. In '*C*', malloc ( ) function is used. The malloc ( ) function simply finds a free block in memory, based on the memory block size requested and then returns a pointer to it, *e.g.*,

PTR = (struct node *) malloc (size of (struct node))

The allocated memory by malloc ( ) function is not free automatically. For this purpose '*C*' has free ( ) function that frees the allocated space, *i.e.*, free (PTR);

In order to access the elements of a structure using a pointer to that structure the [—>] operator is used. For example, the fields info and link can be accessed in any of the following ways after their declaration.

ptr – > info or (* ptr) . info

ptr – > link or (* ptr) . link

Definition : A singly linked list is a linked list in which each node contains only one link field pointing to the next node. In singly linked list, the elements can be traversed only from left to right. The other way is not possible that is why this is also called a one-way list.

Garbage Collection

When an object is no longer used by the program, the heap space it occupies must be recycled so that space is available for new objects that are going to used in future. The operating system of a computer may periodically collect all the deleted space onto the free storage list. Any technique which does this collection is called garbage collection. It is a list of free nodes arranged in the form of stack (Last In First Out). Where **AVAIL** is a pointer variable (like First) containing address of top most free node of the availability list.

To insert a new node into link list, we have to carry a free node from availability list from AVAIL position. Sometimes new nodes are to be inserted but there is no available space *i.e.,* the free storage list is empty (AVAIL = NULL) then this situation is usually called ***overflow.***

4.5 OPERATIONS ON SINGLY LINKED LIST

Let us assume that PTR, CPT, TPT etc. are user defined working pointer variable may contains the address of node.

4.5.1 Creation of a Singly Linked List

Algorithm : CREATION OF A SINGLY LINKED LIST

```
Step 1  PTR = AVAIL
    AVAIL = LINK (AVAIL)
    Read INFO (PTR)
    First = PTR
```

```
Step 2   CH = "Y"
Step 3   Repeat step 4 to 5 while CH = = "Y"
Step 4   CPT = AVAIL
   AVAIL = LINK (AVAIL)
   Read INFO (CPT)
   LINK (PTR) = CPT
   PTR = CPT
Step 5   Read choice <Y/N> Into CH for more Node
Step 6   LINK (PTR) = NULL
Step 7   STOP
```

In step 1 we carry a free node from availability list (AVAIL) and assign its address to PTR, AVAIL Pointer is reset to next top node. Store information into INFO Part of PTR node and assign its address to FIRST, an external Pointer variable. Now Repeat step 4 and 5 to create a new node and insert a next position and linked to previous one, thus every time user is asked before inserting a new node. At last when user input choice 'N', then Last node Link is assigned a NULL address.

EXAMPLE 4.1. Create a singly linked list of elements 5, 10, 25, 20.

Solution: First we take a free node from availability list and assign its address to ptr and read the info field, *i.e.,* 5 and assign its address to FIRST

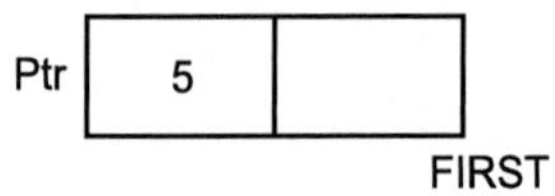

Fig. 4.8.

Now again take a new node cpt from availability list and read info field

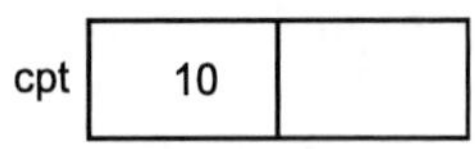

Fig. 4.9.

Now we link cpt to ptr by assign its address to Link (ptr)

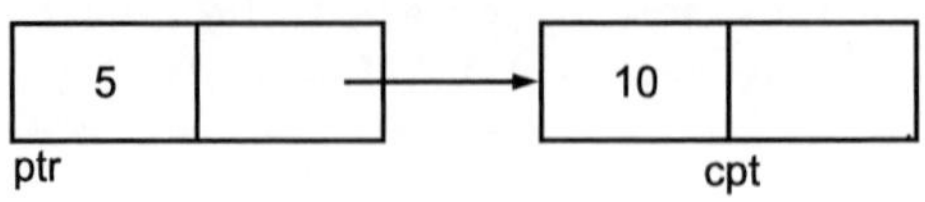

Fig. 4.10.

Similarly node 25 and 20 is added

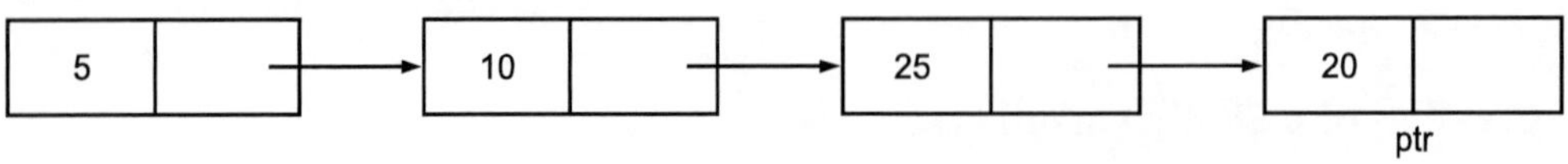

Fig. 4.11.

Now assign a NULL address in last node *i.e.,* in Link (ptr)

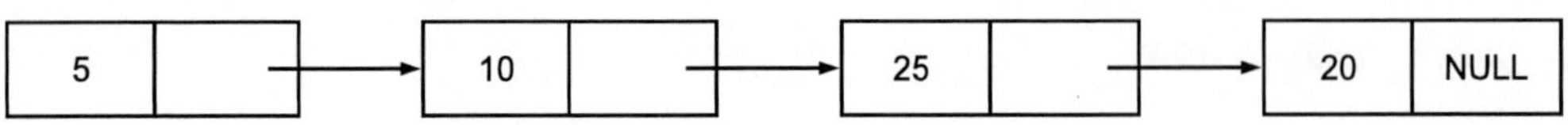

Fig. 4.12.

Implementation in C Language

```
#include<stdio.h>
#include<conio.h>
#include<alloc.h>
struct node
{
   int info;
   struct node *link;
} ;
struct node *first;

void main ( )
{
   void create ( );
   clrscr ( );
   create ( );
   getch ( );
}
void create ( )
(
struct node *ptr, *cpt;
char ch;
ptr = (struct node*) malloc (sizeof (struct node));
printf ("Input first node information");
scanf ("%d', &ptr → info);
first = ptr;
do
{
   cpt = (struct node*) malloc(sizeof(struct node));
   print("Input next node information");
   scanf("%d", &cpt → info);
   ptr → link = cpt;
   ptr = cpt;
printf ("Press <Y/N> for more node");
   ch = getch ( );
}
while (ch = = 'Y');
ptr → link = NULL;
}
```

4.5.2 Traversing of a Singly Linked List

Algorithm : TRAVERSING OF A LINK LIST

1. PTR = FIRST
2. Repeat step (3) to (4) while PTR ¹ NULL
3. Print INFO (PTR)
4. PTR = LINK (PTR)
5. STOP

Implementation :

```
void traverse ( )
{
struct node *ptr;
printf ("Traversing of link list:\n");
ptr = first;
while (ptr! = NULL)
{
printf ("%d\n", ptr -> info);
ptr = ptr -> link;
}
}
```

4.5.3 Insertion in a Singly Linked List

To insert a new node in the existing link list. The insertion can be possible at following any position.

1. At the beginning
2. At the end
3. At after given information node
4. In the sorted information list

Before Insertion, we have to check the **overflow condition**

If AVAIL == NULL then write link list is overflow and stop.

To ***create a new node,*** we use the following step

PTR = AVAIL

AVAIL = LINK (AVAIL)

(*a*) Insertion at the Beginning

Algorithm

Step 1 IF AVAIL = = NULL then write link list is overflow and stop

Step 2 PTR = AVAIL
AVAIL = LINK (AVAIL)
Read INFO (PTR)

Step 3 LINK (PTR) = FIRST
FIRST = PTR

Step 4 STOP

C Implementation :

```
void insert_beginning ( )
{
struct node *ptr;
ptr = (struct node*) malloc (sizeof (struct node));
if (ptr = = NULL)
{
printf ("overflow\n");
return;
}
printf ("Input New Node information");
scanf ("%d", &ptr → info);
ptr → Link = first;
first = ptr;
}
```

EXAMPLE 4.2: Insert 15 at the beginning in the given linked list

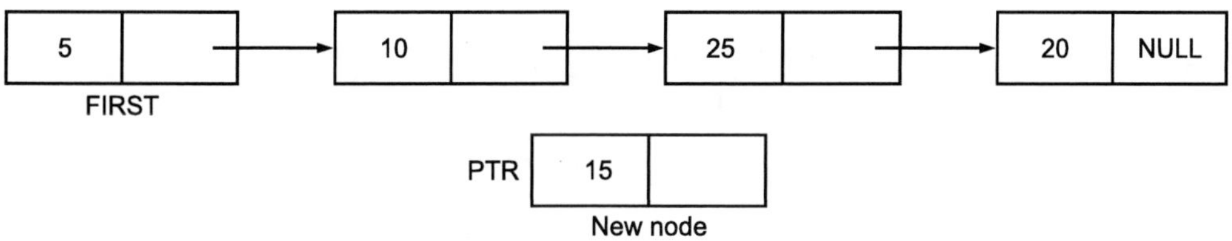

Fig. 4.13.

Assign the address of FIRST in Link (PTR), we get

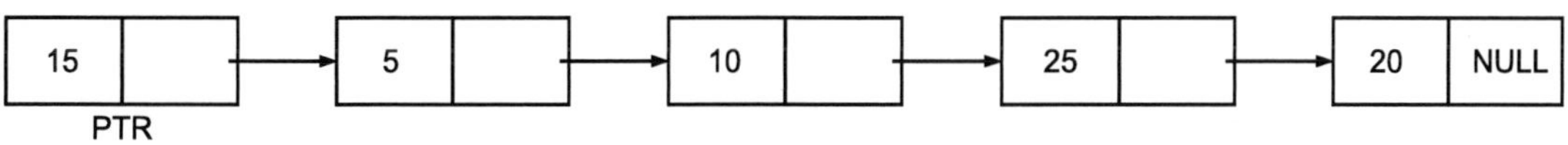

Fig. 4.14.

and now PTR is FIRST node *i.e.*

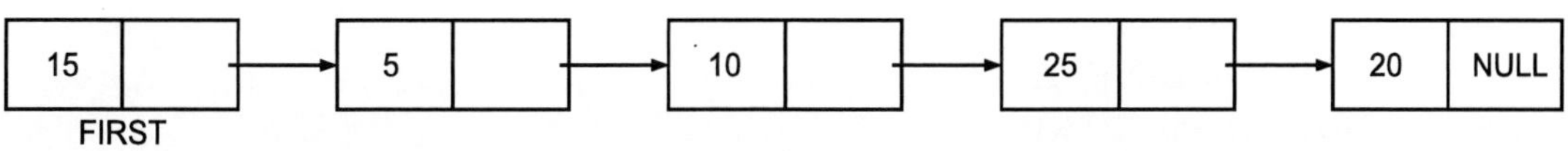

Fig. 4.15.

(*b*) Inserting a node at the End.

Suppose a linked list as

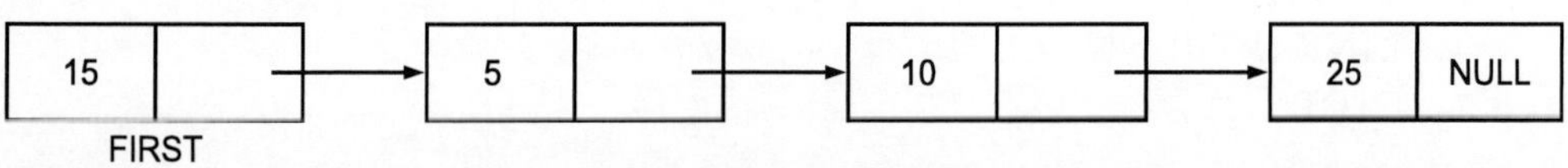

Fig. 4.16.

and we want to insert a node 30 at the end.

Take a pointer CPT and assign the address of FIRST in it. That is now, FIRST is CPT.

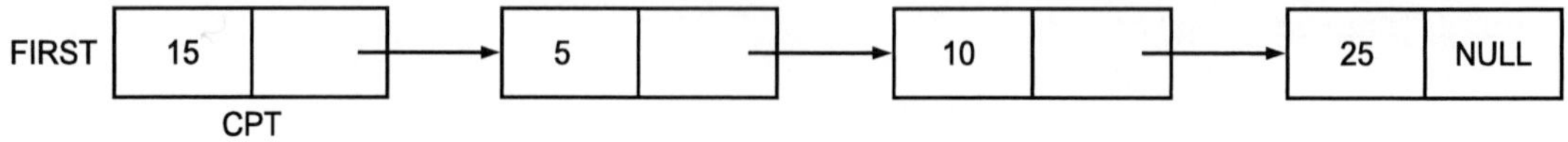

Fig. 4.17.

Now, we move CPT at the last node by assigning Link (CPT) into CPT *i.e.* CPT = Link (CPT), when Link (CPT) = NULL this is last node.

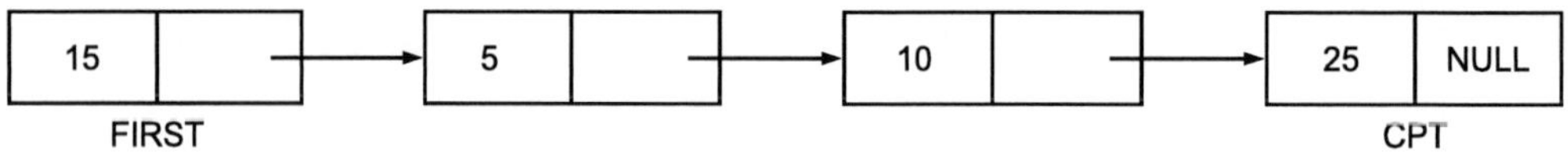

Fig. 4.18.

Now insert node 30 at this position.

Take node 30 as PTR, *i.e.,*

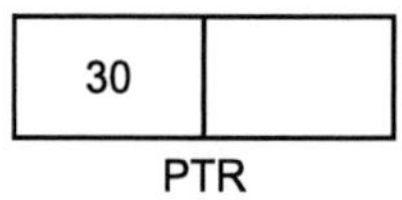

Fig. 4.19.

Assign the address of PTR in the Link (CPT) *i.e.,*

Link (CPT) = PTR

and Link (PTR) = NULL to make it last node.

Thus,

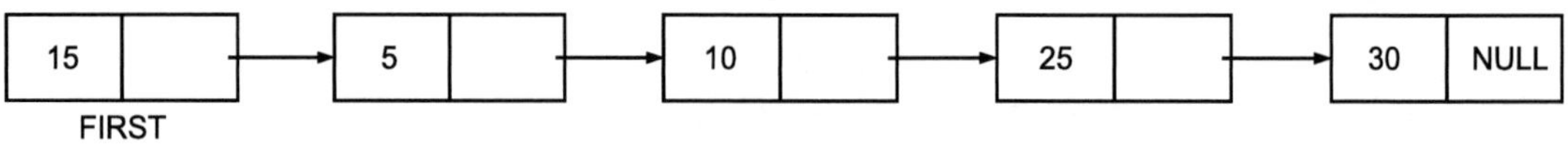

Fig. 4.20 After insert node 30.

Insertion at the End

Algorithm

Step 1 IF AVAIL = = NULL then write link list is overflow and stop

Step 2 PTR = AVAIL
AVAIL = LINK (AVAIL)
Read INFO (PTR)

Step 3 CPT = FIRST

Step 4 Repeat step 5 while LINK (CPT) = = NULL

Step 5 CPT = LINK (CPT)

Step 6 LINK (CPT) = PTR
LINK (PTR) = NULL

Step 7 STOP

C Implementation :

```
void insert_end ( )
{
```

```
struct node *ptr; *cpt;
ptr = (struct node*) malloc (sizeof (struct node));
if (ptr = = NULL)
{
printf ("Link list is overflow\n");
return;
}
printf ("Input New Node information");
scanf ("%d", &ptr → info);
cpt = first;
while (cpt → link! = NULL)
cpt = cpt → link;
cpt → link = ptr;
ptr → link = NULL;
}
```

(*c*) Insertion at after given INFO Node

Algorithm :

Step 1 IF AVAIL = = NULL then write overflow and stop

Step 2 Read DATA as information of node after which insertion will be made

Step 3 PTR = AVAIL
AVAIL = LINK (AVAIL)
Read INFO (PTR)

Step 4 CPT = FIRST

Step 5 Repeat step 5 while INFO (CPT)! = DATA

Step 6 CPT = LINK (CPT)

Step 7 LINK (PTR) = LINK (CPT)
LINK (CPT) = PTR

Step 8 STOP

Suppose we insert a node 40 after node 30 in the linked list as shown in figure.

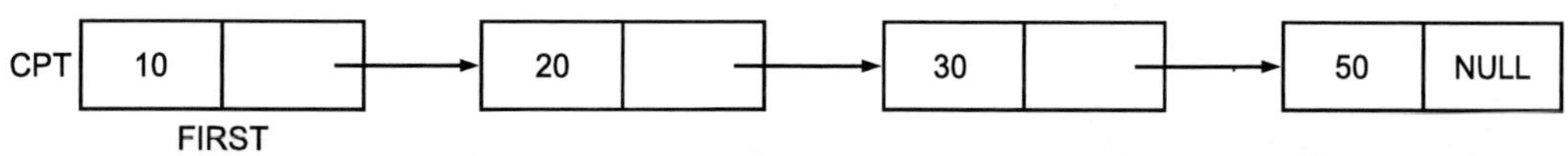

Fig. 4.21.

Make FIRST as CPT and move CPT at the specified position by checking INFO (CPT) Here DATA = 30, so stop moving when INFO (CPT) = DATA

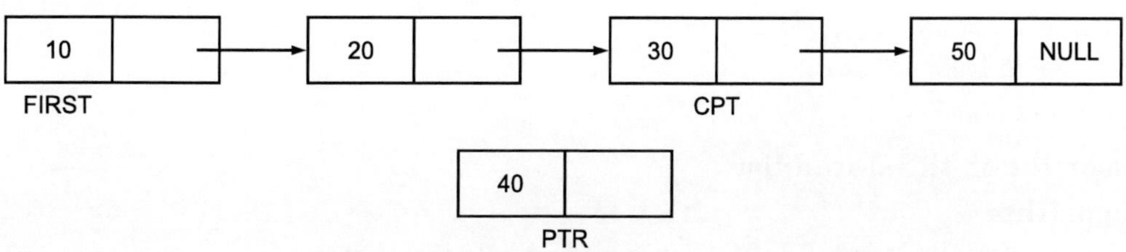

Fig. 4.22.

Now assign

(i) Link (CPT) address in Link (PTR) and

(ii) PTR address in Link (CPT)

That is

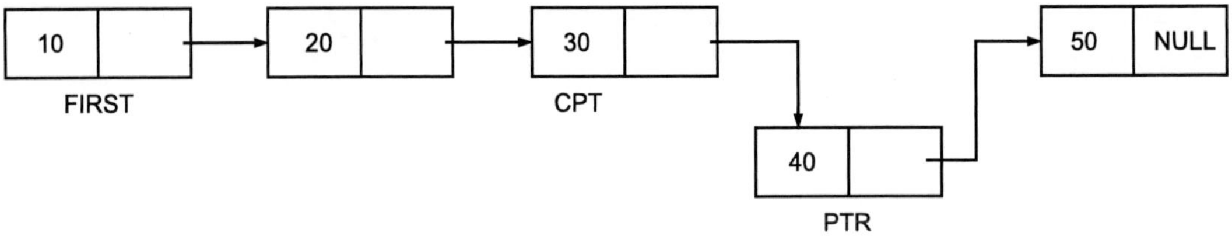

Fig. 4.23.

and final linked list after insertion is:

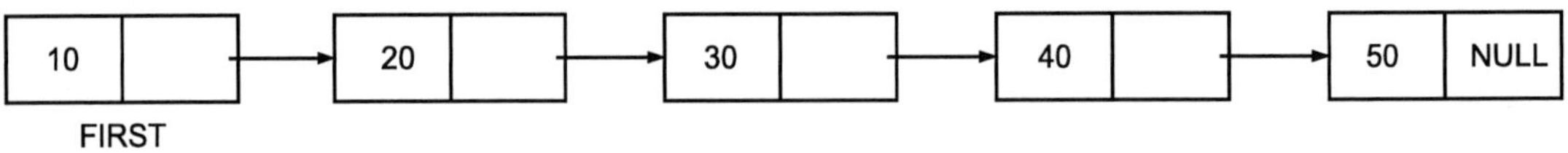

Fig. 4.24.

C Implementation :

```
void insert_given_node ( )
{
struct node *ptr, *cpt;
int data;
ptr = (struct node*) malloc (sizeof (struct node));
if (ptr = = NULL)
{
printf ("overflow\n");
return;
}
printf ("Input New Node information");
scanf ("%d", &ptr → info);
printf ("Input information of node after which insertion will be made");
scanf ("%d", &data);
cpt = first;
while (cpt → info! = data)
cpt = cpt → link;
ptr → link = cpt → link;
cpt → link = ptr;
}
```

***(d)* Insertion in the stored list**

Algorithm

Step 1 IF AVAIL = = NULL then write overflow and stop

Step 2 PTR = AVAIL
AVAIL = LINK (AVAIL)
Read INFO (PTR)

Step 3 CPT = FIRST

Step 4 Repeat step 5 while INFO (CPT) < INFO (PTR)

Step 5 TPT = CPT
CPT = LINK (CPT)

Step 6 LINK (TPT) = PTR
LINK (PTR) = CPT

Step 7 STOP

C Implementation

```
Void insert_sorted_link ( )
{
struct node *ptr, *cpt, *tpt;
ptr = (struct node*) malloc (sizeof (struct node));
if (ptr = = NULL)
{
printf ("overflow\n");
return;
}
printf ("Input new node");
scanf ("%d", &ptr → info);
cpt = first;
while (cpt → info < ptr → info)
{
tpt = cpt;
cpt = cpt → link;
}
tpt → link = ptr;
ptr → link = cpt;
}
```

EXAMPLE :

Let us consider a linked list

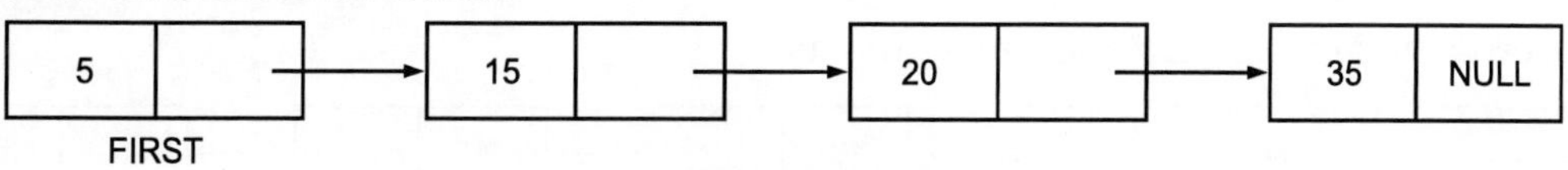

Fig. 4.25.

insert node 18 in this sorted linked list.

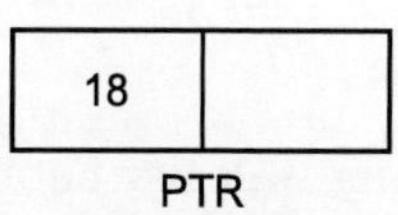

Fig. 4.26.

CPT = FIRST

i.e., now first node is CPT.

If INFO (CPT) < INFO (PTR) then

TPT = CPT and CPT = Link (CPT).

i.e.,

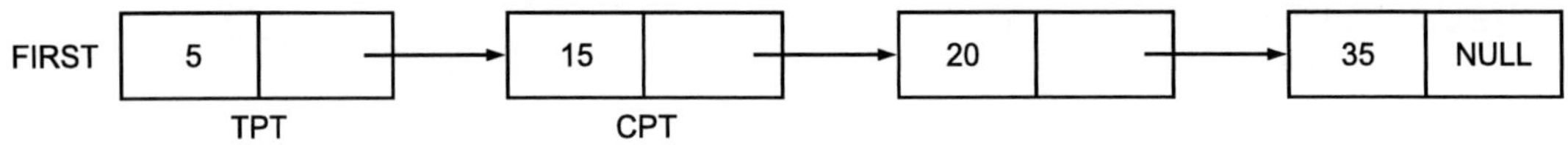

Fig. 4.27.

Now again INFO (CPT) < INFO (PTR) *i.e.,* 15 < 18.

TPT = CPT and CPT = Link (CPT)

i.e.,

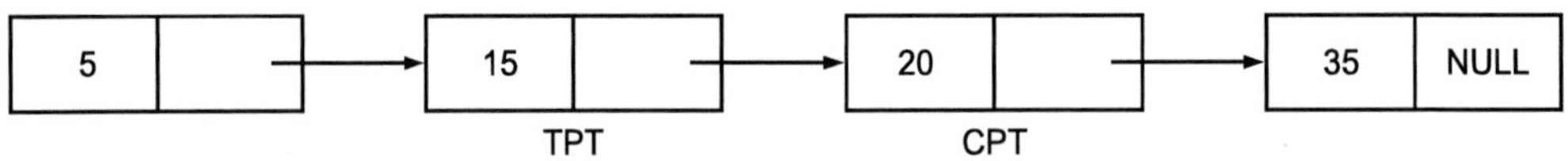

Fig. 4.28.

Now INFO (CPT) $\not<$ INFO (PTR)

So at this position, we insert the node PTR, *i.e.,*

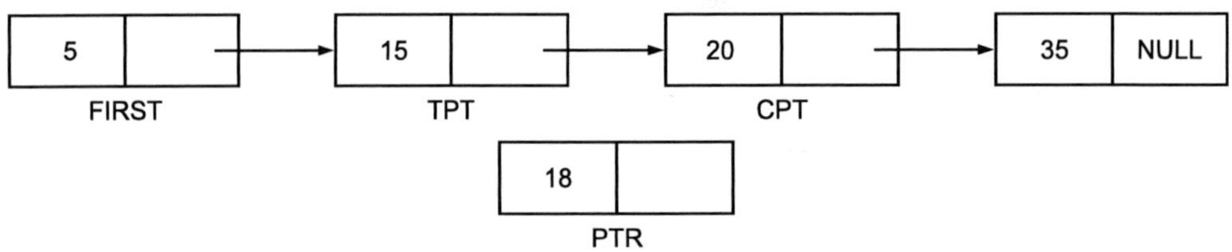

Fig. 4.29.

For insert the node

Link (TPT) = PTR

Link (PTR) = CPT

i.e.,

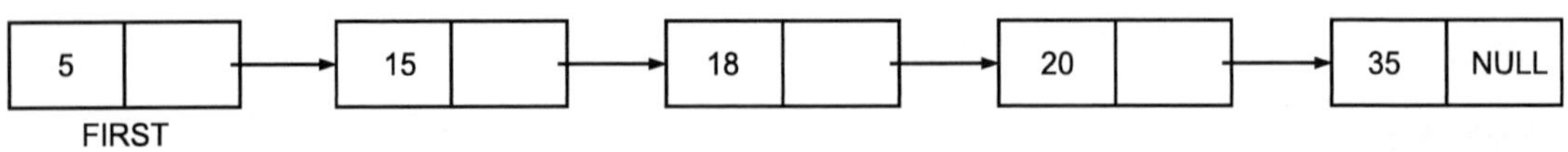

Fig. 4.30.

After insertion, we get final linked list as

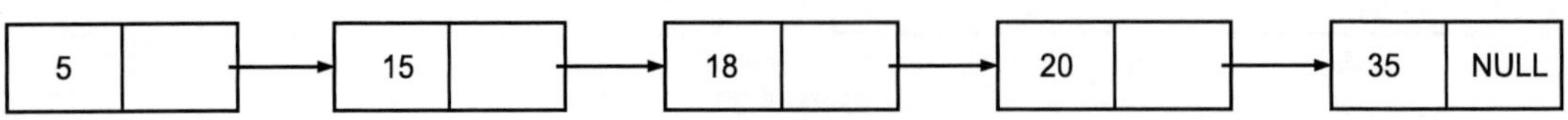

Fig. 4.31. After insertion

4.5.4 Deletion in a Singly Linked List

To delete the node from the existing link list before deleting the node we have to check the underflow condition.

IF FIRST = = NULL then write underflow & stop

After deletion, the deleted node will be inserted back to the availability list at AVAIL position. Let PTR is pointing to the deleted node then following step will be used to insert in the availability list

LINK (PTR) = AVAIL
AVAIL = PTR

Like Insertion, deletion is also possible from following any position.

(*i*) From the beginning
(*ii*) From the end
(*iii*) If information of node is given

(*a*) Deletion from the Beginning

Algorithm

Step 1 IF FIRST = = NULL then write underflow and stop,

Step 2 PTR = FIRST
FIRST = LINK (PTR)

Step 3 LINK (PTR) = AVAIL
AVAIL = PTR

Step 4 STOP

C Implementation :

```
void delete_beginning ( )
{
struct node *ptr;
if (first = = NULL)
{
printf ("underflow\n");
return;
}
ptr = first;
first = ptr → link;
free (ptr);
}
```

Let us consider a linked list

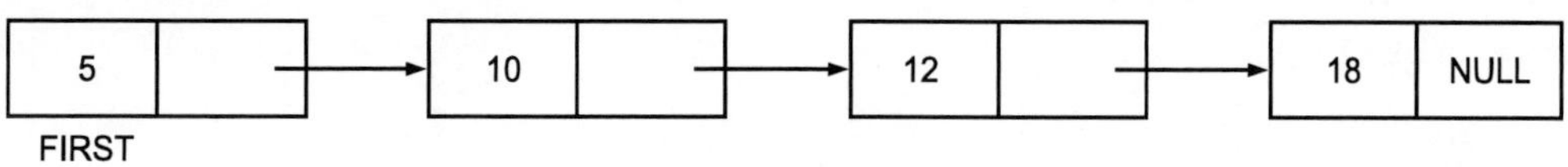

Fig. 4.32.

Delete node 5

Make FIRST as a PTR and Link (PTR) as a FIRST.

i.e., PTR = FIRST
FIRST = LINK (PTR)

Now,

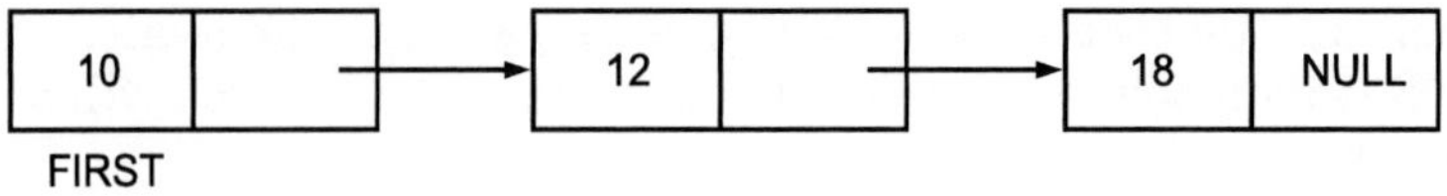

Fig. 4.33.

(*b*) Deletion from the End

Algorithm

Step 1 IF FIRST = = NULL then write underflow & stop
Step 2 PTR = FIRST
Step 3 Repeat step 4 while LINK (PTR)! = NULL
Step 4 CPT = PTR
PTR = LINK (PTR)
Step 5 LINK (CPT) = NULL
Step 6 LINK (PTR) = AVAIL
AVAIL = PTR
Step 7 STOP

C Implementation :

```
void delete_end ( )
{
struct node *ptr, *cpt;
if (first = = NULL)
{
printf ("underflow\n");
return;
}
ptr = first;
while (ptr → link! = NULL)
{
cpt = ptr;
ptr = ptr → link;
}
cpt → link = NULL;
free (ptr);
}
```

(*c*) Deletion of given INFO Node

Algorithm

Step 1 IF FIRST = = NULL then write underflow and stop
Step 2 Read DATA as information of node to be deleted
Step 3 PTR = FIRST
Step 4 Repeat step 5 while INFO (PTR) ≠ DATA

Step 5 CPT = PTR
PTR = LINK (PTR)
Step 6 LINK (CPT) = LINK (PTR)
Step 7 LINK (PTR) = AVAIL
AVAIL = PTR
Step 8 STOP

C Implementation :

```
void delete_given_info ( )
{
struct node *ptr, *cpt;
int data;
if (first = = NULL)
{
printf ("underflow\n");
return;
}
ptr = first;
printf ("Input information of node to be deleted");
scanf ("%d", & data);
while (ptr → info! = data)
{
cpt = ptr;
ptr = ptr → link;
}
cpt → link = ptr → link;
free (ptr);
}
```

4.5.5 Sorting the Link List

To arrange information of the link list in sequence, this sequence may be in ascending order or in descending order.

Algorithm

Step 1 PTR = FIRST
Step 2 Repeat step (3) to (7) while Link (PTR) ≠ NULL
Step 3 CPT = LINK (PTR)
Step 4 Repeat step (5) to (6) while CPT ≠ NULL
Step 5 If INFO (PTR) > INFO (CPT), Then
TEMP = INFO (PTR)
INFO (PTR) = INFO (CPT)
INFO (CPT) = TEMP
Step 6 CPT = LINK (CPT)
Step 7 PTR = LINK (PTR)

Step 8 STOP

C Implementation :

```
void sorting ( )
{
int temp;
struct node *ptr, *cpt;
ptr = first;
while (ptr → link! = NULL)
{
cpt = ptr → link;
while (cpt! = NULL)
{
if (ptr → info > cpt → info)
{
temp = ptr → info;
ptr → info = cpt → info;
cpt ® info = temp;
}
cpt = cpt → link;
}
ptr = ptr → link;
}
}
```

EXAMPLE 4.3: Let us consider a linked list.

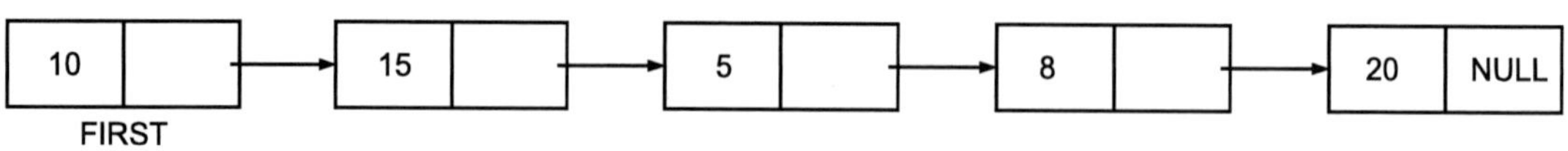

Fig. 4.34.

Arrange information of the linked list in ascending order.

PTR = FIRST and Link (PTR) $\neq$ NULL

CPT = LINK (PTR)

Here, INFO (PTR) < INFO (CPT)

So, CPT = LINK (CPT)

Now,

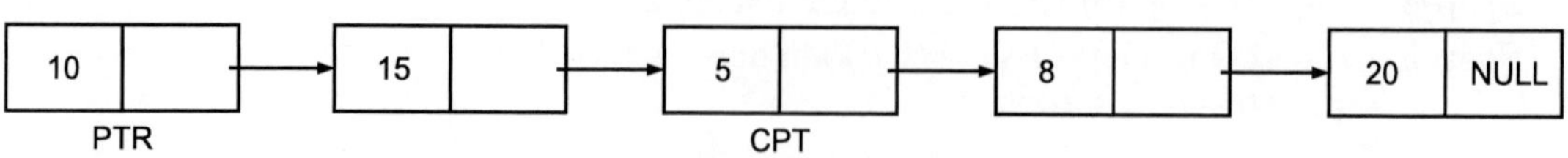

Fig. 4.35.

CPT $\neq$ NULL

Here, INFO (PTR) > INFO (CPT) then exchange it and CPT = LINK (CPT)

i.e.,

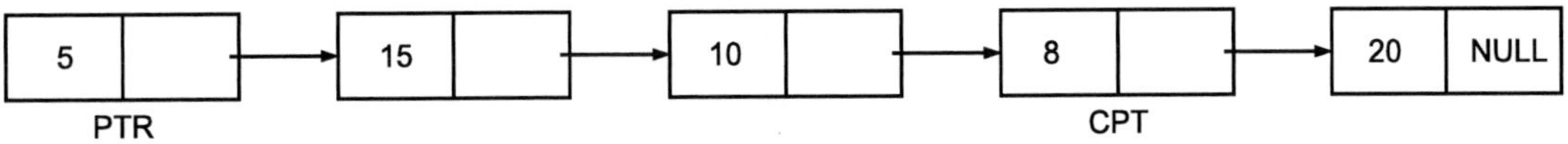

Fig. 4.36.

CPT ≠ NULL

Here, INFO (PTR) < INFO (CPT) No change and CPT = LINK (CPT)

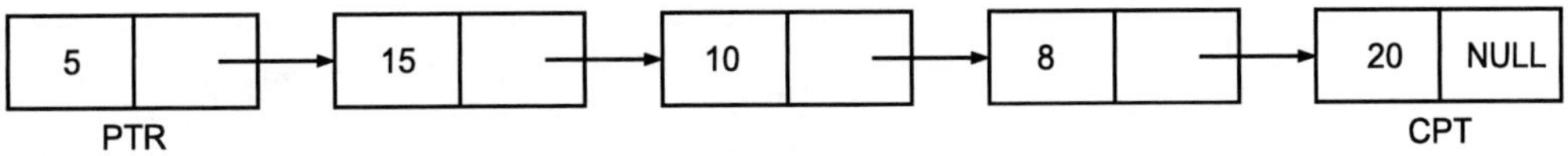

Fig. 4.37.

CPT ≠ NULL

INFO (PTR) < INFO (CPT) No change and CPT = LINK (CPT)

i.e., CPT = NULL

Now, PTR = LINK (PTR)

i.e.,

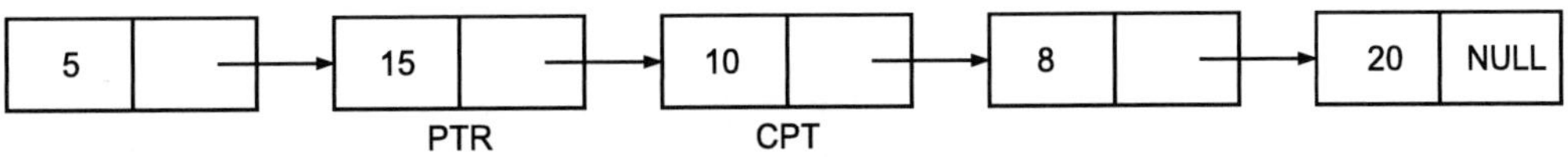

Fig. 4.38.

LINK (PTR) ≠ NULL

and CPT = LINK (PTR)

INFO (PTR) > INFO (CPT). So exchange it and CPT = LINK (CPT)

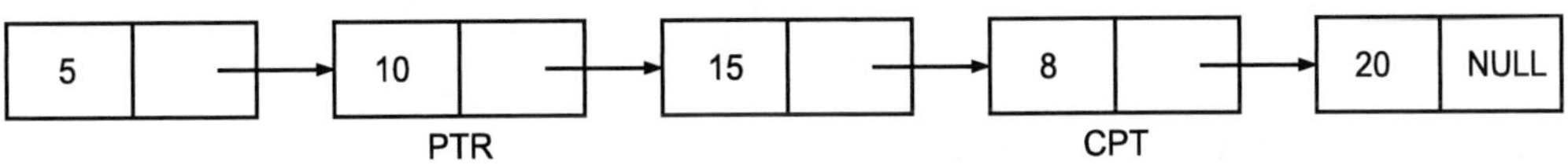

Fig. 4.39.

CPT ≠ NULL

and INFO (PTR) > INFO (CPT)

So exchange it and CPT = LINK (CPT)

i.e.

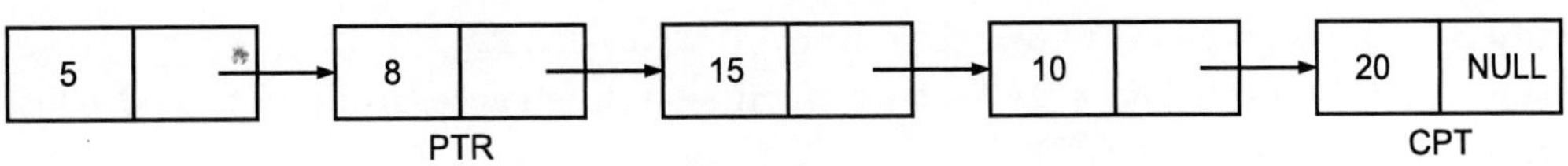

Fig. 4.40.

CPT ≠ NULL

and INFO (PTR) < INFO (CPT) No change and CPT = LINK (CPT)

i.e., now CPT = NULL

PTR = LINK (PTR) and CPT = LINK (PTR)

i.e.,

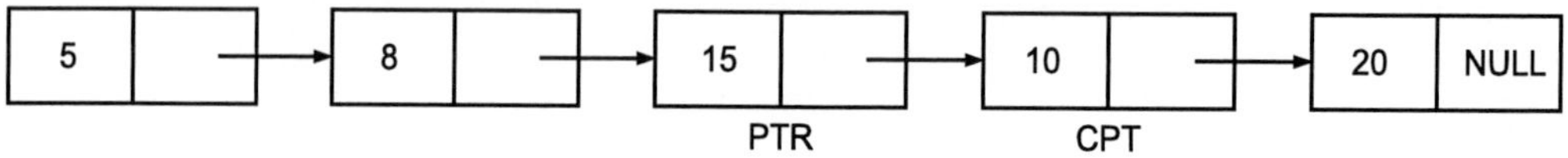

Fig. 4.41.

$$\text{CPT} \neq \text{NULL}$$

INFO (PTR) > INFO (CPT) exchange it and CPT = LINK (CPT)

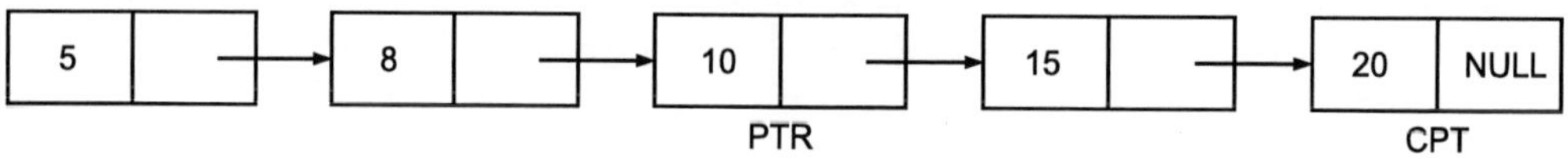

Fig. 4.42.

$$\text{CPT} \neq \text{NULL}$$

INFO (PTR) < INFO (CPT) No change and CPT = LINK (CPT)

So, CPT = NULL and PTR = LINK (PTR) and CPT = LINK (PTR)

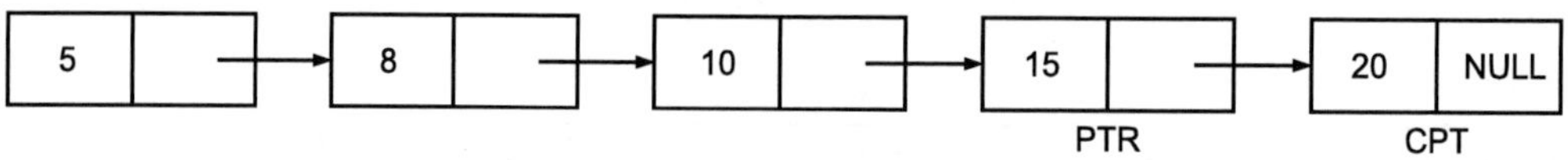

Fig. 4.43.

$$\text{CPT} \neq \text{NULL}$$

INFO (PTR) < INFO (CPT) No change and CPT = LINK (CPT) *i.e.* CPT = NULL
PTR = LINK (PTR) and now LINK (PTR) = NULL. So STOP and final sorted list is

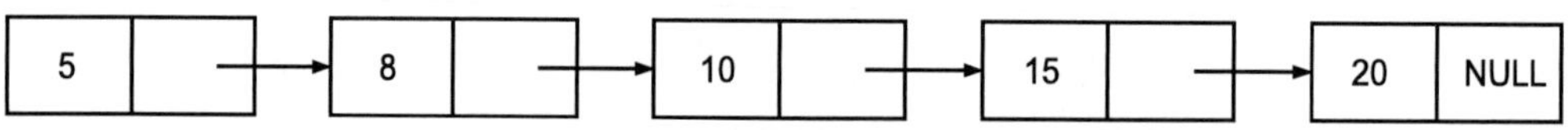

Fig. 4.44.

4.5.6 Searching a Node in Linked List

To locate or search any particular information node is available in the link list or not is known as searching. Basically it resemble with traversing (as linear search).

Algorithm

Step 1 Read DATA as information of node to be search
Step 2 PTR = FIRST
Step 3 Repeat step (4) to (5) while INFO (PTR) ≠ DATA
Step 4 IF INFO (PTR) = DATA then print search is successful and GOTO STEP 7
Step 5 PTR = LINK (PTR)
Step 6 PRINT This info node is not available
Step 7 STOP

C Implementation :

```
void search ( )
{
struct node *ptr;
```

```
int data;
ptr = first;
printf ("Input information of node to be search");
scanf ("%d", & data);
while (ptr! = NULL)
{
if (ptr -> info = = data)
{
printf ("Search is successfull\n");
return;
}
ptr = ptr → link;
}
printf ("This element does not exist\n");
}
```

4.5.7 Concatenation of Two Link Lists

To concate two different size link and make one link list *i.e.*, concatenated link list.

Suppose FIRST1 is pointing towards first node of first link list and FIRST2 is pointing towards first node of second link list.

Algorithm

Step 1 PTR = FIRST 1

Step 2 Repeat step 3 while LINK (PTR) ≠ NULL

Step 3 PTR = LINK (PTR)

Step 4 LINK (PTR) = FIRST2

Step 5 STOP

C Implementation :

```
void concate ( )
{
struct node *ptr;
ptr = first1;
while (ptr → link! = NULL)
ptr = ptr → link;
ptr → link = first2;
}
```

4.5.8 Merging Two Sorted Link Lists

To merge two sorted link list and create a third list in the same sequence of inputted link list.

Algorithm

```
Step 1   PTR 1 = FIRST1
         PTR 2 = FIRST2
Step 2   PTR = AVAIL
         AVAIL = LINK (AVAIL)
         FIRST = PTR
Step 3   If INFO (PTR1) < INFO (PTR2) THEN
         INFO (PTR) = INFO (PTR1)
         PTR1 = LINK (PTR1)
         Otherwise
         INFO (PTR) = INFO (PTR2)
         PTR2 = LINK (PTR2)
Step 4   Repeat Step 5 to Step 7 while PTR1 ! = NULL AND PTR2 ! = NULL
Step 5   CPT = AVAIL
         AVAIL = LINK (AVAIL)
Step 6   If INFO (PTR1) < INFO (PTR2) Then
         INFO (CPT) = INFO (PTR1)
         PTR1 = LINK (PTR1)
         Otherwise
         INFO (CPT) = INFO (PTR2)
         PTR2 = LINK (PTR2)
Step 7   LINK (PTR) = CPT
         PTR = CPT
Step 8   Repeat step 9 while PTR1! = NULL
Step 9     (i) CPT = AVAIL
               AVAIL = LINK (AVAIL)
               INFO (CPT) = INFO (PTR1)
               PTR1 = LINK(PTR1)
          (ii) LINK(PTR) = CPT
               PTR = CPT
         (iii) PTR2 = LINK (PTR1)
Step 10  Repeat step 11 while PTR2! = NULL
Step 11    (i) CPT = AVAIL
               AVAIL = LINK (AVAIL)
               INFO (CPT) = INFO (PTR2)
          (ii) LINK (PTR) = CPT
               PTR = CPT
         (iii) PTR2 = LINK (PTR2)
Step 12  LINK (PTR) = NULL
Step 13  STOP
```

EXAMPLE 4.4. **Let us consider two sorted link lists are:**

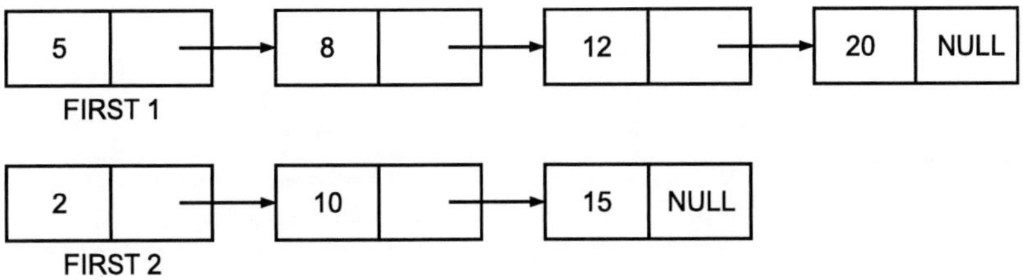

Fig. 4.45.

Merge these link list into third link list.

Initially PTR1 = FIRST1 and PTR2 = FIRST2

INFO (PTR1) = 5

INFO (PTR2) = 2

Here, INFO (PTR2) < INFO (PTR1)

So, create a new node PTR

i.e.,

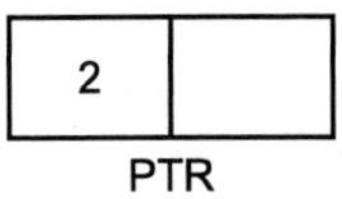

and PTR2 = LINK (PTR)

i.e., now

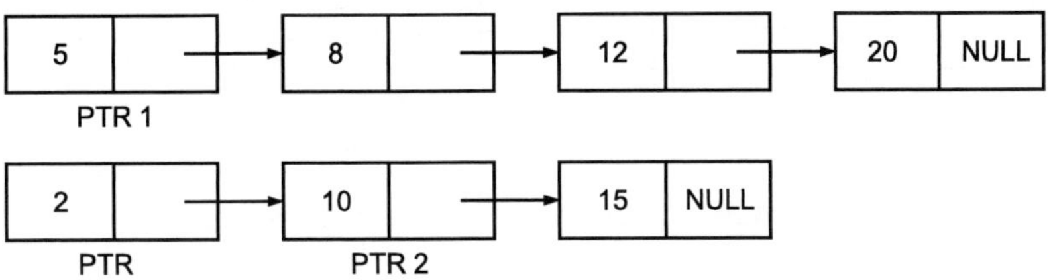

Fig. 4.46.

INFO (PTR1) = 5

INFO (PTR2) = 10

Here, INFO (PTR1) < INFO (PTR2)

Create a new node CPT and INFO (CPT) = INFO (PTR1)

LINK (PTR) = CPT

PTR = CPT

i.e.,

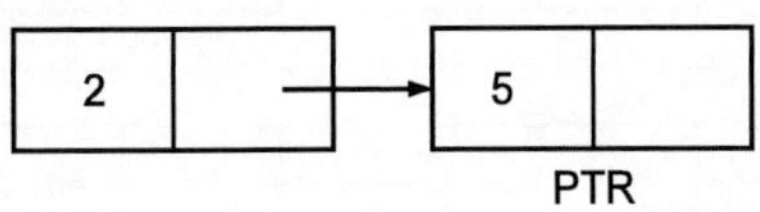

Fig. 4.47.

and PTR1 = LINK (PTR1)

Now,

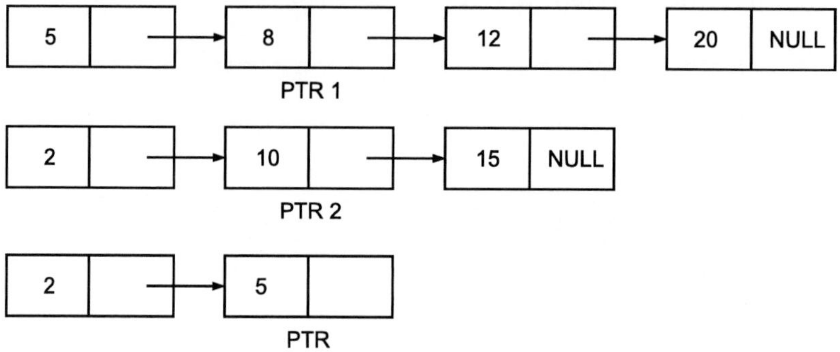

Fig. 4.48.

Here, INFO (PTR1) < INFO (PTR2),

So create new node CPT and INFO (CPT) = INFO (PTR1)

PTR1 = LINK (PTR1)

and LINK (PTR) = CPT

PTR = CPT

Now link lists are:

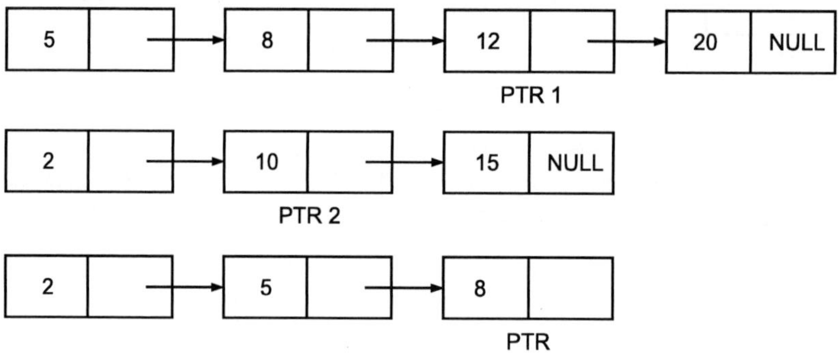

Fig. 4.49.

Here, INFO (PTR1) > INFO (PTR2)

So create new node CPT and INFO (CPT) = INFO (PTR2)

PTR2 = LINK (PTR2)

LINK (PTR) = CPT

PTR = CPT

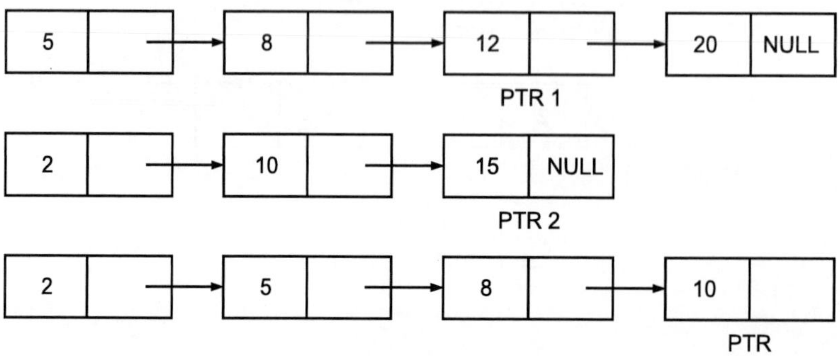

Fig. 4.50.

Now, INFO (PTR1) < INFO (PTR2)

So, INFO (CPT) = INFO (PTR1)

and PTR1 = LINK (PTR1)

LINK (PTR) = CPT

PTR = CPT

i.e. link lists are

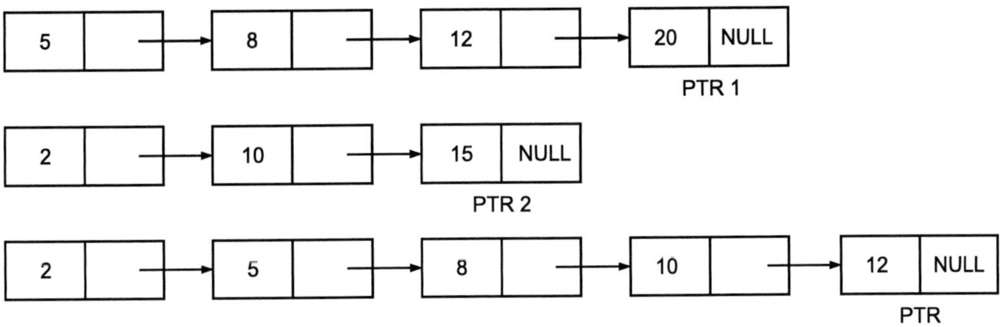

Fig. 4.51.

Now, INFO (PTR1) > INFO (PTR2)

So, create a new node CPT and

INFO (CPT) = INFO (PTR2)

LINK (PTR) = CPT

PTR = CPT

and PTR2 = LINK (PTR2) *i.e.,* PTR 2 = NULL

i.e., now link lists are

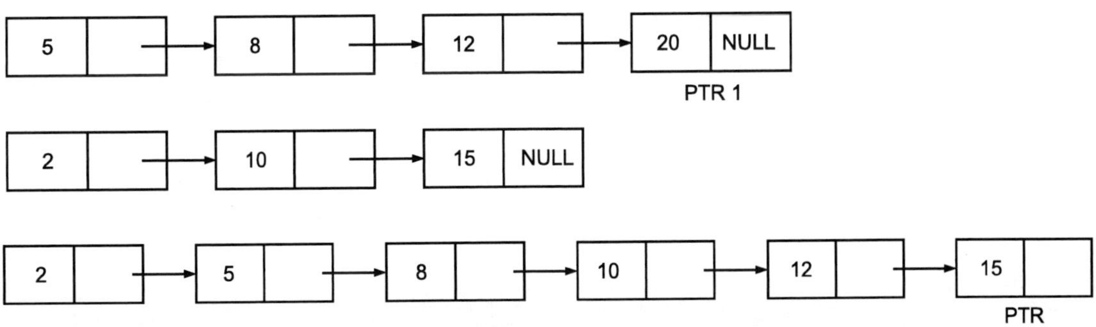

Fig. 4.52.

Here PTR1 ≠ NULL and PTR2 = NULL

So, create a node CPT and

INFO (CPT) = INFO (PTR1)

PTR1 = LINK (PTR1) *i.e.* PTR1 = NULL

and LINK (PTR) = CPT

PTR = CPT

Now, PTR1 = NULL and PTR2 = NULL

So, LINK (PTR) = NULL *i.e.*, final link-list after merging is:

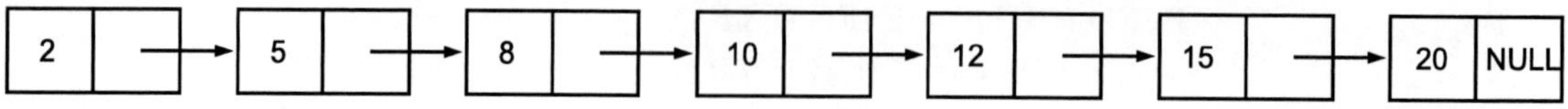

Fig. 4.53.

C Implementation :

```
void merging ( )
{
struct node *ptr1, *ptr2, *ptr, *cpt;
ptr1 = first1;
ptr2 = first2;
ptr = (struct node *) malloc (sizeof (struct node));
first = ptr;
if (ptr1 → info < ptr2 → info)
{
ptr → info = ptr1 → info;
ptr1 = ptr1 → link;
}
else
{
ptr → info = ptr1 → info;
ptr2 = ptr2 → link;
}
while (ptr! = NULL && ptr2! = NULL)
{
cpt = (struct node*) malloc (sizeof (struct node));
if (ptr1 → info < ptr2 → info)
{
cpt → info = ptr1 → info;
ptr1 = ptr1 → link;
}
else
{
cpt → info = ptr2 → info;
ptr2 = ptr2 → link;
}
ptr → link = cpt;
ptr = cpt;
}
while (ptr1! = NULL)
{
cpt = (struct node*) malloc (sizeof (struct node));
```

```
cpt → info = ptr1 → info;
ptr1 = ptr1 → link;
ptr → link = cpt;
ptr = cpt;
}
while (ptr2! = NULL)
{
cpt = (struct node *) malloc (sizeof (struct node));
cpt → info = ptr2 → info;
ptr2 = ptr2 → link;
ptr → link = cpt;
ptr = cpt;
}
ptr → link = NULL
}
```

4.5.9 Reversing a List

To reverse a linear linked list, three pointer fields are used. These are tpt, ptr, cpt which hold the address of previous node, current node and next node in the linear linked list. To begin with the address of the first node, which is held in pointer variable FIRST, is assigned to PTR and the tpt is assigned value NULL.

Algorithm

Step 1 PTR = FIRST

Step 2 TPT = NULL

Step 3 Repeat step 4 while PTR ! = NULL

Step 4 (a) CPT = LINK (PTR)

(b) LINK (PTR) = TPT

(c) TPT = PTR

(d) PTR = CPT

Step 5 STOP

C Implementation :

```
reverse_list ( )
  {
    PTR = FIRST;
    TPT = NULL;
    while (PTR ! = NULL)
    {
      CPT = PTR -> LINK;
      PTR -> LINK = TPT;
      TPT - PTR;
      PTR = CPT;
    }
  }
```

EXAMPLE 4.5. Consider a link list

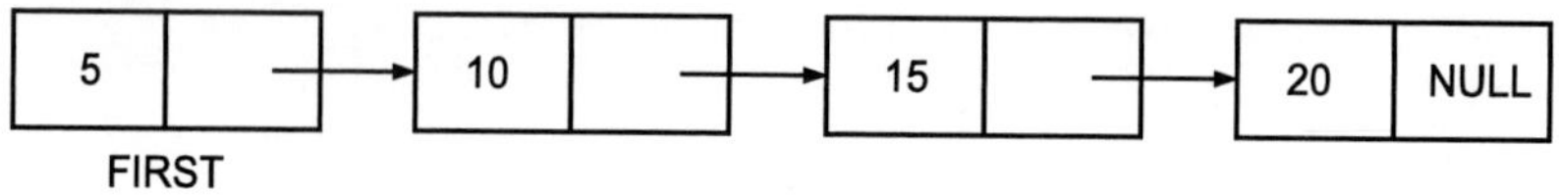

Fig. 4.54.

Reverse this list.

Solution: Initially, PTR = FIRST

TPT = NULL

Here, PTR $\neq$ NULL So,

CPT = Link (PTR)

Link (PTR) = TPT *i.e.* NULL

and TPT = PTR

PTR = CPT

i.e.

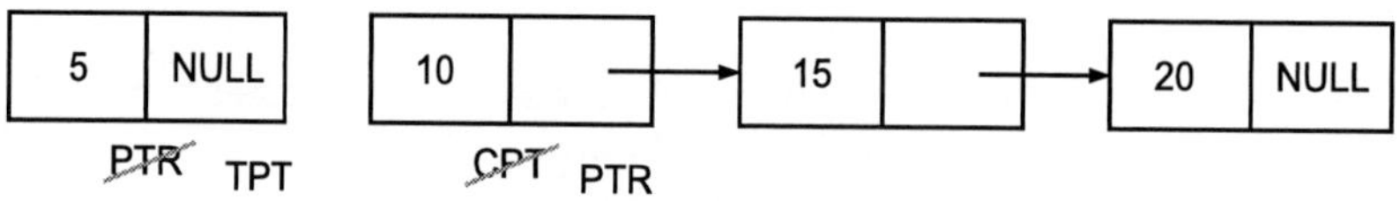

Fig. 4.55.

Now,

PTR is node 10

TPT is node 5

Again, PTR $\neq$ NULL

So, CPT = LINK (PTR) *i.e.* node 15.

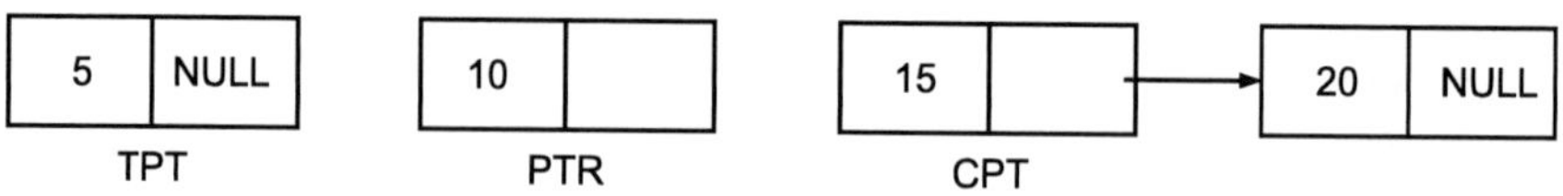

Fig. 4.56.

Now, LINK (PTR) = TPT

i.e.,

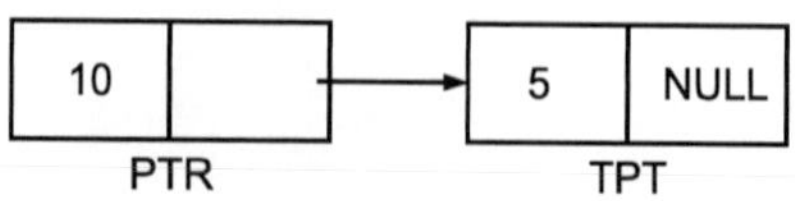

Fig. 4.57.

and TPT = PTR

i.e., node 10 is now TPT

PTR = CPT

i.e., node 15 is now PTR.

Now,

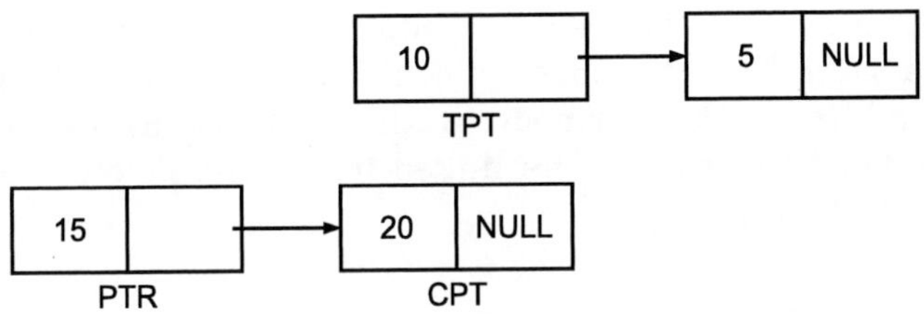

Fig. 4.58.

Again, PTR ≠ NULL

Now, CPT = LINK (PTR) *i.e.,* node 20

LINK (PTR) = TPT

i.e.,

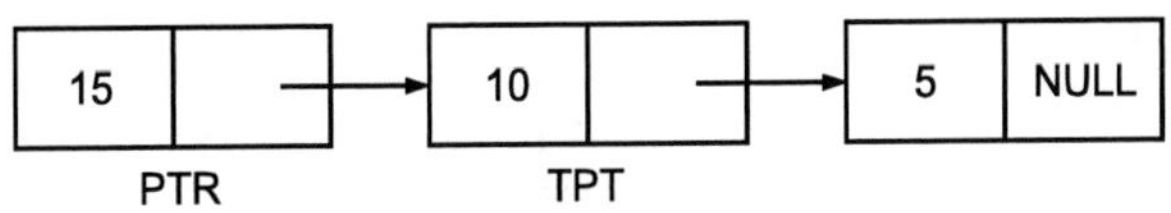

Fig. 4.59.

and TPT = PTR *i.e.,* node 15

PTR = CPT *i.e.,* node 20

Now,

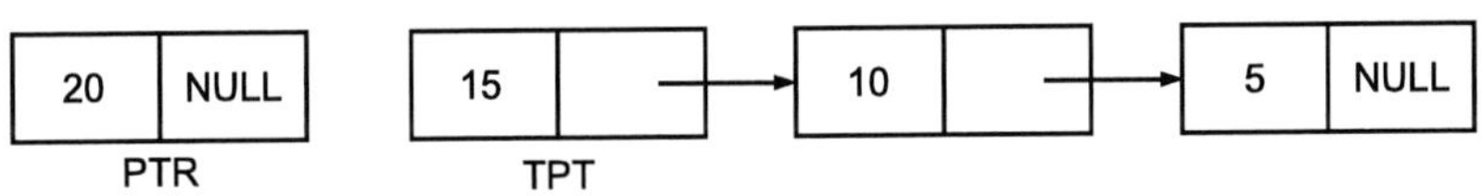

Fig. 4.60.

Again, PTR ≠ NULL

Now, CPT = LINK (PTR) *i.e.,* NULL

So, CPT = NULL

LINK (PTR) = TPT, *i.e.,*

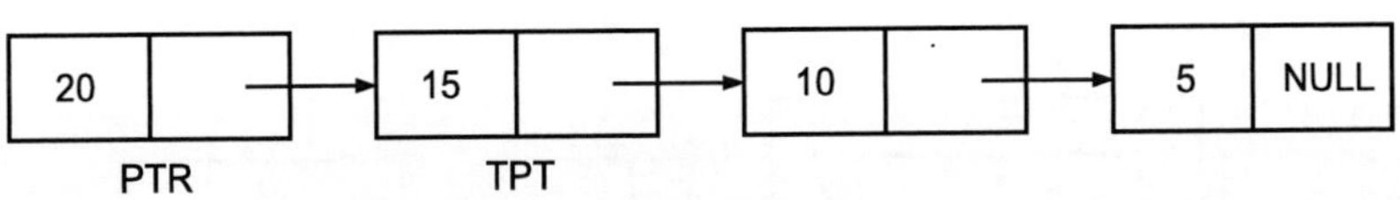

Fig. 4.61.

and TPT = PTR *i.e.,* node 20

PTR = CPT *i.e.,* PTR = NULL

Now, PTR = NULL so STOP and the final link list is

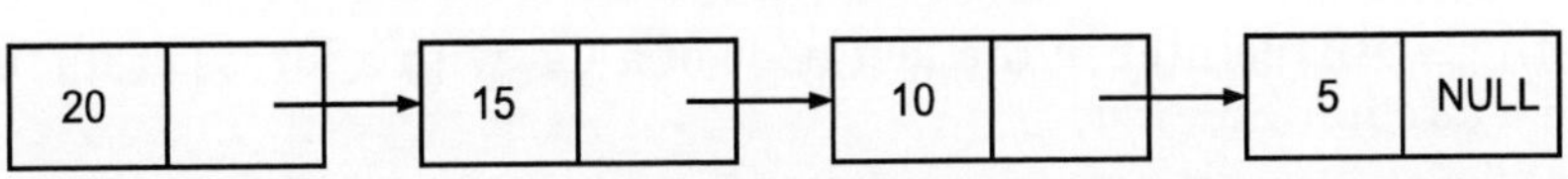

Fig. 4.62.

Header linked list

The header linked list is a special type of linked list where a special **Header node** is present at the beginning of list. The header node contains the address of first node in the linked list. Now, the START pointer does not contain the address of actual first node of the list but it holds the address of header node. In this case, the header node is like a dummy node also known as **sentinel node**. The linked list in fig. below can be shown with the dummy header node as follows:

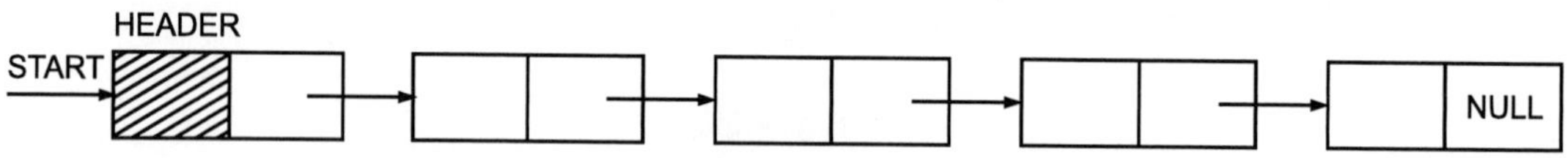

Fig. 4.63.

Observe that the INFO field of header node does not contain any useful information it is just a dummy node. However, in many cases, info field of header node may be used to store length of the linked list, that is the number of nodes other than header node.

(*a*) Insertion in header linked list:

Insertion at beginning uses some different procedure. Inserting a node in the beginning of header list means insert the node after header node.

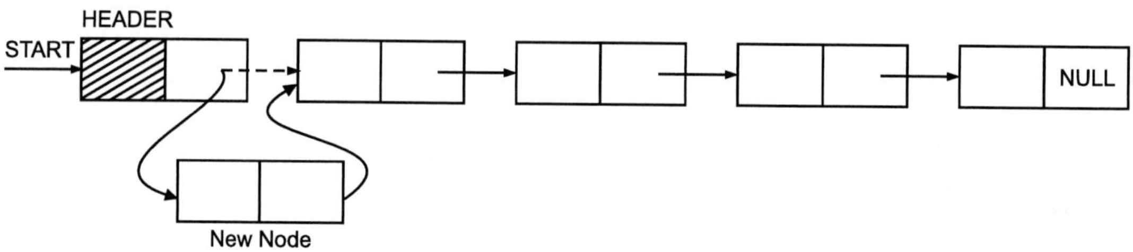

Fig. 4.64.

Other insertion like insertion in last, in middle are similar to one way list where traverse starts from link part of header node.

(*b*) Deletion from header linked list:

Similar to insertion procedure, the deletion of a node is also same as that of one way list except deletion from beginning.

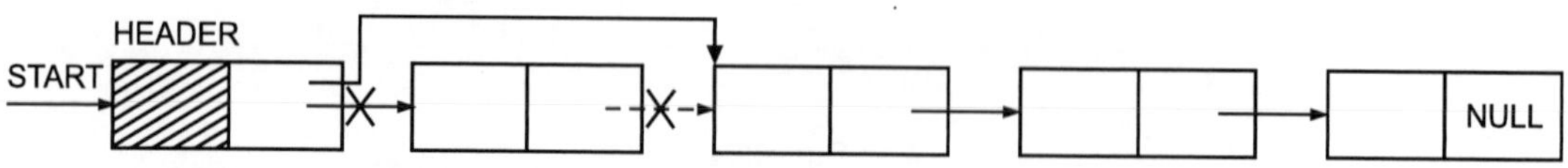

Fig. 4.65.

4.6 CIRCULAR LINKED LIST

If we replace the NULL pointer in the last node of a list with address of its first node, such a list is called a circular link list.

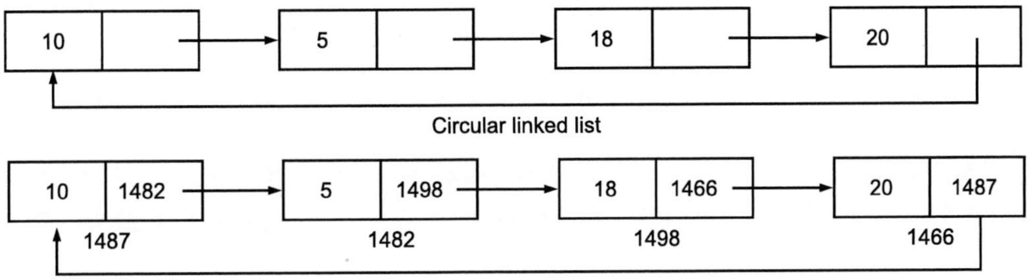

Fig. 4.66.

From any point in such a list it is possible to reach any other point in the list. A circular linked list does not have a first or last node. We must therefore, establish a first and last node by convention.

Advantages of Circular List

Circular lists have certain *advantages* over singly linked lists. In a circular list every node is accessible from a given node. That is, from this given node, all nodes can be reached by merely chaining through the list.

A second advantage concerns the deletion operation. In the deletion algorithm of linear linked list, to find the predecessor requires that a search be carried out by chaining through the nodes from the first node of the list. But this requirement does not exist for a circular list, since the search for the predecessor of node × can be initiated from × itself.

Finally, certain operations on circular lists, such as concatenation and splitting, become more efficient.

Disadvantage of Circular List

The circular linked list required extra care to detect the end of the list. It may be possible to get into an infinite loop. The detection of the end by placing a special node which can be easily identified in the circular list. This special node is often called the **header of the circular list**. This technique has an advantages of its own—the list can never be empty. The header node also indicates the end of the list.

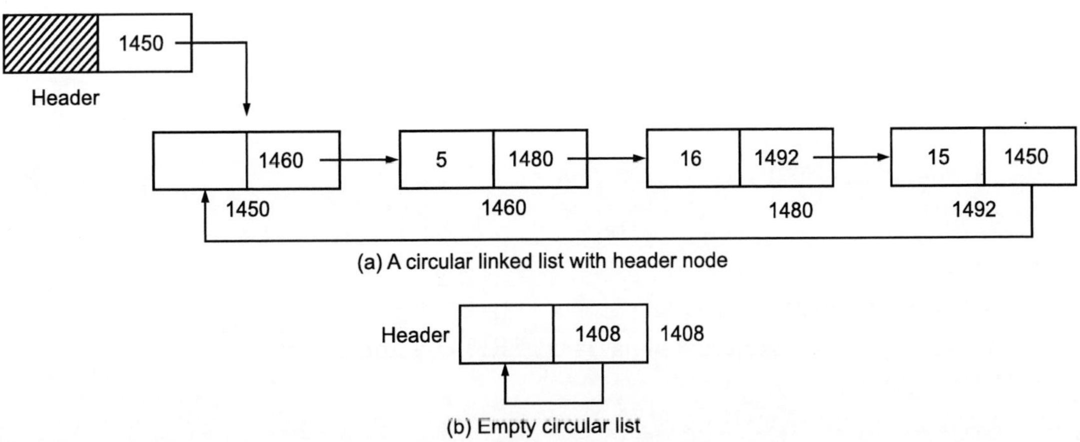

Fig. 4.67.

The representation of a circular link list with a header node is given in figure. The pointer variable Header denotes the address of the header node. Note that the information part in this node is not used, which is illustrated by shading the field.

Circular head lists are mostly used instead of ordinary list because of the following properties :

1. The NULL pointer is not used here and hence all pointers contain valid addresses.
2. Every (except HEADER) node has predecessor, so the first node may not require a special case.

4.6.1 Operations on Circular Linked List

(*a*) Creation of Circular Linked List

1. PTR ← AVAIL
   AVAIL ← LINK (AVAIL)
   Read INFO (PTR)
   FIRST ← PTR
   CH = 'Y'
2. Repeat step 3 to 4 while CH = 'Y'
3. CPT ← AVAIL
   AVAIL ← LINK (AVAIL)
   Read INFO (CPT)
   LINK (PTR) ← CPT
   PTR ← CPT
4. Press (Y/N) for more node
5. LINK (PTR) ← FIRST
6. STOP

(*b*) Traversing of Circular Linked List

1. PTR ← FIRST
   Process INFO (PTR)
   PTR ← LINK (PTR)
2. Repeat step 3 while PTR ! = FIRST
3. Process INFO (PTR)
4. PTR ← LINK (PTR)
5. STOP

(*c*) Insertion

(*i*) At the Beginning

1. If AVAIL = NULL then linked list is OVERFLOW and STOP.
2. PTR ← AVAIL
   AVAIL ← LINK (AVAIL)
   Read INFO (PTR)
3. CPT ← FIRST

4. Repeat step 5 while LINK (CPT) ! = FIRST
5. CPT ← LINK (CPT)
6. LINK (PTR) ← FIRST
   FIRST ← PTR
   LINK (CPT) ← FIRST
7. STOP

(*ii*) At the End

1. If AVAIL = NULL then linked list is overflow and STOP.
2. PTR ← AVAIL
   AVAIL ← LINK (AVAIL)
   Read INFO (PTR)
3. CPT ← FIRST
4. Repeat step 5 while LINK (CPT) ! = FIRST
5. CPT ← LINK (CPT)
6. LINK (CPT) ← PTR
7. LINK (PTR) ← FIRST
8. STOP.

EXAMPLE 4.6. Let us consider a circular link list.

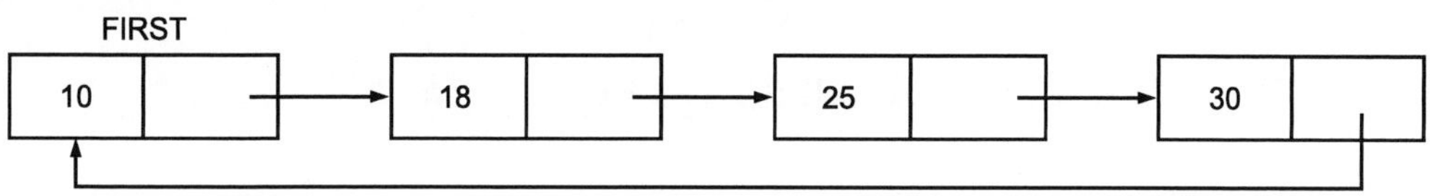

Fig. 4.68.

Insert a new node.

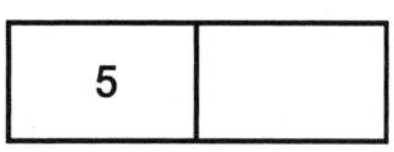

Fig. 4.69.

(a) At beginning, (b) At end.

Solution: (*a*) At beginning:

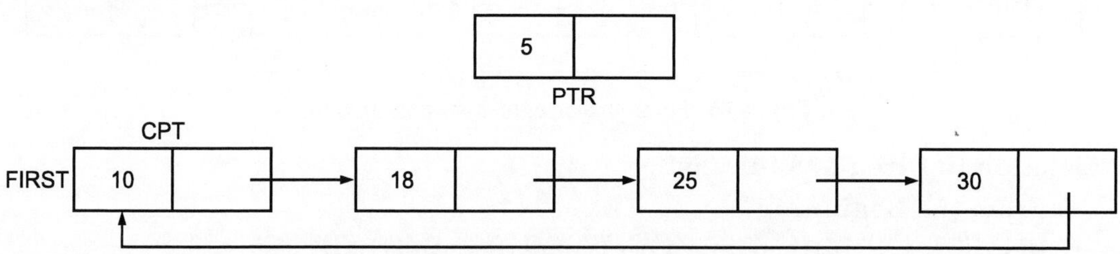

Fig. 4.70.

Repeat CPT = Link (CPT) while LINK (CPT) ≠ FIRST

i.e., Now

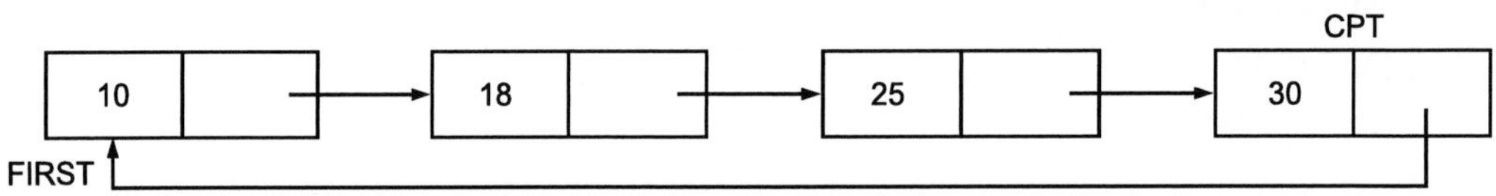

Fig. 4.71.

LINK (PTR) = FIRST
FIRST = PTR
LINK (CPT) = FIRST

i.e.,

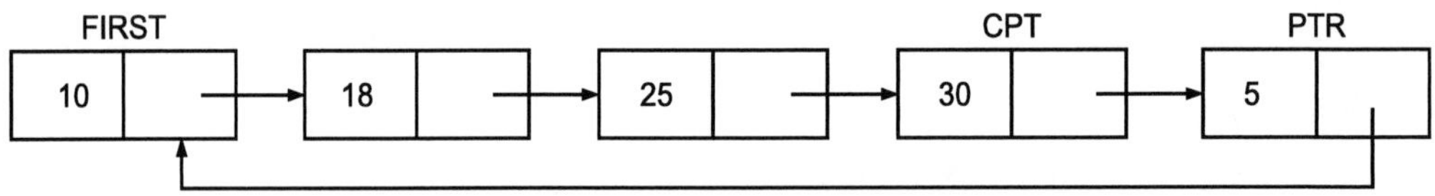

Fig. 4.72. Final linklist after insertion.

(*b*) At end :

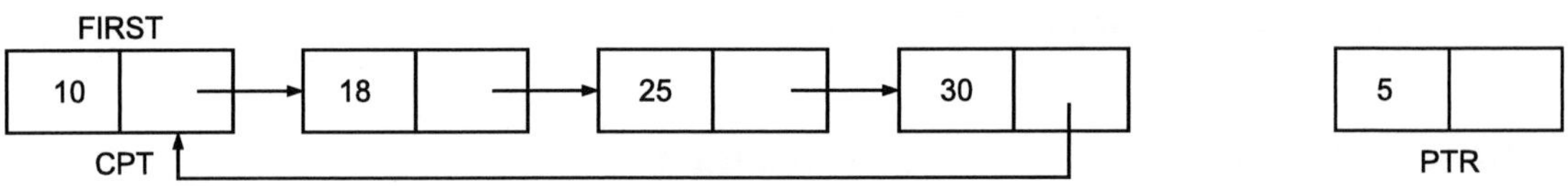

Fig. 4.73.

Repeat CPT = LINK (CPT) till LINK (CPT) $\neq$ FIRST

i.e.,

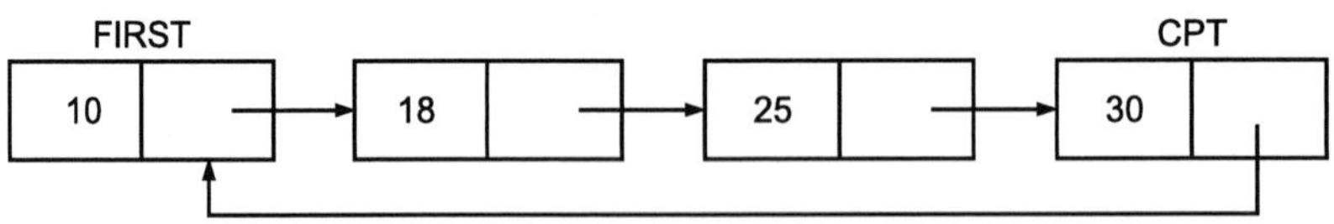

Fig. 4.74.

Then, LINK (CPT) = PTR
LINK (PTR) = FIRST

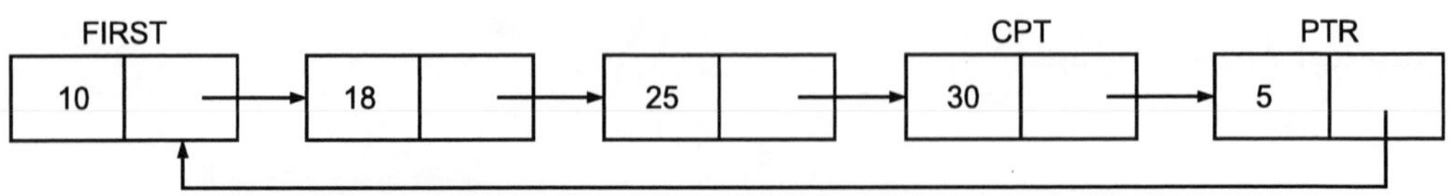

Fig. 4.75. Final linklist after insertion at end.

(*d*) Deletion in Circular Link List :

(*a*) From the Beginning

1. IF FIRST = NULL then linked list is underflow and STOP.
2. CPT ← FIRST.
3. Repeat step 4 while LINK (CPT) ! = FIRST.

4. CPT ← LINK (CPT).
5. PTR ← FIRST
   FIRST ← LINK (PTR)
   LINK (CPT) ← FIRST.
6. LINK (PTR) ← AVAIL
   AVAIL ← PTR
7. STOP.

(*b*) From the End

1. IF FIRST = NULL then linked list is underflow and STOP.
2. CPT ← FIRST.
3. Repeat step 4 while LINK (CPT) ! = FIRST.
4. PTR ← CPT
   CPT ← LINK (CPT).
5. LINK (PTR) ← FIRST.
6. LINK (CPT) ← AVAIL
   AVAIL ← CPT
7. STOP

EXAMPLE 4.7. Let us consider a linked list.

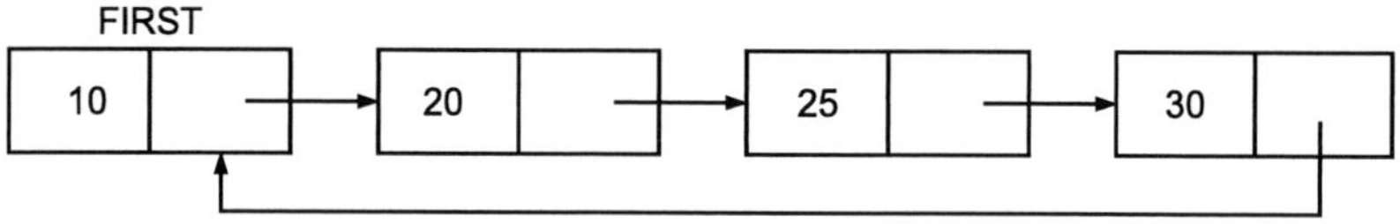

Fig. 4.76.

Delete a node

(*a*) from beginning

(*b*) from the end.

Solution: (*a*) From beginning

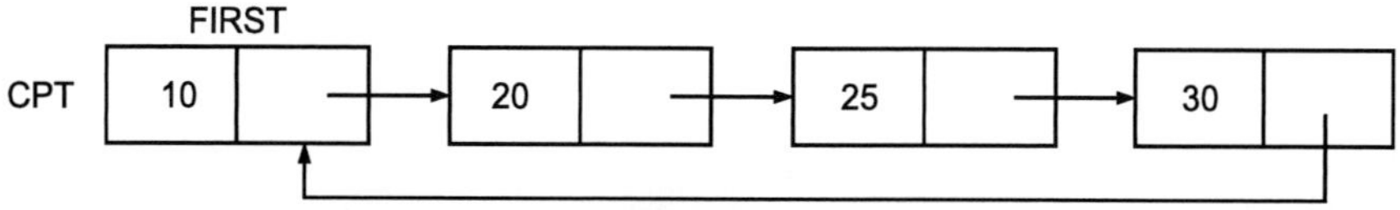

Fig. 4.77.

Repeat CPT = LINK (CPT) till LINK (CPT) ≠ FIRST

Then, PTR = FIRST

and FIRST = LINK (PTR)

LINK (CPT) = FIRST

i.e.,

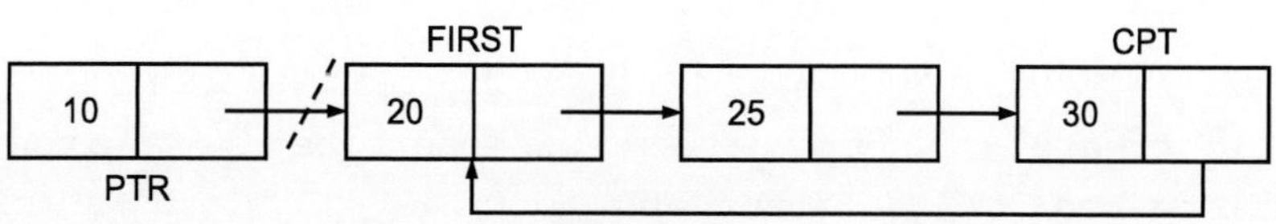

Fig. 4.78.

So, the final link list after deletion is:

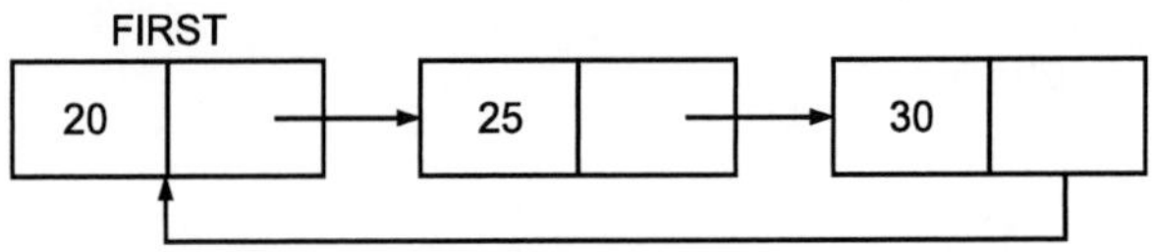

Fig. 4.79.

(*b*) **From end :**

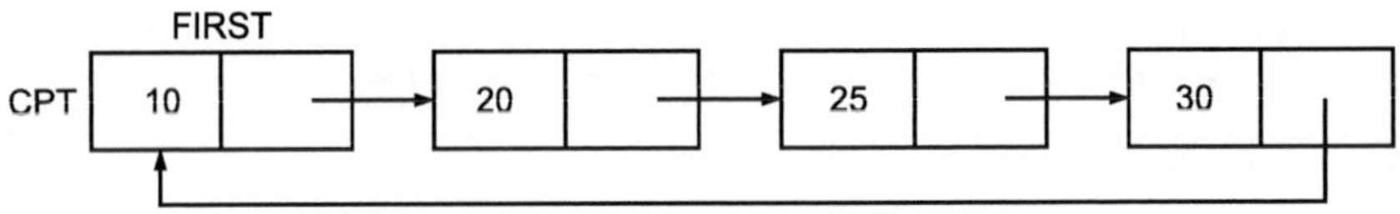

Fig. 4.80.

Repeat (*a*) PTR = CPT, (*b*) CPT = LINK (CPT) till LINK (CPT) ≠ FIRST

i.e.,

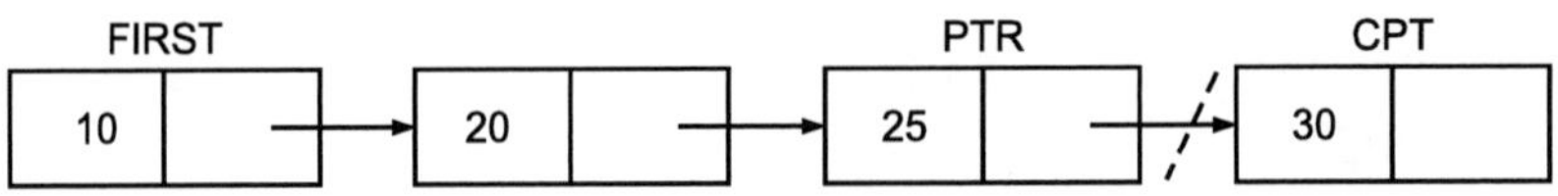

Fig. 4.81.

Now, LINK (PTR) = FIRST

i.e.,

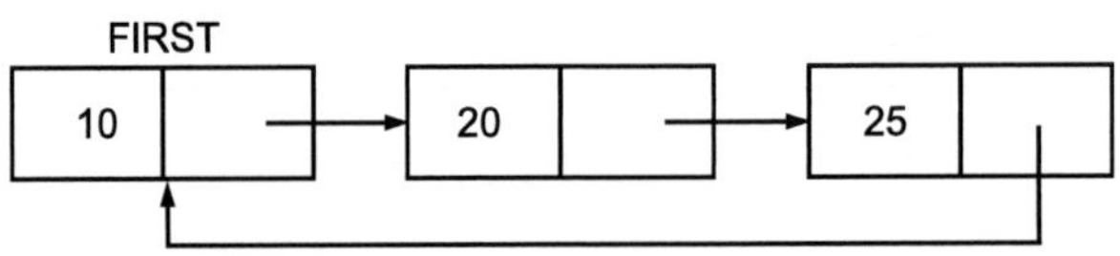

Fig. 4.82. Final linklist after deletion

Implementation in C Language

```
#include <stdio.h>
#include <conio.h>
#include <alloc.h>
struct node
   {
      int info;
      struct node *link;
   };
struct node *first;

void main ( )
   {
      void create ( ), traverse ( ), insert_beg ( ), insert_end ( ),
      delete_beg ( ), delete_end ( );
      clrscr ( );
```

```
        create ( );
        traverse ( );
        insert_beg ( );
        traverse ( );
        insert_end ( );
        traverse ( );
        delete_beg ( );
        traverse ( );
        delete_end ( );
        traverse ( );
        getch ( );
    }

void create ( )
    {
        struct node *ptr, *cpt;
        char ch;
        ptr = (struct node *) malloc (sizeof (struct node));
        printf ("Input first node");
        scanf ("%d" & ptr → info);
        first = ptr;
        do
            {
                cpt = (struct node *) malloc (sizeof (struct node));
                printf ("Input next node");
                scanf ("%d" & cpt → info);
                ptr → link = cpt;
                ptr = cpt;
                printf ("Press <Y/N> for more node");
                ch = getch ( );
            }
            while (ch = = "Y");
            ptr → link = first;
            }

void traverse ( )
    {
        struct node *ptr;
        printf ("Traversing of link list;\n");
        ptr = first;
        while (ptr! = first)
```

```
        {
           printf ("%d\n", ptr → info);
           ptr = ptr → link;
        }
   }

void insert_beg ( )
   {
      struct node *ptr;
      ptr = (struct node *) malloc (sizeof (struct node));
      if (ptr = = NULL)
      { printf ("overflow\n");
         return;
      }
      printf ("Input new node");
      scanf ("%d", & ptr → info);
      cpt = first;
      while (cpt → link != first)
         {
            cpt = cpt → link;
         }
      ptr → link = first;
      first = ptr;
      cpt → link = first;
   }

void insert_end ( )
   {
      struct node *ptr, *cpt;
      ptr = (struct node *) malloc (sizeof (struct node));
      if (ptr = = NULL)
      {
         printf ("overflow\n");
         return;
      }
   printf ("Input new node information");
   scanf ("%d", & ptr → info);
   cpt = first;
   while (cpt → link ! = first);
      cpt = cpt → link;
      cpt → link = ptr;
```

```
        ptr → link = first;
    }

void delete_beg( )
    {
        struct node *ptr, *cpt;
        if (first = = NULL)
        {
            printf ("underflow\n");
            return;
        }
        cpt = first;
        while (cpt → link ! = First)
            cpt = cpt → link;
        ptr = first;
        first = ptr → link;
        cpt → link = first;
        free (ptr);
    }

void delete_end( )
    {
        struct node *ptr, *cpt;
        If (first = = NULL)
            {
                printf ("underflow\n");
                return;
            }
        cpt = first;
        while (cpt → link ! = first)
            {
                ptr = cpt;
                cpt = cpt → link;
            }
        ptr → link = first;
        free (cpt);
    }
```

4.7 DOUBLY LINKED LIST

In one-way or singly linked list we can traverse only in one direction *i.e.*, in forward direction only because in this link list each node containing only one link which stores the address of next successor node.

The doubly or two-way linked list uses double set of pointers, one pointing to the next node and the other pointing to the preceding node. That is, a doubly linked list is one in which all nodes are linked together by multiple links which help in accessing both the successor and predecessor node for any arbitrary node within the list.

Every nodes in the doubly linked list has three fields.

LPT	INFO	RPT

Fig. 4.83.

LPT will point to the node in the left side (or previous node) *i.e.*, LPT will hold the address of the previous node RPT will point to the node in the right side (or next node) that is RPT will hold the address of the next node. INFO field store the information of the node.

A doubly linked list can be shown as follows :

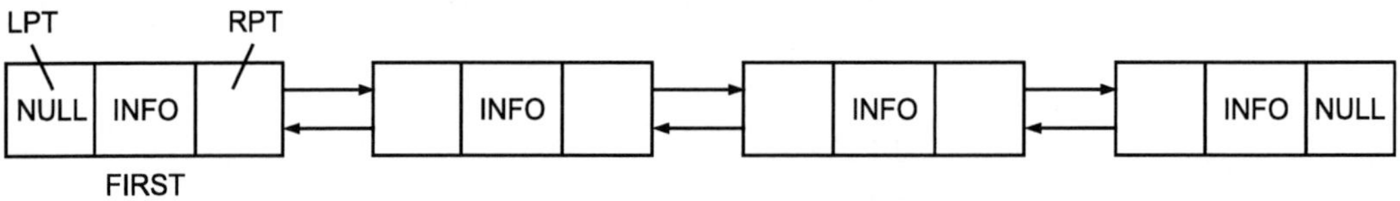

Fig. 4.84. Doubly linked list.

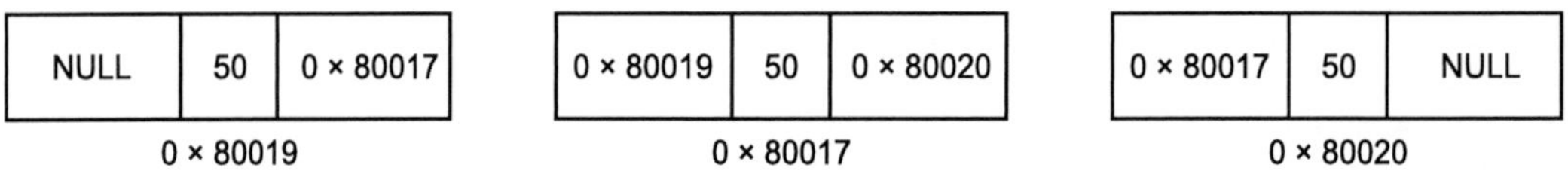

Fig. 4.85. Doubly linked list.

The structure defined for doubly linked list is:

```
struct node
   {
      int info;
      struct node *rpt;
      struct node *lpt;
   }
```

4.8 OPERATIONS OF DOUBLY LINK-LIST

(*a*) Creation of Doubly Linked List :

1. PTR ← AVAIL
   AVAIL ← RPT (AVAIL)
   READ INFO (PTR)
   LPT (PTR) ← NULL
   FIRST ← PTR.
2. STORE 'Y' to CH.
3. Repeat step 4 to 5 while CH = 'Y'.

LPT	INFO	RPT

4. (*a*) CPT ← AVAIL
   AVAIL ← RPT (AVAIL)
   Read INFO (CPT).
   (*b*) RPT (PTR) ← CPT
   LPT (CPT) ← PTR
   PTR ← CPT.
5. Input choice <Y/N> for more node.
6. RPT (PTR) ← NULL.
7. STOP.

Traversing of two-way linked list :

(*a*) Forward Traversing.

(*b*) Backward Traversing.

(*b*) Forward Traversing

1. PTR ← FIRST.
2. Repeat step 3 to 4 while PTR ! = NULL.
3. Process INFO (PTR).
4. PTR ← RPT (PTR).
5. STOP.

(*c*) Backward Traversing

1. PTR ← FIRST.
2. Repeat step (3) while RPT (PTR) ! = NULL.
3. PTR ← RPT (PTR)
4. Repeat step (5) to (6) while PTR ! = NULL.
5. Process INFO (PTR).
6. PTR ← LPT (PTR).
7. STOP.

Insertion of Doubly Linked List

(*a*) At Beginning

1. If AVAIL = NULL then write OVERFLOW and STOP.
2. PTR ← AVAIL
   AVAIL ← RPT (AVAIL)
   Read INFO (PTR).
3. RPT (PTR) ← FIRST
   LPT (FIRST) ← PTR
   LPT (PTR) ← NULL
   FIRST ← PTR
4. STOP

(*b*) At End

1. If AVAIL = NULL then write OVERFLOW and STOP.

2. PTR ← AVAIL
   AVAIL ← RPT (AVAIL)
   Read INFO (PTR).
3. CPT ← FIRST
4. Repeat Step (5) while RPT (CPT) ≠ NULL
5. CPT ← RPT (CPT)
6. RPT (CPT) ← PTR
   LPT (PTR) ← CPT
   RPT (PTR) ← NULL
7. STOP

(*c*) After given Node Information

1. If AVAIL = NULL then write OVERFLOW and STOP.
2. PTR ← AVAIL
   AVAIL ← RPT (AVAIL)
   Read INFO (PTR).
3. Read DATA as information after which insertion will be made.
4. CPT ← FIRST.
5. Repeat step 6 while INFO (CPT) ≠ DATA.
6. CPT ← RPT (CPT).
7. TPT ← RPT (CPT)
   RPT (CPT) ← PTR
   LPT (PTR) ← CPT
   RPT (PTR) ← TPT
   LPT (TPT) ← PTR.
8. STOP.

(*d*) Insertion in the Sorted Doubly Linked List

1. If AVAIL = NULL then write OVERFLOW and STOP.
2. PTR ← AVAIL
   AVAIL ← RPT (AVAIL)
   Read INFO (PTR).
3. CPT ← FIRST.
4. Repeat step 5 to 6 while INFO (CPT) < INFO (PTR).
5. CPT ← RPT (CPT).
6. TPT ← LPT (CPT).
7. RPT (TPT) ← PTR
   LPT (PTR) ← TPT
   RPT (PTR) ← CPT
   LPT (CPT) ← PTR.
8. STOP.

EXAMPLE 4.8. **Let us consider a doubly link list as:**

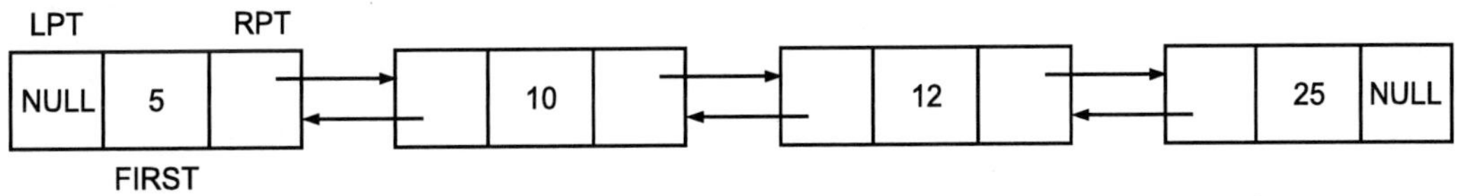

Fig. 4.86.

(*a*) **Insert 8 at beginning.**

(*b*) **Insert 30 at end.**

(*c*) **Insert 15 at node 12.**

Solution: (*a*) Insert 8 at beginning

Create a node PTR whose INFO (PTR) = 8

RPT (PTR) = FIRST LPT (FIRST) = PTR

LPT (PTR) = NULL

FIRST = PTR

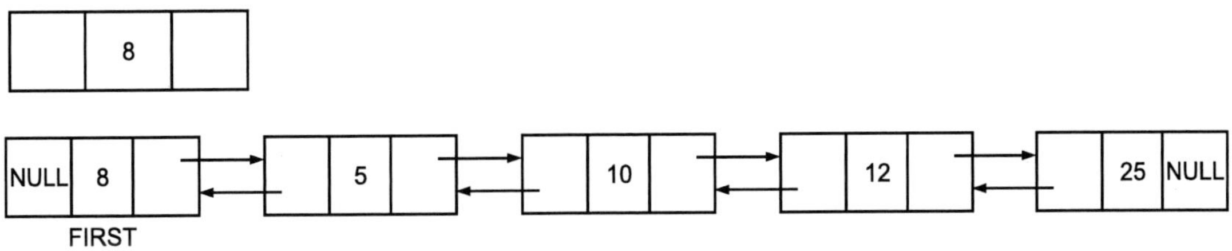

Fig. 4.87.

(*b*) Insert note 30 at end

Create a new node as PTR

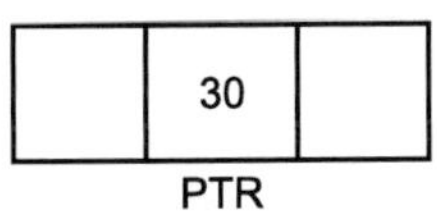

CPT = FIRST i.e.

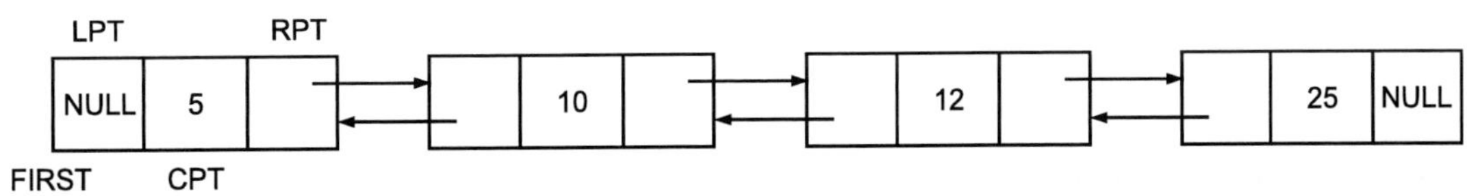

Fig. 4.88.

Repeat CPT = RPT (CPT) till RPT (CPT) ≠ NULL

i.e. now

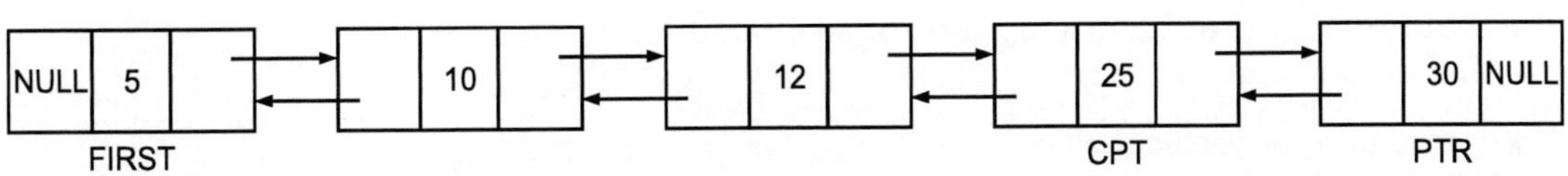

Fig. 4.89.

RPT (CPT) = PTR

LPT (PTR) = CPT

RPT (PTR) = NULL

(c) Insert node 15 after node 12 :

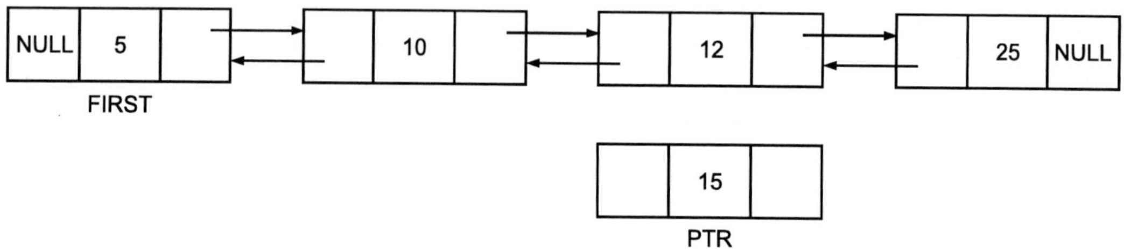

Fig. 4.90.

Here, DATA = 12

i.e. after which insertion will be made.

Now CPT = FIRST

Repeat CPT = RPT (CPT) till.

INFO (CPT) $\neq$ DATA

i.e. now

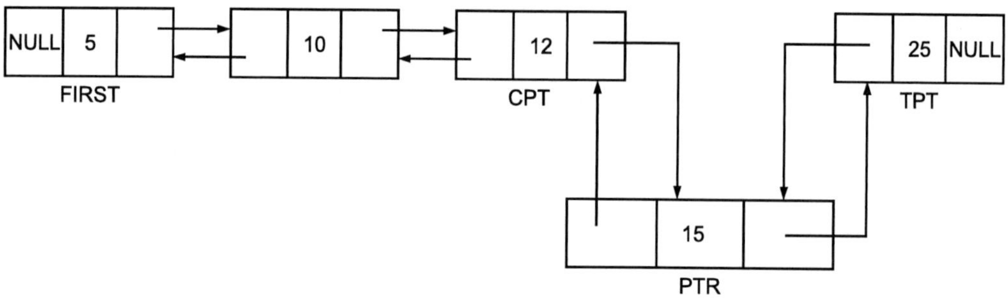

Fig. 4.91.

i.e.

TPT = RPT (CPT)

RPT (CPT) = PTR

LPT (PTR) = CPT

RPT (PTR) = TPT

LPT (TPT) = PTR

Now, the final list is

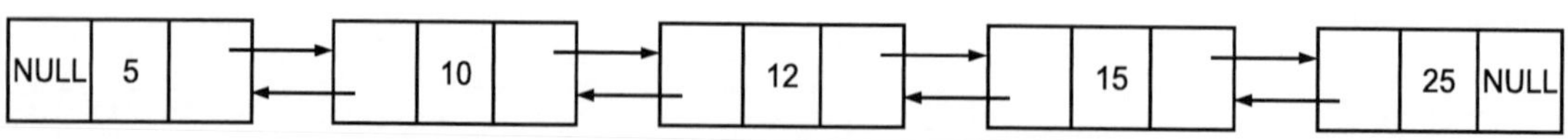

Fig. 4.92. Final doubly linklist after insertion.

C Implementation of Two way/Doubly Linked List :

```
# include <stdio.h>
# include <conio.h>
# include <alloc.h>
struct node
  {
     int info;
```

```
        struct node *lpt;
        struct node *rpt;
    };
struct node *first;
void main ( )
    {
        void create ( ); void ftraverse ( );
        void btraverse ( );
        create ( );
        btraverse ( );
        ftraverse ( );
        getch ( );
    }
void create ( )
    {
        struct node *ptr, *cpt;
        char ch;
        ptr = (struct node *) malloc (size of (struct node));
        printf ("Input first node information");
        scanf ("%d", & ptr → info);
        ptr → lpt = NULL;
        first = ptr;
        do
            {
                cpt = (struct node *) malloc (size of (struct node));
                pintf ("Input next node information");
                scanf ("%d", & cpt → info);
                ptr → rpt = cpt;
                cpt → lpt = ptr;
                ptr = cpt;
                printf ("Press <Y/N> for more node");
                ch = getch ( );
            }
                while (ch= = 'Y');
                ptr → rpt = NULL;
    }
void ftraverse ( )
    {
        struct node *ptr;
        printf ("forward Traversing :\n");
        ptr = first;
        while (ptr ! = NULL)
```

```
        {
          printf ("%d \n", ptr → info);
          ptr = ptr → rpt;
        }
    }
  void btraverse ( )
    {
        struct node * ptr ;
        printf ("Backward traversing :\n");
        ptr = first ;
        while (ptr → rpt! = NULL)
          ptr = ptr → rpt ;
        while (ptr! = NULL)
        {
          printf ("%d \n", ptr → info);
          ptr = ptr → lpt;
        }
    }
```

C Implementation of Insertion at the beginning

```
void insert_beg ( )
  {
    struct node *ptr;
    ptr = (struct node *) malloc (size of (struct node));
    If (ptr = = NULL)
      {printf ("OVERFLOW");
      return;
      }
    printf ("Input new node");
    scanf ("%d", & ptr → info);
    ptr → rpt = first;
    first → lpt = ptr;
    first = ptr;
    printf ("New node is insert\n");
  }
```

C Implementation of Insertion at the end

```
void insert_end ( )
  {
    struct node *ptr, *cpt;
    ptr = (struct node *) malloc (size of (struct node));
    If (ptr = = NULL)
      {
```

```
            printf ("OVERFLOW");
            return;
         }
      printf ("Input new node information");
      scanf ("%d", & ptr → info);
      cpt = first;
      while (cpt → rpt! = NULL)
         cpt = cpt → rpt;
         cpt → rpt = ptr;
         ptr → lpt = cpt;
         ptr → rpt = NULL;
      printf ("Insertion is done\n");
   }
```

C Implementation of Insertion after given node

```
void insert_given_node ( )
   {
      struct node *ptr, *cpt, *tpt;
      int m;
      ptr = (struct node *) malloc (size of (struct node));
      IF (ptr = = NULL)
         {
            printf ("OVERFLOW");
            return;
         }
      printf ("Input new node information");
      scanf ("%d", & ptr → info);
      printf ("Input node information after which insertion");
      scanf ("%d", & m);
         cpt = first;
      while (cpt → info! = m)
         cpt = cpt → rpt;
         tpt = cpt → rpt;
      cpt → rpt = ptr;
      ptr → lpt = cpt;
      ptr → rpt = tpt;
      tpt → lpt = ptr;
      printf ("Insertion is done\n");
   }
```

C Implementation of Insertion in the stored list

```
void insert_sorted ( )
   {
```

```
    struct node *ptr, *cpt, *tpt;
    ptr = (struct node *) malloc (size of (struct node));
    if (ptr = = NULL)
       {
          printf ("OVERFLOW");
          return;
       }
    printf ("Input new node information");
    scanf ("%d", & ptr → info);
    cpt = first;
    while (cpt → info < ptr → info)
       {
          tpt = cpt;
          cpt = cpt → tpt;
       }
       tpt → rpt = ptr;
       ptr → lpt = tpt;
       ptr → rpt = cpt;
       cpt → lpt = ptr;
    printf ("Insertion is done\n");
}
```

Deletion

(a) From beginning

1. If FIRST = NULL then write UNDERFLOW and STOP.
2. PTR ← FIRST
   FIRST ← RPT (PTR)
   LPT (FIRST) ← NULL.
3. RPT (PTR) ← AVAIL.
   AVAIL ← PTR
4. STOP.

(b) From end

1. If FIRST = NULL then write UNDERFLOW and STOP.
2. PTR ← FIRST.
3. Repeat step 4 while RPT (PTR) ≠ NULL.
4. PTR ← RPT (PTR).
5. CPT ← LPT (PTR)
   RPT (CPT) ← NULL.
6. RPT (PTR) ← AVAIL
   AVAIL ← PTR.
7. STOP.

(c) If specific node is given :

1. If FIRST = NULL then write UNDERFLOW and STOP.
2. Read DATA as information of node to be deleted.
3. PTR ← FIRST.
4. Repeat step (5) while INFO (PTR) ≠ DATA.
5. PTR ← RPT (PTR).
6. CPT ← LPT (PTR)
   TPT ← RPT (PTR)
   RPT (CPT) ← TPT
   LPT (TPT) ← CPT.
7. RPT (PTR) ← AVAIL
   AVAIL ← PTR
8. STOP.

C implementation

(a) Deletion from beginning

```
void delete_beg ( )
   {
      struct node *ptr;
      if (first = = NULL);
         {
            printf ("Underflow\n");
            return;
         }
      ptr = first;
      first = ptr → rpt;
      first → lpt = NULL;
      free (ptr);
      printf ("Deletion is done\n");
   }
```

(b) Deletion from end :

```
void delete_end ( )
   {
      struct node *ptr, *cpt;
      if (first = = NULL)
         {
            printf ("Underflow\n");
            return;
         }
      ptr = first;
      while (ptr → rpt! = NULL)
         {
```

```
            cpt = ptr;
            ptr = ptr → rpt;
         }
      cpt → rpt = NULL;
      free (ptr);
      printf ("Deletion is done\n");
   }
```

(c) Deletion at the given information :

```
void delete_given_node ( )
   {
      struct node *ptr, *cpt, *tpt;
      int m;
      IF (first = = NULL)
         {
            printf ("Underflow");
            return;
         }
      printf ("Insertion of the node to be deleted");
      scanf ("%d", & m);
      ptr = first;
      while (ptr → info ! = m)
         ptr = ptr → rpt;
      cpt = ptr → lpt;
      tpt = ptr → rpt;
      cpt → rpt = tpt;
      tpt → lpt = cpt;
      free (ptr);
      printf ("Deletion is done\n");
   }
```

EXAMPLE 4.9. **Let us consider a link list as**

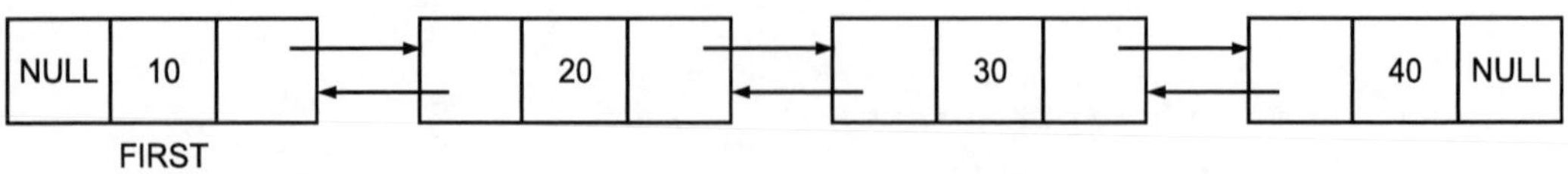

Fig. 4.93.

(a) Delete node 10

(b) Delete node 40

(c) Delete node 30.

Solution: (a) Delete node 10 :

Make PTR = FIRST

and FIRST = RPT (PTR)

LPT (FIRST) = NULL

i.e.

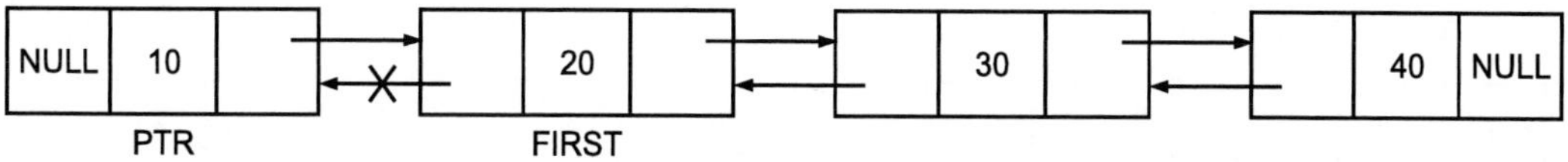

Fig. 4.94.

Thus, final list is

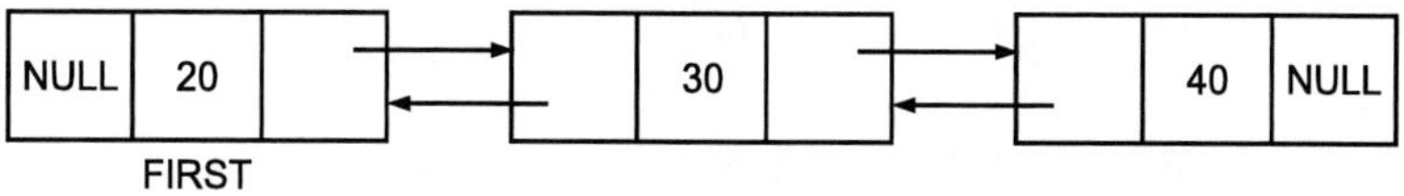

Fig. 4.95.

(*b*) Delete node 40 :

Make PTR = FIRST

Repeat PTR = RPT (PTR) till RPT (PTR) $\neq$ NULL

i.e., now

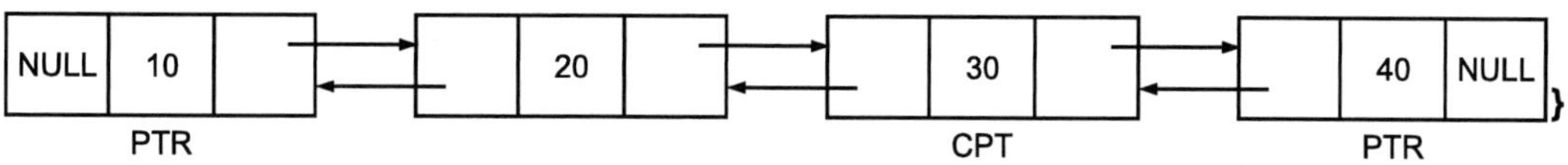

Fig. 4.96.

Now, CPT = LPT (PTR)

Make RPT (CPT) = NULL

i.e.,

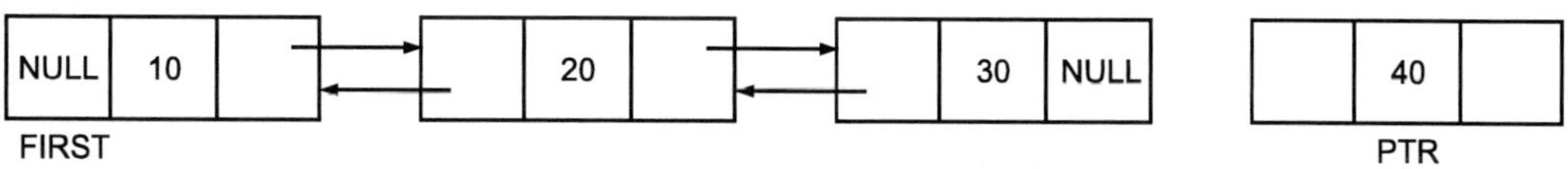

Fig. 4.97.

Now, delete PTR node *i.e.,* the final list is

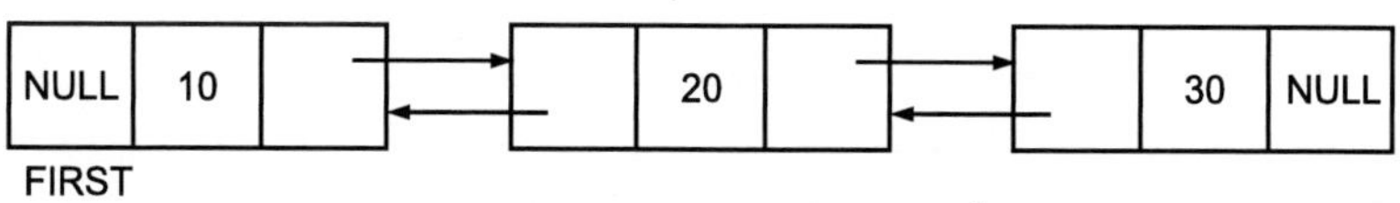

Fig. 4.98.

(c) Delete node 30 :

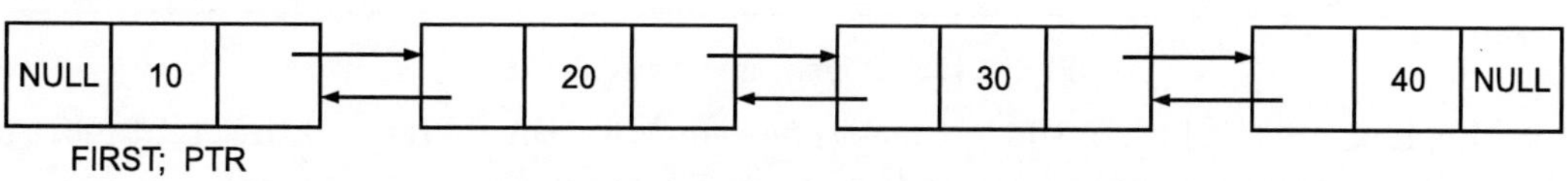

Fig. 4.99.

Here, DATA = 30

Now, PTR = FIRST

Repeat PTR = RPT (PTR)

till INFO (PTR) = DATA

i.e., now

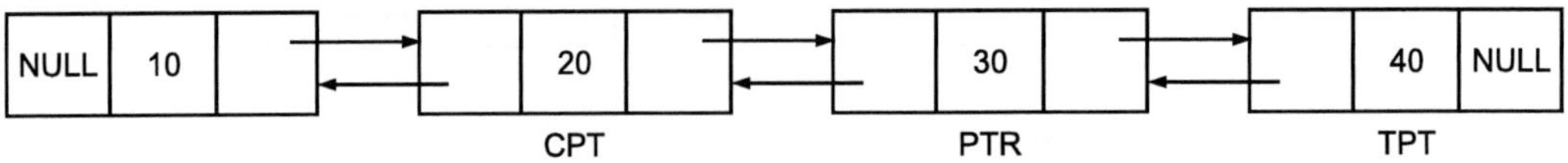

Fig. 4.100.

Now, CPT = LPT (PTR)

TPT = RPT (PTR)

Then, RPT (CPT) = TPT

and LPT (TPT) = CPT

i.e.,

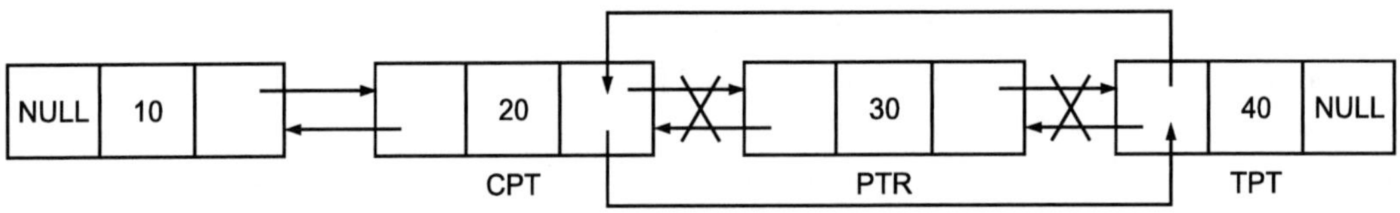

Fig. 4.101.

Then delete node 30.

Thus the final link list is

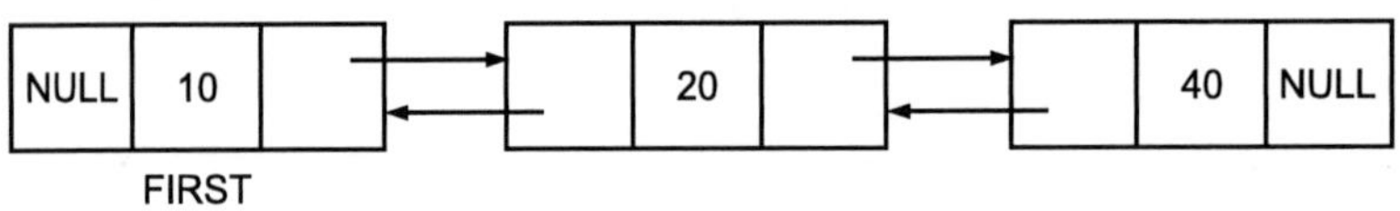

Fig. 4.102.

4.9 DOUBLY CIRCULAR LINKED LIST

A doubly linked circular list means that the last node points to the first node and there are two links (one for pointing to the predecessor node, another for pointing to the next node) between the nodes of the linked list. The list is shown in the figure is a doubly linked circular list.

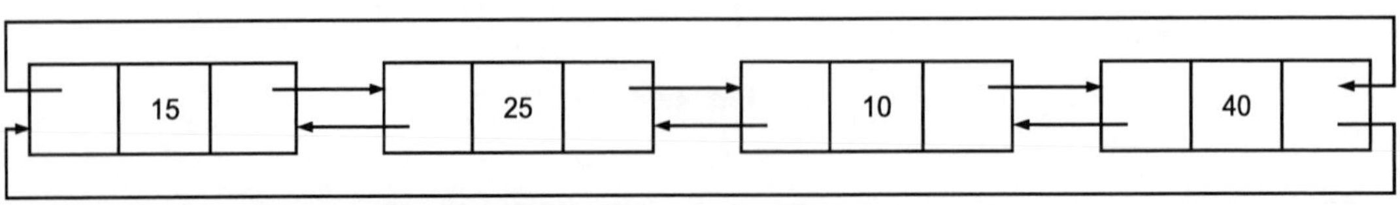

Fig. 4.103. Doubly linked circular list.

In this list, the left link of the left most node contains the address of the right most node and the right link of the rightmost node contains the address of the left most node.

The aim of considering circular doubly linked list is to simplify the insertion and deletion operations performed on doubly linked list.

A circular doubly linked list with a head node is shown in the figure.

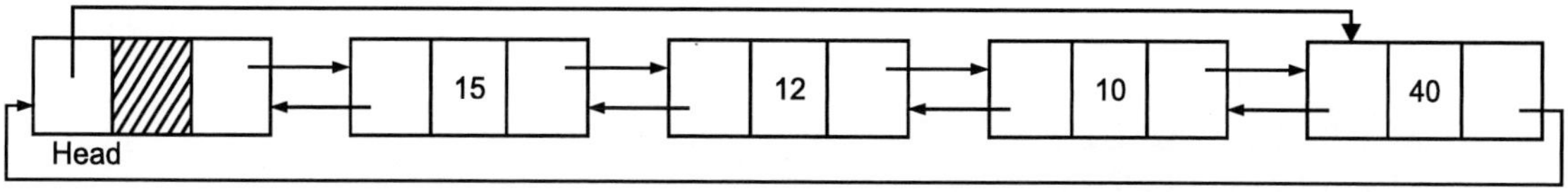

Fig. 4.104 Circular doubly linked list with header node.

In this list, the left link of a header node contains address of the last node and right link of last node contains address of the header node. An empty list with a header node can be represented as:

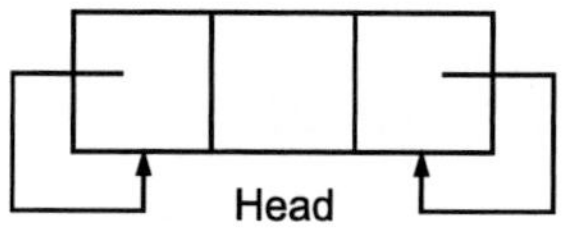

Fig. 4.105.

where the left link and right link of a header node points to itself.

4.9.1 Operations on Circular Doubly Linked List

(*a*) Creation of circular doubly linked list :

1. PTR ← AVAIL
   AVAIL ← RPT (AVAIL)
   AVAIL ← LPT (AVAIL)
   Read INFO (PTR)
   LPT (PTR) ← PTR
   RPT (PTR) ← PTR
   FIRST ← PTR
2. STORE 'Y' to CH.
3. Repeat step 4 to 5 while CH = 'Y'.
4. (*a*) CPT ← AVAIL
   AVAIL ← RPT (AVAIL)
   Read INFO (CPT).
   (*b*) RPT (PTR) ← CPT
   LPT (CPT) ← PTR
   PTR ← CPT.
5. Input choice <Y/N> for more node.
6. RPT (PTR) ← FIRST
   LPT (FIRST) ← PTR.
7. STOP.

(*b*) Inserting a node :

(*i*) At the beginning

1. If AVAIL = NULL then write "overflow and STOP".
2. PTR ← AVAIL
   AVAIL ← RPT (AVAIL)
   AVAIL ← LPT (AVAIL)

Read INFO (PTR).
3. TPT ← LPT (FIRST).
4. RPT (PTR) ← FIRST
   LPT (FIRST) ← PTR.
5. LPT (PTR) ← TPT
   RPT (TPT) ← PTR
6. FIRST ← PTR.
7. STOP.

(*ii*) At the end

1. If AVAIL = NULL then write "OVERFLOW" and STOP.
2. PTR ← AVAIL
   AVAIL ← LPT (AVAIL)
   AVAIL ← RPT (AVAIL)
   Read INFO (PTR).
3. TPT ← LPT (FIRST).
4. RPT (TPT) ← PTR
   LPT (PTR) ← TPT.
5. RPT (PTR) ← FIRST
   LPT (FIRST) ← PTR.
6. STOP.

EXAMPLE 4.10. Consider a circular doubly linked list as

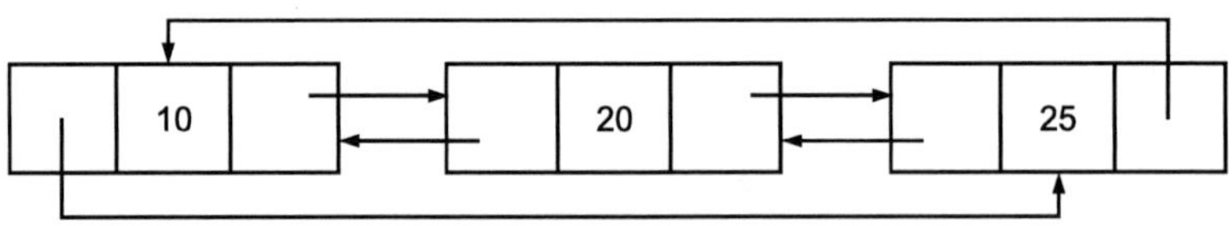

Fig. 4.106.

(*a*) Insert node 5 at beginning.

(*b*) Insert node 30 at end.

Solution: (*a*) Insert node 5 at beginning :

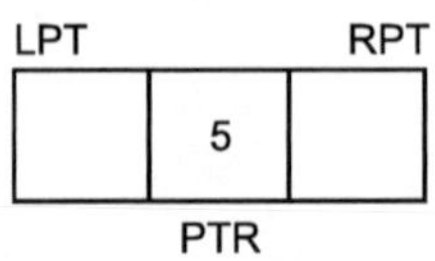

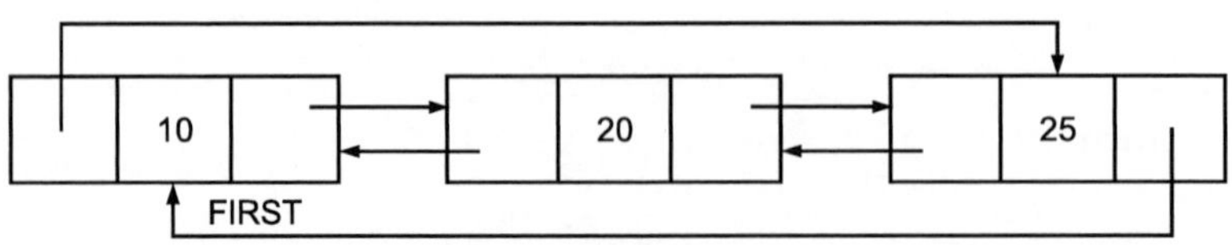

Fig. 4.107.

Make LPT (FIRST) as TPT *i.e.*, node 25 as TPT.

Now RPT (PTR) = FIRST and LPT (FIRST) = PTR

and LPT (PTR) = TPT and RPT (TPT) = PTR

i.e.,

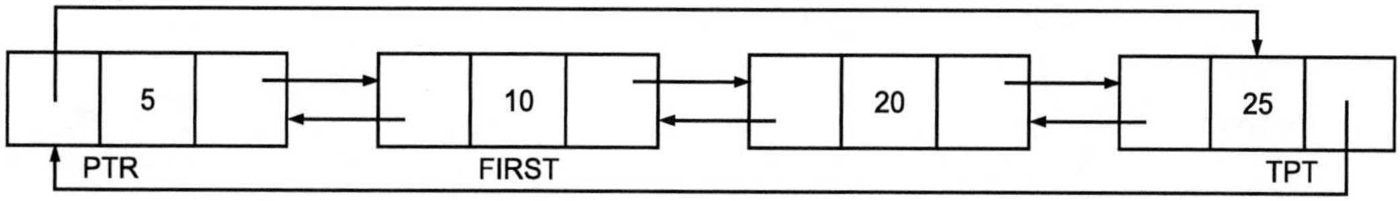

Fig. 4.108.

Make PTR as FIRST, thus final list after insertion is:

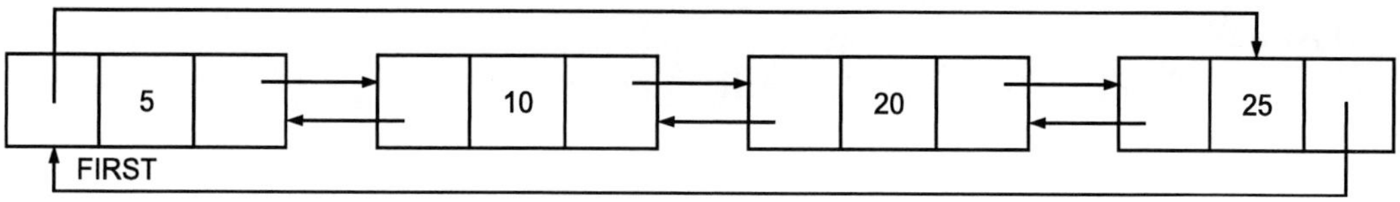

Fig. 4.109.

(*ii*) Insert node 30 at end :

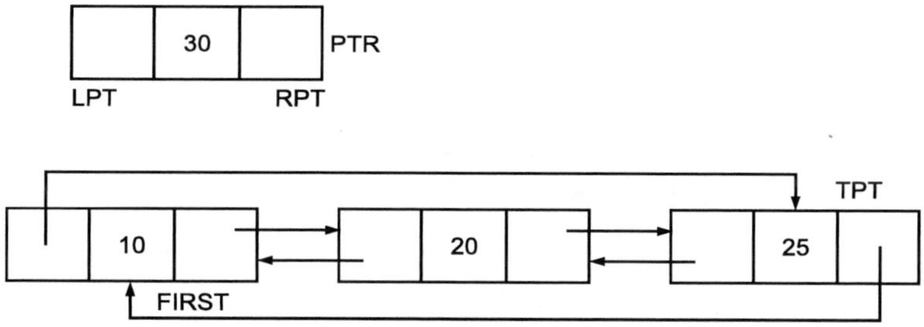

Fig. 4.110.

Make LPT (FIRST) as TPT *i.e.*, node 25 is now TPT

Now, RPT (TPT) = PTR

LPT (PTR) = TPT

and RPT (PTR) = FIRST

LPT (FIRST) = PTR

That is,

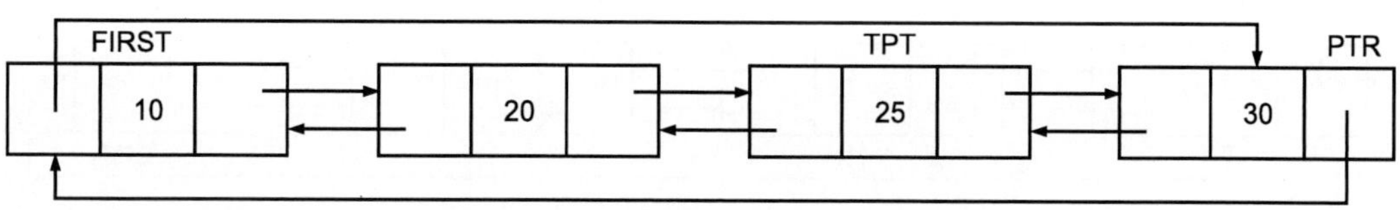

Fig. 4.111. Final list after insertion.

(*b*) Deletion :

(*a*) At the beginning

1. If FIRST = NULL then write "UNDERFLOW" and STOP.
2. TPT ← FIRST

PTR ← RPT (FIRST)
CPT ← LPT (FIRST).

3. LPT (PTR) ← CPT
   RPT (CPT) ← PTR.
4. FIRST ← PTR.
5. LPT (TPT) ← AVAIL
   RPT (TPT) ← AVAIL
   AVAIL ← TPT.
6. STOP.

(*b*) At end

1. If FIRST = NULL then write "underflow" and STOP.
2. PTR ← LPT (FIRST)
   CPT ← LPT (PTR).
3. LPT (FIRST) ← CPT
   RPT (CPT) ← FIRST.
4. LPT (PTR) ← AVAIL
   RPT (PTR) ← AVAIL
   AVAIL ← PTR.
5. STOP.

EXAMPLE 4.11. Let us consider a list as:

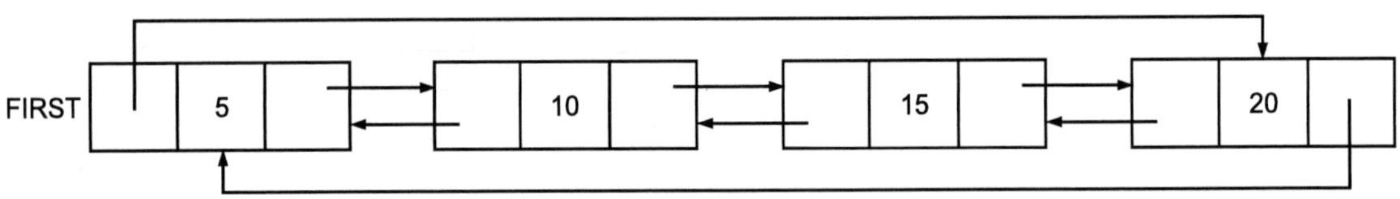

Fig. 4.112.

(*a*) Delete node at beginning.
(*b*) Delete node at end.

Solution: (*a*) Delete node at beginning :

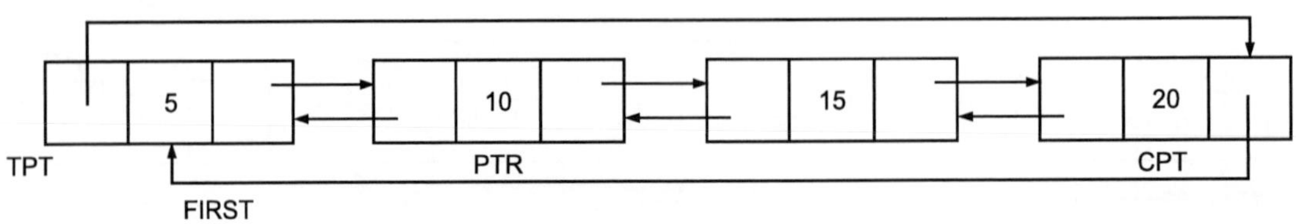

Fig. 4.113.

Firstly, PTR = RPT (FIRST)
CPT = LPT (FIRST)

Now, LPT (PTR) = CPT
and RPT (CPT) = PTR

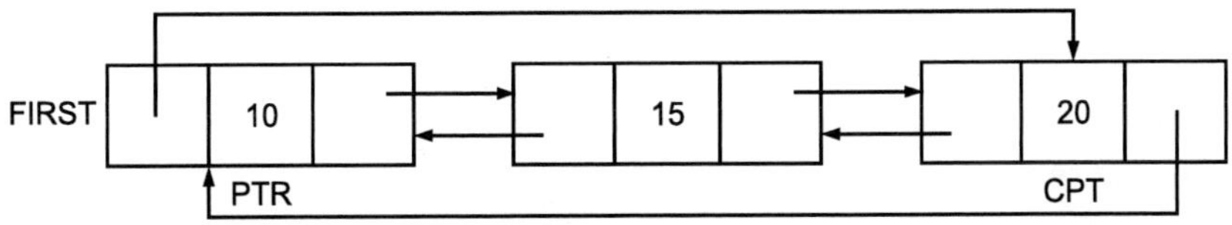

Fig. 4.114.

and make FIRST = PTR and delete TPT node.

That is final list is

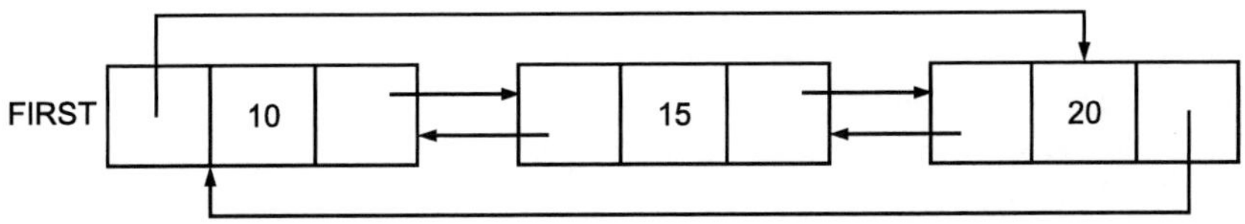

Fig. 4.115.

(*b*) Delete node at end :

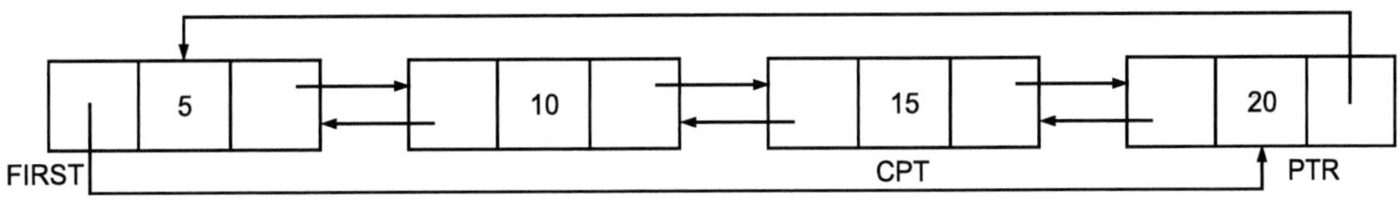

Fig. 4.116.

$$
\begin{aligned}
\text{PTR} &= \text{LPT (FIRST)} \\
\text{CPT} &= \text{LPT (PTR)}
\end{aligned}
$$

Now,

$$
\begin{aligned}
\text{LPT (FIRST)} &= \text{CPT} \\
\text{RPT (CPT)} &= \text{FIRST}
\end{aligned}
$$

i.e.

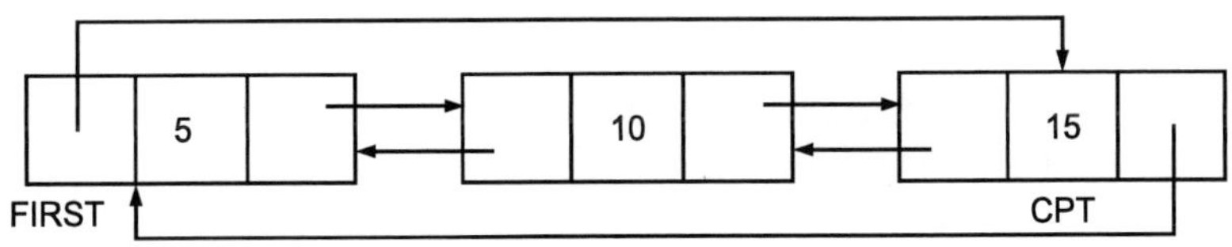

Fig. 4.117.

and delete node PTR.

4.10 LINKED LISTS AND POLYNOMIALS

We can maintain polynomials like $6x^4 + 2x^3 + 5x^2 + 9x + 16$ with the help of linked list. In the linked representation of polynomials, each node should consist of three elements, namely coefficient, exponent and a link to the next term. The coefficient field hold the value of the coefficient of a term, the exponent field contains the exponent value of that term and the link field contains the address of the next term in the polynomial while maintaining the polynomial it is assumed that the exponent of each successive term is less than that of the previous term.

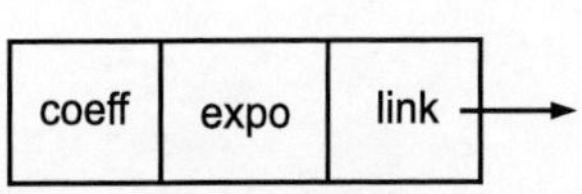

Fig. 4.118.

For example, the following polynomial of degree 4

3 * power (x, 4) + 8 * power (x, 2) + 6 * power (x, 1) + 8 * power (x, 0)

i.e., $3x^4 + 8x^2 + 6x + 8$

can be represented as linked list as

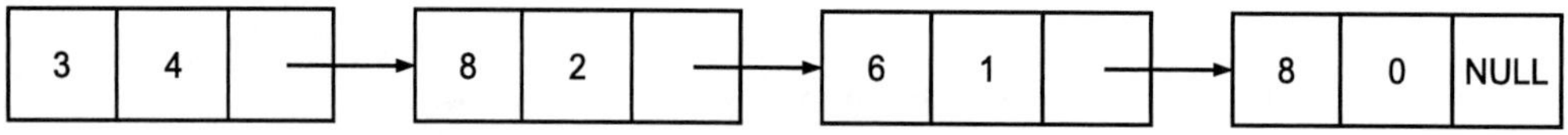

Fig. 4.119.

The link coming out of the last node is NULL pointer.

Consider the example of representing a term of a polynomial in the variables x, y and z. A typical node is represented as below, which consists of five sequentially allocated fields. The first three fields represent the power of the variables x, y and z respectively. The fourth and fifth fields represent the coefficient of the term in the polynomial and the address of the next term in the polynomial. For example, term $5x\,y^2\,z^3$ would be represented as:

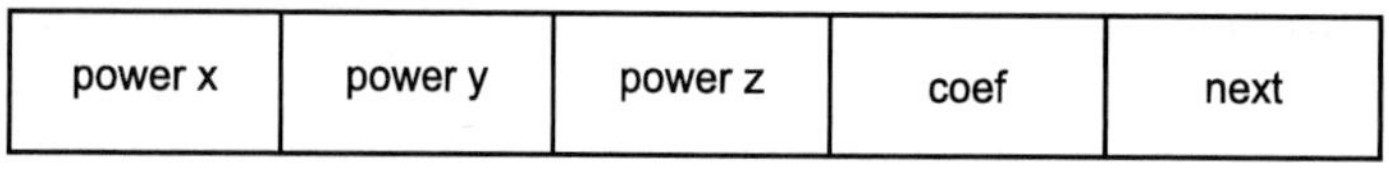

i.e.,

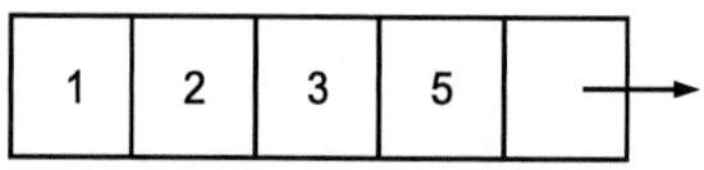

we can view a polynomial such as $3x^2 + 2xy^2 + 5y^3 + 7yz$

we can replace each term of the polynomial with node of the linked list as.

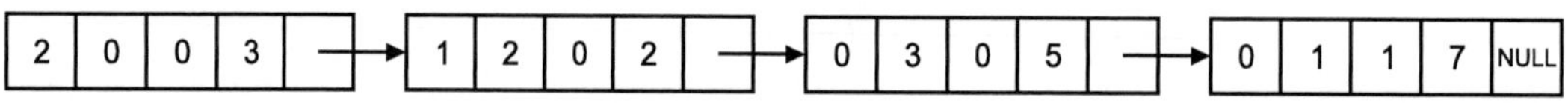

Assume that the nodes in the list are to be stored such that the terms with power of variable x are stored first in decreasing order of the power, then terms with power of variable y are stored in decreasing order of the power, then terms with power of variables z are stored in decreasing order of the power *i.e.*, $3x^2 + 2xy^2 + 5y^3 + 7yz$

The primary reason for representing polynomials as linked list is that only the non-zero terms are required to be represented. This is advantages as frequently we encounter polynomials in which many coefficients are zero.

Another reason for choosing linked list for polynomials is that polynomial grow and shrink as operations are performed on them. As a result, representation of polynomials has become a important application of link lists.

Creation and Traversing of Polynomial :

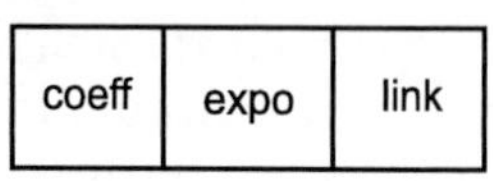


```
# include <stdio.h>
# include <conio.h>
# include <alloc.h>
struct node
  {
    int coff;
    int expo;
    struct node *link;
```

```
   };
struct node *first;
void main ( )
   {
      void create ( ), traverse ( );
      create ( );
      traverse ( );
      getch ( );
   }
void create ( )
   {
      struct node *ptr, *cpt;
      char ch;
      ptr = (struct node *) malloc (size of (struct node));
      printf ("Input first term coefficient and exponent);
      scanf ("%d %d", & ptr → coff, & ptr → expo);
      first = ptr;
   do
      {
         cpt = (struct node *) malloc (size of (struct node));
         printf ("Input next term");
         scanf ("%d %d", & cpt → coff, & cpt → expo);
         ptr → link = cpt;
         ptr = cpt;
         printf ("Press <Y/N> for more node");
         ch = getch ( );
      }
         while (ch = = 'Y');
         ptr → Link = NULL;
      }
void traverse ( )
   {
      struct node *ptr;
      printf ("terms of polynomials:\n");
      ptr = first;
      while (ptr ! = NULL)
         {
            printf ("coefficient : %d\n", ptr → coff);
            printf ("exponent : %d\n", ptr → expo);
            ptr - ptr → link;
         }
         getch ( );
   }
```

4.10.1 Addition of Two Polynomials of Single Variable

Let us consider the addition of two polynomials of single variable $5x^4 + 6x^3 + 2x^2 + 10x + 4$ and $7x^3 + 3x^2 + x + 7$. We can visualise this as follows:

$$\begin{array}{r} 5x^4 + 6x^3 + 2x^2 + 10x + 4 \\ + 7x^3 + 3x^2 + x + 7 \\ \hline 5x^4 + 13x^3 + 5x^2 + 11x + 11 \\ \hline \end{array}$$

i.e., to add two polynomials, compare their corresponding terms starting from the first node and move towards the end node.

A **general algorithm** for addition of two polynomials with single variable is as follows :

1. Repeat upto step 3 while there are terms in both polynomials yet to be processed.
2. Obtain the values for each term.
3. If the exponents (powers) of the corresponding terms in the two polynomials are equal.
   then the coefficients of two nodes have to be added and place the sum in the resultant polynomial.
   else if the exponent of first polynomial is greater (>) than the corresponding exponent term of the second polynomial then copy the coefficient term of the first polynomial to the coefficient term of the resultant polynomial.
   Also, copy the exponent term of the first polynomial to the exponent term of the resultant polynomial and obtain the next term in the first polynomial.
   else insert the term from second polynomial into sum polynomial and obtain the next term in the second polynomial.
4. Copy, remaining terms from non-empty polynomial into the sum or resultant polynomial.
5. Return the address of resultant polynomial.

C Implementation of Addition of Two Polynomials

```
void addition_poly ( )
  {
    struct node *ptr 1, *ptr 2, *ptr, *cpt;
    ptr = (struct node *) malloc (size of* (struct node));
    ptr 1 = first 1;
    ptr 2 = first 2;
    first 3 = ptr;
    If (ptr 1 → expo > ptr 2 → expo)
      {
        ptr → coff = ptr 1 → coff;
        ptr → expo = ptr 1 → expo;
        ptr 1 = ptr 1 → link;
      }
    else
      if (ptr 2 → expo > ptr 1 → expo)
```

```
        {
          ptr → coff = ptr 2 → coff;
          ptr → expo = ptr 2 → expo;
          ptr 2 = ptr 2 → Link;
        }
    else
      {
        ptr → coff = ptr 1 → coff + ptr 2 → coff;
        ptr → expo = ptr 1 → expo
        ptr1 = ptr1 → link;
        ptr 2 = ptr 2 → link;
      }
    while (ptr 1 ! = NULL && ptr 2 ! = NULL)
      {
        cpt = (struct node *) malloc (size of (struct node));
        if (ptr1 → expo > ptr2 → expo)
          {
            cpt → coff = ptr1 → coff;
            cpt → expo = ptr1 → expo;
            ptr1 = ptr1 → Link;
          }
    else if (ptr2 → expo > ptr1 → expo)
      {
        cpt → coff = ptr 2 → coff;
        cpt → expo = ptr2 → expo;
        ptr2 = ptr2 → Link;
      }
    else
      {
        cpt → coff = ptr1 → coff + ptr2 → coff
        cpt → expo = ptr1 → expo;
        ptr1 = ptr1 → Link;
        ptr2 = ptr2 → Link;
      }
    ptr → Link = cpt;
    ptr = cpt;
  }
while (ptr! = NULL)
  {
    cpt = (struct node *) malloc (size of (struct node));
    cpt → coff = ptr1 → coff;
    cpt → expo = ptr1 → expo;
```

```
            ptr1  =  ptr1  →  Link;
            ptr  →  Link  =  cpt;
            ptr  =  cpt;
        }
    while  (ptr2  !  =  NULL)
        {
            cpt  =  (struct  node  *)  malloc  (size  of  (struct  node));
            cpt  →  coff  =  ptr2  →  coff;
            cpt  →  expo  =  ptr2  →  expo;
            ptr2  =  ptr2  →  Link;
            ptr-Link  =  cpt;
            ptr  =  cpt;
        }
            ptr-Link  =  NULL  ;
        }
```

4.10.2 Multiplication of Polynomials

The multiplication of polynomials is performed by multiplying coefficient and adding the respective power, *e.g.*, the multiplication of polynomials

$8x^4 + 6x^2 + 5x + 2$ and $3x^2 + x + 2$ as

$$
\begin{array}{lllllll}
 & & 8x^4 & + 6x^2 & + 5x & + 2 & \\
 & & \times\ 3x^2 & + x & + 2 & & \\
\hline
24x^6 & & + 18x^4 & + 15x^3 & + 6x^2 & & \\
 & +8x^5 & & + 6x^3 & + 5x^2 & + 2x & \\
 & & +16x^4 & & + 12x^2 & + 10x & + 4 \\
\hline
\multicolumn{7}{l}{24x^6 + 8x^5 + 34x^4 + 21x^3 + 23x^2 + 12x + 4} \\
\hline
\end{array}
$$

To produce the multiplication of two polynomials following steps are performed.

1. Check whether two given polynomials are non-empty. If anyone polynomial is empty then polynomial multiplication is not possible. So exit.
2. Second polynomial is scanned from left to right.
3. For each term of the second polynomial, the first polynomial is scanned from left to right and its each term is multiplied by the term of the second polynomial, *i.e.*, find the coefficient by multiplying the coefficients and find the exponent by adding the exponents.
4. If the product term already exists in the resulting polynomial then its coefficients are added, otherwise a new node is inserted to represent this product term.

Polynomial Expansion of an Integer: Given the integer A, we can form a single linked list representing its polynomial expansion. Each node in the list consists of a coefficient and a next points field in the list. This suggests representing long integers by storing their digits from right to left in a list so that the first node on the list contains the least significant digit (right most) and the last node contains the most significant digits (left most).

To save space, we keep 5 digits in each node. In '*C*' language we can store the maximum integer value 2147483647 and minimum integer value – 2147483646

A list representing the number 101231516237 is as follows:

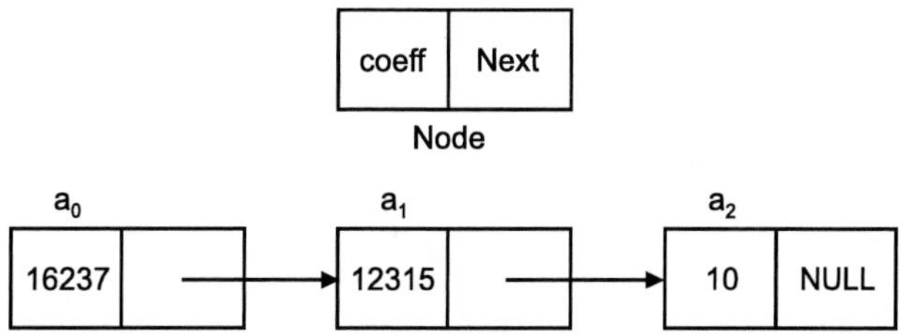

Fig. 4.120.

EXAMPLE 4.12: The algorithm ADDINT adds two multiple-precision integers and returns the address of the firt node of the resultant integer.

Algorithm ADDINT (p, q)

```
Step 1:   Intialize the variables
          MAX = 100000 L;
          ptrp = p;
          ptrq = q;
          ptrs = S;
          carry = 0;
Step 2:   Repeat step(3) to (5) while ((ptrp! = NULL)) and (ptrq! = NULL)
Step 3:   total = info (ptrp) + info (ptrq) + carry;
          carry = total/MAX;
          ptrs = insertlast (ptrs, total % MAX);
Step 4:   If ((ptrp = p) and (ptrq = q)) then
              S = ptrs;
Step 5:   ptrp = next (ptrp);
          ptrq = next (ptrq);
Step 6:   if (ptrp ! = NULL)
              then ptrq = ptrp;
Step 7:   if (ptrq! = NULL) then
          while (ptrq! = NULL)
              {
                  total = info (ptrq) + carry;
                  carry = total/MAX;
                  ptrs = insertlast (ptrs, total% MAX);
                      ptrq = next (ptrq);
              }
Step 8:   If (carry ! = 0)
              ptrs = insertlast (ptrs, carry);
Step 9:   return (S);
```

Algorithm Insertlast (FIRST, S)

```
Step 1:   Create a new node as NODE
```

```
Step 2:    INFO (NODE) = S;
           NEXT (NODE) = NULL;
Step 3:    If (FIRST = NULL) then
               FIRST = NODE;
               Return (FIRST);
           else
               NEXT (FIRST) = NODE;
               Return (NODE);
Step 4:    STOP
```

4.11 GENERALISED LINKED LIST

A generalised link list is a sequence of $n \geq 0$ elements as:

(*i*) A = (a_0, a_1, a_2 ...... a_{n-1}) where a_i is either atomic value or a list.

(*ii*) When a_i is a list, it is called **sub-list.**

A list may be denoted by a parenthesised enumeration of its elements separated by commas. For example,

1. A = ( ) represents null or empty list have zero length.
2. *A* = (2, (3, 4), 8) list having length 3, there are two atomic node and a list containing two atomic nodes.
3. *A* = (*a, A*) is a recursive list of length 2. A corresponds to the infinite list as *A* = (*a*, (*a*, (*a*, (*a* .....)

Now let us consider a list *A* as:

$$A = (5, (2, 8, 3, (9, 6), (\ ), 7), 1)$$

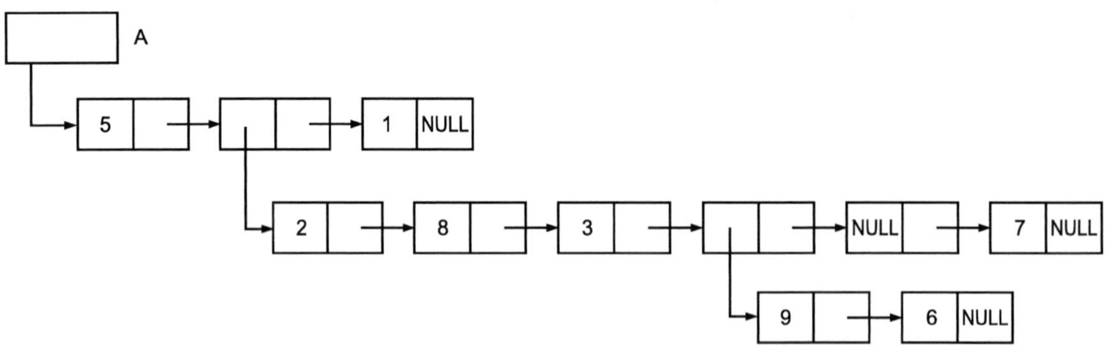

Fig. 4.121.

In generalised list representation of *A*

(*a*) 5 and 1 are at level one.

(*b*) 2, 8, 3, 7 are at level two.

(*c*) 9, 6 are at level three.

EXAMPLE 4.13. Discuss the advantages, if any, of a two-way list over a one-way list for each of the following operations :

(*a*) Traversing the list to process each node.

(*b*) Deleting a node whose location is given.

(c) Searching an unsorted list for a given element.

(d) Searching a sorted list for a given element.

(e) Inserting a node before the node with a given location.

(f) Inserting a node after the node with a given location.

Solution:

(a) No advantage.

(b) The location of the preceding node is needed. The two-way list contains this information whereas in one-way list we must traverse the list.

(c) No advantage.

(d) No advantage.

(e) Two-way list is more efficient.

(f) No advantage.

EXAMPLE 4.14. Whether linked list is linear or non-linear data structure ?

Solution: According to access strategies linked list is a linear one but according to storage linked list is a non-linear data structure.

Exercises

1. What is linked list ? How it is different from array ? Write the different types of linked list.
2. List the advantages of linked lists over arrays.
3. Give any two applications of linked lists.
4. Mention the significance of a circular linked list.
5. List the various operations that can be performed in a linked list.
6. Let LIST be a linked list in memory. Write a procedure which
   (a) Find the number of times a given ITEM occurs in LIST.
   (b) Finds the number of non-zero elements in LIST.
   (c) Adds a given value k to each element in LIST.
   (d) Deletes the last node from LIST.
7. Suppose NAME1 is a list in memory. Write an algorithm which copies NAME1 into a list NAME2.
8. Write a *C* program to reverse a singly linked list.
9. Write a *C* program to create a singly linked list and split it at the middle and make the second half as the first and *vice-versa*. Display the final list.
10. What is doubly linked list. What advantages does a doubly linked list have over linear linked lists?
11. In circular linked lists, how the end-of-list condition will be tested ?
12. How many pointers need to be modified in the following operations ?
    (a) Insert an element in the beginning of a linear linked list.
    (b) Insert an element in the beginning of a doubly linked list.
    (c) Delete an element from the end of a doubly linked list.
13. How a linked list can be used to represent a polynomial of type
    $$9x^2y^2 - 8xy^2 + 10xy + 9y^2.$$
14. What is the advantage of a header node in a linked list ?
15. What is generalised list ? Give some examples.

Chapter 5 Stacks

Chapter Outline

- Introduction
- Definition
- Operations on Stack
- Implementation
- Applications of Stack

5.1 INTRODUCTION

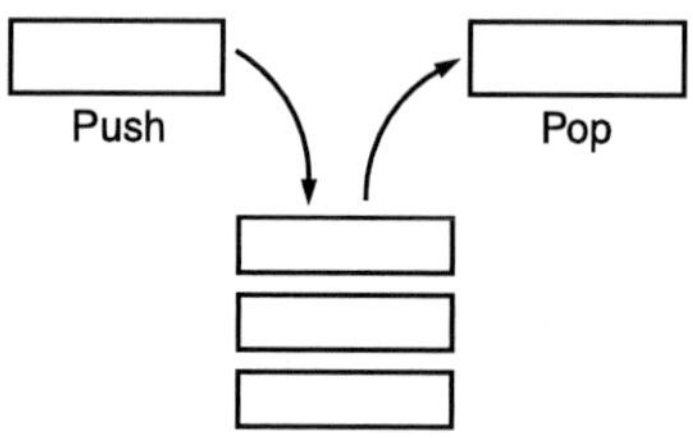

Fig. 5.1.

The linear data structure such as array and a linked list allows us to delete and insert an element at any place in the list, either at the beginning or at the end or even in the middle. However, sometimes it is required to permit the addition or deletion of elements only at one end. That is either at the beginning or at the end. Stack and queue are two types of data structures in which the addition or deletion of an element is done at end, rather than in the middle.

A 'stack' was first proposed in 1955, and then patented in 1957, by the **German Friedrich L. Bauer**. The same concept was developed independently, at around the same time, by the Australian **Charles Leonard Hamblin**.

A stack is a linear data structure in which all insertions and deletions are made at one end, called the top of the stack. It is very useful in many **applications** such as :

- We see stack (pile) of plates in restaurant, stack of books in bookshop. etc.
- Even a packet of papers is also a stack of paper-sheet. A book is also a stack of written papers.
- When anybody takes a plate from a stack of plates, he takes it from the top.

- For the shipment of goods, they have to be loaded into a cargo. During unloading, they are unloaded exactly in the opposite order as they are loaded, *i.e.*, the goods which are loaded last should be unloaded first.

5.2 DEFINITION

A stack is an ordered collection of homogeneous data elements where the insertion and deletion operations occur at only one end. This end is often known as *top of the stack*. Here, the last element inserted will be on top of the stack. Since deletion is done from the same end, last element inserted will be the first element to be removed out from the stack and so on. That is why the stack is also called *Last-in-First-out (LIFO)*.

Initially, when stack is created using arrays the size is fixed, the stack base remains fixed while the stack top changes the elements are inserted or deleted. So, most frequently accessible element in the stack is the top and the last accessible element is the bottom of the stack. Figure 5.1 shows the creation of stack and its operations, *i.e.*, addition or deletion of elements.

Consider an example, assume 5 elements are A, B, C, D and E to be added into the stack one-by-one.

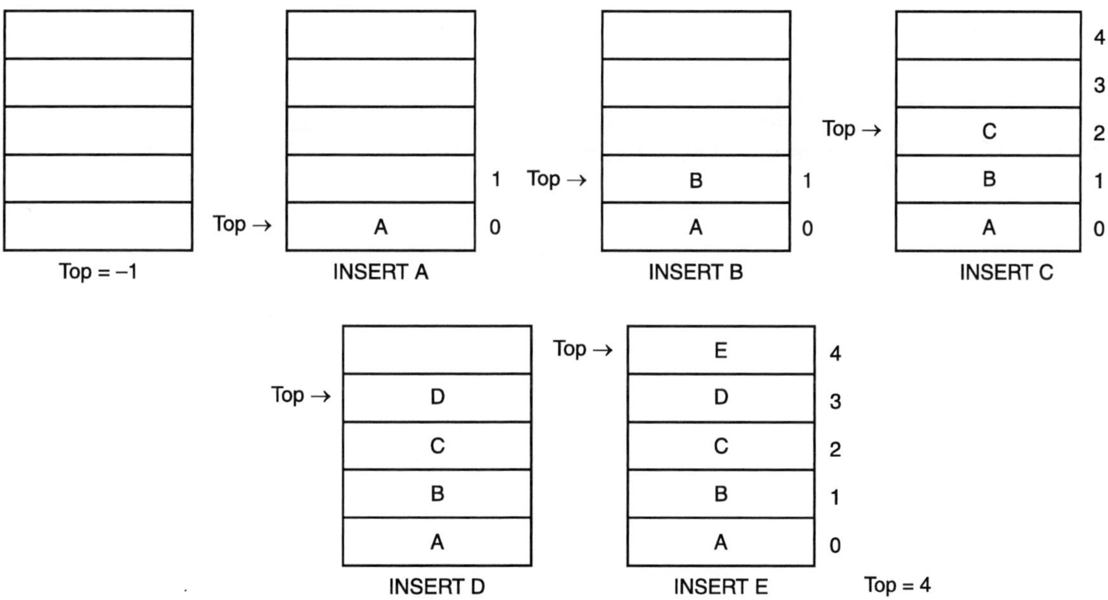

Fig. 5.2. Insertion operations.

Figure 5.2 shows the stack **insertion operations**. Initially the stack is empty and it has one pointer *i.e.*, TOP. It points to bottom of stack (top = – 1). When the elements are inserted into stack pointer Top is incremented and it points to the top most elements in the stack.

Now, figure below shows the stack **deletion operations**, when an element needs to be deleted, it should be deleted from the top. So, the element below the top element become the new top element and so the position of the top most element is decremented.

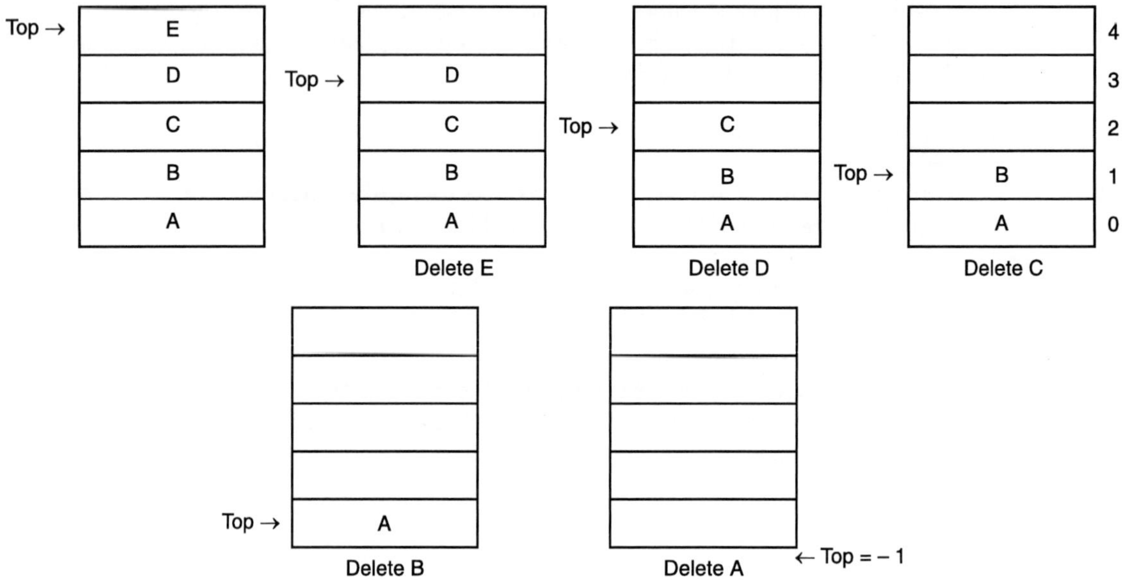

Fig. 5.3. Deletion operation.

5.3 OPERATIONS ON STACK

A stack is generally implemented with two basic operations such as PUSH and POP.

Operation	Description
PUSH	Allows adding an element at the top of the stack.
POP	Allows to remove an element from the top of the stack.

PUSH Operation

The process of adding a new element to the top of stack is called push operation. Pushing an element in the stack invoke adding of element at the top. Suppose, after inserting an element max size is reached, *i.e.*, stack is full and if we want to insert one more new element, it is not possible to insert any new elements. This situation is called ***stack overflow condition***. At this point the stack top is present at the highest location of the stack.

```
IF TOP = MAX SIZE - 1 then STACK IS OVERFLOW
```

POP Operation

The process of deleting an element from the top of stack is called pop operation. After every pop operation the stack is decremented by 1. Finally, when all the elements are deleted, top points to bottom of the stack. When stack is empty, it is not possible to deleted any element and this situation is called ***stack underflow***. Thus, if there is no element on the stack and the Pop operation is performed then this will result into UNDERFLOW condition.

```
IF TOP = -1 then STACK IS UNDERFLOW
```

It is also possible to verify the item placed at the top of the stack without removing it. This operation is called ***PEEP operation***. Thus, **determining the top element of a stack without removing it from the stack is called PEEP operation**.

5.3.1 Algorithms for Push and Pop Operations

(*a*) Push Operation

1. If TOP = MAX-1 then write "STACK OVERFLOW and STOP"
2. READ DATA
3. TOP ← TOP + 1
4. STACK [TOP] ← DATA
5. STOP

(*b*) Pop Operation

1. If TOP < 0 then write "STACK UNDERFLOW and STOP"
2. STACK[TOP] ← NULL
3. TOP ← TOP – 1
4. STOP

C-implementation

/* Operations performed on the stack */

```
#include <stdio.h>
#include <conio.h>
#include <stdlib.h>
#define maxsize 30
int stack [maxsize], top;
void push ( );
void pop ( );
void display ( );
void main ( )
{
clrscr( );
int ch=1; top=-1;
while(ch)
{
printf ("\n1 Push");
printf ("\n2 Pop");
printf ("\n enter your choice");
scanf ("%d", &ch);
switch (ch)
{
case 1:
push ( );
break;
case 2:
pop ( );
break;
```

```
default:
exit (0);
}
}
getch ( );
}
void push ( )
{
int no;
if (top==maxsize-1)
{
printf ("\n stack overflow");
return;
}
top++;
printf ("\n Enter number:\n");
scanf ("%d", &no);
stack[top]=no;
printf("\n Elements in stack are:\n");
display ( );
}
void pop ( )
{
int no;
if (top==-1)
{
printf ("\n stack underflow");
return;
}
else
{
top--;
printf ("\n Elements in stack are:\n");
display ( );
}
}
void display ( )
{
int i;
for (i=0; i<=top; i++)
{
printf("%d\n",stack [i]);
}
}
```

/* Program of reversing a string using stack */

```
#include <stdio.h>
#include <conio.h>
#include <string.h>
#define MAX 20
int top=-1;
char stack [MAX]
char pop ( );
push (char);
main ( )
   {
      clrscr ( );
      char str [20];
      int i;
      printf ("Enter the string :");
      gets (str);
      for (i=0; i<strlen(str); i++)
         push (str[i]);
      for (i=0; i<strlen(str); i++)
         str[i]=pop ( );
      printf ("Reversed string is :");
      puts (str);
      getch ( );
   }
push (char item)
   {
      if (top==MAX-1)
         printf ("Stack overflow\n");
      else
         stack[++top] = item;
   }
char pop ( )
   {
      if (top ==-1)
         printf ("stack underflow \n")
      else
         return stack [top--]
   }
```

5.4 IMPLEMENTATION

A stack can be implemented in anyone of the following two ways :

1. Using array
2. Using linked list.

5.4.1 Array Implementation of Stacks

The easiest way to represent a stack is by using one-dimensional array. A particular variable Top represent the index of last inserted variable.

A stack is an ordered collection of items and C language already contains a data type that is an ordered collection of items such as *array*. A stack and an array are two entirely different things. The number of elements in an array is fixed and is assigned by the declaration for the array. A stack on the other hand, is fundamentally a dynamic object whose size is constantly changing as items are popped and pushed. However, although an array cannot be a stack, it can be home of a stack.

A stack in C may declared as a structure containing two objects :

1. An array to hold the elements of the stack.
2. A variable TOP to hold the topmost element of the stack.

Here the stack is of fixed size. This means that some maximum limit is specified for storing the elements into the stack. Once the maximum limit is reached, it is not possible to store the elements into it.

/* Implementation of stack by Array */

```
#include <stdio.h>
#include <conio.h>
#define MAX 50
int stack [MAX+1], top = 0;
void main ( )
   {
      clrscr ( );
      void create ( ), traverse ( ), push ( ), pop ( );
      create ( );
      printf("\n stack is:\n");
      traverse ( );
      push ( );
      printf("After Push an element the stack is:\n");
      traverse ( );
      pop ( );
      printf("After Pop an element the stack is:\n");
      traverse ( );
      getch ( );
   }
void create ( )
   {
      char ch;
      do
         {
            top ++;
            printf ("Input Element");
            scanf ("%d", &stack[top]);
```

```
            printf ("Press <Y> for more element\n");
            ch = getch ( );
        }
    while (ch=='Y');
    }
void traverse ( )
    {
            int i;
            for (i=top; i>0; --i)
            printf ("%d\n", stack[i]);
        }
    void push ( )
        {
            int m;
            if (top==MAX)
               {
                  printf ("Stack is overflow");
                  return;
               }
            printf ("Input New Element to Insert");
            scanf ("%d", &m);
            top++;
            stack[top]=m;
        }
    void pop ( )
        {
            if (top==0)
               {
                  printf ("Stack is underflow\n");
                  return;
               }
            stack[top]='\0';
            top--;
        }
```

The array implementation has the following **disadvantages** :

1. Memory space is wasted if we store less number of items in the array than the maximum size of the array.
2. Limitation in storing the items into the stack.
3. Insertion and deletion operations in a stack using an array involves more data movements.

Advantages of array implementation is that the implementation of the stack using an array is easy.

EXAMPLE 5.1. **Design a method for keeping two stacks with in a single linear array so that neither stacks overflow until all the memory is used. Write a C function PUSH (X, S) that pushes element *x* onto stack S, where is S one or other of these two stacks. Include all necessary error checks.**

Solution:

```
# define maxsize [10]
struct stack
   {
      int top_1, top_2;
      int A[maxsize]
   };
void int stack (struct stack *S)
   {
      S → top_1 = - 1;
      S → top_2 = maxsize;
   }
void push_1(struct stack *S, int item)
   {
      if (S → top_1 == S → top_2 - 1)
      printf ("\n stack is full");
      return;
   }
   else
      {
         S → top_1 = S → top_1 + 1;
         S → A [S → top_1] = item;
      }
   }
/ For second stack */
void push_2 (struct stack *S, int item)
   {
      if  (S → top_2 = = S → top_1 + 1)
      {
         printf ("\n stack is full");
         return;
      }
      else
      {
         S → top_2 = S → top_2 - 1;
         S → A [S → top_2] = item;
      }
   }
```

5.4.2 Linked List Implementation of a Stack

The stack can be implemented as a linked list in which the top of the stack is represented by the first item in the list. The first element inserted into the stack is pointed out by the second element, the second element by the third and so on. In general the (n–1)th element is pointed out by the nth element.

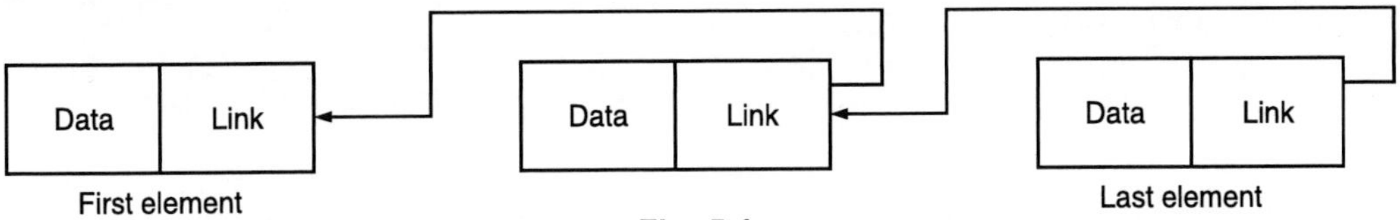

Fig. 5.4.

In linked list implementation, the stack does not need to be of fixed size. There can be any number of elements or nodes in the stack. The second advantage of linked list method is that insertion and deletion operation do not involve more data movements. The other advantage of this method is that memory. Space is not wasted, because memory space is allocated only when the user wants to push an element into the stack. To implement a push, we create a new node in the list and attach it as the new first element. To implement a POP, we advance the Top of the stack to the second item in the list.

/* Implementation of stack by Link-list */

```
#include <stdio.h>
#include <conio.h>
#include <alloc.h>
struct node
      {
         int info;
         struct node *link;
      };
   struct node *top;
   void main ( )
      {
           void create ( ), traverse ( ), push ( ), pop ( );
           create ( );
           printf("\n stack is:\n");
           traverse ( );
           push ( );
           printf("After push the element the stack is:\n");
           traverse ( );
           pop ( );
           printf("After pop the element the stack is:\n");
           traverse ( );
         getch ( );
      }
   void create ( )
      {
```

```
    struct node *ptr, *cpt;
    char ch;
    ptr = (struct node *) malloc (sizeof (struct node));
    printf ("Input first info");
    scanf ("%d", &ptr → info)
    ptr->link = NULL;
do
    {
       cpt = (struct node *) malloc (sizeof (struct node));
       printf ("Input next information");
       scanf ("%d", &cpt → info);
       cpt->Link = ptr;
       prt=cpt;
       printf ("Press <Y/N> for more information");
       ch = getche ( );
    }
    while (ch=='Y')
    top = ptr;
    }
void traverse ( )
       {
          struct node *ptr;
          printf ("Traversing of stack :\n")
          ptr=top;
          while (ptr !=NULL)
             {
                printf ("%d\n", ptr->info);
                ptr=ptr->Link;
             }
       }
    void push ( )
       {
          struct node *ptr;
          ptr = (struct node *) malloc (sizeof (struct
       node));
          if (ptr==NULL)
          {
             printf ("Overflow\n");
             return;
          }
       printf ("Input New node information");
       scanf ("%d", &ptr->info);
```

```
        ptr->link=top;
        top = ptr;
    }
  void pop ( )
    {
        struct node *ptr;
        if (top==NULL)
          {
            printf ("Underflow \n");
            return;
          }
        ptr=top;
        top = ptr->link;
        free (ptr);
    }
```

EXAMPLE 5.2. A programming language provides two functions ALLOCATE(x) and FREE(x) for the maintenance of linked list structures. ALLOCATE(x) allots a node with address *x* for use in the linked list structure and FREE(x) frees the node with address *x* used in the application to the AVAIL list. Assuming the AVAIL list to be maintained as a linked stack. Write procedures to implement the functions ALLOCATE and FREE.

Solution: With the AVAIL list maintained as a linked stack the procedure ALLOCATE(x) performs pop operation *i.e.*, release the top node whose address is x and FREE (x) performs a push operation *i.e.*, inserting the node that has been deleted from the application and whose address is x to the top of the stack.

ALLOCATE(x)

1. If AVAIL = NULL then OVERFLOW
2. x = AVAIL [Allot top node of AVAIL to *x*]
3. AVAIL = LINK [AVAIL]
4. STOP

FREE(*x*)

1. LINK (x) = AVAIL
2. AVAIL = x
3. STOP

5.5 APPLICATIONS OF STACK

(*a*) Checking the Validity of an Arithmetic Expression

Using stack, we can check the validity of an arithmetic expression. We know in a valid expression the parenthesis, brace or bracket must occur in pairs. That is when there is an opening parenthesis; brace or bracket, there should be the corresponding closing parenthesis. Otherwise, the expression is not a valid one.

- Whenever an opening parenthesis is encountered, it is pushed on to the **stack.**

- Whenever a closing parenthesis is encountered, the **stack** is examined.
- If the **stack** is empty, the closing parenthesis does not have an opening parenthesis and the expression is therefore invalid.
- If the STACK is not empty, we pop the **stack** and check whether the popped item corresponds to the closing parenthesis.
- If a match occurs, we continue. Otherwise the expression is invalid.
- When the end of the expression is reached, the **stack** must be empty; otherwise one or more opening parenthesis does not have corresponding closing parenthesis and the expression is invalid.

For example :

[(A + B) – {C + D} – [F + G]

	Symbol Scanned	Stack
(1)	[	[
(2)	(	[, (
(3)	A	[, (
(4)	+	[, (
(5)	B	[, (
(6)	)	[
(7)	-	[
(8)	{	[, {
(9)	C	[, {
(10)	+	[, {
(11)	D	[, {
(12)	}	[
(13)	]	
(14)	-	
(15)	[	[
(16)	F	[
(17)	+	[
(18)	G	[
(19)	]	

In the above example we see the stack is empty at the end, so the expression is valid.

/* Program to check nesting of parenthesis using stack */

```
#include <stdio.h>
#include <conio.h>
#include <string.h>
#define MAX 20
#define true 1
#define false 0
int top = -1;
int stack[MAX];
```

```
push(char);
char pop( );
void main( )
{
            clrscr( );
            char exp [MAX], temp;
            int i, valid=true;
            printf ("Enter an algebraic expression :\n");
            gets (exp);
            for (i=0;i<strlen(exp); i++)
            {
                     if(exp[i]=='('|| exp[i]=='{'||exp[i]=='[')
                           push(exp[i]);
                     if(exp[i]==')' || exp[i]=='}'||exp[i]==']')
                           if(top ==-1) /* stack empty */

                           valid=false;
                           else
                           {
                              temp=pop ( );
                              if (exp[i]==')'&&(temp=='{'||temp=='['))
                                          valid=false;
                              if (exp[i]=='}'&&(temp=='('|| temp=='['))
                                          valid=false;
                              if (exp[i]==']'&&(temp=='('|| temp=='{'))
                                          valid=false;
                           }
            }
            if (top>=0) /*stack not empty */
                     valid=false;
            if (valid==true)
                     printf("Valid expression\n");
            else
                     printf("Invalid expression\n");
getch( );
}
push(char item)
{
            if(top -- (MAX-1))
                     printf ("Stack overflow\n");
            else
            {
```

```
                    top=top+1;
                    stack[top] = item;
            }
}
char pop ( )
{
            if (top == - 1)
                    printf ("stack underflow\n");
            else
                    return (stack [top--]);
}
```

(*b*) Polish Notations

Transforming the expression into a form called Polish notation. An arithmetic expression can be represented in various forms such as *prefix, infix* or *postfix.*

The prefixes "pre-", "in-", and "post-" refer to the relative position of the operator with respect to its operands. If the operator is placed before its two operands then the expression is in **prefix form**, if the operator is placed in the middle it is known as ***infix***, and if it is placed after the two operands it is known as **postfix form**. Let us consider a simple example with an operator "+" and two operands "A" and "B".

Prefix: + AB (operator before its operands)
Infix : A + B (operator in the middle of its operands)
Postfix : AB + (operator after its operands)

Infix form	Prefix form	Postfix form
a * b	* a b	a b *
a + b * c	+ a * b c	a b c * +
(a + b) * c	* + a b c	a b + c *

Reverse Polish notation

Polish notation, named after the polish mathematician **Jan Lukasiewicz**, refers to the notation in which the operator symbol is placed before its two operands.

The fundamental property of Polish notation is that the order in which the operations are to be performed is completely determined by the positions of the operators and operands in the expression. Accordingly, one never needs parentheses when writing, expressions in Polish notation.

Reverse Polish notation refers to the analogous notation in which the operator symbol is placed after its two operands :

AB + CD EF*GH/

Again, one never needs parentheses to determine the order of the operations in any arithmetic expression written in reverse Polish notation. This notation is frequently called postfix (or suffix) notation, whereas prefix notation is the term used for Polish notation, discussed in the preceding paragraph.

The computer usually evaluates an arithmetic expression written in infix notation in two steps. First, it converts the expression to postfix notation, and then it evaluates the postfix expression. In each step, the stack is the main tool that is used to accomplish the given task.

Highest :

- Exponential (^ or ↑)
- Multiplication (* or ×)
- Division (/ or ÷)

Lowest :

- Addition (+) or
- Subtraction (–)

Advantage of using Postfix Notation

Using *infix notation*, one can't tell the order in which operators should be applied by only looking at expression. Whenever an infix expression consists of more than one operators, the precedence rules (BODMAS) should be applied to decide that which operator are evaluated first.

In postfix notation operands appear before the operators, there is no need for operator precedence and other rules.

Convert Infix to Postfix form :

EXAMPLE 5.3. Convert the expression (A + B) / (C – D) to Postfix form.

Solution: (A + B) / (C – D)

$$\underbrace{(AB+)}_{X} \; / \; \underbrace{(CD)-}_{Y}$$

Put X = (AB +)

Y = (CD –)

i.e., X/Y

XY/

Put the values of X and Y, we get

AB + CD –/

This is the required postfix form.

EXAMPLE 5.4. Convert the expression A*B + C/D to postfix form.

Solution: A*B + C/D

We can write the above expression as:

(A*B) + (C/D)

$$\underbrace{(AB*)} + (C/D)$$

X + (C/D)

$$X + \underbrace{(CD/)}_{Y}$$

X + Y

XY +

Put the values of X and Y, we get

AB*CD/ +

This is the required postfix form.

Note : During parenthesising the expression, the operands associated with operator having higher precedence are first parenthesised.

EXAMPLE 5.5. Convert

A + (B*C – (D/E ^ F)*G)*H

into postfix form.

Solution:

A + (B*C – (D/E ^ F)*G)*H

A + (B*C – $\underbrace{(D/EF \wedge)}_{X}$ *G)*H

A + (B*C – (D/X)*G)*H

A + (B*C – $\underbrace{(DX/)}_{Y}$ *G)*H

A + (B*C – (Y*G))*H

A + (B*C – $\underbrace{(YG*)}_{T}$)*H

A + (B*C – T)*H

A + ((B*C) – T)*H

A + ($\underbrace{(B*C)}_{U}$ – T)*H

A + (U – T)*H

A + $\underbrace{(UT-)}_{W}$ *H

A + W*H

A + (W*H)

A + $\underbrace{(WH*)}_{Z}$

A + Z

AZ +

Now put the values

AWH* +

AUT – H*+

ABC*YG*– H*+

ABC*DX/G*– H*+

ABC*DEF^/G*– H*+

This is the required postfix form.

Infix to Postfix Conversion using Stack

We can convert an *infix* expression to *postfix* form using stack. We start to scan the expression from left to right. In an expression, there may be some operands, operators and parenthesis (opening or closing).

- Each time we get an operand it is added to the *postfix* expression.
- When we get an operator, we should check the top of the stack.
- If the operator at the top of the stack has the same or higher precedence than the current operator then we repeatedly pop from the stack and add it to the *postfix* expression, otherwise the current operator is pushed onto the stack.
- When an opening parenthesis is encountered it is pushed onto the stack and when the corresponding closing parenthesis is encountered we repeatedly pop from the stack and add the operators from the stack to the *postfix* expression and the corresponding opening parenthesis is deleted from the stack.

The steps involved in the process discussed above can be written as follows :

1. *Scan the expression from left to right. We shall get a symbol which may be an operand or a parenthesis (opening or closing) or an operator. The symbol is treated as follows.*
2. *Steps :*
   a. *if the symbol is an operand, add it to the postfix expression,*
   b. *if the symbol is an opening parenthesis, push it onto the stack.*
   c. *if the symbol is an operator, then check the top of the stack.*
3. *If the precedence of the operator at the top of the stack is higher or the same as the current operator then repeatedly it is popped and added to the postfix expression otherwise, it is pushed onto the stack.*
4. *If the symbol is a closing parenthesis, then repeatedly pop from the stack and add each operator to the postfix expression until the corresponding opening parenthesis is encountered.*
5. *Remove the opening parenthesis from the stack.*

Let us consider the expression

5 * (6 + 2) – (12/4)

At first we find "5" which is an operand so we add "5" to the *postfix* expression. Next we get "*", which is an operator; it is pushed onto the stack.

Then we get an opening parenthesis, we should push it onto the stack.

Now, we proceed to the next symbol which is "6", an operand so it is treated as the previous operand "5".

Then we get an operator "+", we should add it to the *postfix* expression. Next the operand "2" is added to the *postfix* expression. When we get the closing parenthesis after "2", we repeatedly pop from the stack and add the operators to the *postfix* expression.

Then the corresponding opening parenthesis is removed from the stack. The complete process is shown below.

	Symbol Scanned	Stack	Postfix Expression
1.	5		5
2.	*	*	5

3.	(	*, (	5
4.	6	*, (	5, 6
5.	+	*, (, +	5, 6
6.	2	*, (, +	5, 6, 2
7.	)	*	5, 6, 2, +
8.	-	-	5, 6, 2, +, *
9.	(	-, (	5, 6, 2, +, *
10.	12	-, (	5, 6, 2, +, *, 12
11.	/	-, (, /	5, 6, 2, +, *, 12
12.	4	-, (, /	5, 6, 2, +, *, 12, 4
13.	)	-	5, 6, 2, +, *, 12, 4, /
14.			5, 6, 2, +, *, 12, 4, /, –

EXAMPLE : a+b*c+(d*e+f)*g

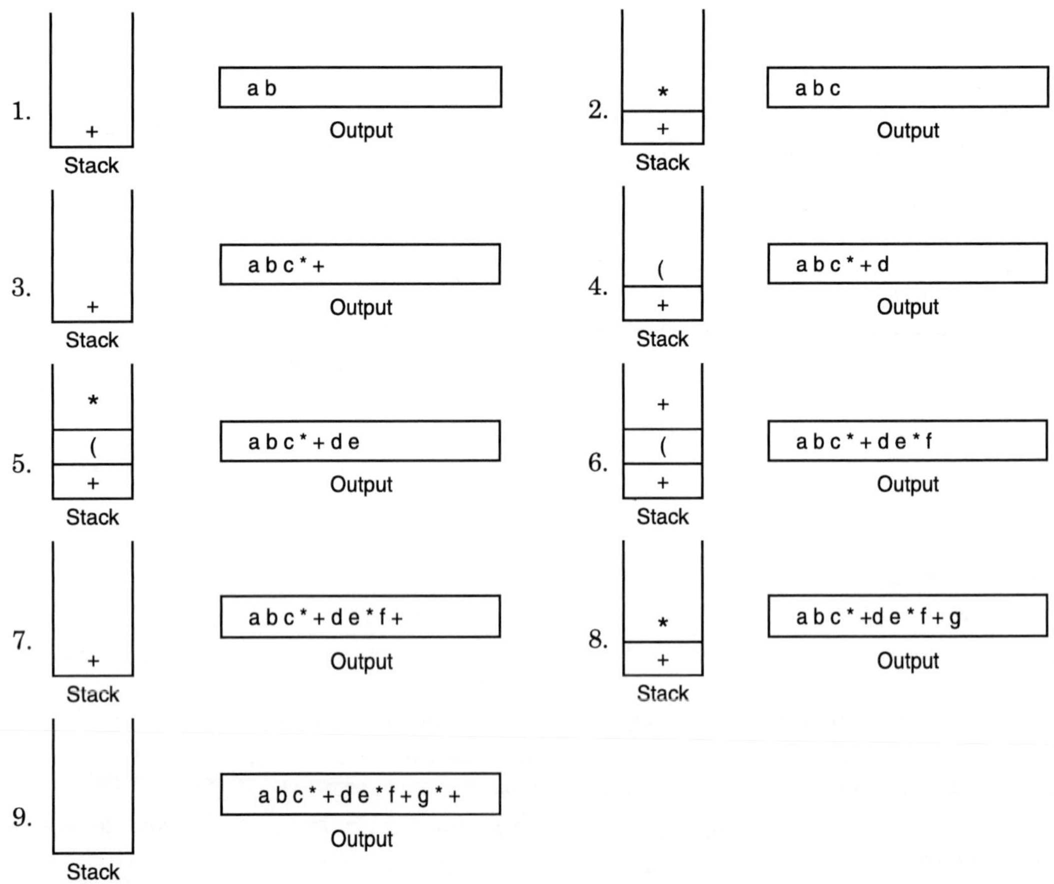

Conversion of parenthesised Infix Expression into postfix by Stack Method

Algorithm

1. Scan the infix exp. from left to right.
2. If scan character is operand than that will display into postfix expression.

3. If scan character is operator than that will push into operator stack.
4. If scan operator is greater than stacked operator then scanned operator will also push into operator stack.
5. If scanned operator is less than or equal to stacked operator then stacked operator will Pop out from stack to postfix expression and scanned operator will push into operator stack.
6. If scanned character is Right parenthesis then till left parenthesis of stack all operators will pop out to postfix expression and left will cancel by right parenthesis.
7. Repeat step(1) to step(6) until end of expression.

/* Convert Infix Expression into Postfix Expression */

```
#include<stdio.h>
#include<conio.h>
#include<string.h>
char stack [30];
int top = -1;
void infix_to_postfix (char*);
void push (char);
char pop ( );
void main ( )
   {
      clrscr ( );
      char infix [30];
      printf ("\n Enter the infix expression");
      gets (infix);
      infix_to_postfix (infix);
      getch ( );
   }
void push (char sym)
   {
      if (top >= 29)
         {
            printf ("\n stack is overflow");
            getch ( );
            return;
         }
      else
         {
            top = top + 1;
            stack [top] = sym;
         }
   }
char pop ( )
   {
```

```
      char i;
   if (top ==-1)
         {
            printf ("\n stack is empty");
            return (0);
         }
      else
         {
            i = stack [top];
            top = top - 1;
         }
      return (i);
   }
int prec (char ch)
   {
      if (ch=='^')
         {
            return (5);
         }
      else if (ch=='*' || ch=='/')
            {
               return (4);
            }
      else if (ch=='+' || ch=='-')
            {
               return (3);
            }
      else
            {
               return (2);
            }
   }
void infix_to_postfix (char infix [ ])
      {
         int length;
         static int index = 0, pos = 0;
         char symbol, temp;
         char postfix [50];
         length = strlen (infix);
while (index < length)
```

```
{
    symbol = infix[index];
    switch (symbol)
        {
            case '(' : push (symbol);
                       break;
            case ')' : temp = pop ( );
                       while (temp != '(')
                         {
                             postfix [pos]=temp;
                             pos++;
                             temp=pop ( );
                         }
                       break;
            case '+' :
            case '-' :
            case '*' :
            case '/' :
            while (prec (stack[top]) >= prec (symbol))
                {
                    temp = pop ( );
                    postfix [pos] = temp;
                    pos++;
                }
            push (symbol);
                break;
            default : postfix [pos++] = symbol;
                      break;
        }
        index++;
    }
while (top>=0)
    {
        temp = pop ( );
        postfix [pos++] = temp;
    }
postfix [pos++] = '\0';
printf("\n Equivalent Postfix expression is:\n");
puts (postfix);
```

```
    return;
}
```

Evaluation of a Postfix Expression using Stack

We can evaluate a postfix expression using stack. Let us consider the *postfix* expression.

5, 6, 2 +, *, 12, 4, /, -

To evaluate the expression we scan the expression from left to right. The steps involved in evaluating a postfix expression are :

1. *If an operand is encountered, push it on STACK.*
2. *If an operator 'op' is encountered.*
   - *Pop two elements of STACK, where A is the top element and B is the next top element.*
   - *Evaluate B op A.*
   - *Push the result on STACK.*
3. *The evaluated value is equal to the value at the top of stack.*

	Symbol Scanned	**Stack**	**Operation (B op A)**
(1)	5	5	
(2)	6	5, 6	
(3)	2	5, 6, 2	
(4)	+	5, 8	[6+2] (A=2, B=6)
(5)	*	40	[5*8] (A=8, B=5)
(6)	12	40, 12	
(7)	4	40, 12, 4	
(8)	/	40, 3	[12/4] (A=4, B=12)
(9)	-	37	[40-3] (A=3, B=40)

EXAMPLE : Evaluate Postfix form : *5 9 8 + 46 * + 7 – **

Push (*5*)

Push (*9*)

Push (*8*)

Push (Pop ( ) + Pop ( ))

Push (*4*)

Push (*6*)

Push (Pop ( ) * Pop ( ))

Push (Pop ( ) + Pop ( ))

Push (*7*)

Push (Pop ( ) - Pop ( ))

Push (Pop ( ) * Pop ( ))

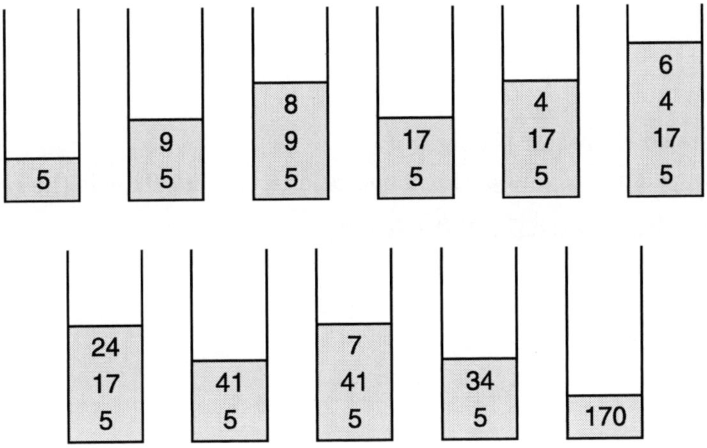

Fig. 5.5.

Algorithm to Evaluate a Postfix Expression by Stack Method

1. Scan postfix expression from left to right.
2. If scanned character is operand then that will push into operand stack.
3. If scanned character is operator then that will operated on top two elements of stack and result will push into operand stack.
4. Repeat step (1) to (3) until end of expression.

EXAMPLE :

(i) 7, 2, – 1, 3, +, / (Postfix)

Push 7 on the stack

Push 2

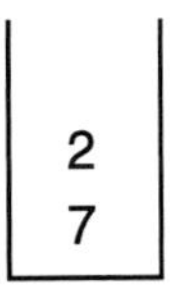

Now scanned character is operator, *i.e.*, '—'. Remove top two elements of stack, *i.e.*, 2 and 7 and perform 7-2, *i.e.*, 5 place this result back on the stack.

Now,

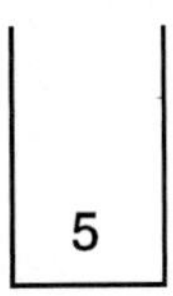

Push 1

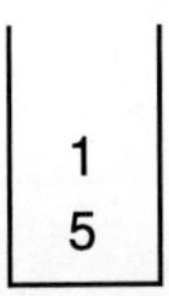

Push 3

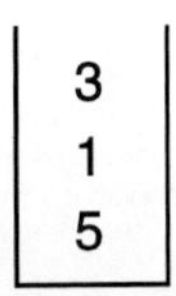

Now scanned character is operator +. Remove top two elements and perform the operation 3 + 1 and place the result back.

Now,

4
5

Again the scanned character is operator /

Remove top two elements and perform operations, *i.e.*, 5/4. This is the required result.

/* Program to Evaluate Postfix Notation */

```
#include<stdio.h>
#include<conio.h>
#include<string.h>
void main( )
  {
     char S[80];
     int i, top=-1, n, x=0, y=0, stack[80];
     clrscr ( );
     printf ("\n Enter the valid postfix notation :");
     gets(S);
     n = strlen(S);
     printf ("The value of the postfix notation is :");
     for (i=0, i<n; i++)
       {
          switch (S[i])
            {
               case '+' :
                  y=stack[top];
                  x=stack[top-1];
                  top = top-1;
                  x=x+y;
                  stack[top] = x;
                  break;
               case '-' :
                  y=stack[top];
                  x=stack[top-1];
                  top = top-1;
                  x=x-y;
                  stack[top] = x;
                  break;
               case '*' :
                  y=stack[top];
                  x=stack[top-1];
                  top = top-1;
```

```
                    x=x*y;
                    stack[top] = x;
                    break;
                case '/' :
                    y=stack[top];
                    x=stack[top-1];
                    top = top-1;
                    x=x/y;
                    stack[top] = x;
                    break;
                default :
                    top = top+1;
                    if (s[i]>=48 && s[i]<=57)
                        x = s[i] - 48;
                        stack[top] = x;
                        x=0;
                    }
        }
        prinft ("%d", stack[top]);
        getch( );
    }
```

Convert Infix to Prefix Form

EXAMPLE 5.6. Convert A/B^C – D into prefix form.

Solution:

A/B^C – D

A/(B^C) – D

$A/\underbrace{(\wedge BC)}_{X} - D$

A/X – D

$\underbrace{(/AX)}_{Y} - D$

Y – D

– YD

Put the values, we get

– /A × D

– /A ^ BC D

Required prefix expression.

EXAMPLE 5.7. Give prefix form for (A – B/C) * (D*E – F)

Solution:

(A – B/C) * (D*E – F)

(A – B/C)) * (CD*E) – F)

$$(A - \underbrace{(/BC)}_{X}) * (\underbrace{(*DE)}_{Y} - F)$$

$$(A - X) * (Y - F)$$

$$\underbrace{(-AX)}_{U} * \underbrace{(-YF)}_{V}$$

$$U * V$$

$$* UV$$

Put the values, we get

$$* - AX - YF$$

$$\boxed{* -A/BC - *DEF}$$

Required Prefix expression.

Infix to Prefix Conversion using Stack :

Algorithm

1. Reverse the input string or start from right of the infix expression.
2. Examine the next element in the input.
3. If it is operand, add it to output string.
4. If it is closing parenthesis, push it on stack.
5. If it is an operator, then
   (i) If stack is empty, push operator on stack.
   (ii) If the top of stack is closing parenthesis, push operator on stack.
   (iii) If it has same or higher priority than the top of stack, push operator on stack.
   (iv) Else pop the operator from the stack and add it to output string, repeat step 5.
6. If it is a opening parenthesis, pop operators from stack and add them to output string until a closing parenthesis is encountered. Pop and discard the closing parenthesis.
7. If there is more input go to step 2.
8. If there is no more input, unstack the remaining operators and add them to output string.
9. Reverse the output string.

EXAMPLE : Suppose we want to convert 2*3/(2-1) +5*(4-1) into Prefix form : Reversed Expression :) 1-4(*5+)1-2(/3*2

Char Scanned	Stack Contents (Top on right)	Prefix Expression (Right to left)
)	)	
1	)	1
-	)-	1
4	)-	14
(	Empty	14-
*	*	14-

5	*	14-5
+	+	14-5*
)	+)	14-5*
1	+)	14-5*1
-	+)-	14-5*1
2	+)-	14-5*12
(	+	14-5*12-
/	+/	14-5*12-
3	+/	14-5*12-3
*	+/*	14-5*12-3
2	+/*	14-5*12-32
	Empty	14-5*12-32*/+

Reverse the output string : +/*23-21*5-41. So, the final Prefix Expression is +/*23-21*5-41.

/* C implementation */

```
#include<stdio.h>
#include<conio.h>
#include<string.h>
#define MAX20
char stack[MAX];
int top=-1;
char pop( );
void push(char item);
int prcd(char symbol)
{
   switch(symbol)
   {
      case '+':
      case '-':
                  return 2;
      case '*':
      case '/':
                  return 4;
      case '^':
      case '$':
                  return 6;
      case '(';
      case ')':
      case '#':
```

```
                    return 1;
    }
return 0;
}
int isoperator (char symbol)
{
    switch (symbol)
    {
        case '+':
        case '-':
        case '*':
        case '/':
        case '^':
        case '$':
        case '(':
        case ')':
                    return 1;
    default:
                return 0;
                }
}
void convertip(char infix[ ], char prefix [ ])
{
int i, symbol, j=0;
char test[MAX];
infix=strrev(infix);
stack[++top]='#';
for(i=0;i<strlen(infix);i++)
    {
        symbol=infix[i];
        if(isoperator(symbol)==0)
            {
                prefix[j]=symbol;
                j++;
            }
        else
            {
                if(symbol==')')
                {
                    push(symbol);
                    }
```

```
            else if(symbol=='(')
              {
                while (stack[top]!=')')
                  {
                    prefix[j]=pop( );
                    j++;
        }
    pop ( );
    }
    else
    {
      if(prcd(symbol)>prcd(stack[top]))
      {
        push(symbool);
      }
    else
      {
        while(prcd(symbol)<=prcd(stack[top]))
        {
          prefix[j]=pop( );
          j++;
        }
      push(symbol);
    }
    }
    }
  }//end of for.
  while(stack[top]!='#')
  {
  prefix[j]=pop( );
  j++;
  prefix[j]='\0';          //null terminate string
  prefix=strrev(prefix);
  }
  }
  int main ( )
  {
    char infix[20], prefix[20];
    clrscr ( );
    printf("Enter the valid infix string:\n");
    gets(infix);
```

```
    convertip(infix, prefix);
    printf("The corresponding prefix string is:\n")
    puts(prefix);
    getch( );
    return 0;
}
void push (char item)
{
    top++;
    stack[top]=item;
}
char pop( )
{
    char a;
    a=stack[top];
    top--;
    return a;
}
```

EXAMPLE 5.8. Consider the following arithmetic expression A, written in postfix notation:

A : 12, 6, 3 –, /, 2, 1, 5, +, *, +

(*a*) Translate A, by inspection and hand, into its equivalent infix expression.

(*b*) Evaluate the infix expression.

Solution: (*a*) Scanning from left to right, translate each operator from postfix to infix notation.

$$\begin{aligned} A &= 12, 6, 3, -, /, 2, 1, 5, +, *, + \\ &= 12, [6 - 3], /, 2, [1 + 5], *, + \\ &= [12 / (6 - 3)], (2 * [1 + 5]), + \\ &= (12 / (6 - 3)) + (2 * (1 + 5)) \end{aligned}$$

(*b*) Now evaluate infix expression

$$\begin{aligned} A &= (12 / (6 - 3)) + (2 * (1 + 5)) \\ &= (12 / 3) + (2 * 6) = 4 + 12 = 16 \end{aligned}$$

Decimal to Binary Conversion

Stack can be use to convert decimal number into binary no.

Algorithm

1. Accept the number N in decimal form.
2. Divide the number N by 2.
3. Store the quotient in N and store the remainder in R.
4. Push the value of R into the stack.
5. Check whether the value in N is not zero. If Yes, go to step 2.
6. If the value in N is zero, retrieve the values from the stack and display them.
7. Stop.

EXAMPLE :

N = 18

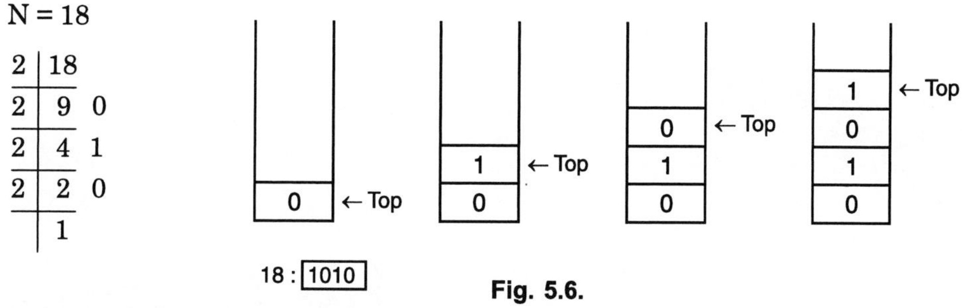

Fig. 5.6.

/* Decimal to binary conversion uses stack */

```
#include <stdio.h>
#include <conio.h>
void main ( )
      {
      clrscr ( );
         int stack[30], dec, rem, top=0;
         printf ("Enter decimal number:");
         scanf("%d", &dec);
         while (dec!=0)
            {
               rem=dec%2;
               top++;
               stack[top]=rem;
               dec=dec/2;
            }
               printf ("The equivalent binary number is");
               for (; top>0; top--)
                  {
                     printf ("%d", stack[top]);
                  }
               getch ( );
            }
```

Exercises

1. What is a stack data structure ? Is stack a linear or non-linear data structure ?
2. List the operations that can be performed on a stack.
3. What are the various applications of a stack structure ?
4. Why an operation to check stack overflow is not implemented on linked stacks ?
5. What is the advantage of using a linked list to implement a stack rather than an array ?
6. What are different notations to represent a mathematical expression ? Which notation is most suitable for use in computers ?

Chapter 6 The Queues

Chapter Outline

- Introduction
- Representation of Queues
- Operation on a Queue
- Implementation of Queue
- Circular Queue
- Double-ended Queue (Deque)
- Applications of Queues
- Difference between Stacks and Queues

6.1 INTRODUCTION

Queues arise quite naturally in the computer for solution of many problems. Perhaps the most common occurrence of a queue in Computer Applications is for the scheduling of jobs. The name "queue" likely comes from the everyday use of the term. A real time example for a queue is people standing in a queue for billing on a stop. The first person in the queue will be the first person to get the service. Similarly the first element inserted in the queue will be the first one that will be retrieved and hence a queue is also called the **First In First Out** or **FIFO** list.

Queue is a linear list which has two ends, one for insertion of elements and other for deletion of elements. The first end is called '*Rear*' and the later is called '*Front*'. *Elements are inserted from Rear End and Deleted from Front End.* Queues are called First-In-First-Out (FIFO) List, since the first element in a queue will be the first element out of the queue. In other words, the order in which the elements enter a queue is the order in which they leave.

It is a homogeneous collection of elements in which new elements are added at one end called ***rear***, and the existing elements are deleted from other end called ***front***. Thus, the two basic operations that are possible in a queue are

1. Insert (or add) an element to the queue (push) or Enqueue.
2. Delete (or remove) an element from a queue (pop) or Dequeue.

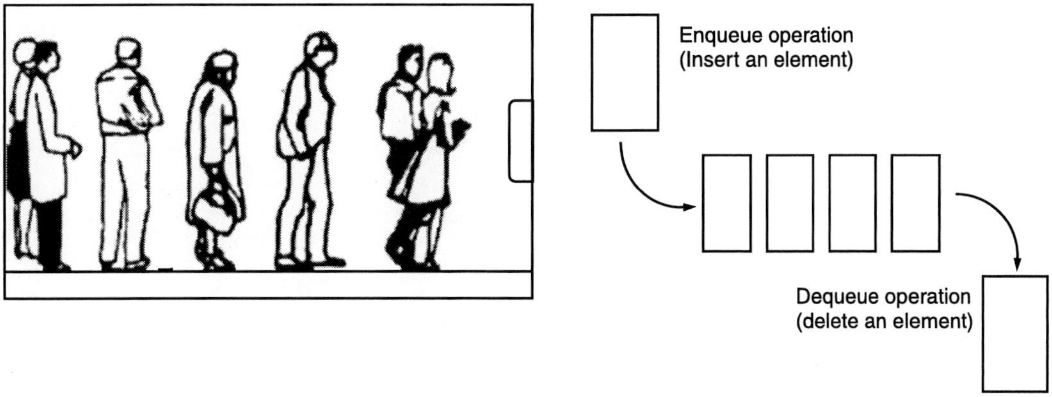

Fig. 6.1.

> A queue is a linear list of elements in which deletions can take place only at one end, called the front, and insertions can take place only at the other end, called the rear.

Push operation will insert (or add) an element to queue, at the rear end, by incrementing the array index. Pop operation will delete (or remove) from the front end by decrementing the array index and will assign the deleted value to a variable. Total number of elements present in the queue is **front-rear+1**, when implemented using arrays.

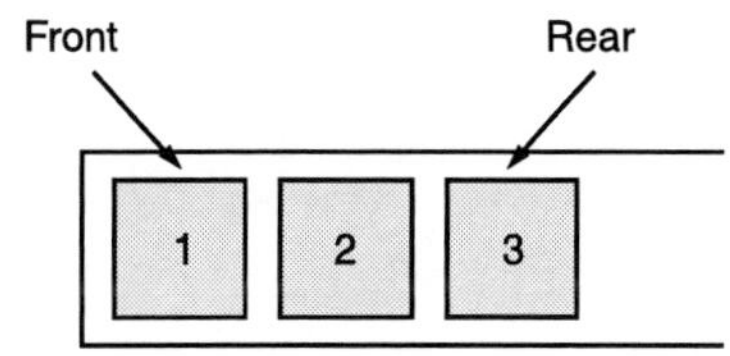

Fig. 6.2.

Example :

Suppose we have an empty queue, with 5 memory cells such as :

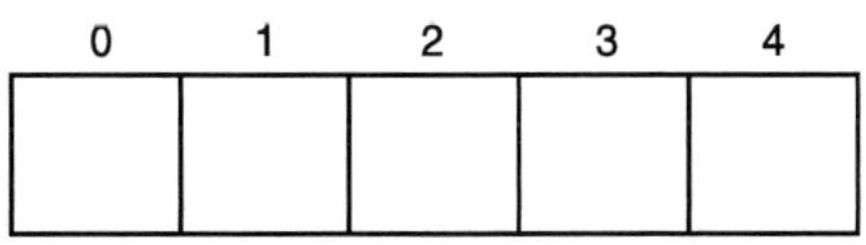

Now Front = – 1

Rear = – 1 *i.e.*, Empty queue.

Insert A :

Now first element A is inserted in queue.

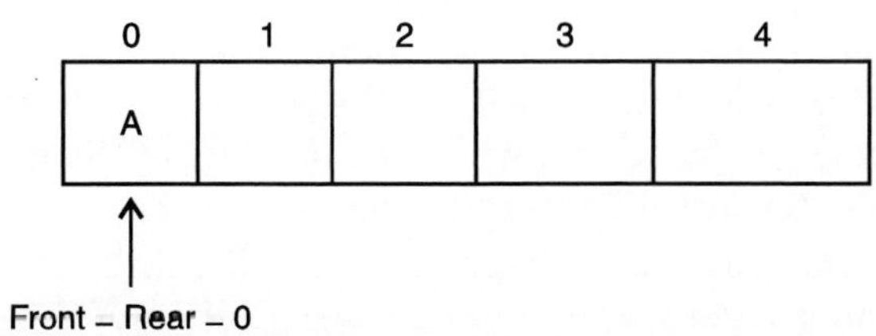

Insert B, C, D :

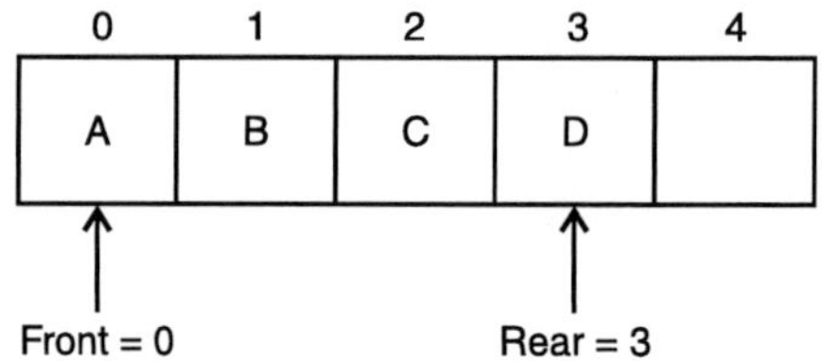

Deletion of an Element :

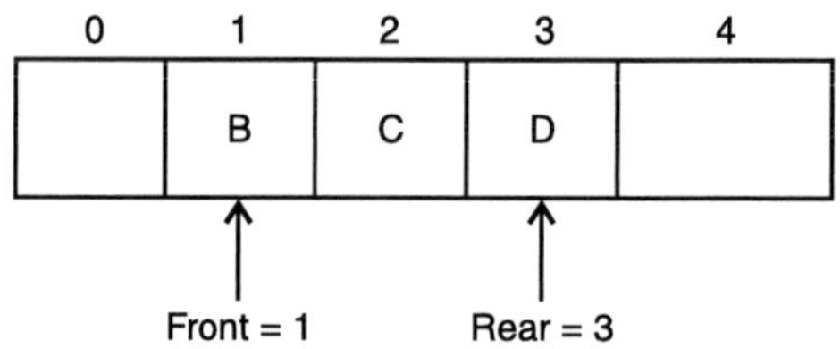

Insertion of E :

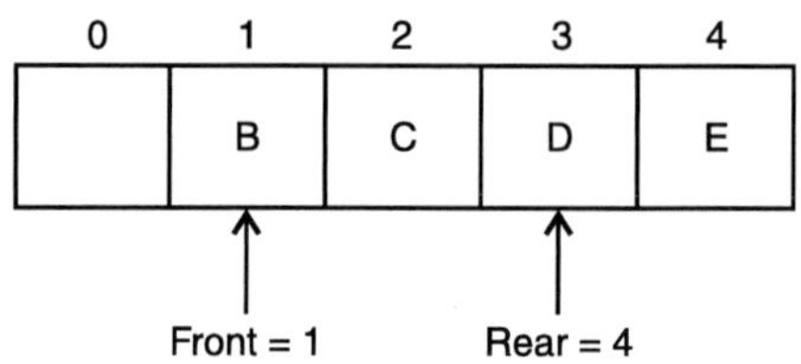

6.2 REPRESENTATION OF QUEUES

Queues, being the linear data structure, can be represented by using both arrays and linked lists.

1. Representation of Queues as an Array

Array is a data structure that stores a fixed number of elements. One of the major limitations of an array is that its size should be fixed prior to using it. But the size of the Queue keeps on changing as the elements are either removed from the front end or added at the rear end. One of the solutions to this problem would be to declare an array with be to declare an array with a maximum size.

Thus, if queue is implemented using arrays, we must be sure about the exact number of elements we want to store in the queue, because we have to declare the size of the array at design time or before the processing starts.

However, although an array cannot be a queue, it can be the home of a queue. That is, an array can be declared large enough for the maximum size of the queue. During the program execution, the queue can grow and shrink with in the space reserved for it. One end of the array is called *start of the queue, i.e.*, front end, while the last end of the queue is rear end and both ends constantly shift as items are deleted and inserted.

A queue in 'C' may be declared as a structure containing three objects : an array to hold the elements of the queue, an integer rear to indicate the position of the end of the current queue and an integer front to indicate the position of the start of queue with in the array.

```
#define MAX 20
struct queue
```

```
        {
            int front;
            int rear;
            int items [MAX];
        };
    struct queue q;
```

2. Linked Representation of Queue

Queue can also be implemented through linked list. The structure of a node will be as

```
        struct node
            {
                int info;
                struct node *link;
            }
        struct node *front, *rear;
```

The info field of the node holds the elements of the queue and link field holds pointer to the neighbouring element in the queue.

Empty queue contains no element and can be indicated by :

front = near = NULL.

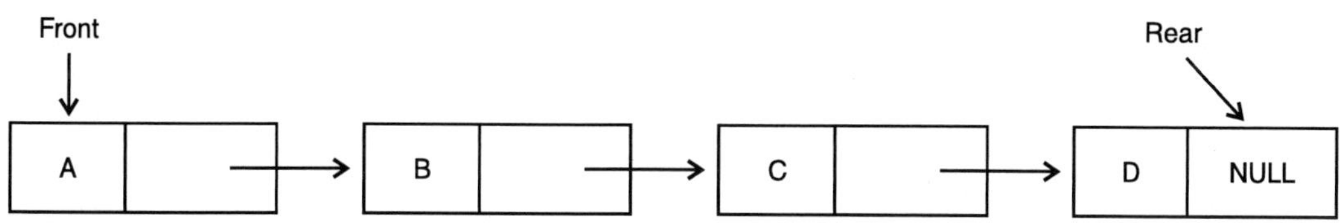

Insertion of E Element

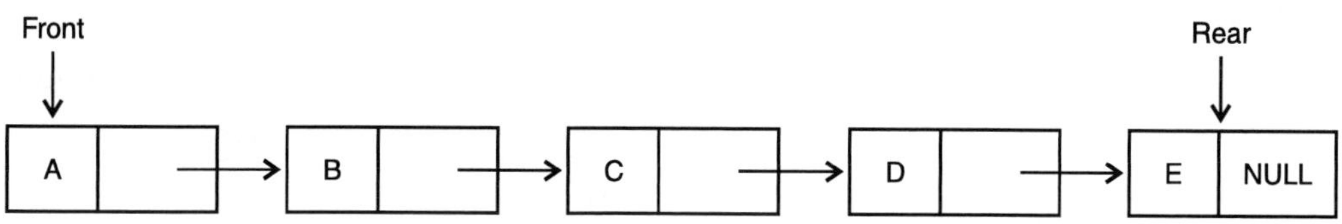

After Deletion of an Element

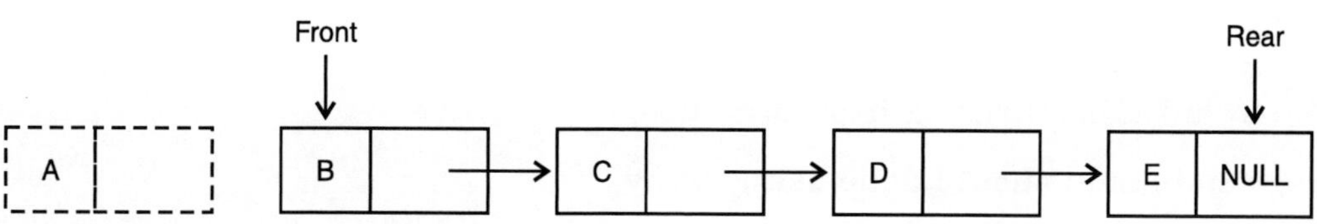

6.3 OPERATIONS ON A QUEUE

Only two basic operations can be performed on queue data structure. These are:

1. Insertion
2. Deletion

Conditions

1. **Overflow Condition :** If REAR = MAX then queue is overflow.

2. **Underflow Condition :** If FRONT = 0 then queue is underflow.
3. **One Element :** If FRONT = REAR then queue containing only one element.

Algorithm for Queue Insertion

1. If REAR = MAX then write "Queue is OVERFLOW & STOP"
2. READ DATA
3. If FRONT = 0 then
   (a) FRONT ← FRONT + 1
   (b) REAR ← REAR + 1
   (c) Q(REAR) ← DATA
   else REAR ← REAR + 1
   Q(REAR) ← DATA

(4) STOP

Note : During the insertion of the first element in the queue we always increment the FRONT by one, *i.e.*, FRONT = FRONT + 1. Afterwards the FRONT will not changed during the entire addition operation.

Algorithm for Deletion from a Queue

1. If FRONT = 0 then write "Queue is UNDERFLOW & STOP"
2. If FRONT = REAR then
   Q(FRONT) ← NULL
   FRONT ← 0
   else
   Q(FRONT) ← NULL
   FRONT ← FRONT + 1
3. STOP

Note : Whenever an element is removed from the queue the value of FRONT is incremented by one.

6.4 IMPLEMENTATION OF QUEUE

We can implement a queue by two ways :

(a) By array (static implementation)
(b) By linked list (Dynamic implementation)

C Program to implement queue using array

```
#include <stdio.h>
#include <conio.h>
#define MAX 100
int q[MAX + 1], front = 0, rear = 0;
void main ( )
   {
   clrscr ( );
      void create ( ); traverse ( ); insert ( ); delet ( );
```

```
        create ( );
        traverse ( );
        insert ( );
        printf("\n After insert an element");
        traverse ( );
        delet ( );
        printf ("\n After deletion");
        traverse ( );
        getch ( );
    }
void create ( )
    {
        char ch;
        front=1;
        do
            {
                rear++;
                printf ("\n Input element in queue:\n");
                scanf ("%d", & q[rear]);
                printf ("Press <Y/N> for more element");
                ch = getch ( );
            }
        while (ch=='Y')
    }
void traverse ( )
    {
        int i;
        printf ("\n elements in the Queue are :\n")
        for (i=front; i<=rear;++i)
            printf ("%d\n", q[i]);
    }
void insert ( )
    {
        int m;
        if (rear ==MAX)
            {
                printf ("Queue is overflow \n");
                return;
            }
        printf ("\n Input new element to insert\n");
        scanf ("%d", &m);
```

```
        rear++;
        q[rear]=m;
    }
void delet ( )
    {
        if (front==0)
            {
                printf ("Queue is underflow\n");
                return;
            }
        if (front==rear)
            {
                q[front] = '\0';
                front = rear = 0;
            }
        else
                {
                    q[front] = '\0'
                    front++;
                }
    }
```

C Program to implement queue using Linked list

```
#include <stdio.h>
#include <conio.h>
#include <alloc.h>
#include <stdlib.h>
struct node
    {
        int info;
        struct node *link;
    };
struct node *front, *rear;
void main ( )
        {
            clrscr ( );
            void insert ( ), delet ( ), display ( );
            int ch;
            while (1)
                {
```

```
            printf ("1. Insert\n");
            printf ("2. Delete\n");
            printf ("3. Display\n");
            printf ("4. Exit\n");
            printf ("Enter your choice :");
            scanf ("%d", &ch);
            switch (ch)
               {
                  case 1 :
                        insert ( );
                        break;
                  case 2 :
                        delet ( );
                        break;
                  case 3 :
                        display ( );
                        break;
                  case 4 :
                        exit (0);
                  default;
                        printf ("Please enter correct choice \n");
                  }
         }
      getch ( );
   }
void insert ( )
   {
      struct node *ptr;
      ptr = (struct node*) malloc (sizeof (struct node));
      int item;
      printf ("Input the element for inserting :\n");
      scanf ("%d", &item);
      prt-> info = item;
      ptr->link = NULL;
      if (front==NULL)                         /* queue is empty*/
            front = ptr;
      else
            rear->link = ptr;
            rear = ptr;
```

```
      }
  void delet ( )
      {
         struct node *ptr;
         if (front==NULL)
            {
               printf ("Queue is underflow \n");
               return;
            }
         if (front==rear)
            {
               free (front);
               rear = NULL;
            }
         else
            {
               ptr = front;
               front = ptr->link;
               free (ptr);
            }
      }
  void display ( )
      {
      struct node *ptr;
      ptr = front;
      if (front==NULL)
         printf("Queue is empty\n");
      else
      {
         printf("\n Elements in the Queue are :\n");
         while(ptr!=NULL)
            {
               printf("%d\n",ptr->info);
               ptr=ptr->link;
            }
         printf("\n");
      }
   }
```

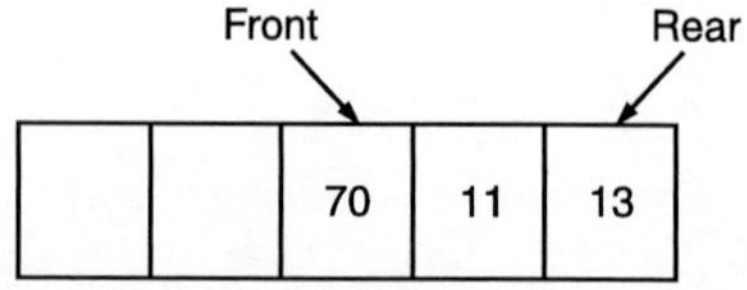

Fig. 6.3.

Suppose a queue Q has maximum size, *i.e.*, MAX = 5, say 5 elements pushed and 2 elements popped (Fig. 6.3).

Now if we attempt to add elements, even though 2 queue cells are free, the elements cannot be pushed. Because in a queue, elements are always inserted at the *rear* end and hence *rear* points to last location of the queue. That is queue is full (overflow condition) though it is empty. This limitation can be overcome if we use **circular queue**.

OTHER QUEUES

There are three major variations in a simple queue. They are:

1. Circular queue
2. Double ended queue (de-queue)
3. Priority queue

6.5 CIRCULAR QUEUE

In circular queues the elements Q[0], Q[1], Q[2] ... Q[n – 1] is represented in a circular fashion. A circular queue is one in which the insertion of a new element is done at the very first location of the queue if the last location at the queue is full.

Suppose Q is a queue array of 6 elements. Push and pop operation can be performed on circular. The following figures will illustrate the same.

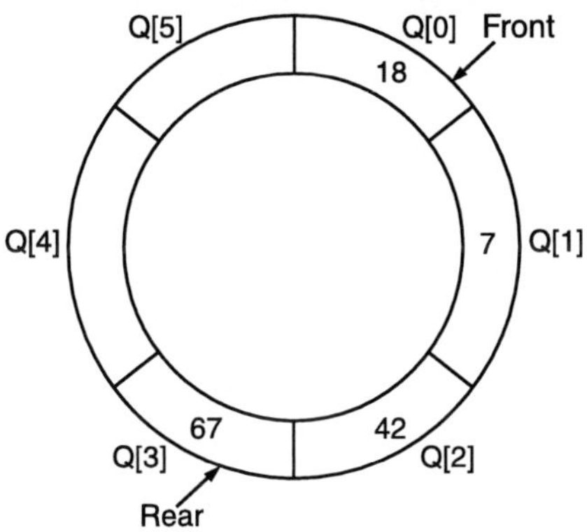

Fig. 6.4. A circular queue after inserting. 18, 7, 42, 67

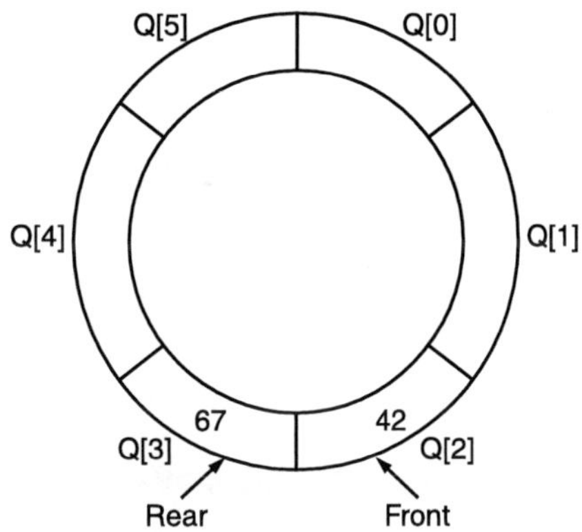

Fig. 6.5. A circular queue after popping. 18, 7

After inserting an element at last location Q[5], the next element will be inserted at the very first location (*i.e.*, Q[0]) that is circular queue is one in which the first element comes just after the last element.

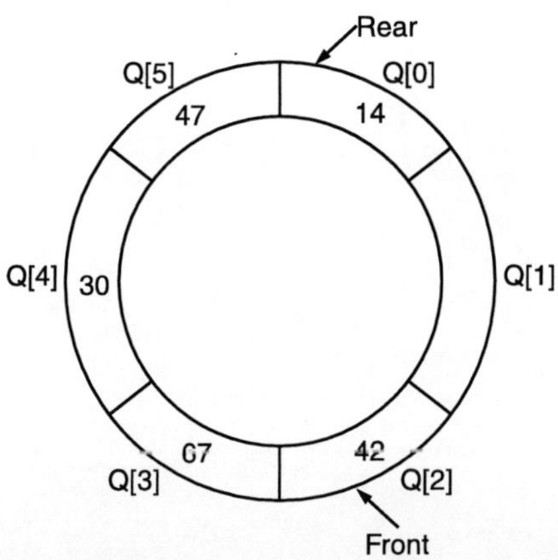

Fig. 6.6. A circular queue after pushing 30, 47, 14.

At any time the position of the element to be inserted will be calculated by the relation

Rear = (Rear + 1) % SIZE

After deleting an element from circular queue the position of the front end is calculated by the relation Front = (Front + 1) % SIZE.

After locating the position of the new element to be inserted, *rear*, compare it with *front*. If (rear = front), the queue is full and cannot be inserted any more.

The pictorial representation of the circular queue and its equivalent array representation is shown in the figure.

(a) After inserting element 15

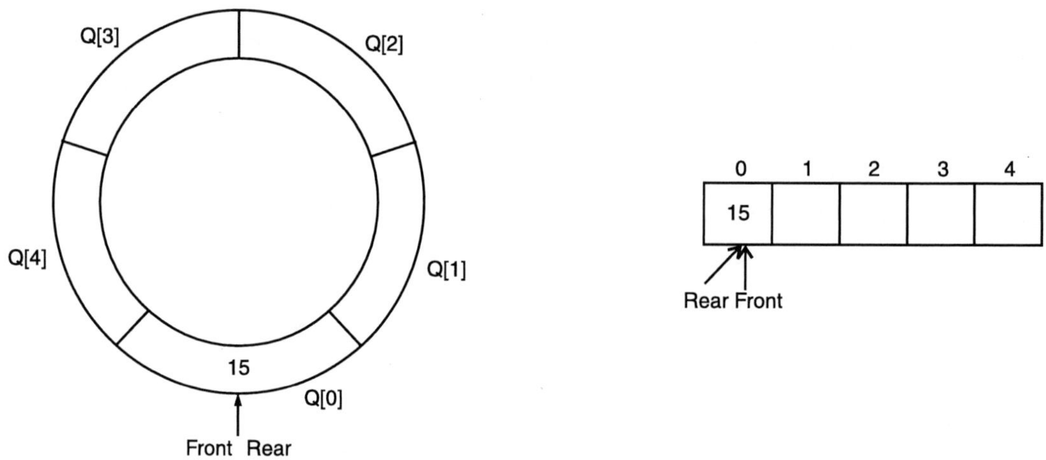

(b) After inserting 10, 20, 30 and 40

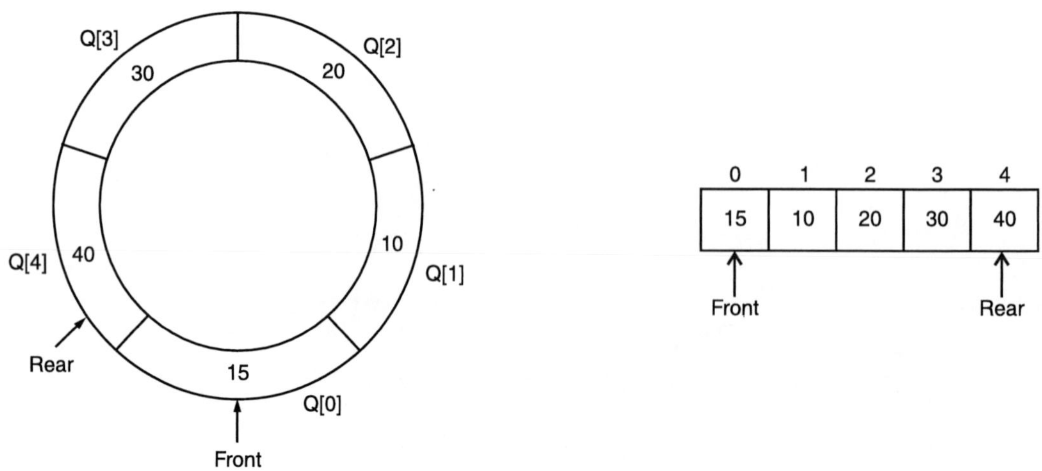

(c) After deleting 15, 10, 20

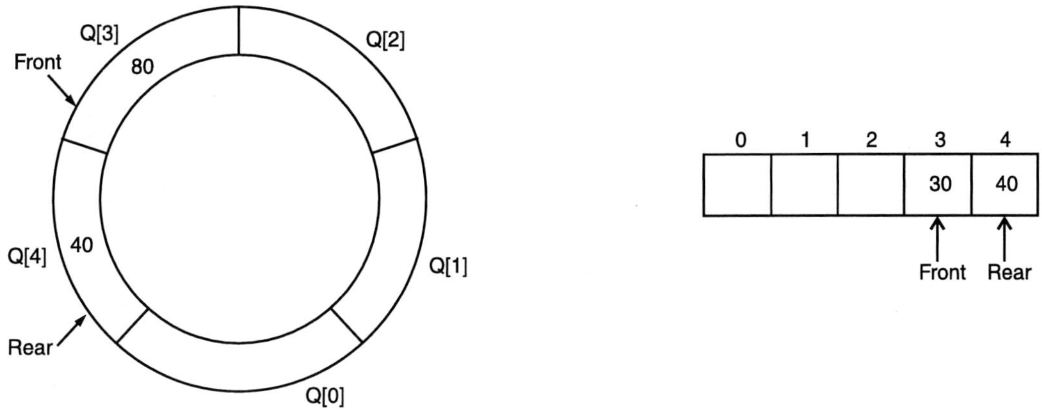

(d) After inserting 80

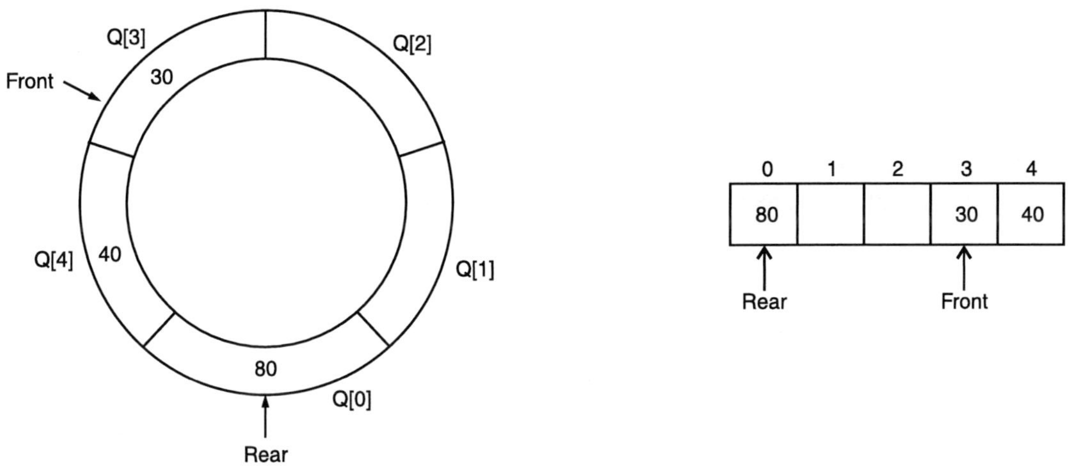

6.5.1 Inserting an Element to Circular Queue

Step 1 : If (FRONT==0 && REAR==MAX-1) | | (Front == rear +1))
then write "Queue is overflow" and stop.

Step 2 : Read data to insert

Step 3 : If (FRONT is equal to – 1)
(a) FRONT = 0
(b) REAR = 0

Step 4 : If (REAR = MAX-1) then REAR = 0
else
REAR = REAR + 1

Step 5 : Cqueue [REAR] = DATA

Step 6 : STOP

6.5.2 Delete Operation in Circular Queue

Step 1 : If FRONT is equal to –1, then write. "Queue is underflow" and stop.

Step 2 : If FRONT = REAR then
(a) Cqueue (FRONT) = NULL
(b) FRONT = –1;
(c) REAR = –1;

Step 3 : If FRONT = MAX-1
(a) Cqueue (FRONT) = NULL
(b) FRONT = 0;
else
(a) Cqueue (FRONT) = NULL
(b) FRONT = FRONT+1

Step 4 : STOP

/* Circular queue using array */

```
#include <stdio.h>
#include <conio.h>
#include <alloc.h>
#include <stdlib.h>
#define MAX 5
int cqueue[MAX];
int front = -1;
int rear = -1;
void main ( )
{
clrscr ( );
void insert ( ), del ( ), display ( );
         int choice;
         while (1)
         {
               printf ("1.Insert\n");
               printf ("2.Delete\n");
               printf ("3.Display\n");
               printf ("4.Quit\n");
               printf ("Enter your choice :");
               scanf ("%d", &choice);
               switch(choice)
               {
               case 1 :
```

```
                    insert( );
                    break;
            case 2 :
                    del( );
                    break;
            case 3 :
                    display( );
                    break;
            case 4 :
                    exit(1);
            default:
                    printf("Wrong choice\n");
            }
        }
}
void insert( )
{
    int item;
    if((front==0 && rear==MAX-1) || ((front==rear+1))
    {
            printf("Queue is Overflow\n")
            return;
    }
    if (front==-1)/*If queue is empty*/
    {
      front = 0;
      rear = 0;
    }
    else
            if (rear==MAX-1) /*rear is at last position of queue*/
                    rear = 0;
            else
                    rear = rear + 1;
    printf("Input the element for insertion :")
    scanf("%d", &item);
    cqueue[rear] = item;
}
void del( )
{
```

```
        if (front == -1)
        {
                printf("Queue Underflow\n");
                return;
        }
        printf ("Deleted element from queue is : %d\n", cqueue[front]);
    if(front == rear)
/* queue has only one element */
        {
                front = -1;
                rear = -1;
        }
        else
                if(front==MAX-1)
                        front = 0;
                else
                        front = front + 1;
}
void display( )
{
        int front_pos = front, rear_pos = rear;
        if(front == -1)
        {
                printf("Queue is empty\n");
                return;
        }
        printf ("Queue elements are:\n")
        if(front_pos <= rear_pos)
                while(front_pos <= rear_pos)
                {
                        printf("%d\n", cqueue[front_pos]);
                        front_pos++;
                }
        else
        {
                        while(fron_pos <= MAX-1)
                        {
                        printf("%d\n", cqueue[front_pos]);
                        front_pos++;
```

```
            }
            front_pos = 0;
            while(front_pos <= rear_pos)
            {
                printf("%d\n", cqueue[front_pos]);
                front_pos++;
            }
        }
        printf("\n");
}
```

EXAMPLE 6.1. Consider the following circular queue capable of accommodating maximum 6 elements.

REAR = 4, FRONT = 2

Queue : _ , L, M, N, _, _.

(Note : We use "—" to denote an empty memory cell)

Describe the queue as the following operations take place

(*a*) Add O. (*b*) Add P, (*c*) Delete two letters (*d*) Add Q, R, S

(*e*) Delete one letter.

Solution:

Initially

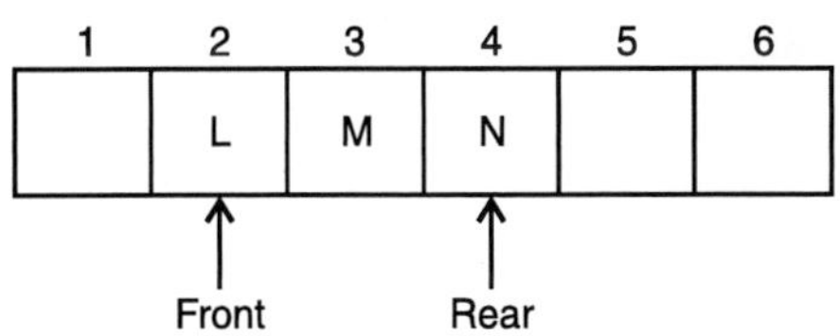

(*i*) Add O

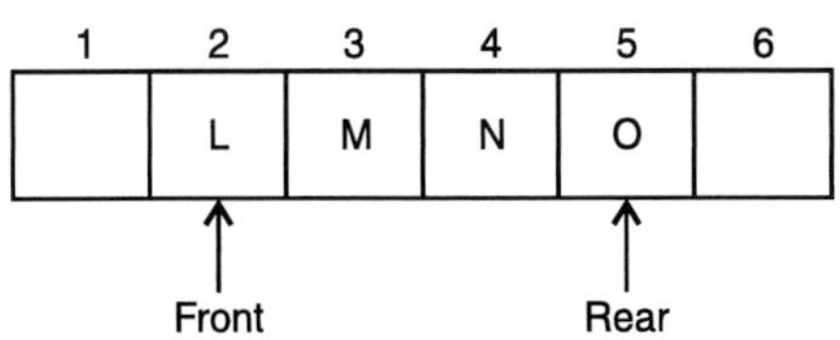

i.e., FRONT = 2
REAR = 5

(*ii*) Add P

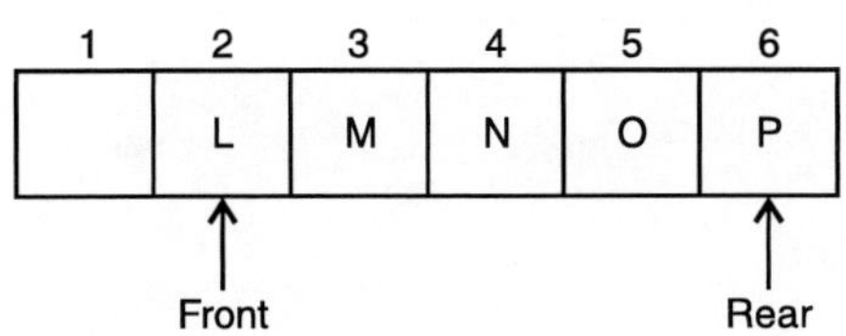

i.e., FRONT = 2
REAR = 6

(*iii*) Delete two letters

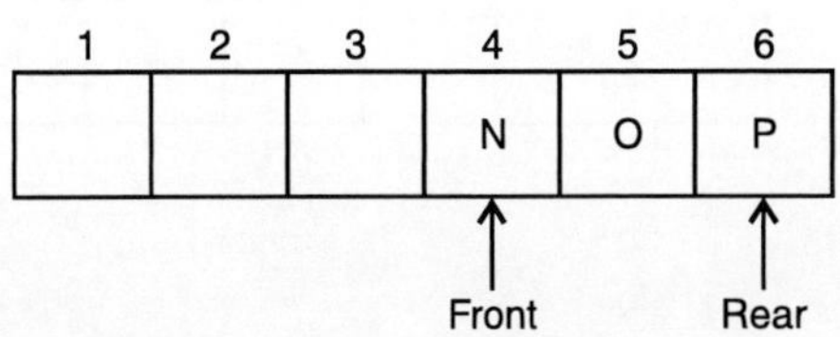

i.e., FRONT = 4
REAR = 6

(iv) Add Q, R, S

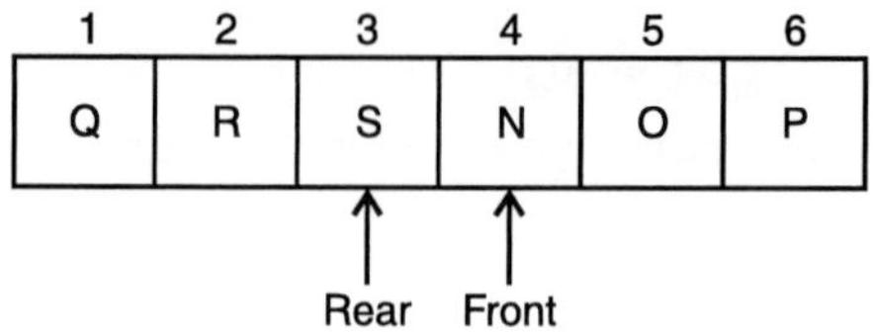

i.e., FRONT = 4
REAR = 3

(v) Delete one letter

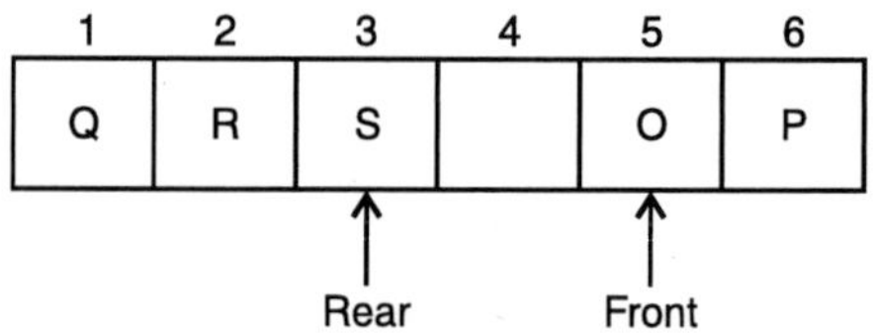

i.e., FRONT = 5
REAR = 3

EXAMPLE 6.2. Consider the following queue characters, where QUEUE is a circular array which is allocated 6 memory cells :

FRONT = 2, REAR = 4 QUEUE :_, A, C, D, _, _

(a) F is added to the queue.
(b) Two letters are deleted.
(c) K, L and M are added.
(d) Two letters are deleted.
(e) S is added to the queue.
(f) One letter is deleted.
(g) R is added to the queue.
(h) One letter is deleted.

Solution:

Initially

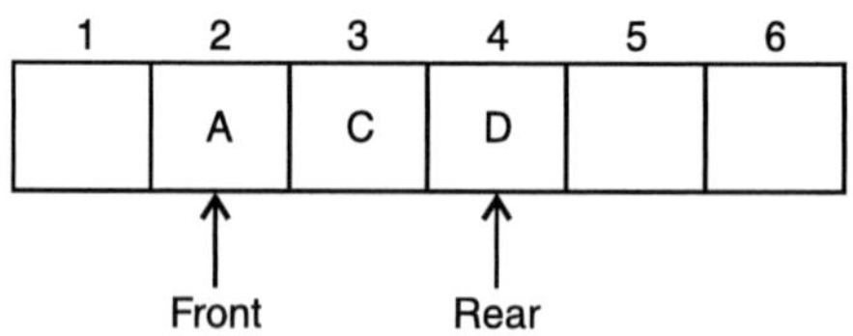

(a) F is added

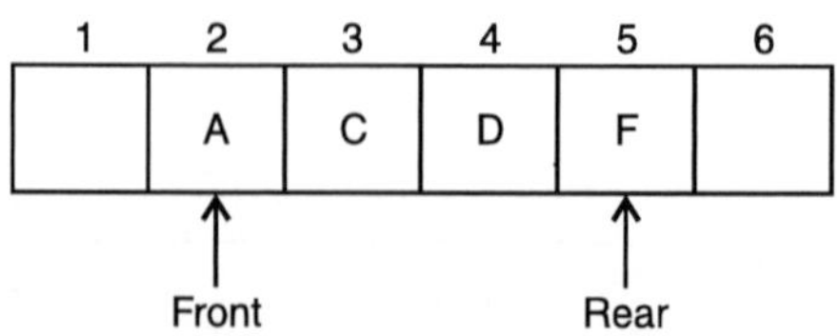

FRONT = 2
REAR = 5

(b) Two letters are deleted

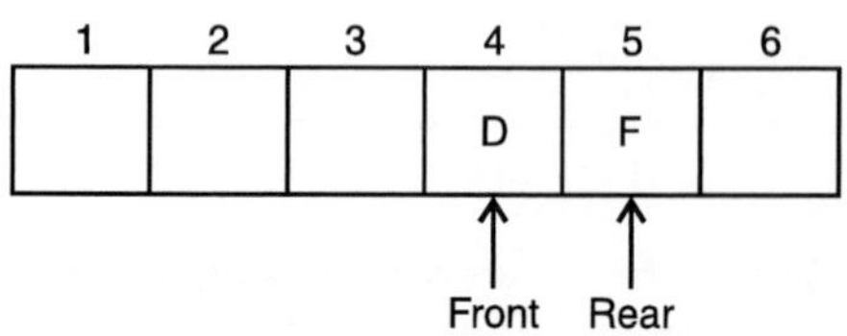

FRONT = 4
REAR = 5

(c) K, L and M are added

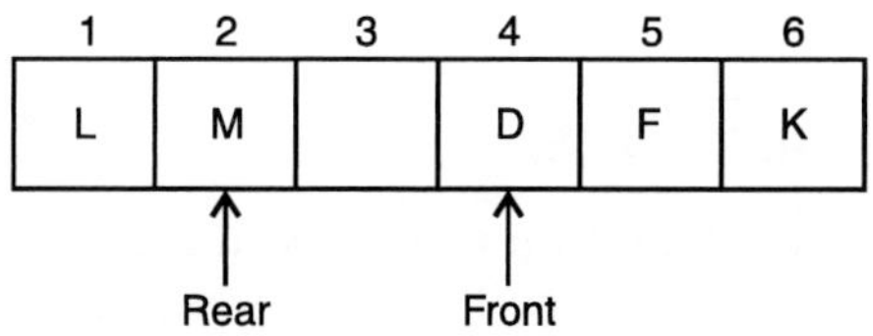

FRONT = 4
REAR = 2

(d) Two letters are deleted

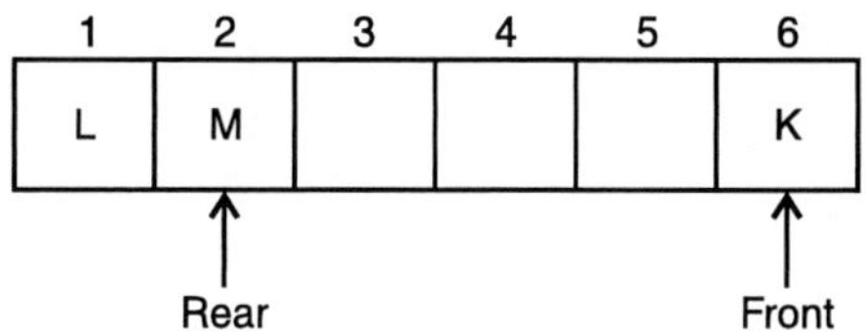

FRONT = 6
REAR = 2

(e) S is added

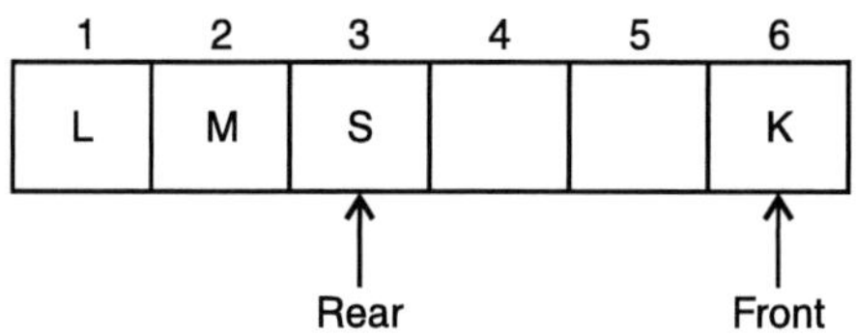

FRONT = 6
REAR = 3

(f) One letter is deleted

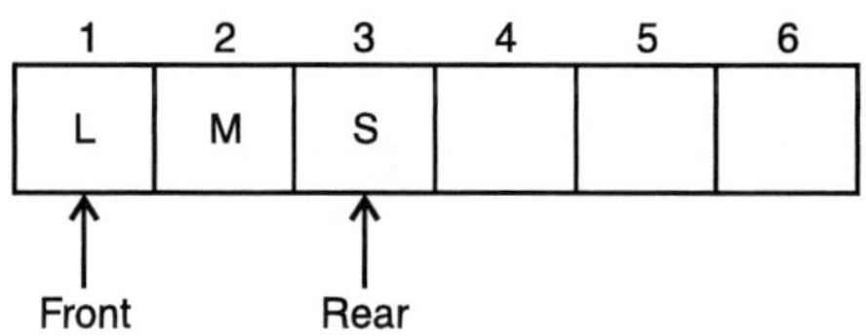

FRONT = 1
REAR = 3

(g) R is added

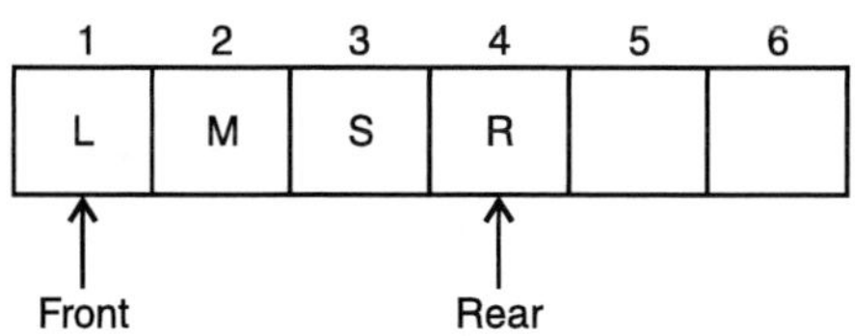

FRONT = 1
REAR = 4

(h) One letter is deleted

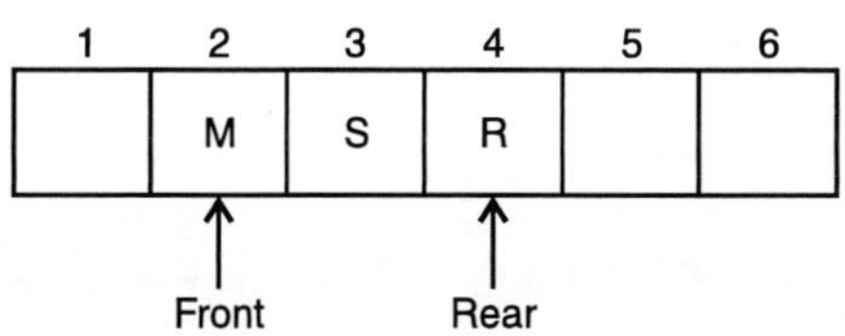

FRONT = 2
REAR = 4

EXAMPLE 6.3. Suppose the data structure is stored in a circular array with N memory cells. Find the number of elements in a queue in terms of FRONT and REAR.

Solution: If FRONT ≤ REAR then no. of elements in queue is:

$$\boxed{\text{REAR} - \text{FRONT} + 1}$$

and if REAR < FRONT then FRONT – REAR – 1 is the number of empty cells. So no. of elements in the queue is N – (FRONT – REAR – 1) = N + REAR – FRONT + 1.

Using arithmetic modulo N, we need only one formula as:
(REAR – FRONT + 1) mod N.

6.6 DOUBLE-ENDED QUEUE (DEQUE)

Another type of queue is called deque. In a deque, both insertion and deletion operations are performed at either end of the queues. That is, we can insert a element from the rear end or the front end. Also deletion is possible from either end.

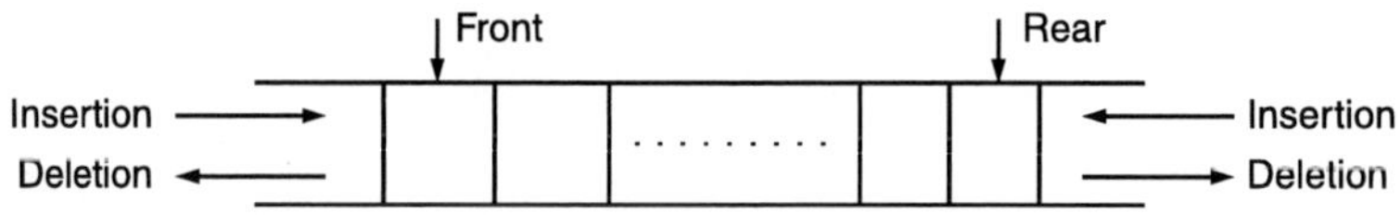

Fig. 6.7. Structure of a deque.

This deque can be used both as a stack and as a queue. There are various ways by which this deque can be represented. The most common ways of representing this type of deque are

(*a*) Using a doubly linked list
(*b*) Using a circular array.

6.6.1 Types of Deques

Dequeue can be of two types :

1. Input-restricted deque.
2. Output-restricted deque.

In **Input-restricted deque**, element can be added at only one end but we can delete the element from both ends. An **output-restricted deque** is a deque where deletions take place at only one end but allows insertion at both ends.

(a) Input-restricted deque. (b) Output-restricted deque.

Fig. 6.8.

The two possibilities that must be considered while inserting or deleting elements into the queue are :

- When an attempt is made to insert an element into a deque which is already full, an *overflow* occurs.
- When an attempt is made to delete an element from a deque which is empty, *underflow* occurs.

The possible **operation** performed on deque is:

1. Add an element at the rear end
2. Add an element at the front end.
3. Delete an element from the front end.
4. Delete an element from the rear end.

Only 1st, 3rd and 4th operations are performed by input-restricted deque and 1st, 2nd and 3rd operations are performed by output-restricted deque.

Algorithms for Inserting an Element

Let Q be the array of MAX elements, *front* (or left) and *rear* (or right) are two array index (pointers), where the addition and deletion of elements occurred. Let DATA be the element to be inserted. Before inserting any element to the queue *left* and *right* pointer will point to the -1.

Insert an Element at the Right side of the Dequeue

1. Input the DATA to be inserted
2. If ((left == 0 && right == MAX-1) | | (left == right + 1))
   (*a*) Display "Queue Overflow"
   (*b*) Exit
3. If (left == – 1)
   left = 0
   right = 0
4. Else
   if (right == MAX – 1)
   left = 0
   else
   right = right + 1
5. Q[right] = DATA
6. Exit

Insert an Element at the Left side of the Dequeue

1. Input the DATA to be inserted
2. If ((left == 0 && right == MAX-1) | | (left == right + 1))
   (*a*) Display "Queue Overflow"
   (*b*) Exit
3. If (left == – 1)
   left = 0
   right = 0
4. Else
   if (left == 0)
   left = MAX – 1
   else
   left = left – 1
5. Q[left] = DATA
6. Exit

Algorithms for Deleting an Element

Let Q be the array of MAX elements, *front* (or left) and *rear* (or right) are two array index (pointers), where the addition and deletion of elements occurred. DATA will contain the element just deleted.

an Element from the Right side of the Dequeue

1. If ((left == –1)

Display "Queue Underflow"
Exit

2. DATA = Q[right]
3. If (left == right)
   left = – 1
   right = – 1

   Else
   if (right == 0)
   right = MAX – 1
   else
   right = right – 1
4. Exit

an Element from the Left side of the Dequeue

1. If ((left == –1)
   Display "Queue Underflow"
   Exit
2. DATA = Q[left]
3. If (left == right)
   left = – 1
   right = – 1
4. Else
   if (left == MAX – 1)
   left = 0
   Else
   left = left + 1
5. Exit

C implementation :

(1) Insertion at Rear (or right) end

```
void insert_rear ( )
     {
        int m;
        if (rear == MAX)
           {
              printf ("Queue is overflow \n");
              return;
           }
              printf ("Input data to insert");
              scanf ("%d",&m);
              rear++;
```

```
                deque[rear] = m;
            }
```

(2) Deletion from the front (left) end

```
void delete_front ( )
      {
         if (front == 0)
            {
               printf ("Underflow \n");
               return;
            }
               deque[front] = '\0';
               front++;
            }
```

(3) Insertion at front end

```
void insert_front ( )
      {
         int m;
         if (front == 1)
            {
               printf ("Overflow \n");
               return;
            }
               printf ("Input element to insert");
               scanf ("%d",&m);
               front --;
               deque[front] = m;
            }
```

(4) Deletion from the rear end

```
void delete_rear ( )
      {
         if (front == 0)
            {
               printf ("Underflow \n");
               return;
            }
               deque[rear] = '\0';
               rear--;
            }
```

To implement various operations on Double ended Queue

Algorithm :

Step 1 : Start the process

Step 2 : Initialise and declare variables

Step 3 : Enter the Choice

Step 4 : If choice is ENQUEUE at FRONT then
(*a*) Check if dequeue is full
(*b*) else check for FRONT at first position
(*c*) else decrement the FRONT position.

Step 5 : If choice is ENQUEUE at REAR then
(*a*) Check if dequeue is full
(*b*) else check for REAR at last position
(*c*) else increment the Rear position.

Step 6 : If choice is DEQUEUE at FRONT then
(*a*) Check if dequeue is empty
(*b*) else check for deque contains only one element
(*c*) else increment the FRONT position.

Step 7 : If choice is DEQUEUE at REAR then
(*a*) Check if dequeue is empty
(*b*) else check for deque contains only one element
(*c*) else decrement the REAR position.

Step 8 : Stop the process.

/* C implementation */

```
/* Program of Dqueue*/

#include <stdio.h>
#include <conio.h>
#include<stdlib.h>
#define MAX 8
void addqatbeg ( int *, int, int *, int * ) ;
void addqatend ( int *, int, int *, int * ) ;
int delqatbeg ( int *, int *, int * ) ;
int delqatend ( int *, int *, int * ) ;
void display ( int * ) ;
int count ( int * ) ;
void disp( );
void main( )
{
```

```
int arr[MAX] ;
int front, rear, i, n ;
clrscr( ) ;
/* initialize data members */
front = rear = -1 ;
for ( i = 0;i<MAX;i + + )
arr[i] = 0;

int ch, data;
disp();
while (1)
{
printf("\n Enter the Choice");
scanf("%d", &ch);
switch(ch)
{
case 0:
disp();
break;
case 1:
printf("\n Enter the Element");
scanf("%d", &data);
addqatbeg ( arr, data, &front, &rear ) ;
break;
case 2:
printf("\n Enter the element");
scanf("%d", &data);
addqatend ( arr, data, &front, &rear ) ;
break;
case 3:
i = delqatbeg ( arr, &front, &rear ) ;
printf ( "\nItem extracted: %d", i ) ;
break;
case 4:
i = delqatend ( arr, &front, &rear) ;
printf ( "\nItem extracted: %d", i ) ;
break;
case 5:
n = count ( arr);
```

```
printf('Number of elements in dequeue is %d",n);
break;
case 6:
display(arr);
break;
case 7:
exit(0);
default:
printf("Please enter right choice....");
}
}
}
void disp( )
{
printf("\nRepresentation of Dequeue");
printf("\n\**************************\n");
printf("\n\t 1.Enqueue at front");
printf("\n\t 2.Enqueue at rear");
printf("\n\t 3.Dequeue at front");
printf("\n\t 4.Dequeue at rear");
printf("\n\t 5.Size of the queue");
printf("\n\t 6.view");
printf("\n\t 7.Exit");
}
/* adds an element at the beginning of a deque */
void addqatbeq ( int *arr, int item, int *pfront, int *prear)
{
int i, k, c ;
if (*pfront == 0 && *prear == MAX - 1)
{
printf ( "\nDeque is full.\n" ) ;
return ;
}
if ( *pfront == -1 )
{
*pfront = *prear = 0 ;
arr[*pfront] = item ;
return ;
```

```
}
if ( *prear != MAX - 1 )
c = count ( arr ) ;
k = *prear + 1 ;
for (i = 1 ; i <= c ; i++)
{
arr[k] = arr[k - 1] ;
k--;
}
arr[k] = item ;
*pfront = k ;
( *prear)++ ;
}
else
{
( *pfront )-- ;
arr[*pfront] = item ;
}
}
/* adds an element at the end of a deque */
void addqatend ( int *arr, int item, int *pfront, int *prear)
{
int i, k ;
if ( *pfront == 0 && *prear == MAX - 1 )
{
printf ( "\nDeque is full.\n" ) ;
return ;
}
if ( *pfront == -1 )
{
*prear = *pfront = 0 ;
arr[*prear] = item ;
return ;
}
if ( *prear == MAX - 1 )
{
k = *pfront - 1 ;
for ( i = *pfront - 1 ; i < *prear ; i++)
{
```

```
k = i ;
if ( k == MAX - 1 )
arr [k] = 0;
else
arr[k] = arr[i + 1] ;
}
(*prear)-- ;
(*pfront)-- ;
}
(*prear)++ ;
arr[*prear] = item ;
}

/* removes an element from the *pfront end of deque */
int delqatbeg ( int *arr, int *pfront, int *prear)
{
int item ;
if ( *pfront == -1 )
{
printf ( "\nDeque is empty.\n" ) ;
return 0 ;
}
item = arr[*pfront] ;
arr[*pfront] = 0 ;
if ( *pfront == *prear )
*pfront = *prear = -1 ;
else
( *pfront)++;
return item ;
}

/* removes an element from the *prear end of the deque */
int delqatend ( int *arr, int *pfront, int *prear )
{
int item ;
if ( *pfront == -1 )
{
printf ( "\nDeque is empty.\n" ) ;
return 0 ;
}
```

```
item = arr[*prear] ;
arr[*prear] = 0 ;
( *prear )-- ;
if ( *prear == -1 )
*pfront = -1 ;
return item ;
}
/* displays elements of a deque */
void display ( int *arr)
{
int i ;
printf ("\n front->" ) ;
for ( i = 0 ; i < MAX ; i++ )
printf ("\t%d", arr[i] ) ;
printf (" <-rear" ) ;
}
/* counts the total number of elements in deque */
int count ( int *arr )
{
int c = 0, i ;
for ( i = 0 ; i < MAX ; i++)
{
if ( arr[i] != 0 )
c++ ;
}
return c ;
}
```

/* Input and output restricted dequeue using array */

```
#include <stdio.h>
#include <conio.h>
#include <malloc.h>
#define MAX 5
int dequ[MAX];
int left =-1;
int right=-1;
main ( )
{
        int choice;
        printf("1.Input restricted dequeue\n")
```

```
        printf("2.Output restricted dequeue\n")
        printf("Enter your choice :")
        scanf("%d",&choice);
        switch(choice)
        {
        case 1 :
                input_que( );
                break;
        case 2 :
                output_que( );
                break;
        default :
                printf("Wrong choice\n")
        }
}
input_que( )
{
        int choice;
        while (1)
        {
              printf("1.Insert at right\n");
              printf("2.Delete from left\n");
              printf("3.Delete from right\n");
              printf("4.Display\n");
              printf("5.Quit\n");
              printf("Enter your choice :");
              scanf("%d", &choice);
              switch(choice)
             {
                case 1 :
                input_right( ),
                break;
        case 2 :
                delete_left( );
                break;
        case 3 :
                delete_right( );
                break;
        case 4 :
```

```
                display_queue( );
                break;
        case 5 :
                exit( );
        default:
                printf("Wrong choice\n")
        }
    }
}
output_que( )
{
        int choice;
        while (1)
        {
              printf("1.Insert at right\n");
              printf("2.Delete at left\n");
              printf("3.Delete from left\n");
              printf("4.Display\n");
              printf("5.Quit\n");
              printf("Enter your choice :");
              scanf("%d", &choice);
              switch(choice)
             {
                case 1 :
                insert_right( );
                break;
        case 2 :
                insert_left( );
                break;
        case 3 :
                delete_left( );
                break;
        case 4 :
                display_queue( );
                break;
        case 5 :
                exit( );
        default:
                printf("Wrong choice\n")
```

```
        }
    }
}
    insert_right( )
    {
            int item;
            if(left==0 && right==MAX-1) || (left==right+1))
    {
            printf("Queue Overflow\n");
            return;
    }
    if (left==-1) /* if queue is initially empty */
    {
            left = 0;
            right = 0;
    }
    else
      if (right==MAX-1) /* right is at last position of queue */
              right = 0;
         else
              right = right+1;
      printf("Input the element for adding in queue :");
      scanf("%d", &item);
      dequ[right] = item;
}
insert_left ( )
{
            int item;
            if(left==0 && right==MAX-1) || (left==right+1))
    {
            printf("Queue Overflow\n");
            return;
    }
    if (left==-1) /* if queue is initially empty */
    {
            left = 0;
            right = 0;
    }
    else
```

```
        if (left == 0)
                left=MAX-1;
        else
                left = left-1;
        printf("Input the element for adding in queue :");
        scanf("%d", &item);
        dequ[left] = item;
}
delete_left ( )
{
        if (left ==-1)
        {
                printf("Queue Underflow\n");
                return;
        }
        printf("Element deleted from queue is :%d\n", dequ[left]);
        if(left==right)/*Queue has only one element */
        {
                left =-1;
                right =-1;
        }
        else
                if(left == MAX-1)
                        left = 0;
                else
                        left = left+1;
}
delete_right ( )
{
        if (left ==-1)
        {
                printf("Queue Underflow\n");
                return;
        }
        printf("Element deleted from queue is :%d\n", dequ[right]);
        if(left==right)/*Queue has only one element */
        {
                left =-1;
                right =-1;
```

```
        }
        else
                if(right == 0)
                        right = MAX-1;
                else
                        right = right-1;
}
display_queue( )
{
        int front_pos = left, rear_pos = right;
        if(left==-1)
        {
                printf("Queue is empty\n");
                return;
        }
        printf("Queue elements :\n");
        if(front_pos <= rear_pos)
        {
                while(front_pos <= rear_pos)
                {
                          printf("%d",dequ[front_pos]);
                          front_pos++;
                }
        }
        else
        {
                while(front_pos <= MAX-1)
                {
                          printf("%d",dequ[front_pos]);
                          front_pos++;
                }
                front_pos = 0;
                while(front_pos <= rear_pos)
                {
                          printf("%d",dequ[front_pos]);
                          front_pos++;
                }
        }
```

```
        printf("\n");
}
```

EXAMPLE 6.4. **Consider the following deque of characters where DEQUE is a circular array which is allocated six memory cells :**

LEFT = 2 RIGHT = 4 DEQUE : _, A, C, D, _, _

Describe the deque while the following operations take place

(*a*) F is added to the right of the deque.
(*b*) Two letters on the right are deleted.
(*c*) K, L and M are added to the left of the deque.
(*d*) One letter on the left is deleted.
(*e*) R is added to the right of the deque.
(*f*) S is added to the left of the deque.

Solution:

Initially

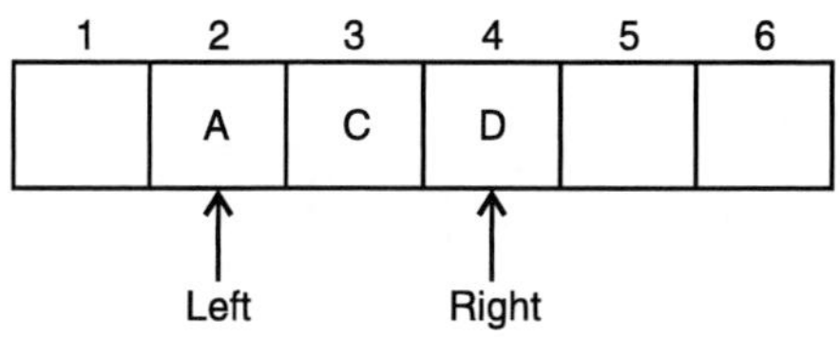

(*a*) F is added on the right, yielding

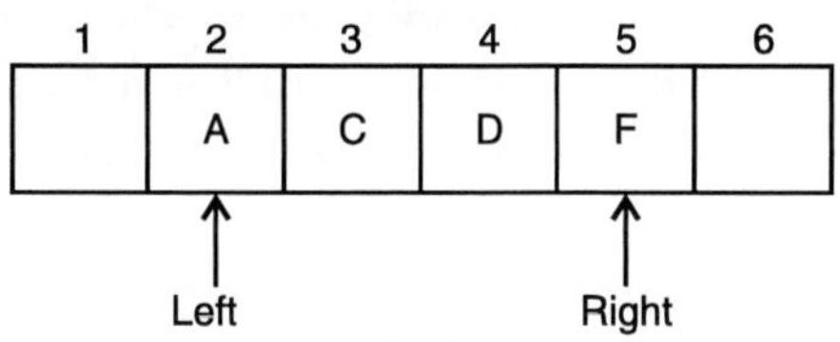

LEFT = 2
RIGHT = 5

(*b*) Two letters on the right are deleted

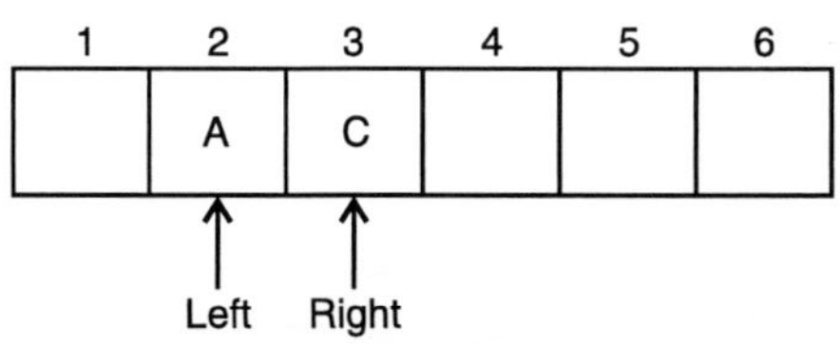

LEFT = 2
RIGHT = 3

(*c*) K, L, M are added on the left

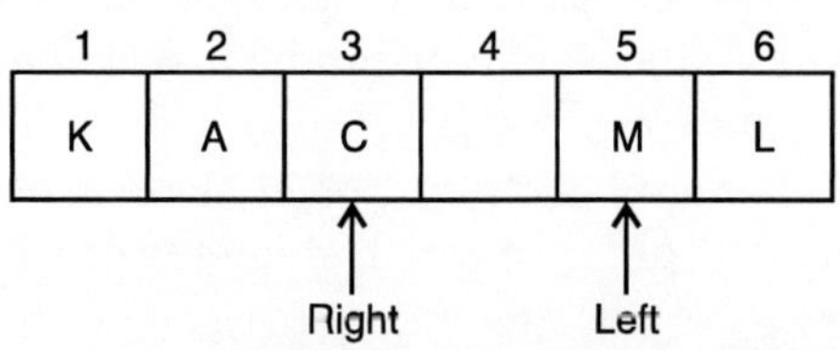

LEFT = 5
RIGHT = 3

(*d*) One letter on the left is deleted

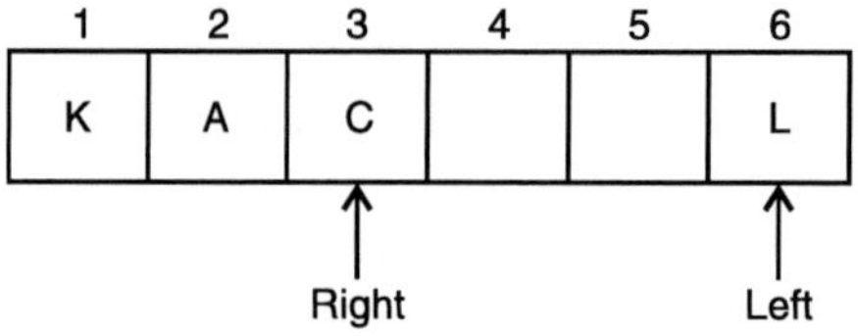

LEFT = 6
RIGHT = 3

(*e*) R is added to the right

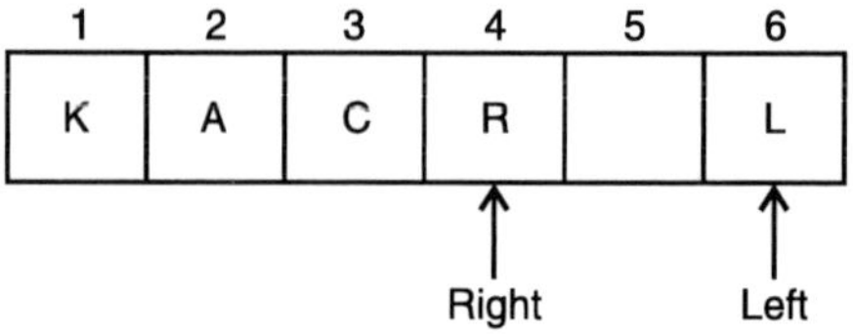

LEFT = 6
RIGHT = 4

(*f*) S is added to the left

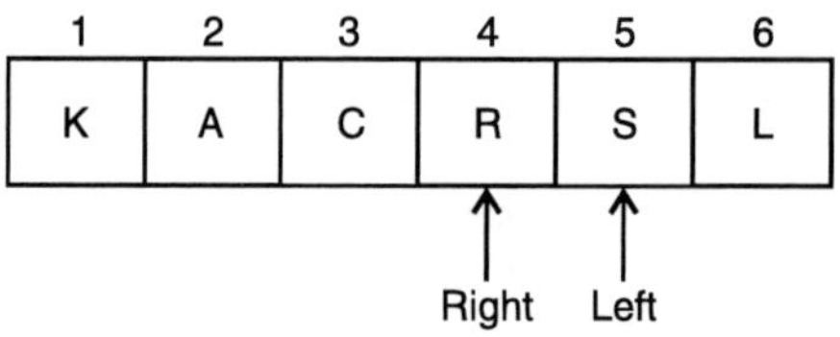

LEFT = 5
RIGHT = 4

EXAMPLE 6.5. Consider a deque maintained by a circular array with N memory cells

(*a*) Suppose an element is added to the deque. How is LEFT or RIGHT changed ?

(*b*) Suppose an element is deleted. How is LEFT or RIGHT changed ?

Solution:

(*a*) If the element is added on the left, then LEFT is decreased by 1 (mod N). On the other hand, if the element is added on the right, then RIGHT is increased by 1 (mod N).

(*b*) If the element is deleted from the left, then LEFT is increased by 1 (mod N). However if the element is deleted from the right, then RIGHT is decreased by 1 (mod N). In the case that LEFT = RIGHT before the deletion (that is, when the deque has only one element), then LEFT and RIGHT are both assigned NULL to indicate that the deque is empty.

6.7. PRIORITY QUEUE

A priority queue is another type of queue structure in which elements can be inserted or deleted based on the priority.

Definition : A priority queue is a data structure in which each element has been assigned a value called the *priority of the element* and an element can be inserted or deleted not only at the ends but at any position on the queue.

A priority queue is a collection of elements such that each element has been assigned an explicit or implicit priority and such that the order in which elements are deleted and processed comes from the following rules :

(a) An element of higher priority is processed before any element of lower priority.

(b) Two elements with the same priority are processed to the order in which they were inserted to the queue.

The priority of element is decided by its value is known as **implicit priority** and priority number is assigned with each element is known as **explicit priority**.

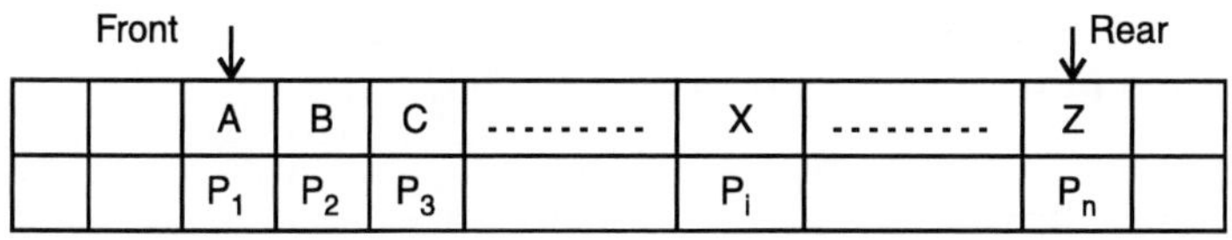

Fig. 6.9. View of priority queue

Here, an element X of priority P_i may be deleted before an element which is at the position pointed out by FRONT.

Types of Priority Queues

There are two types of priority queues :

1. Ascending priority queue
2. Descending priority queue.

In **ascending priority queue** elements can be inserted in an order. But, while deleting elements from the queue, always a small element only to be deleted first whereas, in **descending priority queue** can also elements are inserted in any order but while deleting elements from the queue always a largest element to be deleted first.

An *ascending priority queue* is a collection of items into which items can be inserted arbitrarily and from which only the smallest item can be removed. A *descending priority queue* is a collection of items into which items can be inserted arbitrarily and from which only the largest item can be removed.

6.7.1 Representation of the Priority Queue

There are various ways of maintaining a priority queue. These are:

- One way linked list.
- Multiple queues, one for each priority.
- Maximum or minimum heap.

The simplest way to represent a priority queue is an unordered liner list. Suppose that we have n elements in this queue and to delete the element with maximum key. If the list is represented sequentially, insertions are most easily performed at the end of this list. Hence, the insert time is 0(1). A deletion requires a search for an element with the largest key, followed by its deletion. So each deletion takes 0(n) time.

An alternative is to use an ordered linear list. The elements are in non-decreasing order if sequential representation is used. The deletion time for this one is 0(1) but insertion time is 0(n).

In case of heap, both insertions & deletions can be performed in $0(\log_2 n)$ time.

One way List Representation

In this representation each node of the linked list will have three fields.

1. an information field INFO
2. a priority number PRN
3. a link number LINK.

A node x precedes a node y in the list when x has higher priority than y or when both have same priority but x was inserted to the list before y.

Priority numbers will operate in the usual way, *i.e.*, the lower the priority number, the higher the priority.

The first element in the list is the element with maximum priority. So it is very simple to *delete an element* from priority queue. It takes only 0(1) time.

Algorithm :

1. ITEM ← INFO (FIRST)
2. Delete first node from the list
3. Process ITEM
4. STOP

Adding an element to priority queue is much more complicated than deleting an element from the queue, because we need to find the correct place to insert the element.

Suppose we adds an ITEM with priority number N to a priority queue than

(*a*) Traverse the list until finding a node X whose priority number exceeds N.

(*b*) Insert the node ITEM in front of node X.

(*c*) If no such node is found, insert ITEM as the last node of the list.

So, insertion needs 0(n) time.

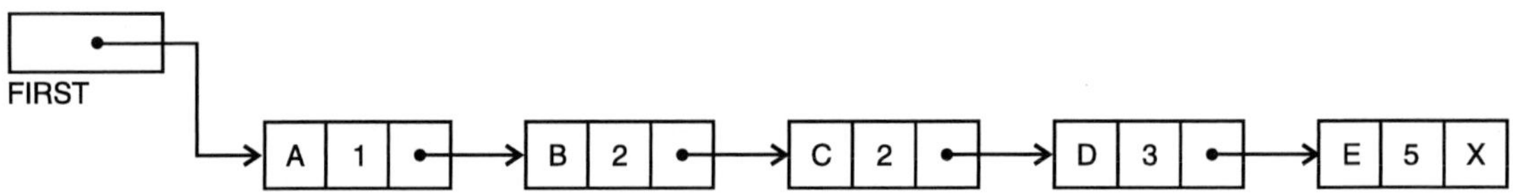

Fig. 6.10.

Figure 6.10 shows the schematic diagram of a priority queue with 5 elements. This figure tell us that B is inserted before C because B and C have the same priority number and B appears before C in the list.

Multiple Queue Representation

Figure 6.11 shows the scheme to represent multiple queues in a same array queue[ ] of size n. Each queue i is allocated a predefined space bounded by array indices. In order to preserve multiple queues, there should be multiple Front and multiple Rear. So, here a vector front and rear denotes the indices for each queue.

Queue 1 indices from front [0] to rear [0]

Queue 2 indices from front [1] to rear [1]

Queue 3 indices from front [2] to rear [2]

Queue 4 indices from front [3] to rear [3]

Queue 5 indices from front [4] to rear [4]

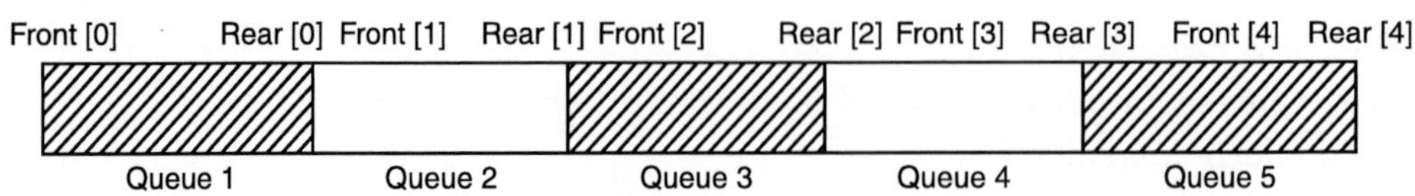

Fig. 6.11. Representation of 5 queues by a single array.

The most common use of multiple queues is implementation of priority queue. In this representation, one queue is maintained for each priority number.

In order to process an element of the priority queue, element from the first non-empty highest priority number queue is accessed. In order to add a new element to the priority queue, the element is inserted in an appropriate queue for given priority number. The first queue has highest priority number elements, second queue has next higher priority number elements and so on.

Heap Representation of a Priority Queue

A heap is a complete binary tree with additional property—the root node is either smallest or largest from its children. If root node of the heap is smallest from its children. It is called *min heap*. If root node of a the heap is largest from its children it is known as *max heap*.

A priority queue having highest priority for lower number can be represented using min heap and a priority queue having highest priority for higher number can be represented using *max heap*.

6.8 APPLICATIONS OF QUEUES

Some of the applications of queues are as follows :

1. There are several algorithms that use queues to solve problems easily. For example, BFS, traversing of a binary tree etc.
2. Round-robin technique for processor scheduling is implemented using queues.
3. When the jobs are submitted to a networked printer, they are arranged in order of arrival. *i.e.*, jobs sent to a printer are placed on a queue.
4. Every real-life line is a queue. For example, lines at ticket counters at cinema halls, railway stations, bus stands etc., are queues because the service, *i.e.*, ticket is provided on first come first served (FIFO) basis.

6.9 DIFFERENCE BETWEEN STACKS AND QUEUES

Stacks	Queues
1. Stack is an ordered list where in all insertions and deletions are performed at one end called top. 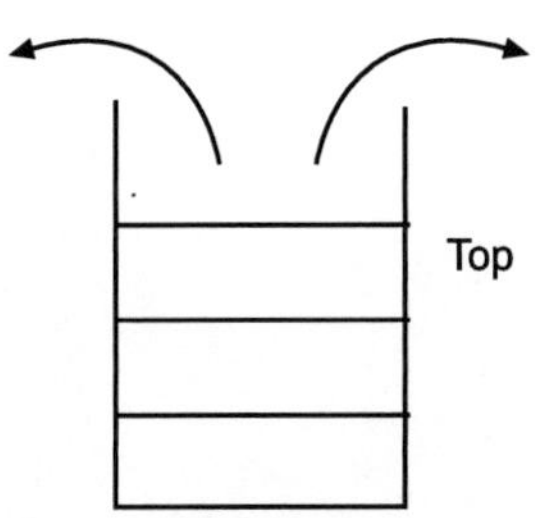 	1. Queue is an ordered list where in insertions are performed at rear end and deletions are performed at front end. 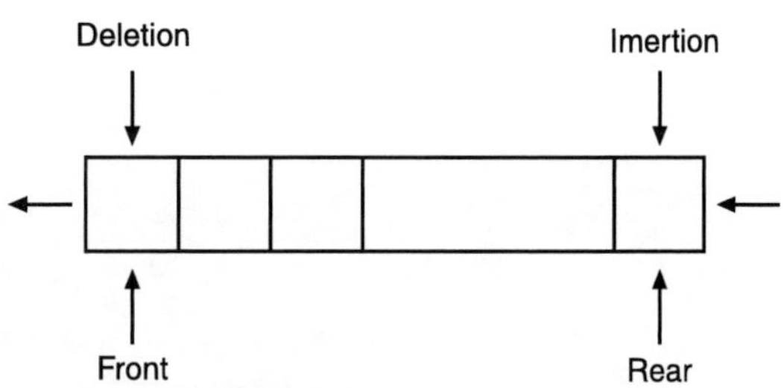
2. Stack is called LIFO structure.	2. Queue is called FIFO.
3. Associated with stacks there is one variable called top.	3. Associated with queues there are two variables called front and rear.
4. Stack is full can be represented by the condition Top = MAX-1	4. Queue is full can be represented by the condition rear = MAX-1.
5. Stack is empty is represented by the condition, Top =– 1.	5. Queue is empty is represented by the condition front > rear.

6. To insert an element into the stack top is incremental by 1.	6. To insert an element into the queue rear is incremented by 1.
7. To delete an element from the stack Top = Top-1.	7. To delete an element from the queue, front is incremented by 1.

/* Program of priority queue using linked list */

```
#include <stdio.h>
#include <conio.h>
#include <stdlib.h>
#include <malloc.h>
void insert ( ), del ( ), display ( );
struct node
{
        int priority;
        int info;
        struct node *link;
}*front = NULL:
void main ( )
{
        clrscr ( );
        int choice;
        while(1)
        {
                printf("1.Insert\n");
                printf("2.Delete\n");
                printf("3.Display\n");
                printf("4.Quit\n");
                printf("Enter your choice :");
                scanf("%d", &choice);
                switch(choice)
                {
                   case 1:
                           insert ( );
                           break;
                   case 2:
                           del ( );
                           break;
                   case 3:
                           display ( );
                           break;
```

```
                        case 4:
                                exit (1 );
                        default :
                                printf ("Wrong choice\n");
                    }
        }
getch ( );
}
void insert( )
{
            struct node *tmp, *q;
            int added_item, item_priority;
            tmp = (struct node*) malloc(sizeof(struct node));
            printf("Input the item value to be added in the queue :");
            scanf("%d", &added_item);
            printf("Enter its priority :");
            scanf("%d", &item_priority);
            tmp->info = added_item;
            tmp->priority = item_priority;
            /*Queue is empty or item to be added has priority
      more than first item */
            if(front == NULL || item_priority < front->priority)
            {
                  tmp->link = front;
                  front = tmp;
            }
            else
            {
                  q = front;
                  while(q->link!=NULL&&q->link->priority<= item_priority)
                          q=q->link;
                  tmp->link = q->link;
                  q->link = tmp;
            }
}
void del( )
{
            struct node *tmp;
            if(front == NULL)
                  printf("Queue underflow\n")
```

```
        else
        {
            tmp = front;
            printf("Deleted item is %d\n", tmp->info);
            front = front->link;
            free(tmp);
        }
}
void display( )
{
        struct node *ptr;
        ptr = front;
        if(front == NULL)
            printf("Queue is empty\n");
        else
        {
            printf("Queue is :\n");
            printf("Priority item\n");
            while(ptr != NULL)
            {
              printf(%5d  %7d\n", ptr ->priority, ptr->info);
                     ptr = ptr->link;
            }
        }
}
```

Exercises

1. What do you mean by queue ? What are the various operations that can be performed on queue ?
2. What are the different types of queues ?
3. How is the queue different from the stack ?
4. How do you represent a queue in computer's memory ?
5. When do you get error message "Queue overflow" and "Queue underflow" ? Explain with example.
6. What is circular queue ? How is it different from ordinary queue.
7. What is deque ? What are the different operations can be performed on double ended queue.
8. What is the need for the deque ?
9. How elements are inserted and deleted in a circular queue ? Explain with examples.
10. Define input restricted queue and output restricted queue with suitable example.
11. What is a priority queue ? Write a C program to implement priority queues using arrays.
12. Write a program to implement the concept of multiple queues in C with example.

PART—B

7. **Trees:** Basic terminology, sequential and linked representations of trees, traversing a binary tree using recursive and non-recursive procedures, inserting a node, deleting a node, brief introduction to threaded binary trees, AVL trees and B-trees.
8. **Heaps:** Representing a heap in memory, operations on heaps, application of heap in implementing priority queue and heap sort algorithm.
9. **Graphs:** Basic terminology, representation of graphs (adjacency matrix, adjacency list), traversal of a graph (breadth-first search and depth-first search), and applications of graphs.
10. **Hashing & Hash Tables:** Comparing direct address tables with hash tables, hash functions, concept of collision and its resolution using open addressing and separate chaining, double hashing, rehashing.
11. **Searching & Sorting:** Searching an element using linear search and binary search techniques, Sorting arrays using bubble sort, selection sort, insertion sort, quick sort, merge sort, heap sort, shell sort and radix sort, complexities of searching & sorting algorithms.

Chapter

7 Recursion

7.1 INTRODUCTION

Chapter Outline

- Introduction
- Why Use Recursion
- Principles of Recursion
- Advantages and Disadvantages of Recursion
- Comparison of Iteration and Recursion
- Recursion through Stack
- Recursion in C
- Tail Recursion
- The Towers of Hanoi Problem
- Direct and Indirect Recursion
- Simulating Recursion
- Removal of Recursion

Recursion is a process of expressing a function in terms of itself. We can also define recursion as a process in which a function calls itself with reduced input and has a base condition to stop the process, *i.e.,* in order to solve a problem recursively, two conditions must be satisfied. First, the problem must be written in a recursive form, and second, the problem statement must include a stopping condition.

When a function is called, the return address, the values of local and formal variables are pushed onto the stack, a block of memory of contiguous locations, set aside for this purpose. After this the control enters into the function. Once the return statement is encountered, control comes back to the previous call, by using the return value present in the stack, and it substitutes the return value to the call. If the function does not return any value, control goes to the statement that follows the function call.

```
/* Calculating factorial of an integer
using recursion/
#include<studio.h>
void main()
{
int n;
long int factorial(int n);
printf("Enter the number n=");
scanf("%d",&n);
```

```
printf("factorial is=%d", factorial (n));
getch ( );
}
long int factorial (int n)
{
if(n<=1)
return(1);
else
return(n*factorial(n-1));
}
```

Definition

A function is said to be *recursively defined,* if a function containing either a Call statement to itself or a Call statement to a second function that may eventually result in a Call statement back to the original function.

A recursive function must have the following properties:

1. There must be certain criteria, called **base criteria** for which the function does not call itself.
2. Each time the function does call itself (directly or indirectly), the argument of the function must be closer to a base value.

7.2 WHY USE RECURSION

Programming in interative method helps the programmer to perform certain tasks repeatedly for a known number of times or as long as the desired condition is true using the **for loop** or **while loop** or **do-while loop.** But our interest is to make a function execute repeatedly interms of itself. So this mechanism is called **recursion.**

Recursion is a very powerful technique to write a complicated program in a easy way. This is an important facility available in most of the structured programming languages such as Pascal, C, C++ etc.

7.3 PRINCIPLES OF RECURSION

Recursion is implemented through the use of functions. A function that contains a function call to itself or a function call to a second function which eventually calls the first function, is known as a **recursive function.**

Two important **conditions** must be satisfied by any recursive function:

1. Each time a function calls itself it must be closer, in some sense to a solution.
2. There must be a decision criterion for stopping the process or computation.

For designing the good recursive program we must make certain *assumptions* such as:

1. **Base case:** Base case is the terminating condition for the problem while designing any recursive function.
2. **If conditions:** If condition in the recursive algorithm defines the terminating condition.
3. When a recursive program is subjected for execution, the recursive function calls will not be executed immediately.
4. The initial parameter input value pushed on to the stack.

5. Each time a function is called a new set of local variable and formal parameters are again pushed on to the stack and execution starts from the beginning of the function using changed new value. This process is repeated till a base condition (stopping condition) is reached.
6. Once a base condition or stopping condition is reached the recursive function calls poping elements from stack and returns a result to the previous values of the function.
7. A sequence of returns ensures that the solution to the original problem is obtained.

7.4 ADVANTAGES AND DISADVANTAGES OF RECURSION

There are number of *advantages* using recursion is:

1. We can create a simple and easy version of programs using recursion.
2. Always recursion will be written in the name of recursive definition. It can be translated into C code very easily.
3. We can avoid initialisation of variable inside the functions, but iterative solutions are required to be initialised.
4. Some specific applications are meant for recursion such as Binary tree traversal, tower of Hanoi etc. can be easily understood.

There are quite number of *disadvantage* using recursion are:

1. It *occupies lot of memory:* When function is called outside or called within, the function stores formal parameters local variables and returns address to confirm function are working well. Apart from this, it stores function variables separately.
2. It *consumes more time to get desired result:* After matching the base condition, the function should restore the most recently saved parameters, local variables, and return address. This operation spents lot of time during pushing and poping the necessary items from the stack.
3. *Function execution is slower than iterative method* because of the overhead of calling functions repeatedly.

Therefore, recursion is *expensive* in terms of processor time and memory usage.

7.5 COMPARISON OF ITERATION AND RECURSION

Any **iterative function** consists of four parts:

1. **Initialisation:** The decision parameter is set to an initial value or more precisely, the loop variable is initialised, other pre-defined variables are also initialised.
2. **Decision:** The decision parameter is used to determine whether further looping is necessary. The loop variable is compared and based on the outcome, the loop is executed again.
3. **Computation:** Necessary computation is performed within the loop.
4. **Update:** The decision parameter is updated and a transfer is made to the next iteration. The loop variable is incremented/decremented.

For a **recursive function.**

1. **Initialisation:** The variables in the *form of arguments* are passed on to the function.
2. **Decision:** The argument values are used to determine whether further recursive calls are required.
3. **Computation:** Necessary computation is performed using the local variables and the parameters at the current depth.
4. **Update:** The update is done so that the variables can be used for further recursive calls.

Iteration	Recursion

1. It is a process of executing or group of statements repeatedly until some specified condition is satisfied. Here counter controlled and body of the loop terminates when the termination condition is failed.
For example :
`int fact (int n)`
`{`
`int i; prod = 1;`
`for (i = 1; i < n; i++)`
`prod = prod * i;`
`return (prod);`
`}`
	1. It is the name gives to the technique of defining something in terms of itself. It is terminated, when a base condition is met. The terminal condition is gradually reached, by invocation of the same function. Each time function is called itself till the base condition is reached.
	For example :
	`int fact (int n)`
	`{`
	`if (n == 0)`
	`return 1;`
	`return (n*fact n-1);`
	`}`

2. This process involves some clear cut processing sections such as initialisation, decision, computation an updation (of iteration counter).	2. There will be an exclusive stopping condition within the recursive body of the function.
3. It uses repetitions structures such as *for loop, while-loop* and *do-while loop* and explicitly uses repetation structures.	3. It uses selection structures such as *if-statement,* if-else or switch statement and achieves repetition through function calls itself repeatedly.
4. These functions exert very fast and occupy less memory and it can be easy.	4. In recursive, every time the function is called, all the local variables, formal parameters and return address are pushed into the stack. So it occupies lot of memory (stack) and it occupies more memory because more number of times pushing and poping items. Therefore, recursion, is *expensive* in terms of *processor time* and *memory usage.*
5. There are some applciations in which iteration is not best suited such as tower of Hanoi, tree traversals etc. These functions are very difficult to design, takes more programmers time and more error prone.	5. There are some applications in which recursion is best suited for designing algorithms such as Tower of Hanoi, Binary Tree traversals etc. In this case, these functions are more efficient and can be understood easily.

7.6 RECURSION THROUGH STACK

Some programming languages like 'C' provide the facility of recursive function. So we can use it very easily. But in other language also we can provide recursion technique with stack implementation whenever the situation arises for recursion implementation. Even we can convert recursive function in 'C' into non-recursive function through stack implementation. We will see this approach in graph traversal where we use Depth first traversal using recursion as well as stack implementation. We have a need to do following things for using recursion through stack implementation.

1. One stack for each parameter of function.
2. Only one stack for return address.

3. One stack for each local variable of function.

We have a need to take right decision where to use recursion function or stack implementation because most of the places using stack implementation will be complex and takes more execution time but in some situations it can be more useful than recursion also.

7.7 RECURSION IN C

EXAMPLE : C program to print array elements in an reverse order (Recursion).

```
#include<stdio.h>
#include<conio.h>
void reverse(int a[], int n)
    {
        if(n == -1)
            return;
        printf("%d\n",a[n]);
        reverse(a,n-1);
    }
void main()
    {
        clrscr ( );
        int i,n,a[20];
        printf("Enter the maximum number of elements in array\n");
        scanf("%d",&n);
        printf("Enter %d elements\n",n);
        for(i=0;i<n;i++)
            scanf("%d",&a[i]);
        printf("The reversed array is :\n");
        reverse(a, n-1);
        getch();
    }
```

EXAMPLE : C program to find product of elements of the given array (Recursion).

```
#include<stdio.h>
#include<conio.h>
int prod(int a[],int n)
    {
        if(n == 0)
            return(a[n]);
        else
            return(a[n] * prod (a, n-1));
    }
void main()
```

```
        {
        clrscr ( );
            int i, n, a[20], result;
            printf("Enter the maximum number of elements in array\n");
            scanf("%d",&n);
            printf("Enter elements");
            for (i=0; i<n; i++)
                scanf("%d",&a[i]);
            result = prod(a, n-1);
            printf("Product of the elements in the array is %d", result);
        getch ( );
        }
```

EXAMPLE : C program to multiply two natural numbers. (Recursion)

```
    #include<stdio.h>
    #include<conio.h>
    int evalmul (int i, int j)
        {
        if(i == 0, j == 0)
            {
                return 0;
            }
        if(j == 1)
            {
                return i;
            }
                return (evalmul (i, j-1) + i);
        }
    void main()
        {
            clrscr ( );
            int i, j result;
            printf("Enter the two elements:\n");
            scanf("%d %d", &i, &j);
            result = evalmul(i, j);
            printf("The multiplication of (%d, %d)=%d", i, j, result);
        getch ( );
        }
```

EXAMPLE : C program to find largest and smallest of n elements in an array (Recursion).

```
    #include<stdio.h>
```

```
#include<conio.h>
int largest (int a[], int n)
    {
        int lar;
        if(n == 0)
            return (a[n]);
        lar = largest (a, n-1);
        if (a [n] > lar)
            return(a[n]);
        return lar;
    }
int minimum (int a[], int n)
    {
        int small;
        if(n == 0)
            return (a[n]);
        small = minimum (a, n-1);
        if (a [n] < small)
            return (a[n]);
        return small;
    }
void main()
    {
        clrscr ( );
        int, n, a [20], lar, small;
        printf("Enter the maximum number of elements in array:\n");
        scanf ("%d", &n);
        printf ("Enter elements\n");
        for (int i=0; i<n; i++)
            scanf("%d", &a[i]);
        lar = largest (a, n-1);
        small = minimum (a, n-1);
        printf ("largest in the given array is %d\n", lar);
        printf ("Smallest in the given array is %d\n", small);
    getch ( );
    }
```

Greatest Common Divisor (GCD) of Two numbers

The greatest common divisor of two integers *m* and *n* is the largest common divisor of *m* and *n* and is denoted by GCD (*m*, *n*). This divides both *m* and *n* such that the remainder is zero.

Euclid's Method: To find the greatest common divisor of two positive integers with the help of Euclid's algorithm is as follows:

The recursive definition for finding an GCD of two numbers is

$$\text{GCD}(m, n) = \begin{cases} GCD(n,m) & \text{if } n > m \\ m & \text{if } n = 0 \\ GCD(n, m \bmod n) & \text{if } m > n \end{cases}$$

EXAMPLE : C program to find GCD of two positive numbers.

```
#include<stdio.h>
#include<conio.h>
int gcd (int m, int n)
     {
          if (n == 0)
               {
                    return m;
               }
          if (m < n)
               {
                    return gcd (n, m)
               }
          return (gcd (n, m%n));
     }
void main ()
     {
          clrscr ( );
          int m, n, result;
          printf ("Enter two numbers");
          scanf ("%d %d", &m, &n);
          result = gcd (m, n)
          printf("The GCD (%d, %d)= %d\n", m, n, result);
     getch ( );
     }
```

EXAMPLE 7.1. Write a recursive algorithm to implement the following function

$$A(m, n) = \begin{cases} n+1 & \text{if } m = 0 \\ A(m-1,1) & \text{if } n = 0 \\ A(m-1), A(m, n-1) & \text{else} \end{cases}$$

Solution:

int A (m, n)

1 if $m == 0$
   return $n + 1$

2. If $n == 0$
   return call A $(m - 1, 1)$

3. Call A $((m - 1), A\ (m, n - 1))$

4. STOP

EXAMPLE 7.2. Let a and b are positive integers. Suppose a function Q is defined recursively as follows:

$$Q(a, b) = \begin{cases} 0 & \text{if } a < b \\ Q(a-b, b)+1 & \text{if } b \le a \end{cases}$$

(*i*) Find the value of $Q(2, 3)$ and $Q(14, 3)$

(*ii*) Find Q (5861, 7)

Solution:

(*i*) Q (2, 3)

Here $a = 2 \quad b = 3$ *i.e.*, $2 < 3$

$\therefore \quad Q(2, 3) = 0$

Now Q (14, 3) here $a = 14 \quad b = 3$ *i.e.*, $b < a$

So

$$\begin{aligned} Q(a, b) &= Q(a-b, b) + 1 = Q(14-3, 3) + 1 \\ &= Q(11, 3) + 1 = [Q(11-3, 3) + 1] + 1 \\ &= [Q(8, 3) + 1] + 1 = [Q(8-3, 3) + 1 + 1 + 1] \\ &= Q(5, 3) + 3 = [Q(5-3, 3) + 1] + 3 \\ &= Q(2, 3) + 4 = 0 + 4 = 4 \end{aligned}$$

(*ii*) Q (5861, 7)

We know $Q(a-b, b) + 1$ equivalent to $\dfrac{a}{b}$

$$\therefore \quad Q(5861, 7) = \frac{5861}{7} = 837$$

EXAMPLE 7.3. A recursive function f is as follows:

```
int f(int x)
{
 if (x < 2)
return 1;
else
return f(x – 1) + f (x – 2);
}
```

find value of *f*(5)

Solution:

$$f(x) = \begin{cases} 1 & \text{if } x < 2 \\ f(x-1) + f(x-2) & \text{if } x \ge 2 \end{cases}$$

$$\begin{aligned} f(5) &= f(4) + f(3) = f(3) + f(2) + f(2) + f(1) \\ &= f(2) + f(1) + f(1) + f(0) + f(1) + f(0) + f(1) \\ &= f(1) + f(0) + 1 + 1 + 1 + 1 + 1 + 1 = 1 + 1 + 6 = 8 \end{aligned}$$

7.8 TAIL RECURSION

All recursive functions contain a call to themselves. However, there are a number of different types of recursive function, based on how many recursive calls there are, and where in the function they occur. The simplest type of recursion is called ***tail recursion***.

As we know the process in which function calls itself in function body is called as recursion. But when this called function is the last executed statement in the function body then it is called ***tail recursion***. Here we take the return value as one parameter of function itself. We use stack to maintain all the functions but here it will not append new function in stack but it will overwrite the value of previous function with the current one. So we can say the function call time and stack implementation time will be reduced and it will give better performance.

In *tail recursion,* there is a single recursive call that occurs at the very end of the function implementation. In other words, when the recursive call is made, there are no more statements to be executed in the function. The single recursive call is the last thing to happen when the function is executed. The function tail() is defined as:

```
void tail (int n)
{
if (n > 0)
     printf("The value of n is %d",n);
tail (n - 1);
}
}
```

is an example of tail recursion.

```
void nonTail (int n)
{
if (n > 0)
{
non-Tail (n - 1);
printf("Non tail recursive");
non-Tail (n - 2);
}
```

The non-Tail() function is not tail recursion for two reasons; there are two recursive calls, whereas tail recursion features only a single recursive call; and there are statements following a recursive call, whereas in tail recursion the recursive call must be the final statement in the function.

The reason these restrictions are made on the definition of tail recursion is that if they apply to a function it can easily be converted into a loop.

EXAMPLE :

```
\* Find the factorial of any number with use of tail recursion */
#include <stdio.h>
#include <conio.h>

void main()
     {
          int n, val;
          printf("Enter the number");
```

```
        scanf("%d",&n);
        if (n < 0)
            printf ("Factorial not possible\n");
        else
            if(n == 0)
                printf ("factorial of zero is 1\n');
            else
                {
                    val = fac (n, 1)
                    printf ("factorial of %d" = %d\n, n, val);
                }
    }

fac (int n, int fact)
    {
        if (n == 1)
            return fact;
        else
            fac (n - 1, n * fact);
    }
```

7.9 THE TOWERS OF HANOI PROBLEM

Another complex recursive problem is that of towers of Hanoi. The problem has a historical basis in the ritual of the ancient Tower of Brahma. This is a problem which is not specified in terms of recursion and we have to see how we can use recursion to produce a logical and elegant solution. In this problem, there are *n* disks of different sizes and there are three poles *A*, *B* and *C*. All the *n* disks are placed on a pole *A* in such a way that a larger disk is always below a smaller disk. The other two poles are initially empty. The aim is to move the *n* disks to the second pole (*C*) using the third pole (*B*) as a temporary storage.

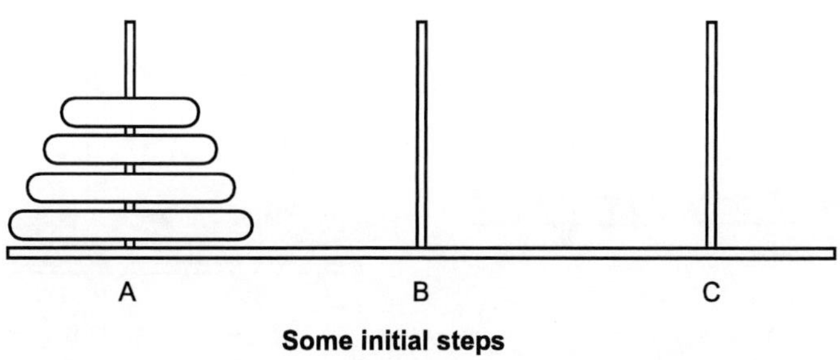

Fig. 7.1 Some initial steps

The rules for the movement of disks are as follows:

- Only one disc could be moved at a time.
- A larger disc must never be stacked above a smaller one.
- One and only one extra needle could be used for intermediate storage of disks.

In the great temple at Benares beneath the dome which marks the centre of the world, rests a brass plate in which are fixed three diamond needles, each a cubit high and as thick as the body of a bee. On one of these needles, at the creation, God placed sixty-four disks of pure gold, the largest disk resting on the brass plate, and the others getting smaller and smaller upto the top one. This is the **Tower of Bramah**. Day and night unceasingly the priests transfer the disks from one diamond needle to another according to the fixed and immutable laws of Bramah, which require that the priest on duty must not move more than one disk at a time and that he must place this disk on a needle so that there is no smaller disk below it. When the sixty-four disks shall have been thus transferred from the needle on which at the creation God placed them to one of the other needles, tower, temple, and Brahmins alike will crumble into dust, and with a thunderclap the world will vanish.

Sometimes we will write $A \to B$ to denote the instruction. "Move top disk from pole A to pole B".

The solution to the Towers of Hanoi problem for $n = 3$ appears in Fig. 7.2.

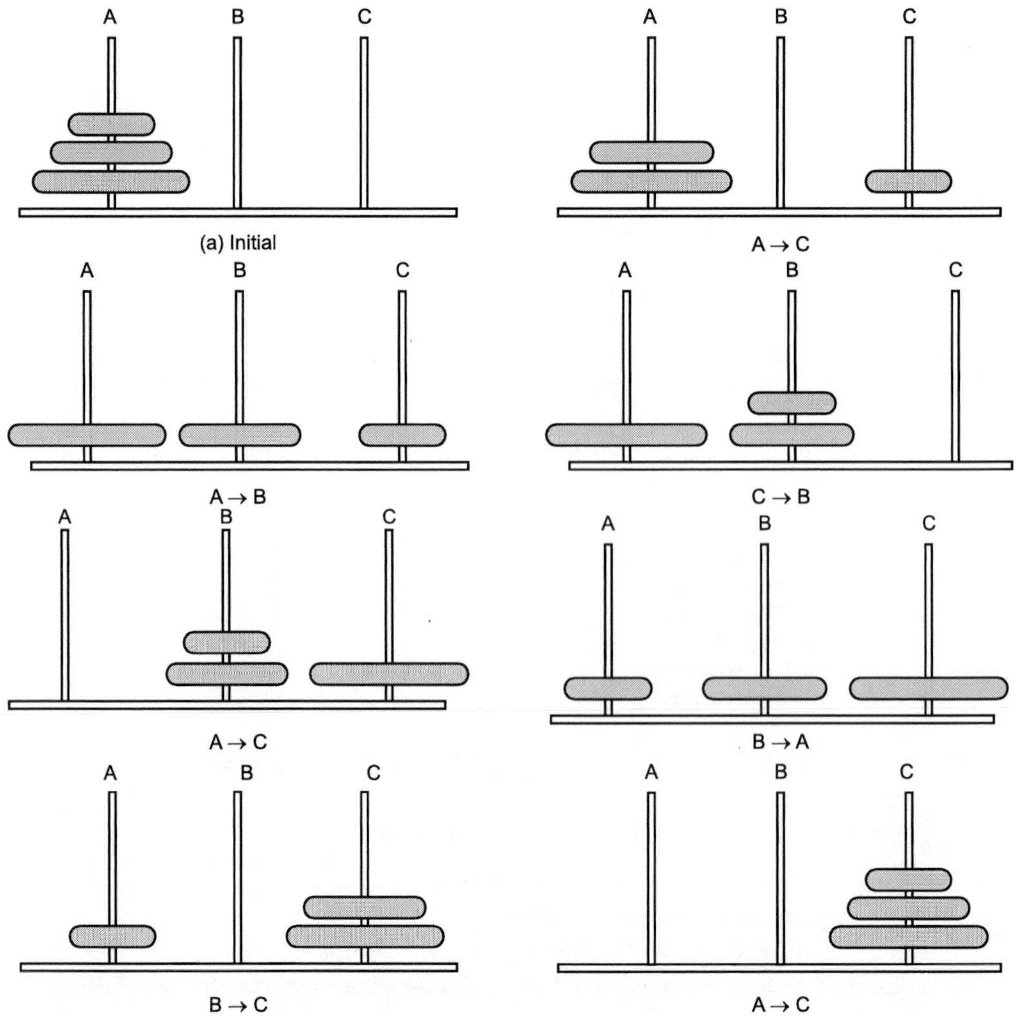

Fig. 7.2.

For $n = 3$.

1. $A \rightarrow C$
2. $A \rightarrow B$
3. $C \rightarrow B$
4. $A \rightarrow C$
5. $B \rightarrow A$
6. $B \rightarrow C$
7. $A \rightarrow C$

We also give the solution to the Towers of Hanoi problem for $n = 1$ and $n = 2$.

For $n = 1$ $A \rightarrow C$

For $n = 2$ $A \rightarrow B$ $A \rightarrow C$ $B \rightarrow C$

Note that for 1 disk it takes 1 move to transfer 1 disk from A to C.

For 2 disks it will take 3 moves : $2M + 1 = 2(1) + 1 = 3$.

For 3 disks, it will take 7 moves : $2M + 1 = 2(3) + 1 = 7$.

For 4 disks, it will take 15 moves : $2M + 1 = 2(7) + 1 = 15$.

For $n = 4$ disks consist of the following 15 moves:

$A \rightarrow B$, $A \rightarrow C$, $B \rightarrow C$, $A \rightarrow B$, $C \rightarrow A$, $C \rightarrow B$, $A \rightarrow B$, $A \rightarrow C$, $B \rightarrow C$, $B \rightarrow A$, $C \rightarrow A$, $B \rightarrow C$, $A \rightarrow B$, $A \rightarrow C$, $B \rightarrow C$.

Number of Disks (n)	*Number of Moves*
1.	$2^1 - 1 = 2 - 1 = 1$
2.	$2^2 - 1 = 4 - 1 = 3$
3.	$2^3 - 1 = 8 - 1 = 7$
4.	$2^4 - 1 = 16 - 1 = 15$
5.	$2^5 - 1 = 32 - 1 = 31$
6.	$2^6 - 1 = 64 - 1 = 63$

So, the formula for finding the number of steps it takes to transfer n disks from A to C is $\boxed{2^n - 1}$.

We use the technique of **recursion** to develop a general solution. The solution to the Towers of Hanoi problem for $n > 1$ disks may be reduced to the following sub-problems :

1. Move the top $n - 1$ disks from A to B.
2. Move the top disk from A to C *i.e.*, $A \rightarrow C$.
3. Move the top n - 1 disks from B to C.

Now introduce the general notation.

Tower (N, BEG, AUX, END)

To denote a procedure which moves the top n disks from the initial pole BEG to the final pole END using the pole AUX as an auxiliary pole.

Tower (N, BEG, AUX, END)

1. If $N = 1$ then write BEG $\rightarrow$ END and EXIT.
2. (Move $N - 1$ disks from BEG to AUX) call Tower ($N - 1$, BEG, END, AUX).

3. Write BEG → END.
4. (Move $N-1$ disks from AUX to END) call Tower ($N-1$, AUX, BEG, END).
5. EXIT.

The recursive solution for
Tower (3, A, B, C) as follows

Here $N = 3$
BEG = A
AUX = B
END = C

1. Call Tower (2, A, C, B)
2. Then $A \to C$.
3. Then call Tower (2, B, A, C)

That is,

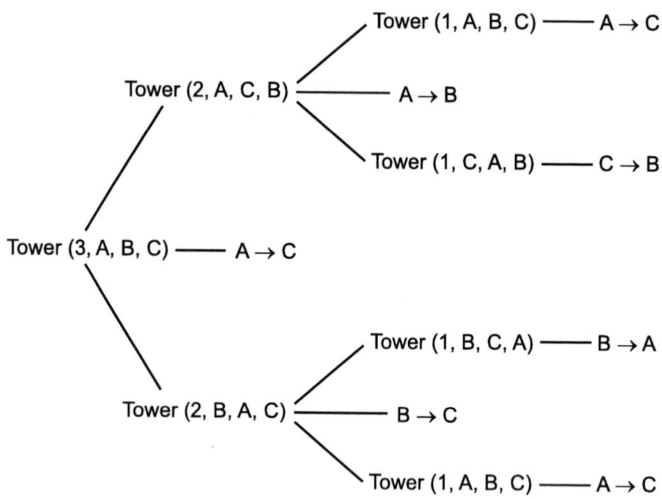

Fig. 7.3.

Observe that the recursive solution for $n = 3$ disks consists of following 7 moves:
$A \to C, A \to B, C \to B, A \to C, B \to A, B \to C, A \to C$.

```
/* Towers of Hanoi problem using recursion */
#include<stdio.h>
#include<conio.h>

void main()
     {
     clrscr( );
          int n;
          char A = 'A', B = 'B', C = 'C';
          void hanoi (int, char, char, char);
          printf ("Enter Number of Disks:");
          scanf("%d",&n);
```

```
        printf("\n\n Tower of Hanoi problem with %d disks\n",n);
        printf ("Sequence is:\n");
        hanoi (n, A, B, C);
        printf ("\n");
        getch ( );
    }
void hanoi(int n, char A, char B, char C)
    {
        if (n!=0)
            {
                hanoi (n-1, A, C, B);
                printf ("Move disk %d from %c to %c\n",n,A,C);
                hanoi (n-1, B, A, C);
            }
    }
```

7.10 DIRECT AND INDIRECT RECURSION

Recursion is said to be **direct** if a subprogram calls itself, it is **indirect** if there is a sequence of more than one sub-program call that eventually calls the first sub-program such as when function *A* calls function *B*, which in turn calls *A*.

7.11 SIMULATING RECURSION

Simulating recursion means to simulate the recursive mechanisms with using non-recursive techniques. Many commonly used programming languages such as FORTRAN, COBOL and many machine languages do not allow recursive programs. Problems such as the Towers of Hanoi and prefix-to-post-fix conversion, whose solutions can be derived and stated quite simply using recursive techniques, can be programmed in these languages by simulating the recursive solution using more elementary operations. So the recursive programs can be programmed into non-recursive techniques by simulating recursive techniques.

In *C* language which supports recursion, a recursive solution to a problem is often more expensive than a non-recursive solution, both in terms of time and space.

So reduce the expense of design and then simulate to the non-recursive solution to use in actual practice. However, in this conversion some of the part of recursive solution will be superfluous in the non-recursive solution of particular application and so reduce the amount of work also.

7.12 REMOVAL OF RECURSION

Suppose *P* is a directly recursive function. We can translate *P* into a non-recursive function by including instructions in *P* to accomplish the following tasks. These steps involve insertion of statement labels and goto statements as well as other constructions that will make the result appear messy.

General Rules:

1. Declare a stack (or stacks) that will hold all local variables, parameters, called by value and flags to specify where *P* was called. As the first executed statement of *P*, initialise

the stack(s) to be empty by setting the counter to 0. The stack(s) and the counter are to be treated as global variables, even through they are declared in *P*.

2. To enable each recursive call to start at the beginning of the original function *P*, the first executable statement of the original *P* should have a label attached to it.

The following steps should be done at each place inside *P* where *P* calls itself.

3. Make a new statement label L_i (if this is the ith place where *P* is called recursively) and attach the label to the first statement after the call to *P*.
4. Push the integer *i* into the stack.
5. Push all local variables and parameters called by value onto the stack.
6. Set the dummy parameters called by value to the values given in the new call to *P*.
7. Replace the call to *P* with a goto to the statement label at the start of *P*.

At the end of *P*, the following steps should be done:

8. If the stack is empty then the recursion has finished, make a normal return.
9. Otherwise pop the stack to restore the values of all local variables and parameters called by values.
10. Pop an integer *i* from the stack and use this to go to the statement labelled L_i.

By mechanically following the preceding steps we can remove direct recursion from any function.

Exercises

1. What do you mean by recursion? Discuss advantages and disadvantages of recursion.
2. Define principles of recursion in detail.
3. Differentiate iteration and recursion.
4. Define tail recursion with example.
5. What do you mean by Towers of Hanoi problem?
6. Write a short note on:

   (*a*) Simulating recursion

   (*b*) Removal of recursion

❒❒❒

Chapter 8

Trees

8.1 INTRODUCTION

Have you ever thought how does the operating system manage our files ? Why do we have a hierarchial file system ? How do files get saved deleted under hierarchical directories? Well, we have answers to all these questions in this section through a hierarchical data structure called **Trees.**

Linked list usually provide greater flexibility than arrays, but they are linear structures and it is difficult to use them to organise a hierarchial representation of objects. Although stacks and queues reflect some hierarchy, they are limited to only one-dimension. To overcome this limitations, we create a new data structure called *tree*.

Tree is a data structure which allows you to associate a parent-child relationship between various pieces of data and thus allows us to arrange our records, data and files in a hierarchical fashion. Consider a Tree representing your family structure. Let us say that we start with your grand-parent; then come to your parent and finally, you and your brothers and sisters. So, familiar examples of trees are genealogies and organisation charts. Trees are used to help analyse electrical circuits and to represent the structure of mathematical formulas. Trees also arise naturally in many different areas of computer science. For example, trees are used to organise information in database systems and to represent the syntactic structure of source programs in compilers.

Unlike natural trees, a tree data structure is depicted upside down with the root at the top and

Chapter Outline

- Introduction
- Terminologies Used
- Binary Tree
- Representation of a Binary Tree
- Types of Binary Tree
- Strictly Binary Tree
- Complete Binary Tree
- Almost Complete Binary Tree
- Traversing in Binary Tree
- Non-recursive Functions for Traversals
- Creation of Binary Tree Using Travesals
- Converting Algebraic Expression into Binary Tree
- Extended Binary Tree
- Huffman Tree
- Threaded Binary Trees
- Traversing in Threaded Tree
- Insertion and Deletion in a Threaded Binary Tree
- Finding Inorder Successor and Inorder Predecessor of a Node in Treaded Binary Tree
- General Tree
- General Tree Represented as BInary Tree
- Forest to Bianry Tree Conversion

the leaves at the bottom. The *root* is a node that has no parent; it can have only child nodes. *Leaves,* on the other hand, have no children, or rather, their children are empty structures.

Formally, a *tree* can be defined *recursively* in the following manner.

1. A single node by itself is a tree. This node is also the root of the tree.
2. Suppose n is a node and $T_1, T_2, ..., T_k$ are trees with roots $n_1, n_2, ..., n_k$, respectively. We can construct a new tree by making n be the parent of nodes, $n_1, n_2, ..., n_k$. In this tree n is the root and $T_1, T_2, ..., T_k$ are the *sub-trees* of the root. Nodes $n_1, n_2, ..., n_k$ are called the *children* of node n.

Sometimes, it is convenient to include among trees the **null tree,** a "tree" with no nodes, which we shall represent by $\wedge$.

Examples of Trees

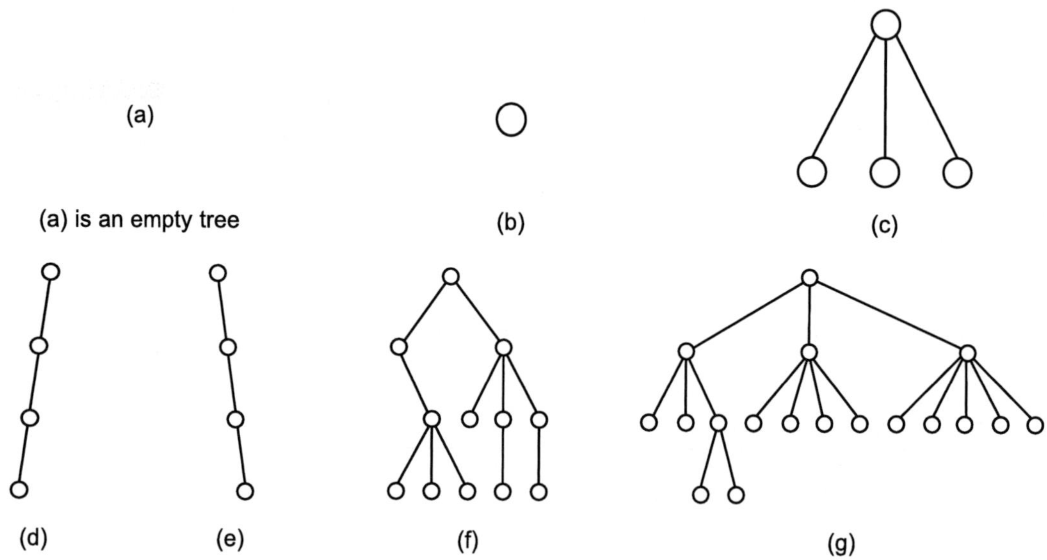

Fig. 8.1.

Fig. 8.1. contains examples of trees. Each node has to be reachable from the root through a unique sequence of arcs, called a *path.* The number of arcs in a path is called the *length* of the path. The *level* of a node is the length of the path from the root to the node plus 1, which is the number of nodes in the path. The *height* of a non-empty tree is the maximum level of a node in the tree. The empty tree is a legitimate tree of height 0 (by definition), and a single node is a tree of height 1. This is the only case in which a node is both the root and a leaf. The level of a node must be between 1 (the level of the root) and the height of the tree.

A **binary tree** is either empty, or it consists of a node called the **root** together with two binary trees called the **left sub-tree** and the **right sub-tree** of the root. The definition of a tree does not impose any combination on the number of children of a given node. This number can vary from 0 to any integer.

8.2 TERMINOLOGIES USED

1. **Node:** We will use the term *node,* rather than vertex, with binary tree. This is the main component of any tree structure. It stores the actual data along with linkes to other nodes. Fig. shows the structure of a node.

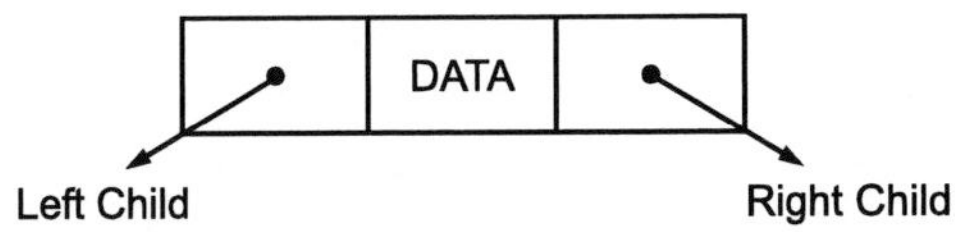

Fig. 8.2.

2. **Parent:** The parent of a node is the immediate-predecessor of that node. In Fig. A is the parent of nodes B and C.

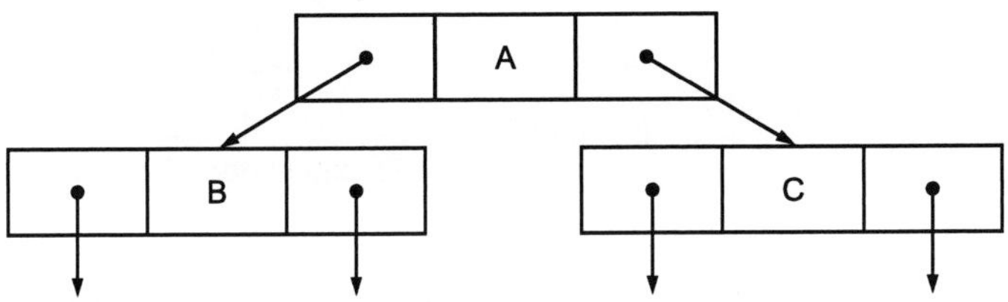

Fig. 8.3.

3. **Child:** The immediate successors of a node are called *child nodes.* A child which is placed at the left side is called the **left child** and a child which is placed at the right side is called the **right child.** In the above, Fig. B and C are the child nodes of A.
4. **Root:** A root is a specially designated node in a tree. It is a node which has no parent. There can be only one root in a tree. In Fig. 8.4, the node A is the root of the tree.

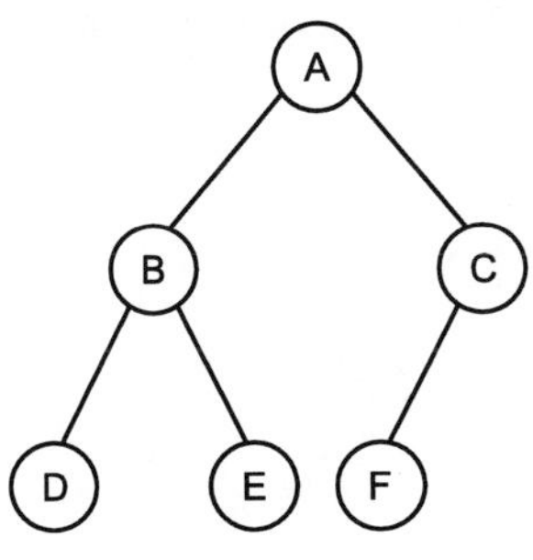

Fig. 8.4.

5. **Leaf Node (External node):** A leaf node which does not have any child node. In Fig. 8.4 nodes D, E and F are leaf nodes.
6. **Siblings:** The child nodes of a given parent node are called *siblings.* In the Fig. 8.4, D and E are siblings.
7. **Branch or Edge:** It is a connecting line between two nodes. A node can have more than one edge. In Fig. 8.4, the node A has two branches. Actually, a branch is a pointer to a node in the tree.
8. **Path:** Each node has to be reachanable from the root through a unique sequence of edges called a Path. The number of edges in a path is called the **length of path.**

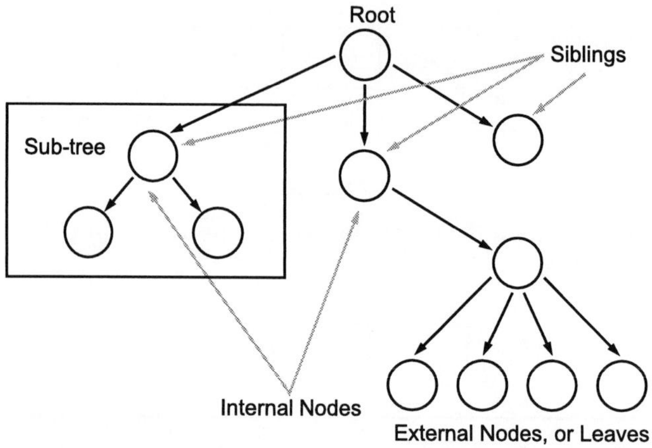

Fig. 8.5.

9. **Rooted Tree:** A **Rooted Tree** is one where we designate one node as the root. In computer science we normally reserve the term **tree** to refer to **rooted trees**. The more general structure is a **free tree.** In a rooted tree, any node is the root of a **subtree** consisting of it and the nodes below it. There is exactly one path between the root and each of the other nodes in the tree. Each node except the root has exactly one node above it in the tree, (*i.e.*, it's **parent**), and we extend the family analogy talking of **children, siblings** or **grandparents.**
10. **Level of a Node:** The level of a node (vertex) is the number of edges along the unique path between it and the root. The level of the root is defined as zero. If a node is at level I then its child is at level I + 1 and its parent is at level I – 1. This rule is common for all nodes except the root node.

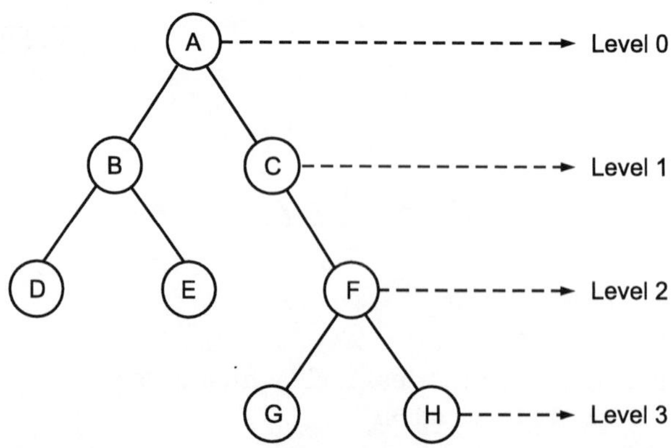

Fig. 8.6.

Here the root node A is at level 0.

Nodes B and C are at level 1.

Node D, E and F are at level 2.

Nodes G and H are at level 3.

11. **Height:** The height of a rooted tree is the maximum level of a node + 1 in the tree. The height of an empty tree is zero. The height of a tree containing a single node is 1.

The formla used to find the height of a tree is $h = I_{max} + 1$

where I_{max} is the maximum level of the tree and h is the height of the tree.

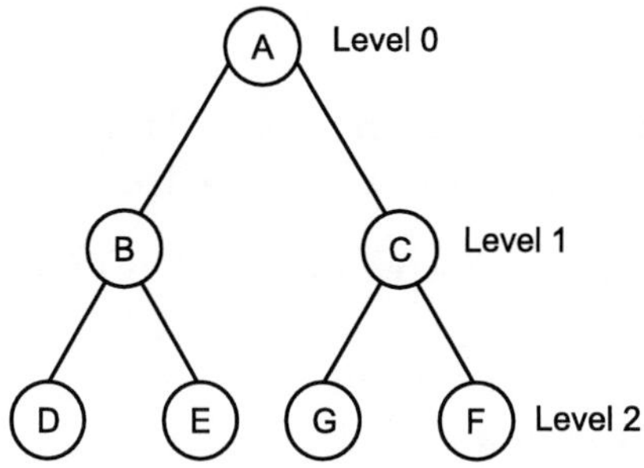

Fig. 8.7. So height of the tree is 3.

12. **Ordered Tree:** An ordered tree is a rooted tree in which the order of the children at every node is specified. If each node *must* have a specific number of children appearing in a specific order, then we have an **M-ary** tree. The simplest type of M-ary tree is the **binary tree**. A binary tree is a tree where each node has exactly zero, one or two children, *i.e.,* each parent can have no more than 2 children.

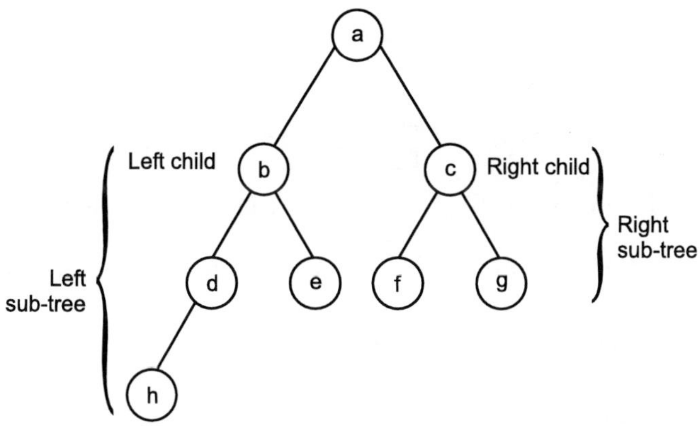

Fig. 8.8.

8.3 BINARY TREE

A binary tree is made of nodes, where each node contains a "left" pointer, a "right" pointer, and a data element. The "root" pointer points to the topmost node in the tree. The left and right pointers recursively point to smaller "sub-trees" on either side. A null pointer represents a binary tree with no elements–the *empty tree*. The formal recursive definition is: a **binary tree** is either empty (represented by a null pointer), or is made of a single node, where the left and right pointers point to a **binary tree.**

Properties of Binary Trees

1. The number of external nodes is 1 more than the number of internal nodes. It is easy to see this if we start removing external nodes with their internal parent, one pair at a time. At the end of this process, only the root with its two external children will remain.
2. The number of external nodes is at least h + 1, where h is the height of the tree, and at most 2h. The later holds for a **full binary tree,** which is a tree where internal nodes completely fill every level.

3. The number of internal nodes is at least h and at most $2^h - 1$.
4. The total number of nodes in a binary tree is at least $2^h + 1$ and atmost 2h + 1 – 1.
5. The height, h, of a binary tree with n nodes is at least log n + 1 and at most n.
6. A binary tree with n nodes has exactly n – 1 edges.

EXAMPLE 8.1. Prove that in a binary tree the number of leaves is equal to the number of nodes with two children + 1.

Solution: Let n_0 be the number of leaf nodes

n_1 be the number of nodes with degree 1

n_2 be the number of nodes with degree 2

$$n = \text{total number of nodes}$$

i.e.,
$$n = n_0 + n_1 + n_2 \quad ...(1)$$

We know the number of branches of binary tree is $n - 1$. It can also be observed that branches may emerge from the node that has degree one or the node that has degree two. Hence, $2n_2 + n_1$

So,
$$n - 1 = 2n_2 + n_1$$
$$n = 2n_2 + n_1 + 1 \quad ...(2)$$

From equation 1 and 2 we get $n_0 + n_1 + n_2 = 2n_2 + n_1 + 1$

$$\boxed{n_0 = n_2 + 1}$$

i.e., Number of leaves is equal to the number of nodes with degree 2 + 1.

EXAMPLE 8.2. A binary tree of height *h*, *h* ≥ 0 has at least *h* and at most $2^h - 1$ elements in it.

Solution: Since, there must be at least one element at each level, the number of elements is at least h.

As each element can have atmost two children, the number of elements at level i is at most $2^i - 1$, $i > 0$. For $h = 0$, the total number of elements is 0, which equals $2^0 – 1$. For $h > 0$ the number of elements cannot exceed.

$$\sum_{i=1}^{h} 2^{i-1} = 2^{1-1} + 2^{2-1} + 2^{3-1} + ... + 2^{h-1}$$

$$= 2^0 + 2^1 + 2^2 + ... 2^{h-1} = 1 + 2 + 2^2 + ... 2^{h-1}$$

$$= \frac{1\,(2^{h-1})}{2-1} = 2^{h-1}.$$

EXAMPLE 8.3. The height of a binary tree that contains *n*, *n* ≥ 0, elements is atmost *n* and atleast $\lceil \log_2 (n+1) \rceil$

Solution: Since there must be at least one element at each level, the height cannot exceed '*n*'.

We know that a binary tree of height h can have no more $2^h - 1$ elements.

Thus, $n \le 2^h - 1$

$n + 1 \leq 2^h$

Taking log both sides

$\log_2(n + 1) \leq \log_2{}^{2h}$

$\log_2(n + 1) \leq h \, . \log_2 2$

$\log_2(n + 1) \leq h \, . 1$

i.e., $h \geq \log_2(n+1)$

Since h is an integer, therefore we have $h \geq \lceil \log_2 (n + 1) \rceil$

8.4 REPRESENTATION OF A BINARY TREE

Binary tree can be represented in any of the following two ways:

1. Using an array (Linear representation)
2. Using a linked list (Link representation)

1. **Array Representation:** A binary tree can be stored represented using an array, which is called *sequential representation* (array representation). This type of representation is static *i.e.,* a block of memory for an array is to be allocated before going to store the actual tree in it.

In this representation, the nodes of the tree are stored level-by-level, starting from the 0th level. The root node of the tree is stored in the first memory location of allocated memory. The following rules are used to decide the location of any node.

> 1. The root node is at location 1.
> 2. For any node with index i, $1 < i \leq n$.
>   (a) Parent (i) = $\lfloor i/2 \rfloor$
>   (b) Left child = [i] = 2i
>   (c) Right child [i] = 2i + 1.

For example: The array representation of the binary tree.

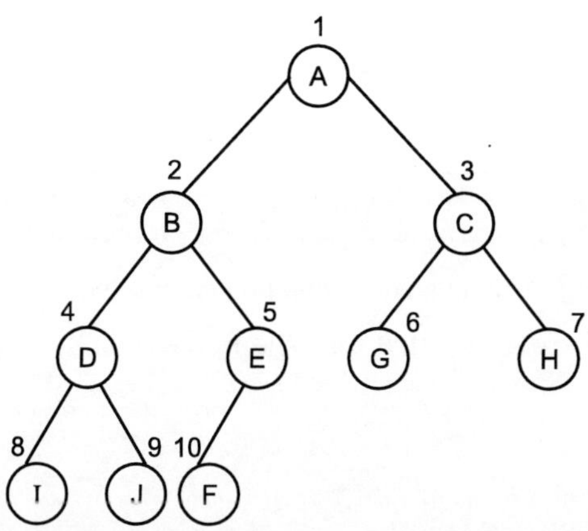

Fig. 8.9.

is

1	2	3	4	5	6	7	8	9	10
A	B	C	D	E	G	H	I	J	F

Advantages

1. Data are stored without any pointer to its successor or ancestor.
2. Any node can be accessed from any other node by calculating the index.
3. Programming languages which do not support dynamic memory allocation, use only this type of representation for a tree.
4. Simplicity.

Disadvantages

1. Array representation is not suitable for normal binary tree but it is only ideal for complete binary tree.
2. Here most of the array entries are empty *i.e.,* unnecessarily more memory is wasted.
3. There is no way possible to enhance the tree structure.
4. Additions and deletions of nodes are inefficient because of the data movements in the array.

To overcome all these problems by using linked list representation.

2. Linked Representation of a Binary Tree: The binary tree can be represented using dynamic memory allocation of a node in a linked list form. In a linked list allocation technique a node in a tree contains three fields:

1. Info (data): It contains the info value.
2. Left link (left): It contains address of the left sub-tree.
3. Right link (Right): It contains address of the right sub-trees.

Left	Info	Right

when a node has no children, the corresponding pointer field are NULL.

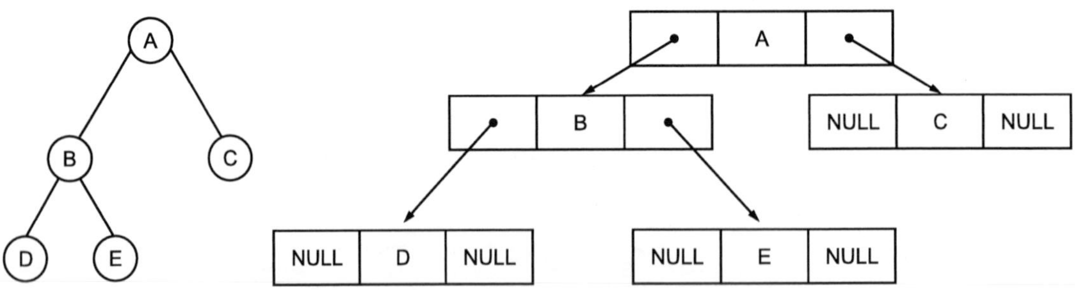

Fig. 8.10. Linked representation of a binary tree.

In this, we should declare structure for tree node.

```
struct node
    {
        int info;
        struct node *left;
        struct node *right;
    }
```

Here first member info is for information field of node, second member is for left chld of node which points to the structure itself, it contains the address of left child. If node has no left child then it should be NULL. Third member is for right child of node which also points to the structure itself, it contains the address of right child. If node has no right child it should be NULL.

This type of representation is dynamic *i.e.,* block of memory required for the tree need not be allocated before. They are allocated only on demand.

Advantages

1. Insertions and deletions involve no data movement.
2. No wastage of memory.
3. Enhancement of the tree is possible.

Disadvantages

1. In this, pointer fields are involved which occupy more space than just data fields.
2. Programming languages which do not support dynamic memory allocation have difficulty in implmenting the binary tree.

8.5 TYPES OF BINARY TREE

A tree is a **binary tree** if and only if:

1. It has a root node, which may not have any child nodes **(0 child nodes, NULL tree).**
2. Root node may have 1 or 2 child nodes. Each such node forms a binary tree itself.
3. Number of child nodes can be **0, 1, 2, ... not more than 2.**
4. There is a unique path from the root to every other node.

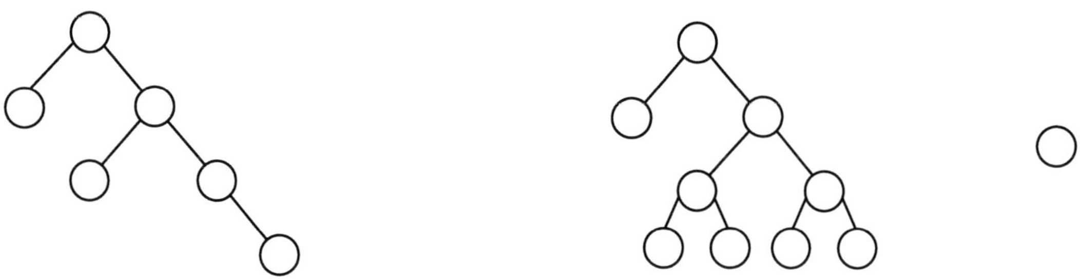

Fig. 8.11.

A binary tree is a **strictly binary tree** if and only if each node has **exactly two** child nodes or **no** nodes.

Fig. 8.12.

Full and Complete Binary Tree

A binary tree of height h that contains exactly $2^h - 1$ elements is called a full binary tree.

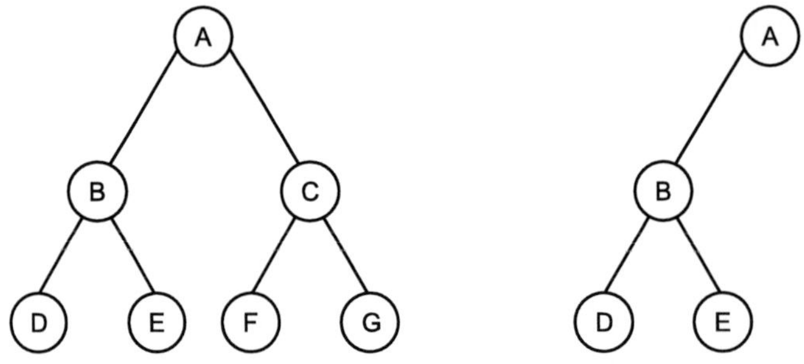

(i) Full binary tree of height 3 (ii) Not a full binary tree

Fig. 8.13.

A binary tree is a **full binary tree** if and only if:

(*a*) Each non-leaf node has **exactly two** child nodes.

(*b*) All leaf nodes are at the **same level.**

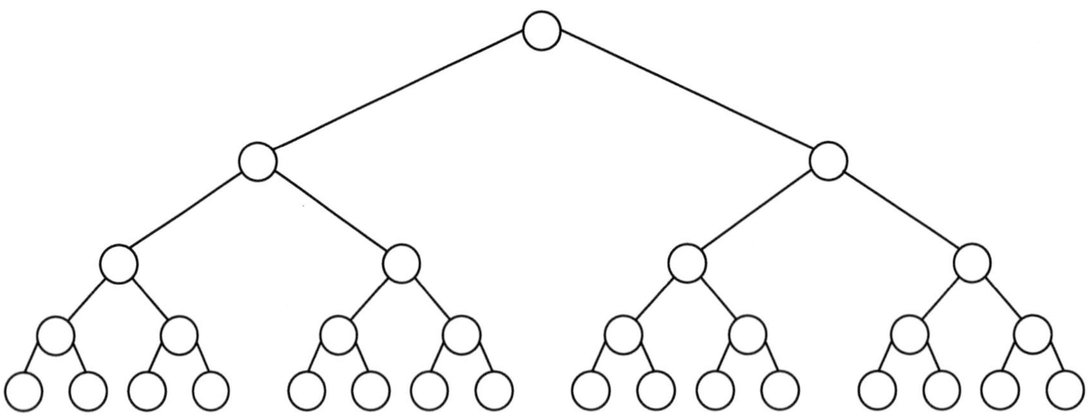

Fig. 8.14.

A binary tree is a **complete** binary tree (**of height** h**,** we take root nodes as 0) if and only if:

(*a*) Level 0 to $h - 1$ represent a **full binary tree** of height $h - 1$.

(*b*) One or more nodes in level $h - 1$ may have 0, or 1 child nodes.

(*c*) If j, k are nodes in level $h - 1$, then j has more child nodes than k if and only if j is to the left of k, ***i.e.*, the last level (*h*) can be missing leaf nodes, however the ones present must be shifted to the left.**

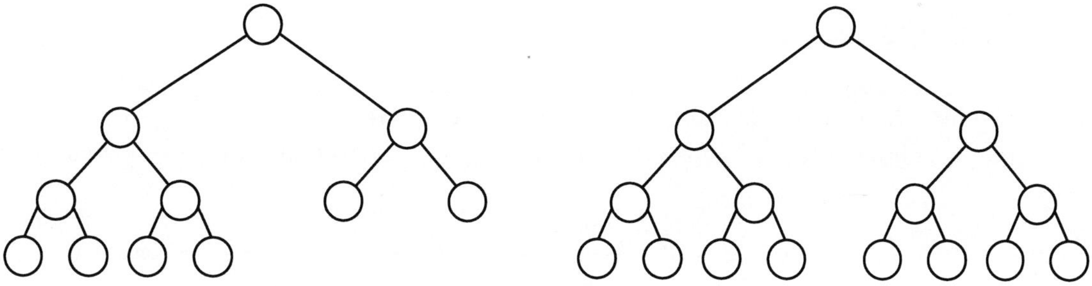

Fig. 8.15.

Examples of Trees:

- A free tree

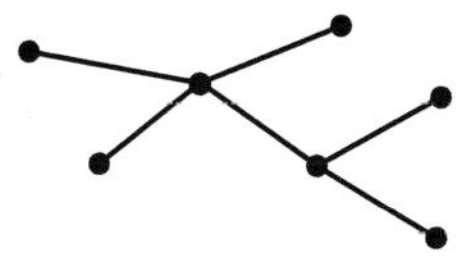

- A binary tree

- A rooted binary tree

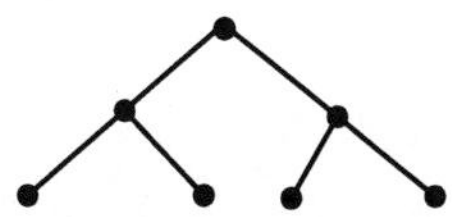

- A forest

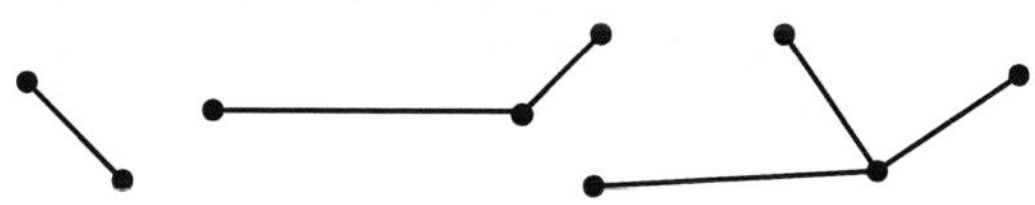

8.6 STRICTLY BINARY TREE

If every non-leaf node in a binary tree has non-empty left and right sub-trees, the tree is termed as strictly binary tree, *e.g.*,

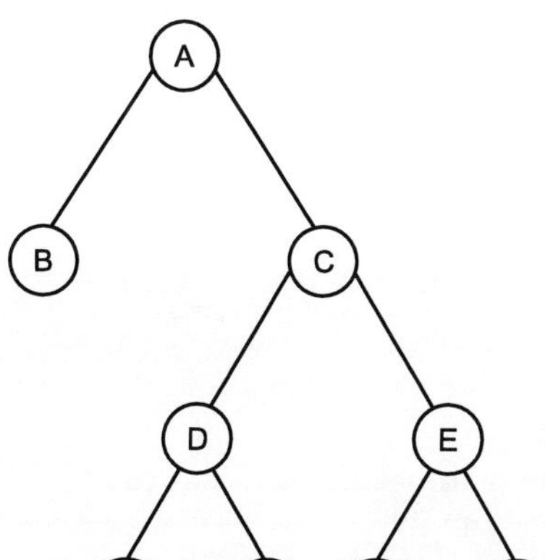

Fig. 8.16 Strictly binary tree

A strictly binary tree with n leaves always contains $2n–1$ node.

8.7 COMPLETE BINARY TREE

As we know, the depth of a binary tree is the maximum level of any leaf in the tree. This equals the length of the longest path from the root to any leaf.

A complete binary tree of depth *d* is the strictly binary tree all of whose leaves are at level *d*.

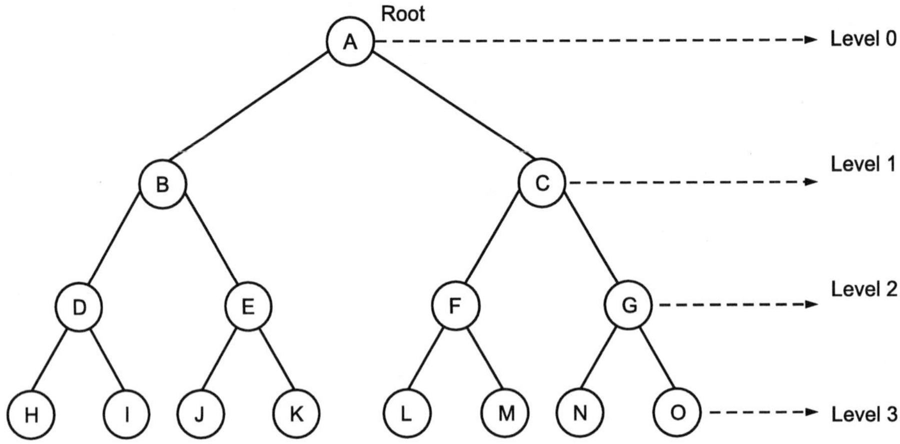

Fig. 8.17. Complete binary tree of depth 3.

If a binary tree contains *m* nodes at level I, it contain at most 2*m* nodes at level I + 1. Since a binary tree can contain atmost one node (root) at level 0, it can contain atmost 2^I nodes at level I.

A complete binary tree of depth *d* is the binary tree of depth *d* that contains exactly 2I nodes at each level I between 0 and d.

Thus, the total number of nodes in a complete binary tree of depth *d* equals the sum of the number of nodes at each level between 0 and *d i.e.,*

Number of nodes at level 0 is $2^0 = 1$

Number of nodes at level 1 is $2^1 = 2$

Number of nodes at level 2 is 2^2.

.....

.....

Number of nodes at level *d* is 2^d.

The total number of nodes in complete binary tree is

$$2^0 + 2^1 + 2^2 + \ldots\ldots 2^d = \sum_{j=0}^{d} 2^j$$

By induction, it can be shown that this sum equals $\mathbf{2^{d+1} - 1}$. Since all leaves in such a tree are at level *d*, the tree contains 2^d leaves and therefore $2^d - 1$ non-leaf nodes.

The significance of a complete binary tree is that it is the binary tree with the maximum number of nodes for a given depth.

8.8 ALMOST COMPLETE BINARY TREE

A binary tree of depth *d* is an **almost complete binary tree,** if

1. Any node at level less than d – 1 has two childrens.
2. For any node 'x' in the tree with a right descendant at level d, x must have a left child and every left descendant of x is either a leaf at level d or has two childrens.

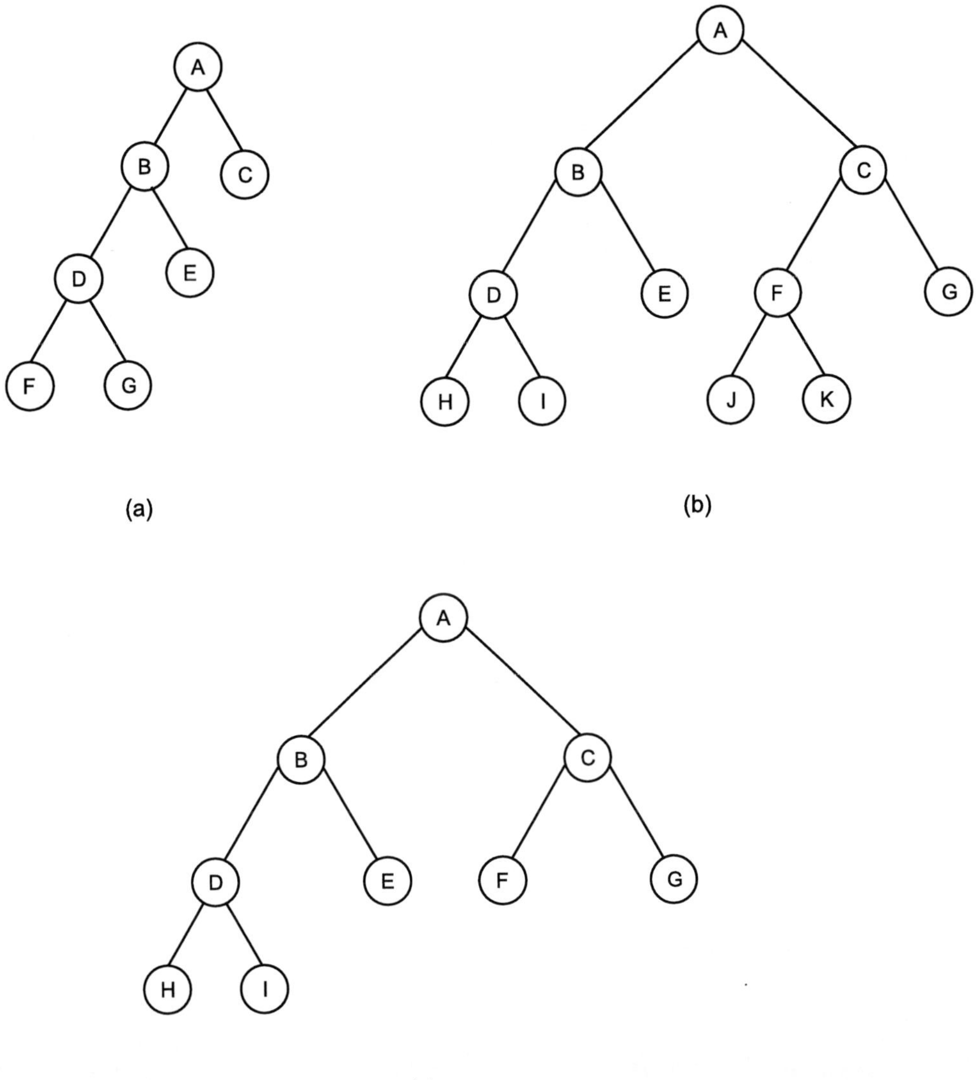

Fig. 8.18.

Fig. 8.18(*a*) is not almost complete binary tree because it contains leaf nodes at level 1, 2 and 3. So violating condition 1.

Fig. 8.18(*b*) satisfies condition 1, since every leaf is either at level 2 or at level 3. However, condition 2 is violated, since A has a right descendant at level 3 (*J*) but also has a left descendant that is a leaf at level 2 (*E*).

Figure 8.18(c) satisfies both conditions 1 and 2 and is therefore an almost complete binary tree.

An *almost complete strictly binary tree* with n leaves has $2n–1$ nodes and an almost complete bianry tree with n leaves that is not strictly binary has $2n$ nodes. There are two distinct almost complete bianry tree with n leaves one of which is strictly binary and one of which is not.

Note that, there is only a single almost complete binary tree with *n* nodes. This tree is strictly binary if and only of *n* is odd.

An almost complete bianry tree of depth d is intermediate between the complete binary tree of depth $d - 1$ that contains $2^d - 1$ nodes and the complete binary tree of depth d, which contains $2^{d+1} - 1$ nodes.

8.9 TRAVERSING IN BINARY TREE

As with lists, we would like to be able to traverse through all nodes in our tree. Tree traversal is the process of visiting each node in the tree exactly once.

Problem: What order should the nodes be visisted in?

- Top-down?
- Left-to-right?
- Bottom-up?

In tree creation we take three parameters node, left and right child, so traversing of binary tree means traversing of node, left sub-tree and right sub-tree. If root is denoted as n, left sub-tree as L and right sub-tree as R, then there will be six combinations of traversals NRL, NLR, LNR, LRN, RNL, RLN. But only three are standard, NLR (node-left-right), LNR (left-node-right) and LRN (left-right-node) traversal. We can see left sub-tree is always traversed before right sub-tree. NLR is called pre-order LNR is inorder NRL is post-order.

These traversals are as:

Preorder (NLR Traversal)

1. Visit the root.
2. Traverse the left sub-tree of root.
3. Traverse the right sub-tree of root.

Inorder Traversal (LNR Traversal)

1. Traverse the left sub-tree of root.
2. Visit the root.
3. Traverse the right sub-tree of root.

Postorder (LRN Traversal)

1. Traverse the left sub-tree of root.
2. Traverse the right sub-tree of root.
3. Visit the root.

That is,

Pre order	Root	Left sub-tree	Right sub-tree
In-order	Left-sub-tree	Root	Right sub-tree
Post-order	Left-sub-tree	Right sub-tree	Root

Now, we will create the function for **preorder traversal.**

```
preorder(ptr)
    {
        struct node *ptr;
        {
            if(ptr != NULL)
            {
                printf("%c", ptr->info);
                preorder(ptr->left);
                preorder(ptr->right);
            }
        }
    }
```

Similarly the function for **inorder traversal** will be.

```
inorder(ptr)
    {
        struct node *ptr;
            {
                if(ptr != NULL)
                    {
                        inorder (ptr->left);
                        printf("%c", ptr->info);
                        inorder(ptr->right);
                    }
            }
    }
```

and the function for **postorder traversal** is

```
postorder(ptr)
    {
        struct node *ptr;
            {
                if(ptr != NULL)
                    {
                        postorder(ptr->left);
                        postorder(ptr->right);
                        printf("%c",ptr->info);
                    }
            }
    }
```

Now apply each traversal in binary tree shown in Fig.

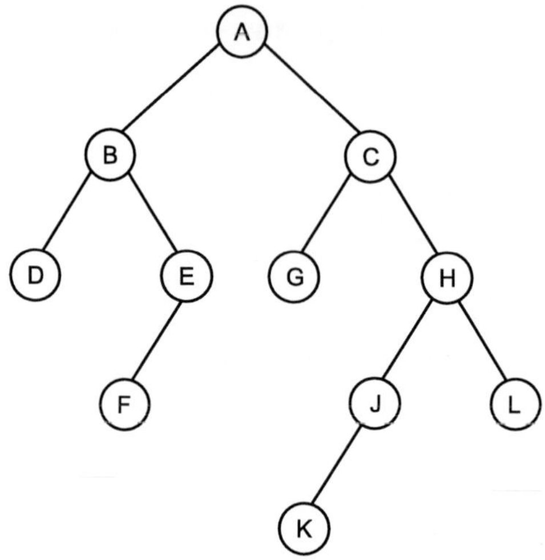

Fig. 8.19.

Pre-order Traversal:

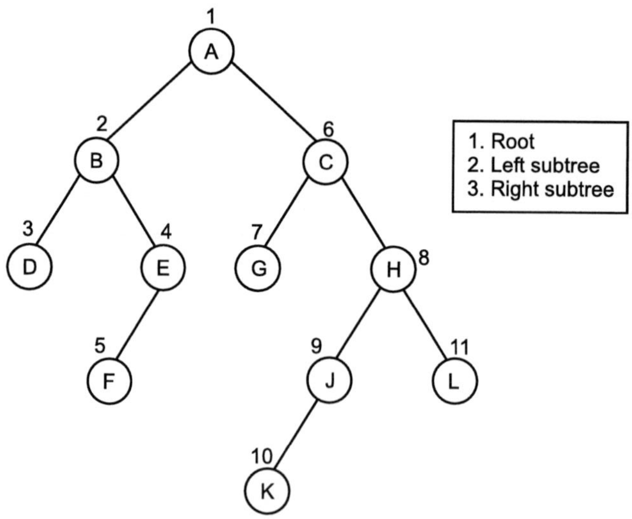

Fig. 8.20.

Here numbers represent the sequence of visited node. So the nodes are visited in pre-order traversal as.

A B D E F C G H J K L

In this A is the root node, so visit it. Once the root node of the tree is visited it goes to the root of the left subtree *i.e.*, node B then left child, *i.e.*, node D. Then the root of right sub-tree *i.e.*, node E then left child *i.e.*, node F.

Now it goes to the root of the right sub-tree of root node, *i.e.*, node C, then left child, *i.e.*, G. Then root of right sub-tree of node C, *i.e.*, node H, then root of left sub-tree of H, *i.e.*, node J, then left child of J, *i.e.*, node K. Then right child of node H, *i.e.*, L.

Inorder Traversal:

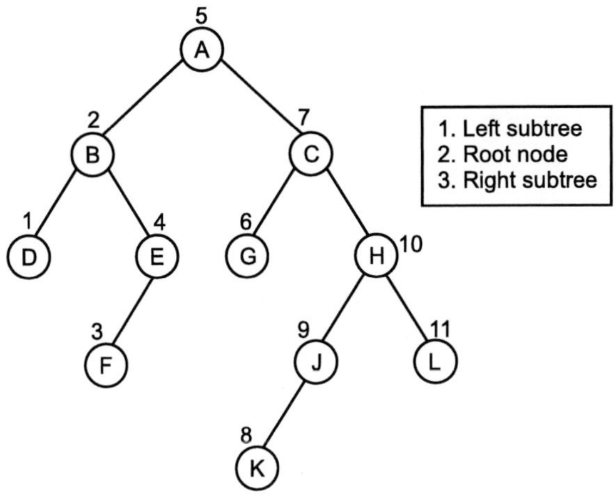

Fig. 8.21.

The nodes are visited in inorder as:

D	B	F	E	A	G	C	K	J	H	L

Postorder Traversal:

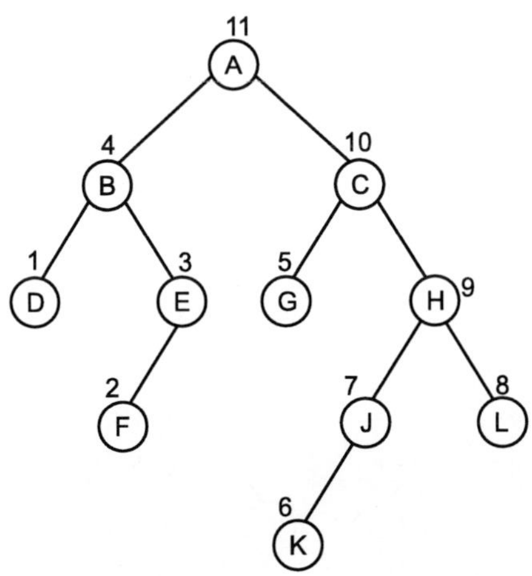

Fig. 8.22.

The nodes are visited in post-order as:

D	F	E	B	G	K	J	L	H	C	A

Example: Apply each traversal techniques in binary tree.

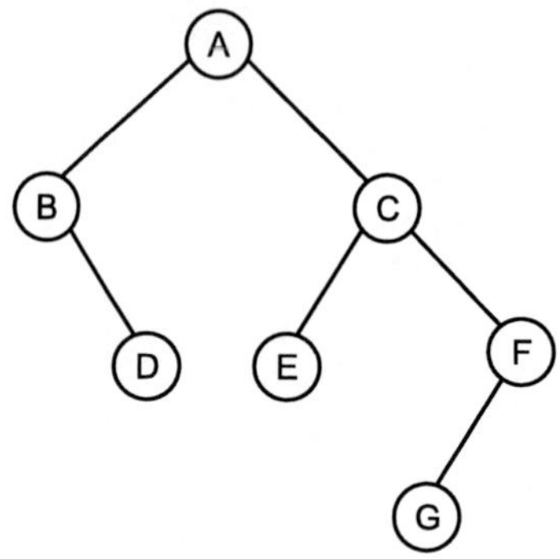

Fig. 8.23.

Pre-order Traversal:

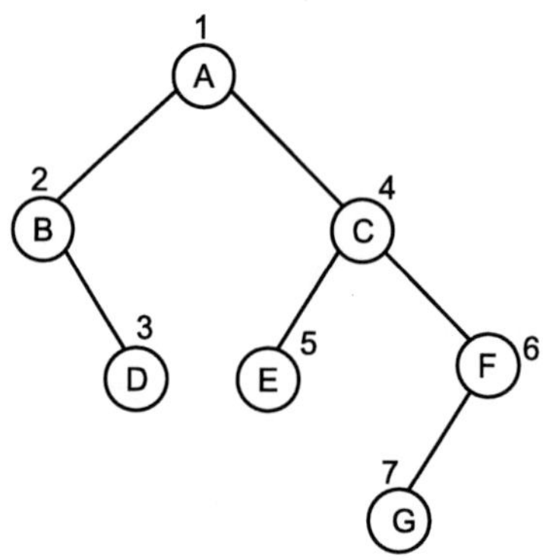

Fig. 8.24.

Pre-order traversing are:

A B D C E F G

In-order Traversal:

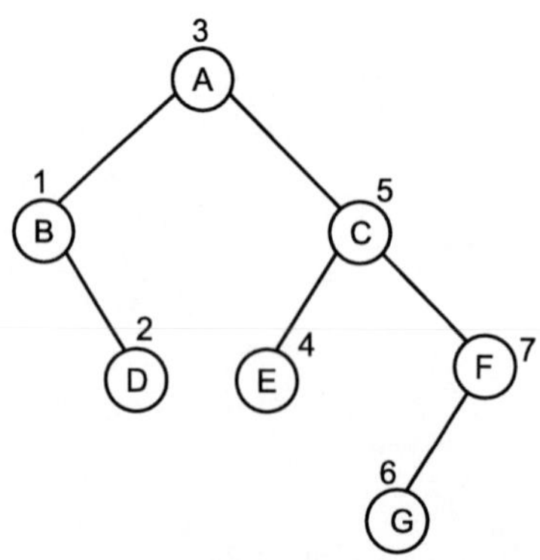

Fig. 8.25.

The nodes are visited in inorder as:

B D A E C G F

Post order Traversal:

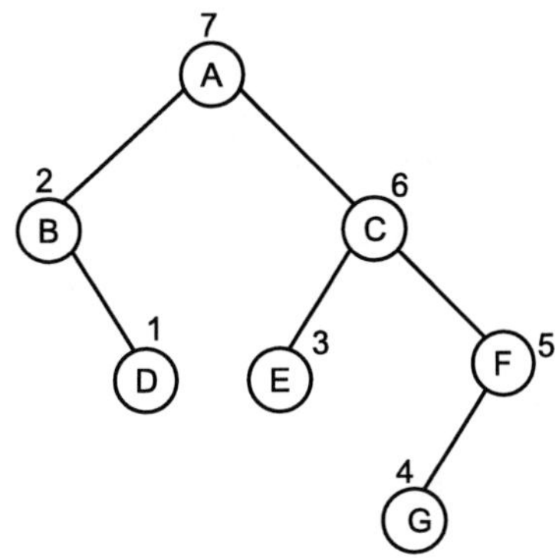

Fig. 8.26.

The nodes are visited in post order as:

D	B	E	G	F	C	A

C Program :

```
/*To write a C-program to create the simple binary tree and recursive
traversal*/
#include<stdio.h>
#include<conio.h>
typedef struct bin
{
int data;
struct bin *left;
struct bin *right;
}node;

void insert (node *, node*);
void inorder (node *);
void preorder (node *);
void postorder (node *);
node *get_node();

void main( )
{
int choice;
char ans = 'n';
node *new,*root;
root = null;
clrscr();
```

```
do
{
printf("\n program for implementing simple binary tree");
printf("\n 1.create");
printf("\n 2.inorder");
printf("\n 3.preorder");
printf("\n 4.postorder");
printf("\n 5.exit");
printf("\n /t Enter Your Choice");
scanf("%d",&choice);

switch(choice)
{
case 1:
root = null;
do
{
new=get_node();
printf("\n enter the element");
scanf("%d",& new->data);
if(root = = null)
     root= new;
else
     insert(root, new);
printf("\n do you want to enter more elements?(y/n)");
ans = getch();
}
while(ans==' Y'|| ans=='y')
clrscr();
break;

case 2:
     if(root ==null)
          printf("tree is not created!");
     else
     inorder(root);
     break;

case 3:
     if(root ==null)
```

```
        printf("tree is not created!");
    preorder (root);
    break;

case 4:
    if(root = = null)
        printf("tree is not created!");
    postorder(root);
    break;
}
}while(choice!=5);
}

node *get_node()
{
node *temp;
temp = (node*) malloc (size of(node));
temp->left = null;
temp->right = null;
return temp;
}

void insert(node *root, node *new)
{
char ch;
printf("\n where to insert left/right of %d", root->data);
ch=getch();
if(ch=='R') || (ch=='r')
    {
        if(root->right == null)
            {
                root->right=new;
            }
        else
            insert(root->right,new);
    }
else
    {
        if(root->left==null)
            {
                root->left=new;
            }
```

```
        else
            insert(root->left,new);
        }
}

void inorder(node *temp)
    {
        if(temp!=null)
            {
                inorder(temp->left);
                printf("%d" temp->data);
                inorder(temp->right);
            }
    }

void preorder(node *temp)
    {
        if(temp !=null)
            {
                printf("%d",temp->data);
                preorder(temp->left);
                preorder(temp->right);
            }
    }

void postorder(node *temp)
    {
        if(temp!=null)
            {
                postorder(temp-> left);
                postorder(temp->right);
                printf("%d",temp->data);
            }
    }
```

8.10 NON-RECURSIVE FUNCTIONS FOR TRAVERSALS

In this, we discuss the implementation of the three standard traversal of binary tree, which were defined recursively in the last section, by means of non-recursive procedures using stack.

(*a*) Pre-order Traversal:

The procedure for traversing a tree in preorder non-recursively is as:

Initially ptr contains the address of root.
1. Push the address of root node on the stack.
2. Pop an address from the stack.
3. If the popped address is not NULL
   Traverse the node
   Push right child of node on stack
   Push left child of node on stack.
4. Repeat steps 2, 3 until the stack is not empty.

Non-recursive function for preorder traversal is

```
nonrec-pre(struct node *ptr)
    {
        stack[++ top]=ptr;
        while(top!=-1)
            {
                ptr=stack[top--];
                if(ptr!=NULL)
                    {
                        printf("%d", ptr->info);
                        stack[++top]=ptr->right;
                        stack[++top]=ptr->left;
                    }
            }
    }
```

Let us consider a binary tree.

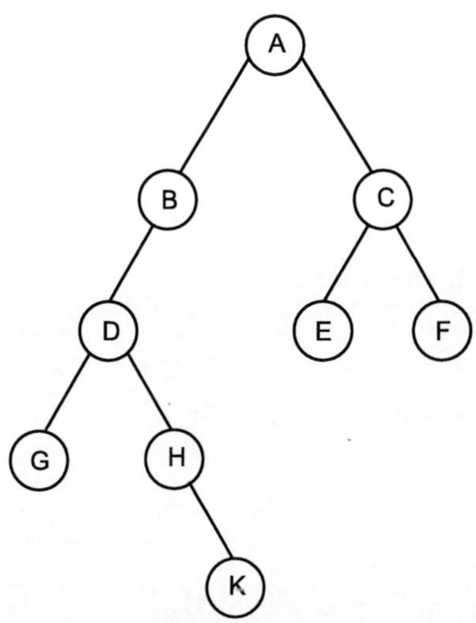

Fig. 8.27.

Initially set ptr = Λ, the root of tree. Push it and then pop it from the stack.

Since PTR ≠ NULL

Traverse the node A.

and push *C* and *B* node on the stack (right and left child). Now pop the top element *B* from the stack.

Stack: | B |
| C |

$$B \neq \text{NULL}$$

Traverse the node B.

and push its right child (No right child) and left child *i.e.,* node *D* on the stack.

Now,

Stack: | D |
| C |

Pop the top element *D* from the stack.

$$D \neq \text{NULL}$$

Traverse the node D.

and push its right child *H* and left child *G* on the stack.

Now, **Stack:** | G |
| H |
| C |

pop the top element *G* from the stack

$$G \neq \text{NULL}$$

Traverse the node G.

and push its right child and left child (No right and left child)

Now, **Stack:** | H |
| C |

pop the top element *H* from the stack.

$$H \neq \text{NULL}$$

Traverse the node *H*.

and push its right child *k* on the stack and its left child (no left child)

Thus, **Stack:** | K |
| C |

Pop the top elemnt *k* from the stack. $k \neq$ NULL

Traverse the node *k*.

and push its right child and left child (No right and left child).

Now, **Stack:** | C |

Pop the top element *i.e, C* from the stack $C \neq$ NULL

Traverse the node *C*.

and push its right child *i.e., F* and left child *i.e., E* on the stack.

Now, **Stack:** | E F |

Pop the top elemnt *i.e., E* from the stack.

Traverse the node *E*.

and push its right and left child (No child).

Pop the top element from the stack *i.e., F*.

Traverse the node *F*.

and push its right and left child (No child).

Now, no element in stack *i.e.,* PTR = NULL and STOP. Thus, final pre order is

| A B D G H K C E F |.

(*b*) Inorder Traversal

The procedure for traversing a binary tree in inorder non-recursively as follows:

1. Repeat step (2) and (3) while stack is not empty or PTR is not equal to NULL.
2. If PTR $\neq$ NULL
   (*a*) push PTR on the STACK.
   (*b*) PRT $\leftarrow$ Left (PTR)
3. If PTR = NULL
   (*a*) POP the element from the STACK.
   (*b*) Traverse the node.
   (*c*) PTR $\leftarrow$ Right (PTR).

The **non-recursive function** for inorder traversal is:

```
void nonrec-inorder (struct node *ptr)
     {
          while (top!=-1 || ptr!=NULL)
               {
                    if(ptr!=NULL)
                         {
                              Stack [++top]=ptr;
                              ptr=ptr->left;
                         }
                    else
                         {
                              ptr=stack[top--];
```

```
                    printf("%d",ptr->info);
                    ptr=ptr->right;
                }
            }
    }
```

Let us consider a binary tree.

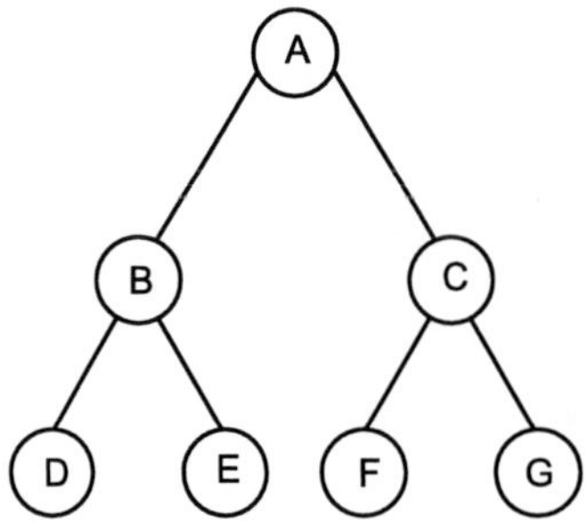

Fig. 8.28.

Initially PTR = Root *i.e.*, node PTR = A.
Push A on the stack.
and now PTR = Left (A) *i.e.*, PTR = B.
Push B on the stack.
and PTR = Left (B) *i.e.*, PTR = D.
Push D on the stack.
and PTR = Let (D) *i.e.*, PTR = NULL.

Now, **Stack:** | D | B | A | and PTR = NULL

Pop the top element from the stack i.e., D.
Traverse the node *D*.
and PTR = Right (*D*) *i.e.*, PTR = NULL.
Again pop the top element from the stack *i.e., B*.
Traverse the node *B*.
and PTR = Right (*B*) *i.e.*, PTR = *E*
Now, PTR ≠ NULL
Push E on the stack.
and PTR = Left (E) i.e., PTR = NULL

Now, **Stack:** | E | A |

pop the top element *i.e.*, E.

Traverse the node E.
and PTR = Right (E) *i.e.*, PTR= NULL
Again pop the top element *i.e.*, A.
Traverse the node A.
and PTR = Right (A) *i.e.*, PTR = C.
PTR ≠ NULL So, push node C on the stack.
Now, PTR = Left (C) *i.e.*, PTR = F.
PTR ≠ NULL So, push node F on the stack.
and PTR = Left (F) *i.e.*, PTR = NULL

Now, **Stack:**

F
C

PTR = NULL.

So, pop the top element *i.e.*, F from the stack.
Traverse the node F.
and PTR = Right (F) *i.e.*, PTR = NULL
Again pop the element *i.e.*, C from the stack
Traverse the node C.
and PTR = Right (C) *i.e.*, PTR = G.
Push G on the stack.

Now, **Stack:**

G

PTR = Left (G) *i.e.*, PTR = NULL.
Pop the element from the STACK.
Traverse the node G.
and PTR = Right (G) *i.e.*, PTR = NULL

Now, stack is empty. So, stop and final in order sequence is | D B E A F C G |.

(c) Post-order Traversal

The procedure for traversing a tree in postorder non-recursively is as:

```
1. Initially ptr has the address of root.
2. Repeat steps (3) to (5) while (ptr! = NULL)
3. Top ← Top + 1
   Stack [Top] ← ptr;
4. If (Right (ptr) ! = NULL)
      {
          Top ← Top + 1
          Stack[Top] ← Right (ptr)
      }
5. ptr ← left (ptr)
6. Pop from the stack.
```

```
7. While (ptr > 0)
      {
         Traverse ptr
         ptr ← stack[Top]
         Top ← Top – 1
      }
8. If ptr < 0 then
      {
         ptr ← right (ptr)
         go to step (2)
      }
9. STOP.
```

The function for non-recursive postorder traversal of tree is as follows:

```
Void nonrec_postorder (struct node * ptr)
    {
        struct node *stack[50];
        int top = - 1;
        int visitd, i visit [10];
    do
        {
            while(ptr!=NULL)
                {
                    printf("%d",ptr->info);
                    stack[++top]=ptr;
                    ptr = ptr->left;
                }
            if(top!=-1)
                {
                    ptr=stack[top--];
                    visitd=0;
                    for(i=0;i<top;i++)
                        if(visit[i]==ptr->right)
                            {
                      visited=1;
                      break;
                   }
                if(visited==1)
                   {
                      printf("%d",ptr->info);
                      ptr=NULL;
```

```
                    }
                else
                    {
                        Stack[++top]=ptr;
                        ptr=ptr->right;
                        visit[ptr]=1;
                    }
                ptr=ptr->right;
            }
        }while(top!=-1 || ptr!=NULL);
    }
```

8.11 CREATION OF BINARY TREE USING TRAVERSALS

In this section, we are creating a binary tree with the help of tree traversals. Here, tree traversals ouput has been given and we need to construct a bianry tree. There are two different ways of creating bianry tree such as:

1. Pre-order and inorder traversals.
2. Post-order and inorder traversals.

8.11.1 Creation of Binary Tree from Pre-order and Inorder Traversals

Here preorder and inorder traversing sequences are given then we need to construct a binary tree. The general procedure for creating a binary tree is as follows:

Step1: Scan the pre-order traversal from left to right.

Step2: For each node scanned locate its position in inorder traversal. Let the scanned node be *X*.

Step 3: The nodex preceding *X* in inorder form its left sub-tree and nodes succeeding it form right sub-tree.

Step 4: Repeat step 1 for each symbol in the preorder.

EXAMPLE: Construct the binary tree given the following traversals.

Pre-order: A B D H E C F G

In-order: D H B E A F C G

Solution: (1) In pre-order traversal root is the first node. So, A is the root node of the binary tree, *i.e.*,

2. Now, we can find the node of left sub-tree and right sub-tree with inorder sequence.

Nodes which are in the left side of the root node in inorder sequence are nodes of left sub-tree and nodes which are in the right side of the root node in inorder sequence are nodes of right sub-tree.

So,

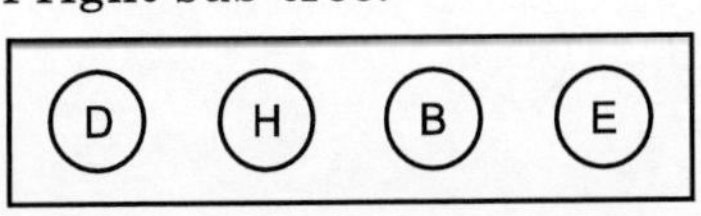

Left sub-tree

and 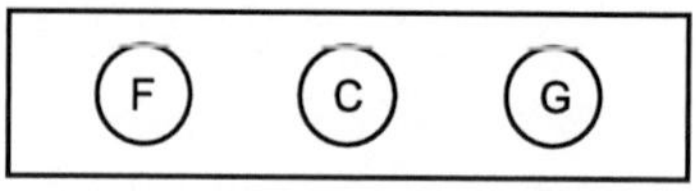

Right sub-tree

3. Now, the left child of the root node will be the first node in the pre-order sequence after root node *A i.e., B*. Hence *B* is the left child of *A*. Similarly the right child of root *A* will be the first node after nodes of left subtree in preorder sequence. Hence *C* is the right child of A.

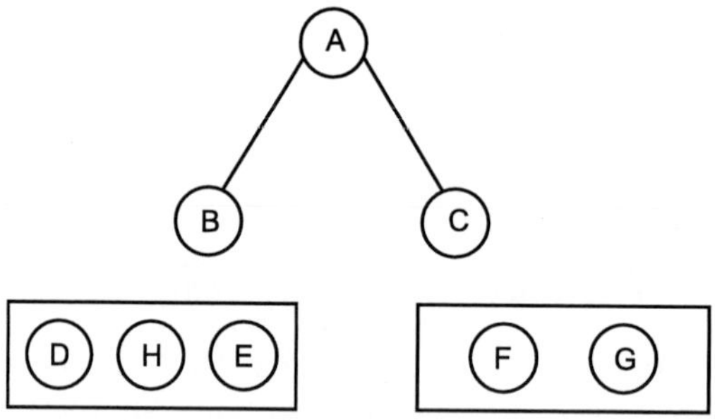

Fig. 8.29.

4. In inorder sequence, *D* and *H* are on the left side of *B* and *E* is on the right side of *B*. So *D* and *H* will form left subtree and *E* will be in the right sub-tree of *B*.

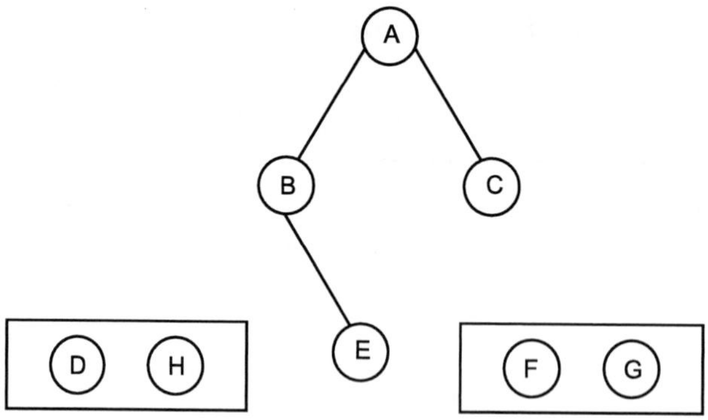

Fig. 8.30.

5. In inorder sequence, *F* is on the left of *C* and *G* is on the right side of *C*. So *F* will be in left subtree of *C* and *G* in right sub-tree of *C*. So,

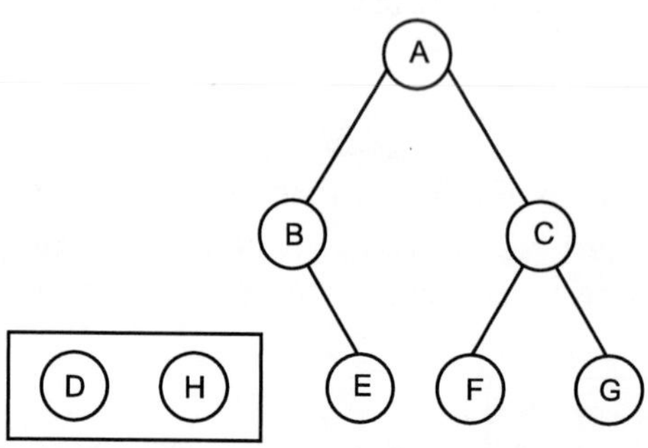

Fig. 8.31.

6. Again, in pre-order traversal, D is coming before H. Hence D is the root of left sub-tree of B. To find out whether H is in left or right sub-tree of D. We look at inorder sequence. H is in right side of D. So, H will be in right sub-tree of D. Thus, final Binary tree is

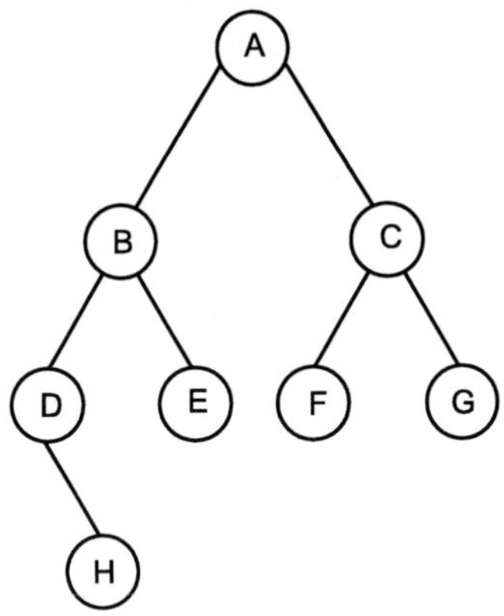

Fig. 8.32.

EXAMPLE 8.4. Draw a binary tree.

Pre-order:	**F**	**A**	**E**	**K**	**C**	**D**	**H**	**G**	**B**
Inorder:	**E**	**A**	**C**	**K**	**F**	**H**	**D**	**B**	**G**

Solution: Root node is (F)

Inorder E A C K (F) H D B G
Left subtree — Right subtree

i.e.,

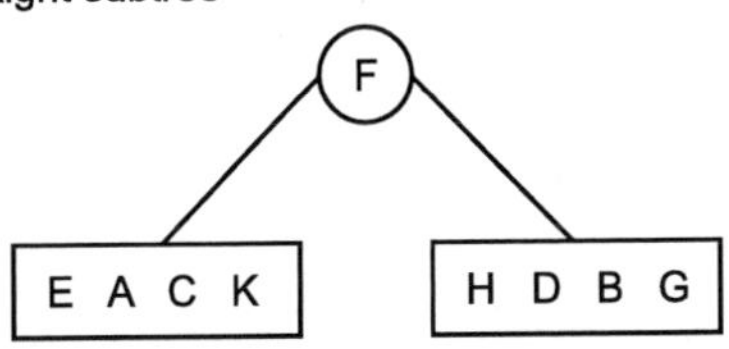

Fig. 8.33.

Now, Inorder E (A) C K F H D B G

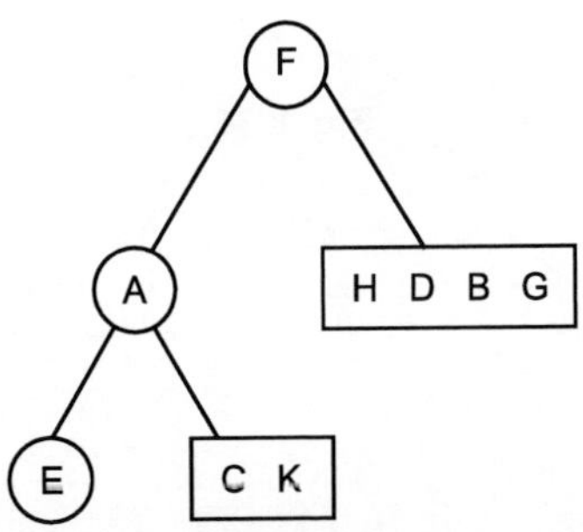

Fig. 8.34.

Now, E A (C) K F H D B G

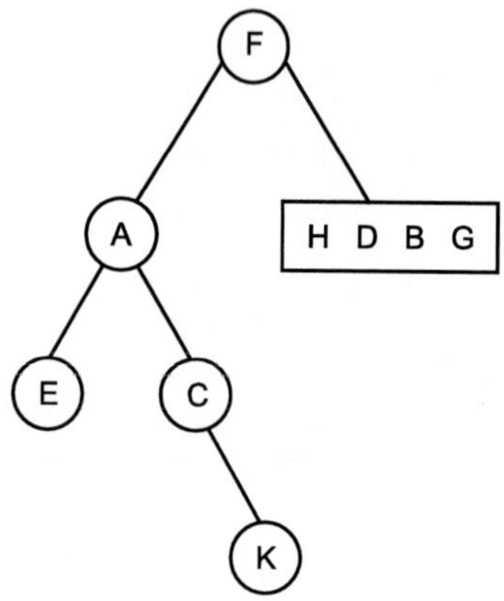

Fig. 8.35.

Now, **D:** Inorder E A C K F H (D) B G

From pre-order

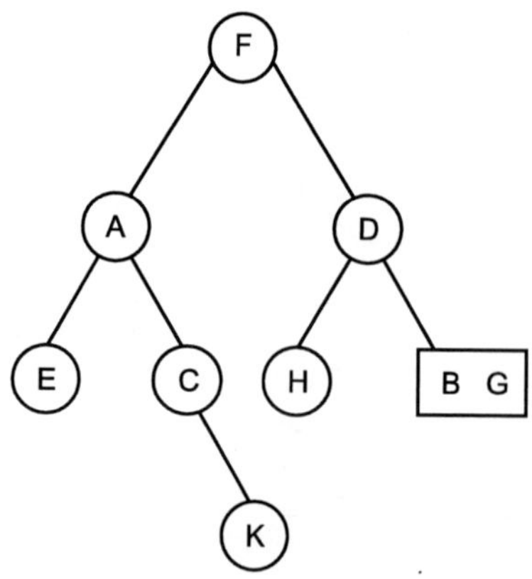

Fig. 8.36.

Now G,

Inorder E A C K F H D B (G)

Thus, final binary tree.

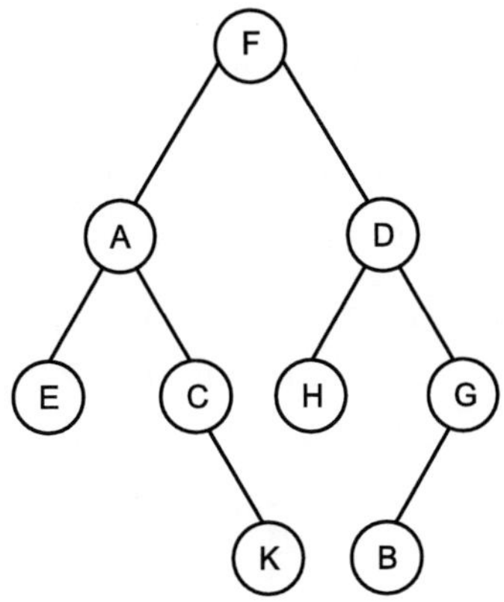

Fig. 8.37.

EXAMPLE 8.5. Create a binary tree.

In-order: B C A E G D H F I J

Pre-order: A B C D E G F H I J

Solution: *A* is the root node (First element in pre-order).

Then Inorder sequence

i.e.,

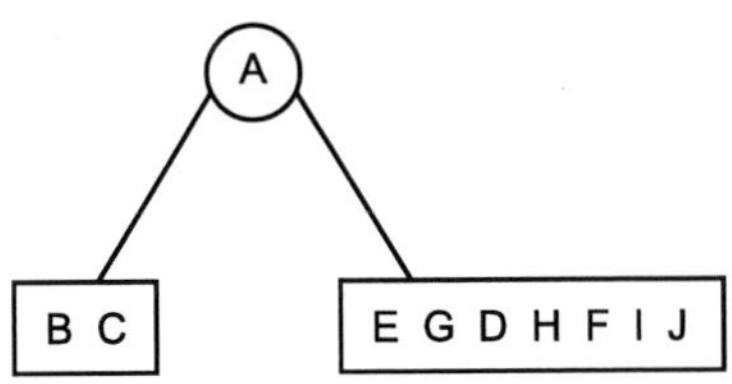

then,

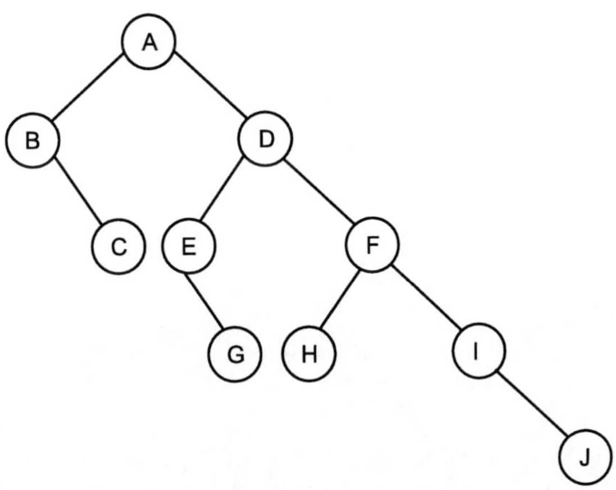

Fig. 8.38. Final Binary tree.

8.11.2 Creation of Binary Tree from Post-order and In-order Traversals

The general method for creating bianry tree using post order and in order traversal is as follows:

1. Scan the post-order traversal from right to left.
2. For each node scanned locate its position in inorder traversal. Let the scanned node be x.
3. The node preceding x in inorder, form its left subtree and nodes succeding it form right sub-tree.

EXAMPLE 8.6. Create a binary tree.

Post-order: H D I E B J F K L G C A

Inorder: H D B I E A F J C K G L

Solution: In postorder traversal root is the last node. Hence A is the root of the binary tree.

Now, from inorder sequence.

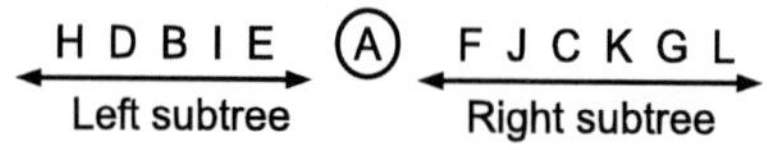

i.e.,

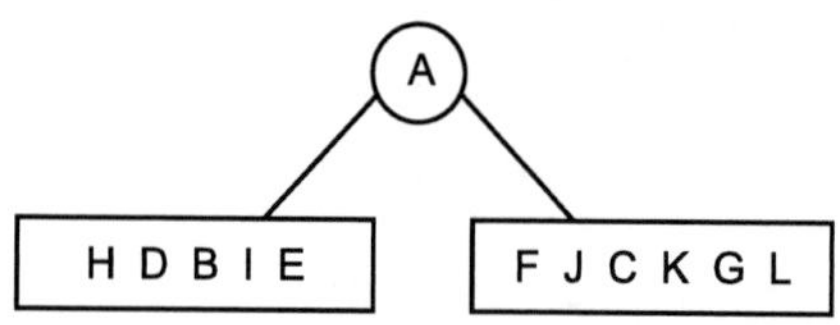

Fig. 8.39.

Now right child of A will be the node which comes just before, node A *i.e.,* node C, and left child of A will be the first node before nodes of right subtree in postorder traversal *i.e.,* node B.

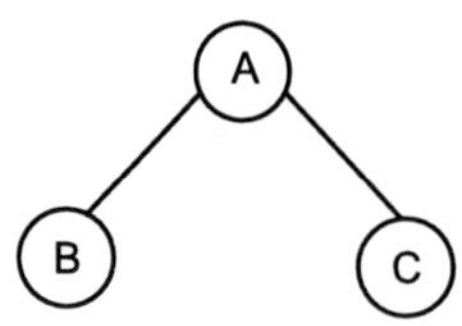

Fig. 8.40.

Now From inorder traversal.

and,

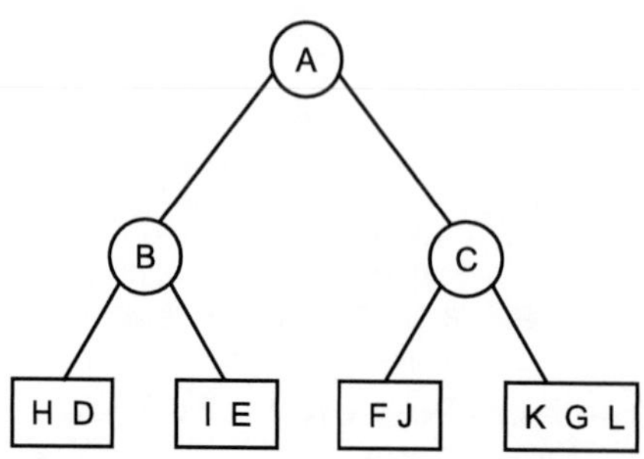

Fig. 8.41.

Now look at post-order traversal, node just before *B* is *E*, so *E* is the right child of *B* and node just before *C* is *G* hence *G* is right child of *C*.

And *D* is the first node before the nodes of right sub-tree of *B*. Hence *D* is the left child of *B* and *F* is the first node before nodes of right subtree of *C*. Hence *F* is left child of *C*.

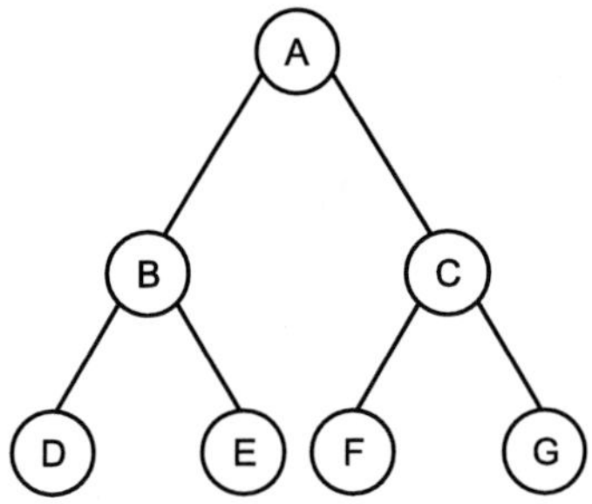

Fig. 8.42.

Now from inorder traversal, *H* is the left of *D* hence it is left child of *D*, *I* is to the left of *E*, hence it is left child of *E*. *J* is to the right of *F* hence it is right child of *F*, *K* is to the left of *G* and *L* to the right of *G* hence *K* is left child of *G* and *L* is right child of *G*.

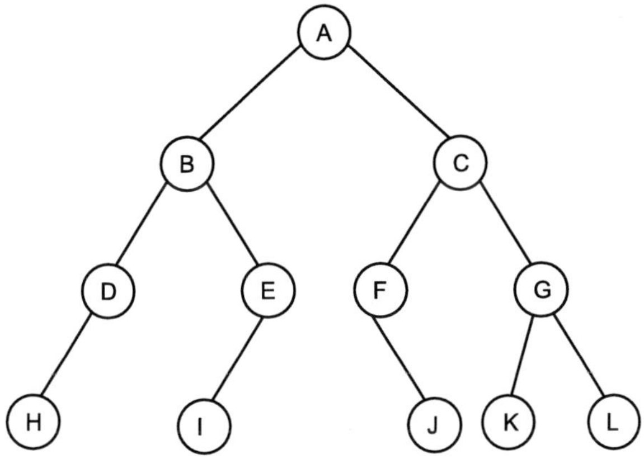

Fig. 8.43.

EXAMPLE 8.7. If the in-order traversal of a binary tree is *B, I, D, A, C, G, E, H, F* and its post-order traversal is *I, D, B, G, C, H, F, E, A*. Determine the binary tree.

Solution:

Post-order: I D B G C H F E (A)

In-order: [B I D] (A) [C G E H F]

A is the last node in post-order traversal hence it is the root node.

Now insert *E* this is the right child of node *A* and *B* is the left child of node *A* because this is the first node before nodes of right subtree in post order *i.e.,*

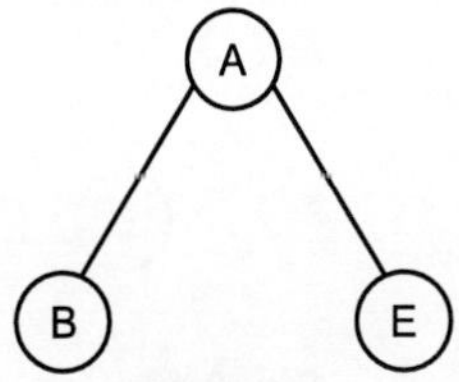

Fig. 8.44.

Now from inorder traversal.

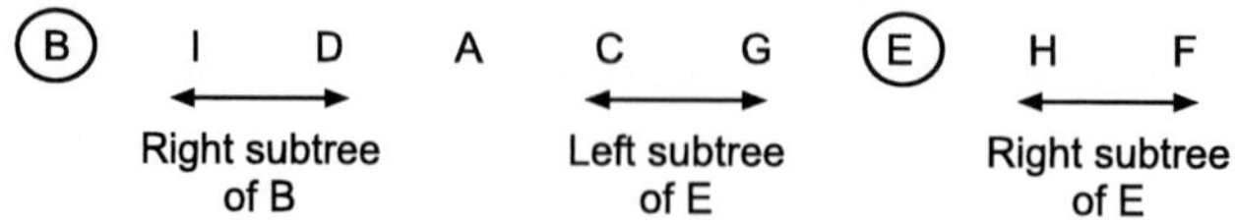

i.e.,

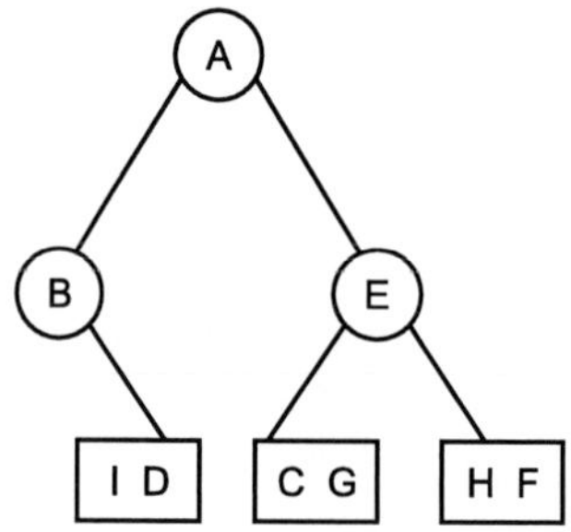

Fig. 8.45.

Now look at postorder traversal node just before *B* is *D*. So *D* is the right child of *B* and *F* is the right child of *E* and *C* is the left child of *E*.

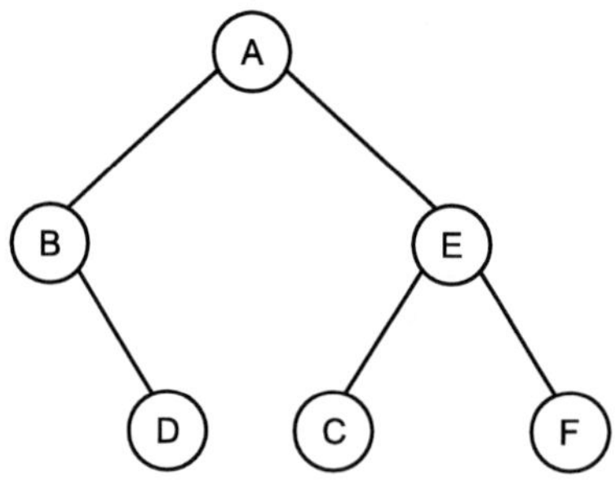

Fig. 8.46.

Now look at inorder traversal, we get

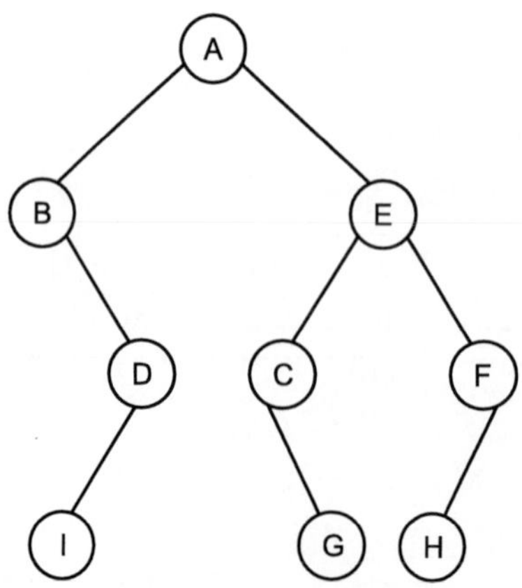

Fig. 8.47.

8.12 CONVERTING ALGEBRAIC EXPRESSION INTO BINARY TREE

The arithmetic expressions represented as binary trees are known as **expression trees**. In this, the root node is operator and the left and right children are operands. In case of unary operators the left child is not present and the right child is the operand.

Fig. shows a tree for the expression.

$$(A + B) * (C + D)$$

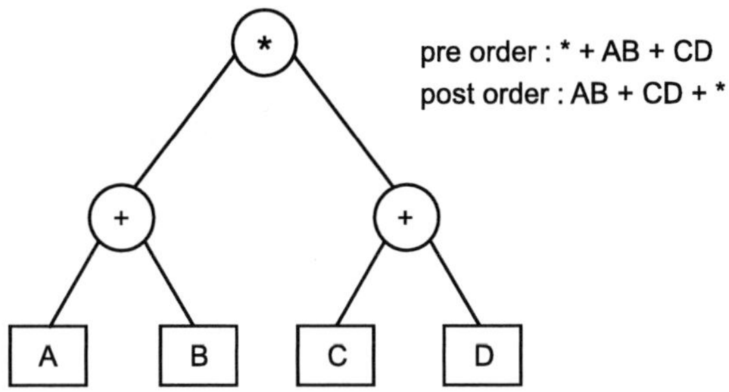

Fig. 8.48.

Generally, nodes appearing in circular shape denote the operators and nodes appearing in square shape are used for operands. Note that once the arithmetic expression tree is created for a given infix expression, the traversing the tree in preorder results prefix expression and traversing the tree in postorder results an postfix expression.

EXAMPLE 8.8. Draw the binary tree to represent the following expression.

$$(5 + 4 * (6 - 7))/(5 + 8)$$

Solution:

1. First consider the inner most parenthesis *i.e.*, (6 – 7)

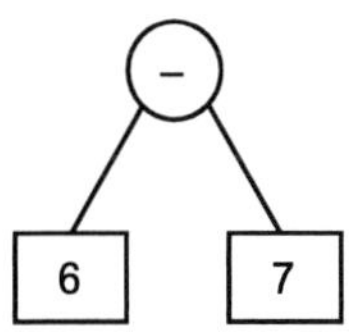

Fig. 8.49.

2. Next consider the next inner most parenthesis *i.e.*,

$$(5 + 4 * (6 - 7))$$

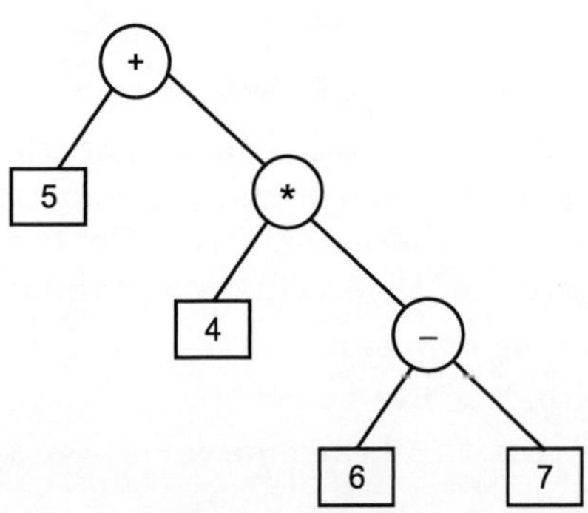

Fig. 8.50.

Now, (5 + 8) as

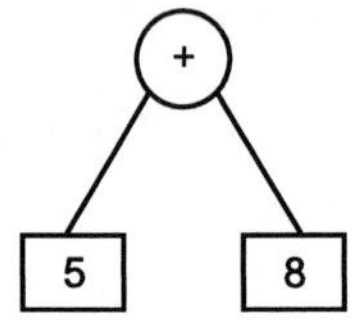

and,

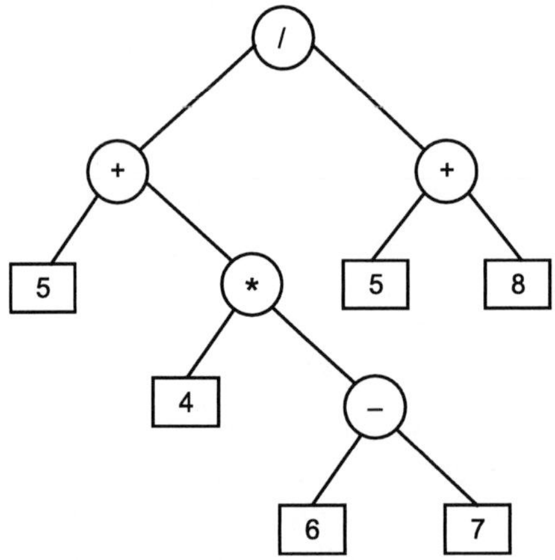

Fig. 8.51.

EXAMPLE 8.9. Construct a bianry tree for the algebraic expression

$$E = (a - b)/((c * d) + e)$$

Solution:

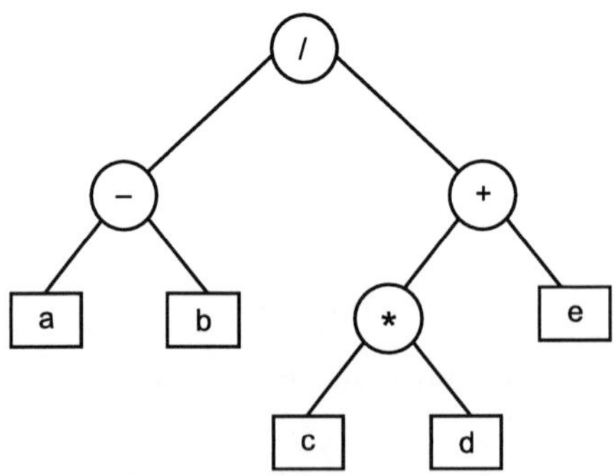

Fig. 8.52.

Note: We get the infix form of algebraic expression when the tree is traversed in inorder. But the parentheses present in the original arithemetic expresion cannot be obtained back from the inorder traversal of the binary tree. This is because the structure of the binary tree itself decides the order of the evaluation of the tree.

EXAMPLE 8.10. For the following expression

$$((a * x + b) * x + e) * x + f$$

Construct the expression tree. Find the equivalent post fix notation.

Solution:

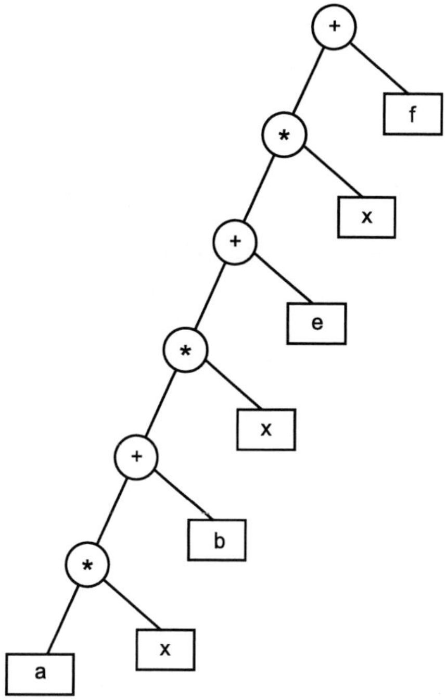

Fig. 8.53.

Post fix notation.

$$\boxed{a\,x * b + x * e + x * f +}$$

8.13 EXTENDED BINARY TREE

A binary tree is called extended binary tree if every node of tree has zero or two children. The extended tree is also known as a **2-tree.**

A binary tree can be converted to an extended binary tree by adding new nodes to its leaf nodes and to the nodes that have only one child. The nodes of the original tree are called **internal nodes** and the new nodes that are added to binary tree, to make it an extended bianry tree are called **external nodes.**

Fig. shows how extended binary tree can be obtained by adding new nodes to the original binary tree.

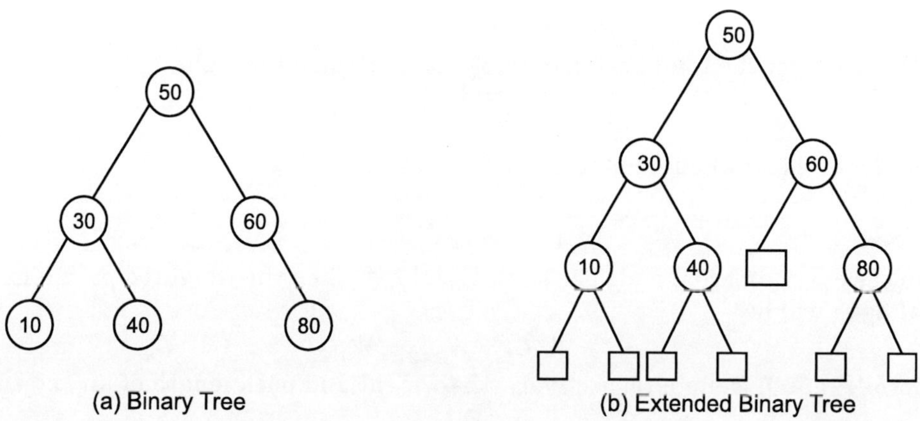

Fig. 8.54.

In Fig. (*b*) square represents the extended (external) nodes in the original binary tree.

Extended binary tree is very useful for representing algebraic expression, because in algebraic expression left and right child represent the operand and parent of children represents the operator.

8.14 HUFFMAN TREE

As we have seen earlier that an extended Binary tree is a binary tree in which every node has zero or two children. The nodes which have two children are called internal nodes and which have no children are called external nodes.

In any 2 Tree the number of external nodes is 1 more than the number of internal nodes *i.e.*,

$$\boxed{E = I + 1}$$

Wher E is the number of external nodes and I is the number of internal nodes.

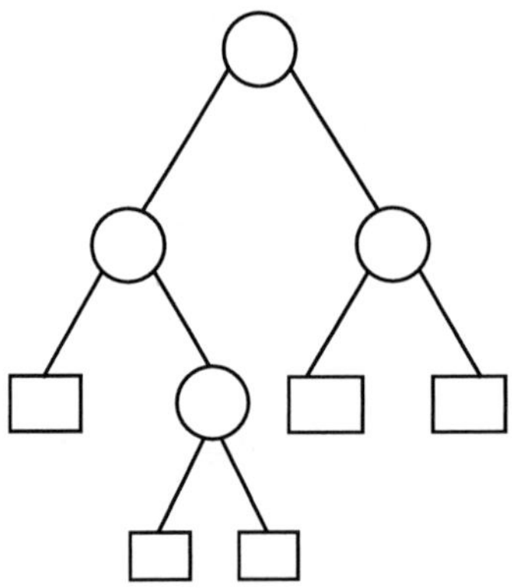

Fig. 8.55.

$$\text{Internal nodes} = 4$$
$$\text{External nodes} = 5$$

We know, the **path length** for any node is the number of minimum nodes traversed from root to that node. With this in mind, we define the External path length. L_E of a 2 Tree to be the sum of all path lengths *i.e.*,

$$L_E = 2 + 3 + 3 + 2 + 2 = 12$$

Similarly total path length for internal nodes are

$$L_I = 0 + 1 + 2 + 1 = 4$$

We can also get the total path length of external node through

$$\boxed{L_E = L_I + 2n}$$

Where n is the total number of nodes in the tree.

Here, $$L_E = 4 + 2 * 4$$
$$= 4 + 8 = 12$$

Suppose every external node has same weight W, then the weighted path length for the external node will be

$$P = W_1L_1 + W_2L_2 + W_3L_3 + \ldots.. + W_nL_n$$

Where W_i and L_i denote respectively, the weight and path length of an external node.

Suppose we create different trees which have same weigths on external nodes then it is not necessary that they have same weighted path length.

Let us take the weights 2, 7, 9, 10 and create three different trees.

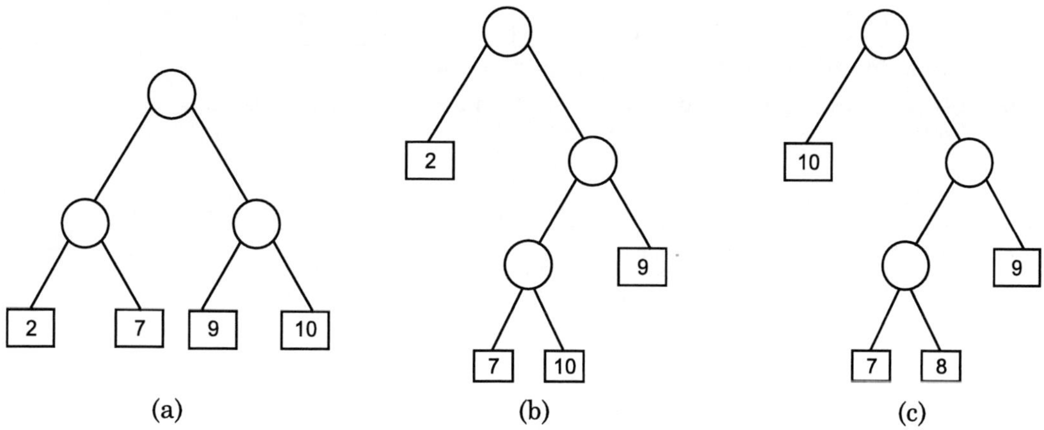

Fig. 8.56.

$$\begin{cases} P_1 = 2*2+7*2+9*2+10*2 = 4+14+18+20 = 56 \\ P_2 = 2*1+7*3+10*3+9*2 = 2+21+30+18 = 71 \\ P_3 = 10*1+7*3+8*3+9*2 = 10+21+24+18 = 73 \end{cases}$$

We can see that three different trees have different path lengths even same tree of (*b*) and (*c*) have different path lengths. So the problem arises for obtaining a unique tree which has minimum path length.

The general problem that we want to solve is as follows. Suppose a list of n weights is given:

$$W_1, W_2, \ldots.., W_n$$

Among all the 2-trees with *n* external nodes and with the given *n* weights, find a tree *T* with a minimum weighted path length. Huffman gave an algorithm to find such a tree.

Huffman Algorithm:

1. Suppose, there are n weights W_1, W_2, ..., W_n.
2. Take two minimum weights among the n given weights. Suppose W_1 and W_2 are first two minimum weights then sub-tree will be:

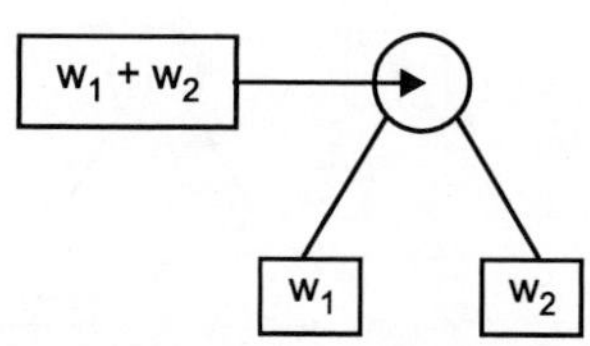

3. Now the remaining weights will be $W_1 + W_2$, W_3, W_4, ... W_n.
4. Create all sub-tree at the last weight.

Thus, Huffman algorithm constructing the tree from the bottom up rather than from the top down approach.

EXAMPLE 8.11. **Suppose *A*, *B*, *C*, *D*, *E*, *F* and *G* are 7 elements with weights as follows:**

Item	**A**	**B**	**C**	**D**	**E**	**F**	**G**
Weights	**15**	**10**	**5**	**3**	**7**	**12**	**25**

Create an extended binary tree by Huffman algorithm.

Solution: Step 1: Taking two nodes with minimum weights *i.e.,* 3 and 5.

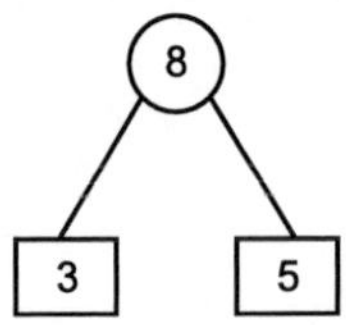

Fig. 8.57.

Now elements in the list are:

15 10 8 7 12 25

Step 2: Taking two minimum weight nodes which are 8 and 7 *i.e.,* now.

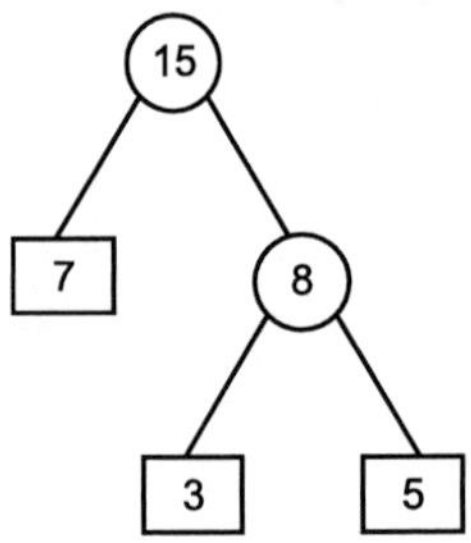

Fig. 8.58.

Now elements in the list are:

15 10 15 12 25

i.e.,

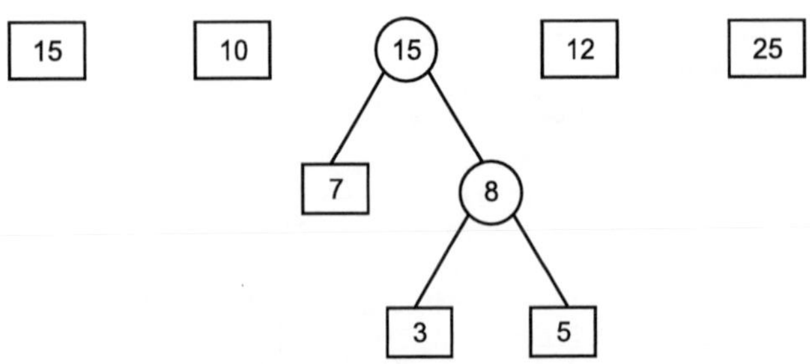

Fig. 8.59.

Step 3: Taking two minimum weighted nodes *i.e.,* 10 and 12.

Now,

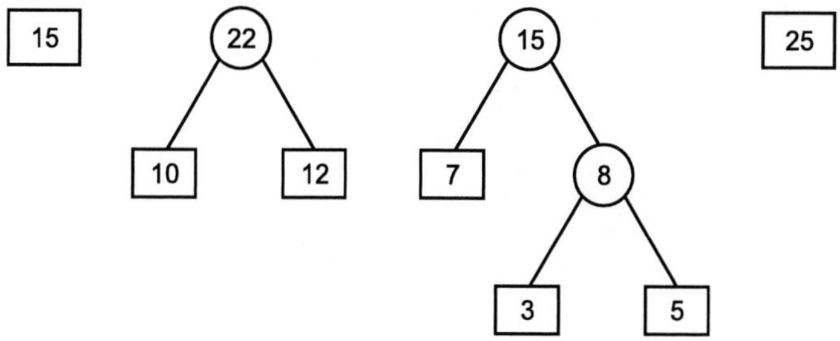

Fig. 8.60.

i.e., elements in the list are

15, 22, 15, 25.

Step 4: Taking two minimum weighted nodes i.e., 15 and 15.

Now,

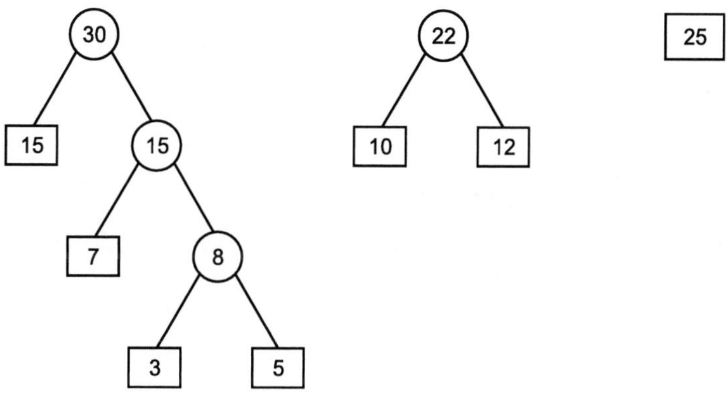

Fig. 8.61.

i.e., elements in the list are

30, 22, 25.

Step 5: Taking two nodes with minimum weights which are 22 and 25.

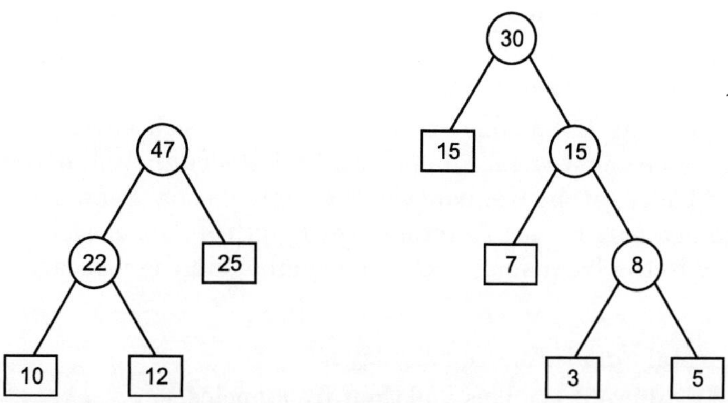

Fig. 8.62.

Now the elements in the list are 47 and 30

Step 6: Taking two nodes with minimum weights which are 30 and 47.

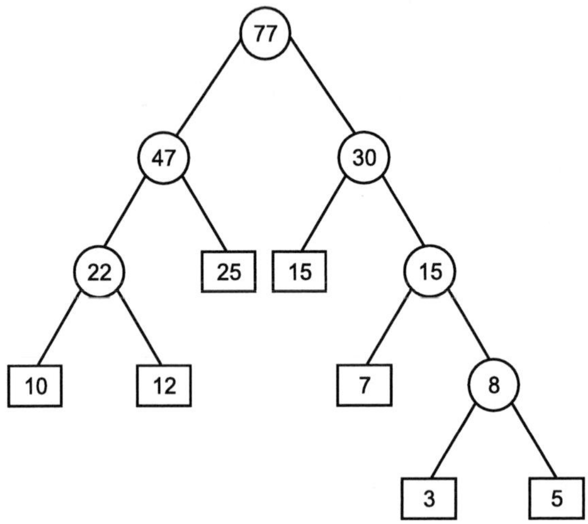

Fig. 8.63 Huffman Tree

8.14.1 Application of Huffman Algorithm

The Huffman algorithm is used to perform the encoding and decoding of a long message consisting of a set of symbols. Suppose we want to send a message there are two options either it sends data as fixed size or to send it as variable length size.

Suppose a collection of n data items A_1, A_2, ..., A_n; are to be coded by means of strings of bits. One way to do this is to code each item by an r-bit string where,

$$2^{n-1} < n \leq 2^r$$

E.g., a 48-character set is frequently coded in memory by using 6-bit strings. One cannot use 5-bit strings since $2^5 < 48 < 2^6$.

Suppose the data items do not occur with the same probability. Then memory space may be conserved by using variable-length strings, were items which occur frequently an assigned shorter strings and items which occur infrequently are assigned longer strings.

Now, we discuss a coding technique using variable-length string that is based on the Huffman Tree for weighted data item.

A **Huffman tree** has to be constructed to perform both encoding and decoding. The Huffman tree has to be constructed to perform both encoding and decoding. The Huffman tree is constructed based on the frequency of the symbols of a given message. The main aim of using the Huffman algorithm is to minimise the length of the encoded message by assigning a shorter bit string to the frequently occurring symbols in the message.

Algorithm

1. Accept the total number of symbols, say *n*.
2. Accept the *n* different symbols and their frequencies.
3. Accept the message to be encoded.
4. Find the two symbols from the given set of symbols such that they occur less frequently.

5. Assign codes (it can be n bits where n can be as minimum as possible) to these chosen symbols such that the codes differ in their last bit positions. One symbol will have 0 as the last bit, the other symbol will have 1 as the last bit.
6. Combine these two chosen symbols by forming a parent node from the nodes corresponds to the chosen symbols. The frequency of the parent node is the sum of the frequencies of the nodes corresponding to the chosen symbols. Now this formed parent node can be treated as a symbol but it will be an internal node.
7. Now choose the second symbol such that its frequency is minimum.
8. Repeat steps 5, 6 and 7 as long as there are some symbols for that node that are not formed.

Once the Huffman tree is formed, the code of any symbol can be formed by starting at that node and climbing up to the root node. Initially the code for that symbol is NULL. If a left branch is climbed, 0 is appended at the beginning of the code. If a right branch is climbed, 1 is appended at the beginning of the code.

Huffman Coding

Huffman follows a bottom-up approach. Procedure which we adopted is as follows :

1. Form the frequency list of all the symbols in the descending order.
2. Locate the two symbols in the list with the lowest frequencies. Frequencies referred as weights.
3. Create a parent node for these two nodes whose weight is equal to the sum of weights of two child nodes.
4. Remove the two child from the list and add newly created parent node to the list alongwith the weight.
5. Repeat thorugh the step (2) until only one node is left in the tree.

EXAMPLE:

Symbol	Frequency
A	19
E	11
I	9
O	7
U	7

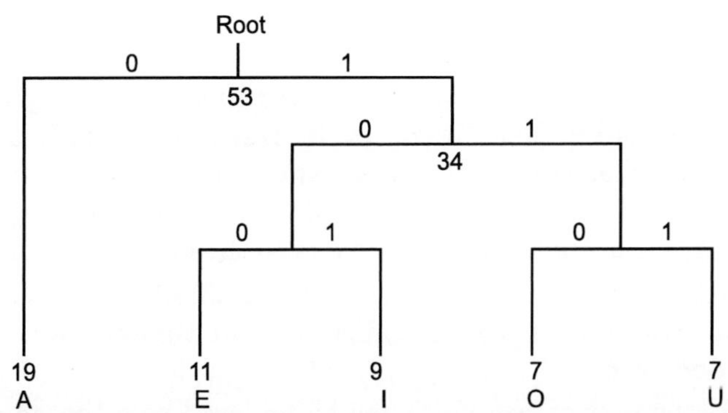

Thus, by using this technique, the codes for all the symbols are given in the following table.

Symbol	*Codes*
A	0
E	100
I	101
O	110
U	111

To perform the decoding operation, just traverse the nodes from the root node towards the leaf nodes of the Huffman tree such that the given code bits are traversed. The important features of the Huffman tree are:

1. The Huffman tree is a binary tree.
2. In the Huffman tree, the most frequently occuring symbols will be the leaf nodes near the root node whereas the least frequently occuring symbols will be farther away from the root node.
3. The most frequently occuring symbols will have smaller code bits whereas, the least frequently occuring symbols will have more code bits.
4. The root node of the Huffman tree is not assigned any code. The left child of the root node is assigned the code 0. The right child of the root node is assigned the node 1.
5. All the symbols of the given message will be leaf nodes of the Huffman tree.
6. The code for the child nodes of the same parent differ only in the last bit. The last-bit is 0 for the left-child node and is 1 for the right child node.

The Huffman tree finds applications wherever the messages have to be transmitted and received with less number of code bits.

8.15 THREADED BINARY TREES

Consider the linked representation of any binary tree, we notice that, half of the entries in the pointer fields LEFT and RIGHT will contain null entries. This space may be more efficiently used by replacing the null entries by special pointers, called **Threads,** which point to the nodes higher in the trees. Such trees are called **Threaded Trees.**

Note: A binary tree with n nodes, when $n \geq 0$ has exactly $n + 1$ null links.

Threads in a binary tree must be distinguished from normal pointers. In the graphical representation of a threaded binary tree, the threads are shown by dotted lines. In computer memory, an extra field, called tag or flag is used to distinguish a thread from a normal pointer. Trees can be threaded using one-way threading or two-way threading. In a one-way threading a therad will appear in the **right field** of the node and will point to the successor node in the inorder traversal of tree; and in two-way threading of tree a thread will also appear in the **left field** of a node and will point to the preceding node in the inorder traversal of tree. So, there are many ways to thread a bianry tree. These are:

1. The right NULL pointer of each node can be replaced by a thread to the successor of that node under in-order traversal called a **right thread,** and the tree will called a **right threaded tree.**
2. The left NULL pointer of each node can be replaced by a thread to the predecessor of that node under in-order traversal called a left-thread and the tree will called a **left-threaded tree**.

3. Both left and right NULL pointers can be used to point to predecessor and successor of that node, respectively, under in-order traversal. Such a tree is called a **fully threaded tree.**

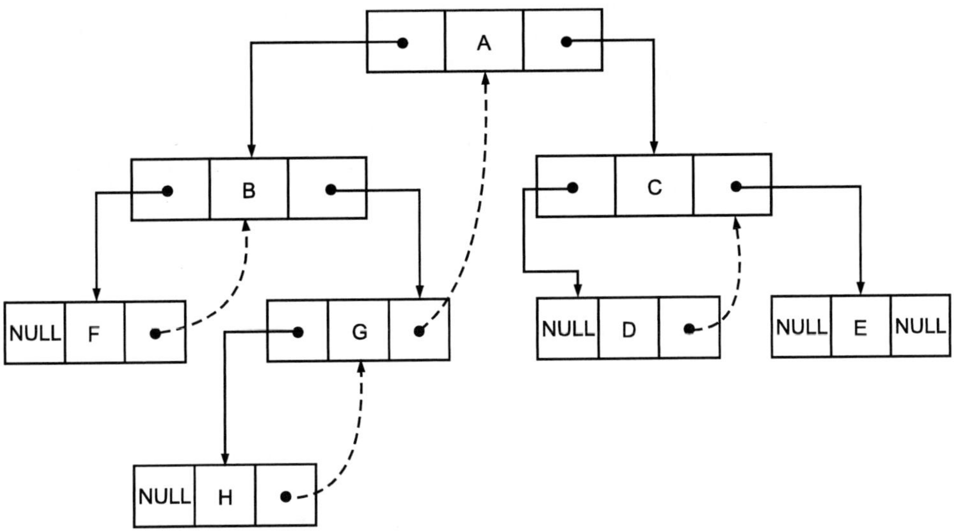

(a) Right threaded binary tree

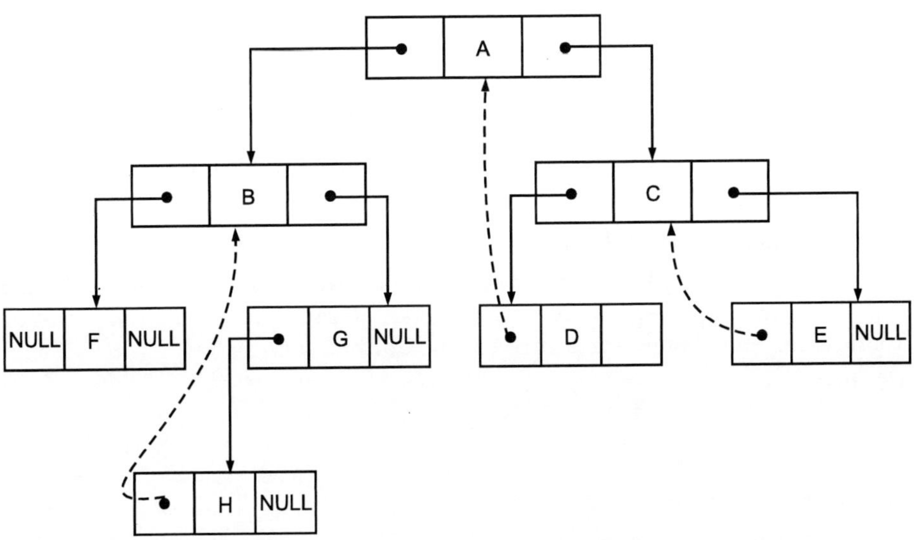

(b) Left threaded binary tree

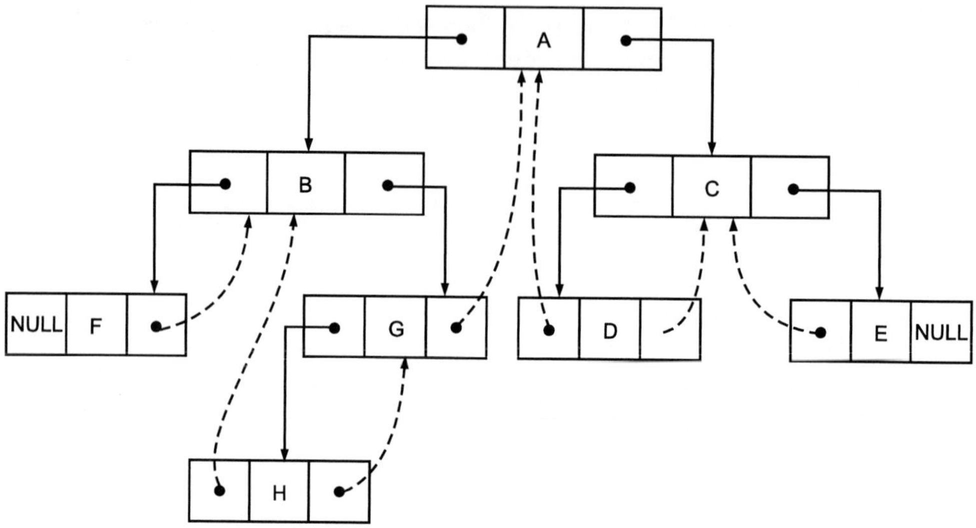

(c) Fully threaded binary tree

Fig. 8.64.

To implement a **right threaded binary tree** an one extra field **rthread** is used. If rthread is assigned to 1 then its corresponding right link represents a thread and if rthread is assigned to 0 then right link represents an ordinary link connecting to the right sub-tree. The *C* node structure can be defined as:

```
struct node
    {
        char info;
        struct node * left;
        struct node * right;
        int rthread;  /*1 indicate threaded link
                              0 indicate ordinary link */
    };
typedef struct node * NODE;
```

Similarly to implement a **left threaded binary tree,** we need one extra field **lthread**. It is assigned to 1 indicate the presence of thread and 0 indicate an ordinary link connecting to the left sub-tree. The *C* node structure can be defined as:

```
struct node
    {
        char info;
        struct node * left;
        struct node * right;
        int lthread;
    }
typedef struct node * NODE;
```

To implement **fully threaded binary tree,** we need two extra field of lthread and rthread as defined earlier are used and the *C* node structure can be defind as:

```
struct node
     {
          char info;
          struct node * left;
          struct node * right;
          int lthread;
          int rthread;
     };
typedef struct node * NODE;
```

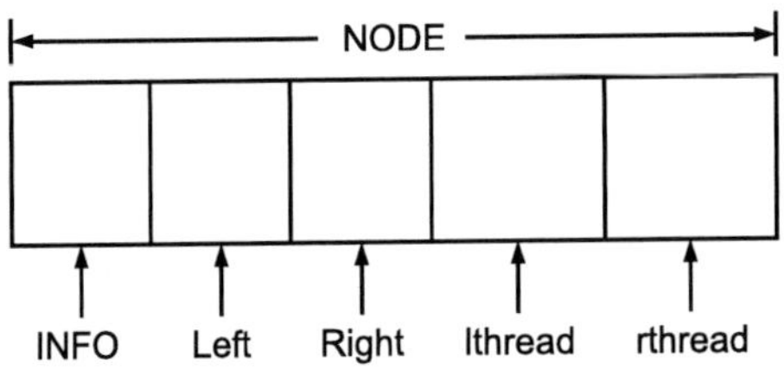

Fig. 8.65 Node of a threaded binary tree

Now, we have seen that if left or right pointer of a node is NULL then it is made a thread pointing to the inorder predecessor or inorder successor of the node. First node in inorder traversal has no predecessor and last node has no successor, so the left pointer of the left most node which is the first node in order traversal and the right pointer of the right most node which is the last node in inorder traversal contains NULL. So we can take a dummy node called the **header node** and we will represent are tree as the left sub-tree of this header node. Left pointer of this header node will point to the root node of our tree. When our tree will be empty then left pointer of this header node will be a thread ponting to itself.

So, now the left most node and right most node of our two-way threaded binary tree will not contain NULL, they will contain threads pointing to this header node.

The structure of a node in a threaded binary tree will be as:

```
typedef enum{thread, link}boolean;
struct node
     {
          struct node * leftptr;
          boolean left;
          int info;
     struct node * rightptr;
          boolean right;
     };
```

Here we have taken two boolean members left and right to differentiate between a thread and link. These members can take values thread or link.

If left = link pointer leftptr points to the left child of the node.
If left = thread pointer leftptr is a thread pointing to inorder predecessor of the node.
If right = link pointer rightptr points to the rightchild of the node.
If right = thread pointer rightptr is a thread pointing to inorder successor of the node.

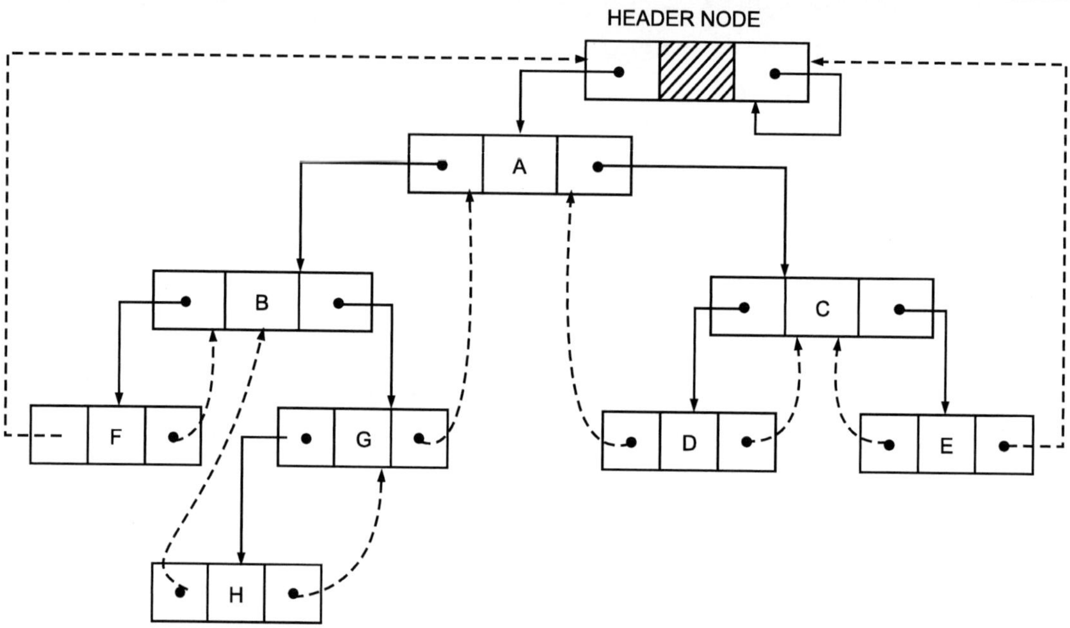

Fig. 8.66.

8.16 TRAVERSING IN THREADED TREE

(*a*) Inorder traversal in threaded binary tree

If the tree is right threaded then we can traverse it in inorder without the use of stack or recursion. In inorder traversal the left most node of the tree is traversed first of all. So first we traverse the leftmost node of the tree and then we find the inorder successor of each node and traverse it. As we know that the right most node of the tree is the last node in inorder traversal and its right pointer is a thread pointing to the header node, hence we wil stop when we reach header node.

```
inorder()
    {
        struct node*ptr;
        if (head->leftptr==head)
            {
                printf(Tree is empty)
                return;
            }
        ptr=head->leftptr;
```

```
            while(ptr->left==link)
                 ptr=ptr->leftptr;
            printf("%d",ptr->info);
   while(1)
        {
            ptr=inorder_sucessor(ptr);
            if(ptr==head)
                 break;
            printf("%d",ptr->info);
        }
   }
```

Pre-order Traversal

In pre-order traversal, we will start traversing from the left child of the header node. If the node has a left child then that left child will be traversed, otherwise if the node has a right child, then that right child will be traversed.

If the node has neither left child nor right child then with the help of right threads we will reach that inorder successor of the node which has a right sub-tree. Now this sub-tree will be traversed as pre-order. We will continue our process until we reach the header nope.

```
   preorder()
        {
             struct node * ptr;
             if(head->leftptr=head)
                  {
                       printf("Tree is empty");
                       return;
                  }
             ptr=head->leftptr;
             while(ptr!=head)
                  {
                       printf("%d", ptr->info);
                       if(ptr->left==link)
                            ptr=ptr->leftptr;
                       else
                            if(ptr->rightptr==link)
                                 ptr=ptr->rightptr;
                            else
                                 {
                                        while(ptr!=head&&ptr->right==thread)
                                             ptr=ptr->rightptr;
                                        if(ptr!=head)
```

```
                    ptr=ptr->rightptr;
                }
        }
}
```

8.17 INSERTION AND DELETION IN A THREADED BINARY TREE

(a) Insertion in a threaded binary search tree

As in binary search tree here also we will find the exact location where we have to insert the new node.

The new node will be inserted as a leaf node so its left and right pointers both will be threads.

```
tmp->leftptr=thread;
tmp->rightptr=thread;
```

Case 1: When the tree is empty.

When the tree is empty the left pointer of the head node is a thread pointing to itself. We will insert the new node as the left child of header node so now the left pointer of the head will be a link pointing to the new node. We know that the left pointer of first node in inorder traversal and the right pointer of last node in inorder traversals are threads pointing to the header node. Here we have only one node which is the first and the last node so its left right pointers will be threads pointing to the header node.

```
head->left=link;
head->leftptr=tmp;
tmp->leftptr=head;
tmp->rightptr=head;
```

Case 2: When new node inserted as the left child of its parent.

The threads of new node will point to its inorder prodecessor and successor. The node which was inorder predecessor of the parent is now the inorder predecessor of this node. The inorder successor of this node is its parent.

```
tmp->lchild=parent->lchild;
tmp->rchild=parent;
```

The parent of new node has a thread in its left pointer pointing to its predecessor, but after insertion its left pointer will be a link pointing to the new node..

```
parent->left=link;
parent->lchild=tmp;
```

Case 3: When new node is inserted as the right child of its parent.

The node which was inorder sucessor of the parent is now the inorder sucessor of this node. The inorder predecessor of this node is its parent. So the left and right threads of the new node will be:

```
tmp->leftptr=parent;
tmp->rightptr=parent->child;
```

The parent of new node had a thread in its right pointer pointing to its successor, but after insertion its right pointer will be a link pointing to the new node.

```
parent->right=link;
parent->rchild=tmp;
```

(b) Deletion from a threaded binary search tree

As in binary search tree here also we have five possibilites while performing delete operation

1. There are no nodes in the tree *i.e.,* the tree is empty.
2. Node to be deleted is not present in the tree.
3. Node to be deleted is leaf node *i.e.,* it has no children.
4. Node to be deleted has only one child.
5. Node to be deleted has two children.

The last three possibilities require the manipulation of pointers so we will discuss them.

Case A: Node to be deleted is leaf node.

If the node to be deleted is the root node of the tree then the tree will become empty after its deletion so then left pointer of head will be a thread pointing to itself.

```
head->left=thread;
head->leftptr=head;
```

If the node to be deleted is a left leaf node then the left pointer of parent will become a thread pointing to its inorder predecessor. Initially its inorder predcessor was its left child but now its inorder predecessor will be that node which was predecessor of its left child.

```
par->left=thread;
par->leftptr=loc->leftptr;
```

If the node to be deleted is a right leaf node then the right pointer of parent will become a thread pointing to its inorder successor. Initially its inorder successor was its right child but now its inorder successor will be that node which was successor of its right child.

```
par->right=thread;
par->rightptr=loc->rightptr;
```

Case B: Node to be deleted has one child.

Delete the node as in binary search tree. Find the inorder successor and inorder predecessor of the node to be deleted.

```
S = inorder_successor(loc);
p = inorder_precessor(loc);
```

If node has right subtree then put the predecessor of the node in the left pointer of its successor.

```
if(loc->right==link)
        s->leftptr=p;
```

If node has left subtree then put the successor of the node in the right pointer of its predecessor.

```
If(loc->left==link)
    p->rightptr=s;
```

Case C: Node to be deleted has two children.

Here first we delete the inorder successor of node and then replace the deleted node with the inorder successor. For deleting inorder successor we will call either case *A* or case *B*.

8.18 FINDING INORDER SUCCESSOR AND INORDER PREDECESSOR OF A NODE IN THREADED BINARY TREE

Inorder successor of a node is the left most node in the right sub-tree of that node. So if the right pointer of a node consists of a link then we can traverse the right sub-tree and find the inorder successor. If the right points is a thread, then that thread will point to the inorder successor.

```
structnode * inorder_successor(struct node * ptr)
    {
        struct node * succe;
        if(ptr->right==thread)
            succe=ptr->rightptr;
        else
            {
                ptr=ptr->rightptr;
                while(ptr->left==link)
                    ptr=ptr->leftptr;
                succe=ptr;
            }
    }
```

We know that **inorder predecessor** of a node is the rightmost node in the left sub-tree of that node. So if the left pointer of a node consists of a link then we can traverse the left subtree and find the inorder predecessor. If the left pointer is a thread, then that thread will point to the inorder predecessor.

```
struct node * inorder_predecessor(struct node * ptr)
    {
        struct node*ptr;
        if(ptr->left==thread)
            pre=ptr->leftptr;
        else
            {
                ptr=ptr->leftptr;
                while(ptr->right==link)
                    ptr=ptr->rightptr;
                pre=ptr;
            }
    }
```

8.19 GENERAL TREE

A tree is a finite set of one or more nodes such that

1. There is specially designated node called root.

2. The remaining nodes are partitioned into $N \geq 0$ disjoint sets $T_1, T_2, \ldots, T_N$. Where each of these sets in a tree $T_1, T_2, \ldots T_N$ are called the sub-trees of the root.

A general tree can have any number of nodes. The children of a node are called as **siblings** of each other. In other words, if a particular node has three children, then the second and third child of that node are the siblings of the first child. Figure shows a general tree.

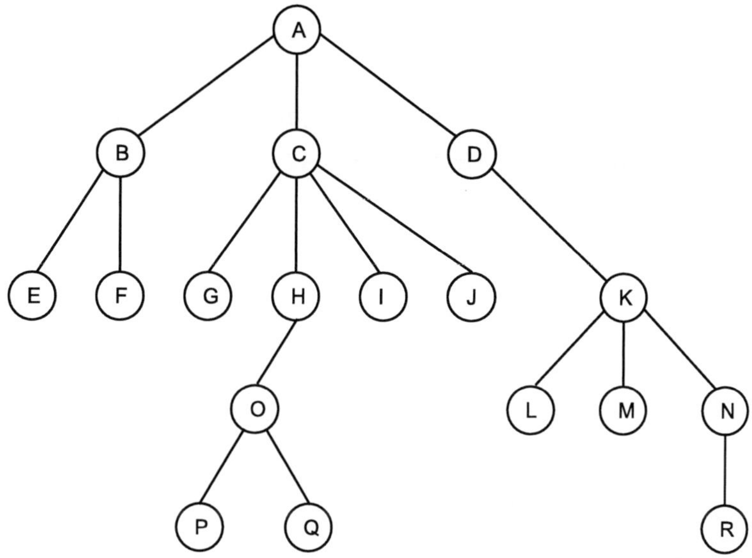

Fig. 8.67. A general tree.

8.19.1 Difference between a Binary Tree and a General Tree

A binary tree T′ is not a special case of a general tree T. Binary trees and general trees are different objects. The two basic differences are as follows:

1. The binary tree may be empty but there such be at least one node in general tree.
2. Every binary tree has at most two sons but general tree may have more than two sons. In binary tree every child is distinguished as a left child or right child but in general tree child cannot be distinguished as left or right. All the childs of a node in general tree are termed as siblings to each other.

8.19.2 Memory Representation of General Tree

Suppose T is a general tree. T will be maintained in memory by means of a linked representation, which uses three parallel arrays. INFO, CHILD (or DOWN) and SIBL (or HORZ) and a pointer variable ROOT as follows. First of all, each node N of T will correspond to a location K such that:

- INFO [K] contains the data at node N.
- CHILD [K] contains the location of the first child of node N.
- The condition CHILD [K] = NULL indicates that N has no children.
- SIBL [K] contains the location of the next sibling of node N.
- The condition SIBL [K] = NULL means that N is the last child of its parent.

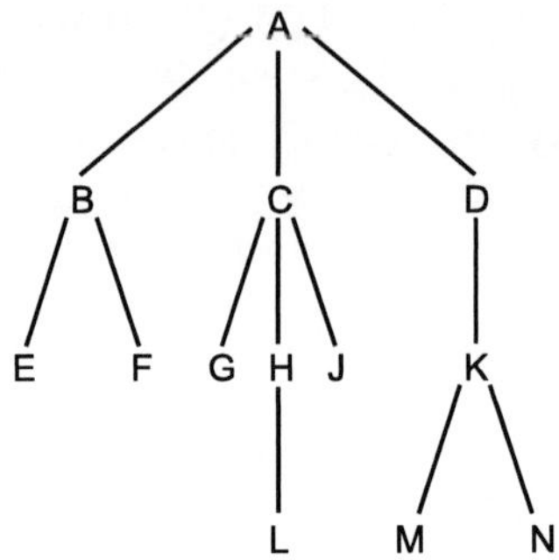

Fig. 8.68. General Tree

The array representation of the tree in Fig. 8.67 is given below:

root: 2

	INFO	CHILD	SIBL
1			
2	A	3	0
3	B	15	4
4	C	6	16
5		13	
6	G	0	7
7	H	11	8
8	J	0	0
9	N	0	0
10	M	0	9
11	L	0	0
12	K	10	0
13		0	
14	F	0	0
15	E	0	14
16	D	12	0

Fig. 8.69. Array representation of the general tree in Fig.

A Node May be Declared as a Dynamic Variable as Below:

```
struct treenode{
    int info;
    struct treenode *father;
    struct treenode *son;
    struct treenode *next;
    };
```

If all traversals are from a node to its sons, the father field may be omitted. Fig. 8.70(*a*) illustrates the representation of the trees of Fig. 8.70(*b*).

```
struct treenode{
     int info;
     struct treenode *son;
     struct treenode *next;
     };
```

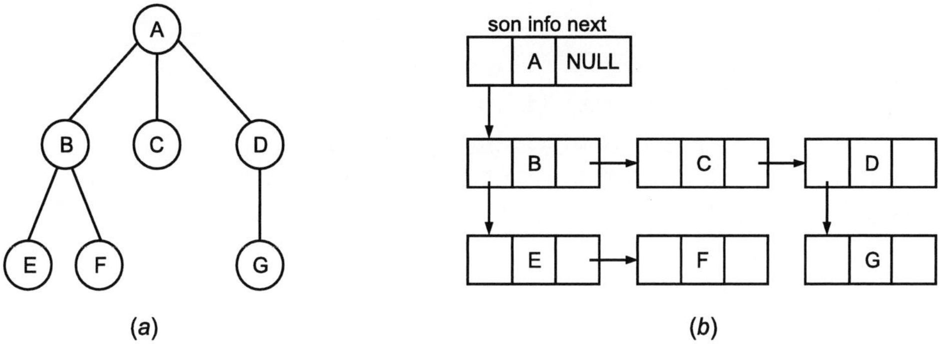

Fig. 8.70.

General tree also supports different ways to traversal tree as similar to binary tree.

1. In-order Traversal.
2. Pre-order Traversal.
3. Post-order Traversal.

The function for each traversal is as follows:

(*a*) In-order Traversal

```
Void inorder(struct treenode *t)
     {
          if (t)
               {
                    inorder(t->son);
                    printf("\t%d",t->info);
                    inorder(t->next);
               }
     }
```

The inorder traversal of general tree in above Fig. is

$$E, F, B, C, G, D, A$$

Pre-order Traversal

```
Void preorder(struct treenode *t)
     {
          if(t)
               {
                    printf ("\t%d",t->info);
                    preorder(t->son);
                    preorder(t->next);
               }
     }
```

Pre-order traversal of general tree in above Fig. is

A, B, E, F, C, D, G

Post-Order Traversal

```
void postorder(struct treenode *ptr)
    {
        if(ptr)
            {
                postorder(ptr->son);
                postorder(ptr->next);
                printf("\t%d",ptr->info);
            }
    }
```

Post-order traversal of general tree is above given Fig. is:

E, F, G, D, C, B, A.

While representing any tree it is necessary to have as many pointers in the node as the number of children that node has. However, each node in a general tree is likely to have different number of children. For example, some nodes may have 2 children, some may have 5 and some may have 4 children. Hence structure of a node of a genaral tree should be declared in such a way that it contains variable number of pointers. This is practically impossible.

To overcome this problem the maximum children occuring in any node in a general tree can be counted and then the structure with those many pointers can be declared. This however would lead to wastage of a large number of pointers. A better solution for this problem is to represent the general trees as binary trees.

8.20 GENERAL TREE REPRESENTED AS BINARY TREE

To represent the general tree as binary tree, two pointers are maintained. One pointer points to the first child of a node and another pointer points to the siblings of the node.

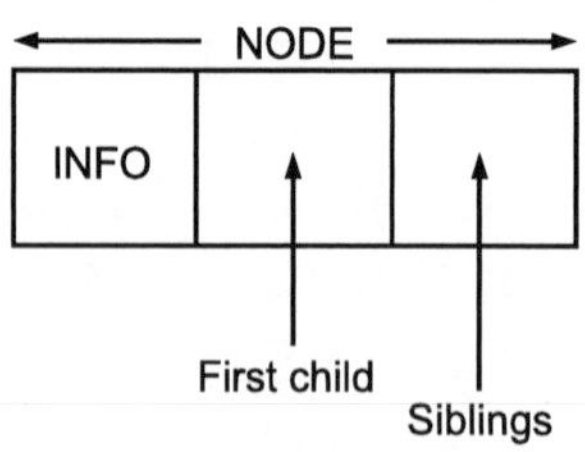

Fig. 8.71.

The structure of the nodes can be defined as follows:

```
struct node
    {
        int info;
        node *firstchild;
        node *siblings;
    };
```

While representing a general tree as a binary tree, the pointer to the first child of a particular node is made the left child of that node and the right child is the pointer to the siblings of the node. Thus while representing a general tree any node has only two children. Hence, general tree can be represented as a binary tree. Fig. 8.71 shows a general tree.

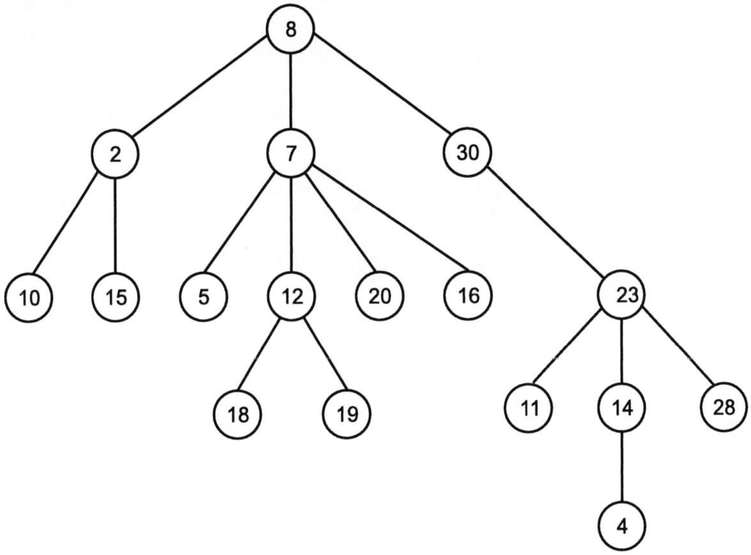

Fig. 8.72. A general tree.

Fig. 8.72 shows the binary presentation of general tree.

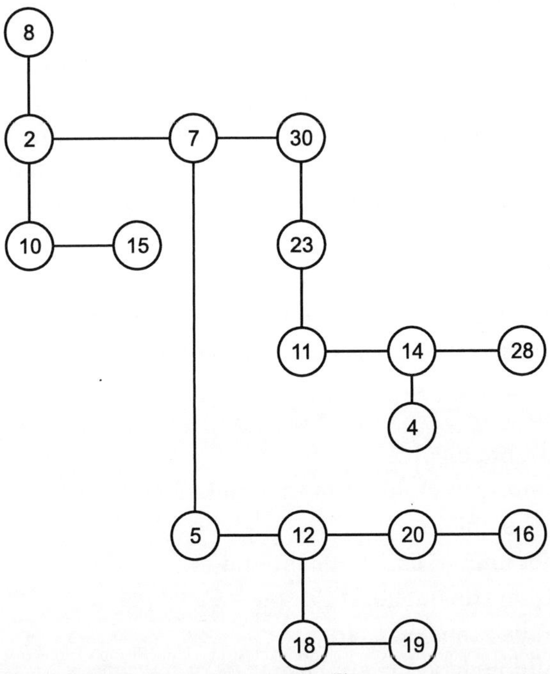

Fig. 8.73. General tree represented as a binary tree.

Though this tree dosn't appear as a binary tree. So, it is rotated by 45° as shown in Fig. 8.74.

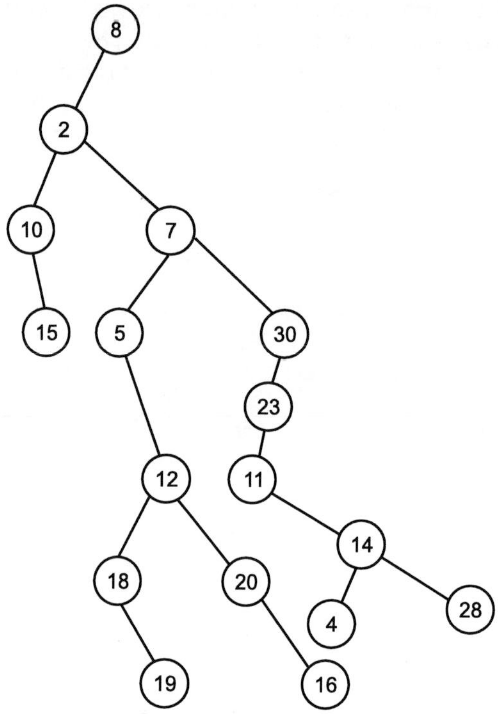

Fig. 8.74. After rotation of 45°

The method used for converting a general tree to a binary tree is as follows:

1. Make the root of the binary tree, the same as the root of the general tree.
2. If X_1, X_2, X_N are the immediate successors of the node X in general tree then make X_1 as the left child of X in equivalent binary tree and X_2 the right child of X_1, X_3 the right child of X_2 ... X_N the right child of X_{N-1} in equivalent binary tree.

Note: It can be observed that for every general tree represented as a binary tree and the right sub-tree of the root is always empty as there are no siblings of the root node.

Advantage of this method are as follows:

1. There is no wastage of memory space.
2. The number of children of the node need not be stored or maintained in the nodes.
3. This implementation is very easy for the languages which support dynamic storage allocation.

Algorithm:

1. (a) Accept the root node value.
   (b) Accept the number of children of this node. Let it be n.
2. Put root node in the queue.
3. While queue is not empty. call steps (4) and (5).
4. Remove a node from the queue, P Parent = P.
5. For i = 1 to n, do the steps (a) to (d).
   (*a*) Get the ith child node value suppose it be 'v'.
   (*b*) Allocate space for this node let it be 's'.
   (*c*) If (i = 0).

(*i*) Put 'v' in a node s.
(*ii*) Put s in the parent node's first child field.
(*iii*) Sibling = s

else

(*i*) Put v in a node 's'.
(*ii*) Make the node 's' to be pointed out by sibling field of the first child node.
(*iii*) Sibling = s.

(*d*) Place 's' in the queue

6. STOP.

Conversion of binary tree to general tree:

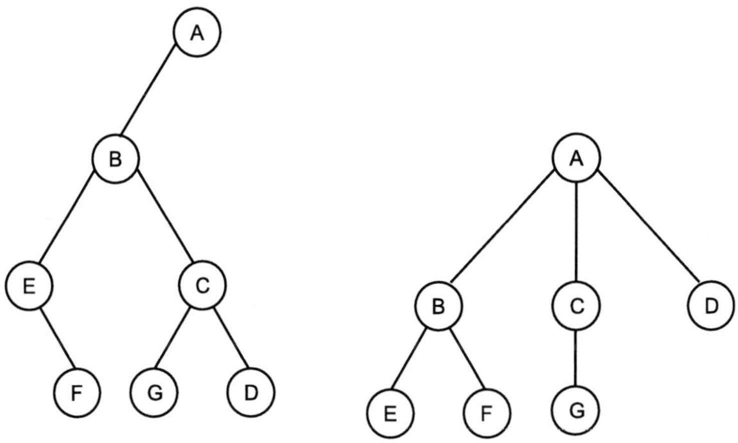

Binary tree General tree

Fig. 8.75.

Method:

1. Make the root of the general tree, the same as root of the binary tree.
2. If X_1 is the left child of the node X in binary tree and X_2 be the right child of X_1, X_3 the right child of X_2 ... X_N the right child of X_{N-1} then make X_1, X_2 ... X_N the children of node X in equivalent general tree.

8.21 FOREST TO BINARY TREE CONVERSION

Forest is a set of several trees that are not linked to each other in any way. Forest can be represented as a binary tree. Initially the left most tree is represented as a binary tree exactly in the same way as explained in the previous section. Then the second tree is made the right child of the root node of the first tree and the third tree is made the right child of the root node of the second tree. Fig. 8.74 shows how a forest can be represented as a binary tree.

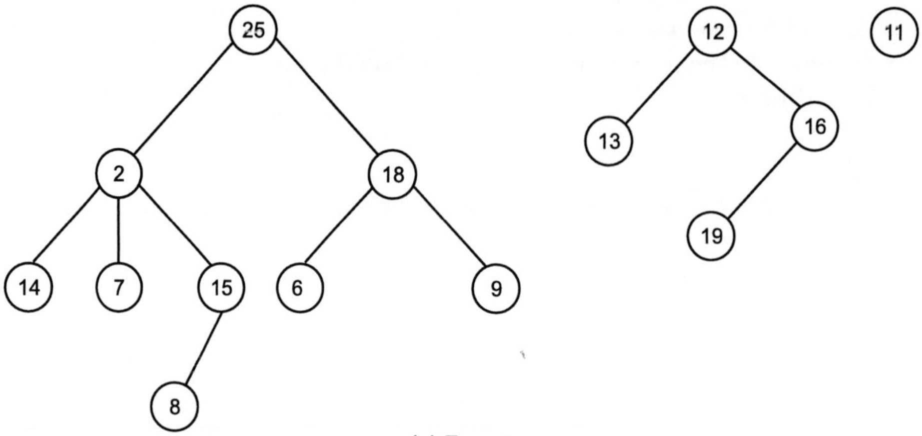

(a) Forest

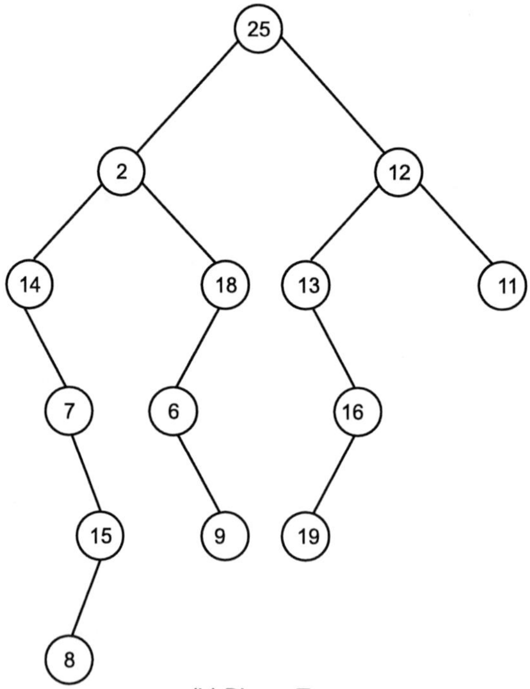

(b) Binary Tree

Fig. 8.76.

Exercises

1. How many binary trees are possible with 4 nodes?
2. Prove that the root of a binary tree is an ancestor of every node in the tree except itself.
3. Write recursive and non recursive algorithms to determine.
   (*a*) The number of nodes in a binary tree.
   (*b*) The depth of a binary tree.
4. Prove that a strictly binary tree with n leaves contains 2n – 1 nodes.
5. Write an algorithm to determine if a binary tree is:

(*a*) Strictly binary tree.

(*b*) Complete binary tree.

(*c*) Almost complete binary tree.

6. Write C functions to create a binary tree given:

   (*a*) The pre-order and inorder traversals of that tree.

   (*b*) The pre-order and postorder traversals of that tree.

7. Write C routines to traverse a binary tree in pre-order and post-order.

8. Prove that the leftmost node at level n in an almost complete strictly binary tree is assigned the number 2^n.

9. Find the binary tree for the following expression.

$$E = (3x - 2y)\ (5a + 3b)^3$$

10. Write a function that finds height of a binary tree.

11. Determine the order in which the vertices of the binary trees will be visited under.

   (i) Pre-order (ii) In-order (iii) Post-order.

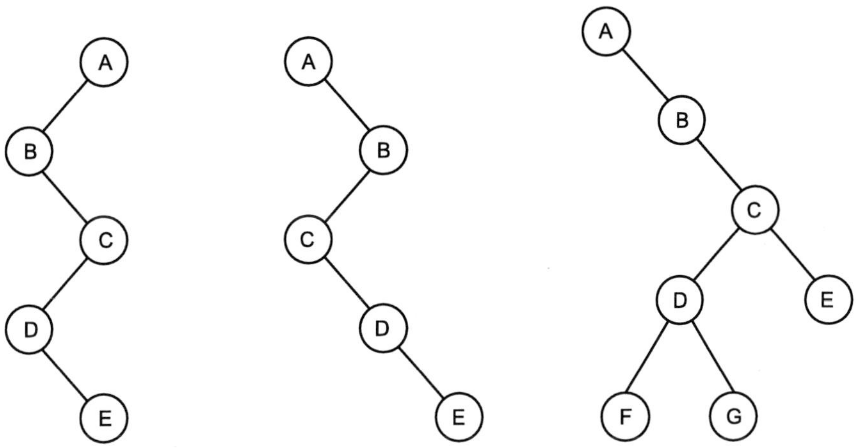

(a) (b) (c)

Fig. 8.77.

12. A binary tree has 9 nodes. The inorder and preorder traversals of tree yield the following sequences of nodes:

In-order:	E	A	C	K	F	H	D	B	G
Pre-order:	F	A	E	K	C	D	H	G	B

   Draw the tree.

13. Suppose a binary tree T is in memory. Write a procedure which deletes all the terminal nodes in T.

14. Suppose T is an extended binary tree. Prove.

   (*a*) $N_E = N_I + 1$, where N_E is the number of external nodes and N_I is the number of internal nodes.

   (*b*) $L_E = L_I + 2n$, where L_E is the external path length, L_I is the internal path length and n is the number of internal nodes.

15. Consider the general tree T in Fig. 8.76 below.

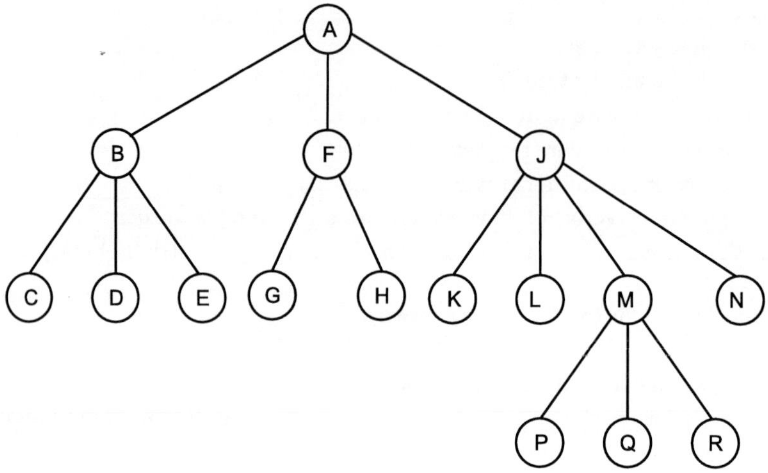

Fig. 8.78.

Find the corresponding binary tree.

16. Write an algorithm to convert a forest into a binary tree.
17. Write a short note on Huffman algorithm explaining various steps with example.
18. Write short note on Threaded Binary tree.

❒❒❒

Chapter 9 Binary Search Tree

Chapter Outline

9.1 INTRODUCTION

Searching a key or a value in an ordinary binary tree is very slow. This searching operation can be made faster by storing the values in the binary tree in an order and using the binary search method to search the values. The corresponding tree is called a **Binary Search Tree (BST)**. The Binary search tree is also called a **Binary Sorted Tree.**

A binary search tree is a binary tree. Such a tree can be represented by a linked data structure in which each node is an object. In addition to a **key** field, each node contains fields **left**, **right** and **p**, that point to the nodes corresponding to its left child, its right child and its parent respectively. The keys in a binary search tree are always stored in such a way as to satisfy the binary-search tree properties.

9.2 BINARY-SEARCH TREE PROPERTIES

A binary search tree may be empty. If it is not empty then it satisfies the following properties :

(1) Every element has a key, *i.e.*, key (x) for element x and no two elements have the same key.

(2) The keys (if any) in the left sub-tree are smaller than the key in the root.

(3) The keys (if any) in the right sub-tree are larger than the key in the root.

(4) The left and right sub-trees are also binary search trees.

Let x be a node in a binary search tree.

- if y is a node in the left sub-tree of x, then

  key [y] ≤ key [x]

- if y is a node in the right sub-tree of x, then

  key [x] ≤ key [y]

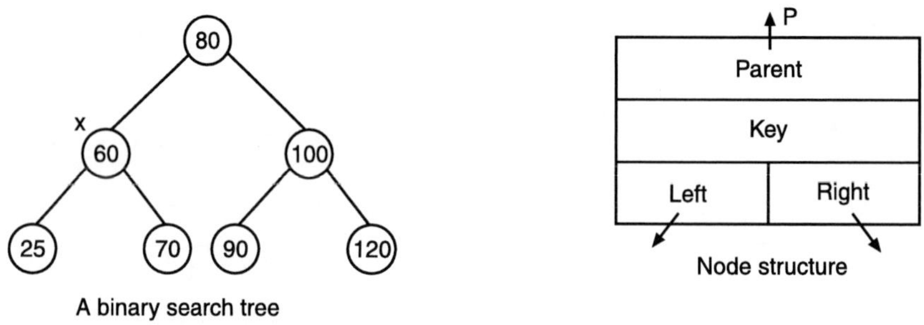

Fig. 9.1.

In this tree key [x] = 60

if y is the node in the left sub tree then key [y] = 25

i.e., key [y] ≤ key [x]

and if y is the node in the right sub tree then key [y] = 70

i.e., key [y] ≥ key [x]

9.3 VARIOUS OPERATIONS ON A BINARY SEARCH TREE

The most common operation performed on a BST is searching for a key stored in the tree. Besides the search operation. Binary search trees can support many operations such as minimum key. Maximum key, successor and predecessor. These operations run in O(h) time, where h is the height of the tree.

(*a*) Searching in a BST

Searching for a data in a binary search tree is much faster than in arrays or linked lists. The TREE-SEARCH (x, k) algorithm searches the tree root at x for a node whose key value equals to k. It returns a pointers to the node if it exists otherwise NIL.

TREE-SEARCH (x, k)

1. If x = NIL or k = key [x]
2. then return x
3. If k < key [x]
4. then return TREE-SEARCH (left [x], k)
5. else return TREE-SEARCH (right [x], k)

Note : Search operation takes time O(h), where h is the height of a BST.

Example :

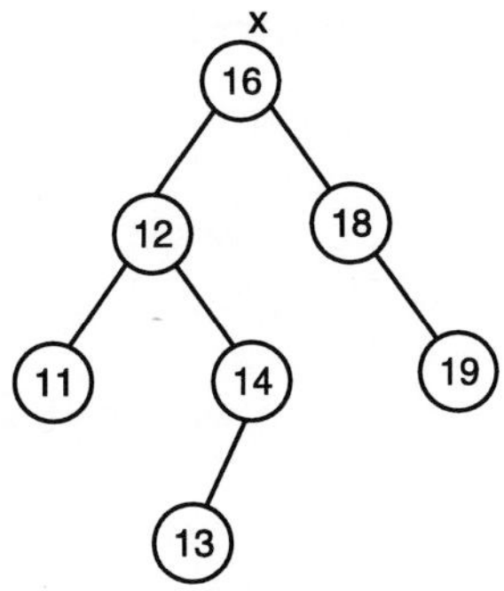

Fig. 9.2.

Suppose TREE-SEARCH (x, 14)

Here x ≠ NIL and key [x] = 16 which is not equal to k.

Here k = 14 and 14 < 16 (True)

then TREE-SEARCH (left (x), k)

i.e., now left [x] becomes x. (12 node)

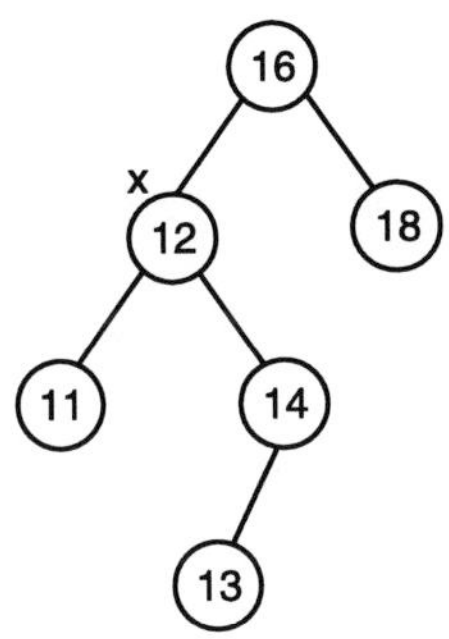

Fig. 9.3.

Now x ≠ NIL and 14 ≠ 12

14 ≮ 12

So, TREE-SEARCH (right [x], 14)

i.e.,

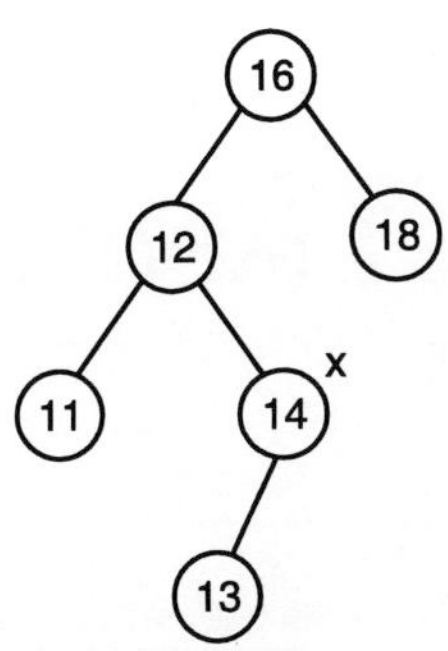

Fig. 9.4.

Now x ≠ NIL and k = key [x]

So return x.

(b) Minimum and Maximum Key Element in BST

The minimum key value element can always be found by following left child pointers from the root until a NIL is encountered.

Tree-Minimum (x)

1. While left [x] ≠ NIL
2. do x ← left [x]
3. return x

and the maximum key value element can always be found by following right child pointers from the root until NIL is encountered.

Tree-Maximum (x)

1. While Right [x] ≠ NIL
2. do x ← Right [x]
3. return x

(c) Successor of a Node x

The following procedure returns the successor of a node x in a BST if it exists and NIL if x has the largest key in the tree.

TREE-SUCCESSOR (x)

1. If right [x] ≠ NIL
2. then return TREE-MINIMUM (right [x])
3. y ← p[x]
4. while y ≠ NIL and x = Right [y]
5. do x ← y
6. y ← p[y]
7. return y

Example :

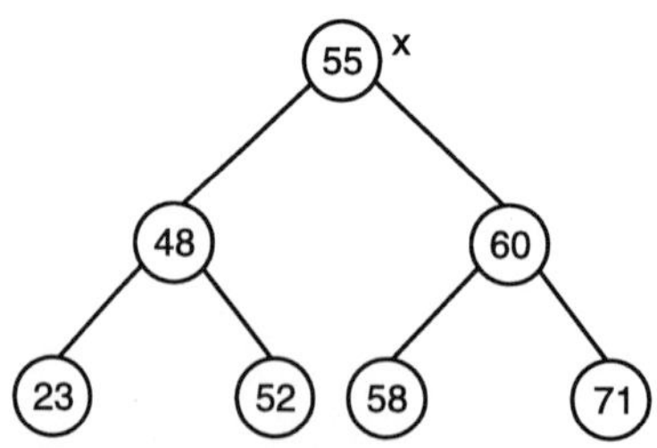

Fig. 9.5.

Suppose we want to find successor of node 55 then, we call TREE-SUCCESSOR (55)

Here right [x] ≠ NIL

So call TREE-MINIMUM (right [x])

i.e., TREE-MINIMUM (60)

Here left (x) ≠ NIL

So left (x) is now x, *i.e.*, x ← left (x)

and return x, *i.e.*, node 58.

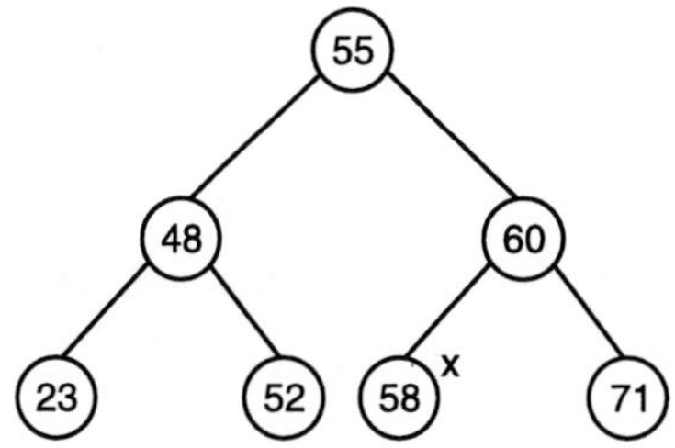

Fig. 9.6.

Thus, node 58 is the successor of node 55.

(*d*) Traversal Operation on BST

All the traversal operations are applicable in binary search trees. The in-order traversal on a binary search tree gives the sorted order of data ascending (increasing) order.

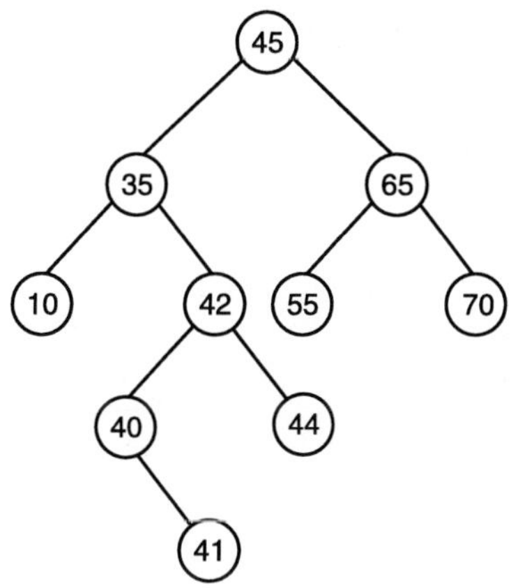

Fig. 9.7.

The in-order traversal of the BST in the above figure is:

10, 35, 40, 41, 42, 45, 55, 65, 70

It gives the increasing order of the numbers in the BST.

(*e*) Insertion of Data into a Binary Search Tree

To insert a new value w into a binary search tree T, we use the procedure TREE-INSERT. The procedure is passed a node z for which key [z] = w left [z] = NIL and Right [z] = NIL.

```
y ← NIL
x ← root [T]
while x ≠ NIL
      do y ← x
            if key [z] < key [x]
                  then x ← left [x]
            else x ← right [x]
P[z] ← y
if y = NIL
      then root [T] ← z
```

```
else if key [z] < key [y]
        then left [y] ← z
    else right [y] ← z.
```

Actually, the sequence of steps to be followed in performing the insertion operation on a binary search tree are as follows :

(1) Start from the root node.

(2) If the data to be inserted is w, compare this with the value of the root node.
  - (i) If they are equal, just stop because this value already exists.
  - (ii) If they are different and if the data to be inserted is less than the value of the root node, choose the left sub-tree.
  - (iii) If the data to be inserted (w) is greater than the value of the root node, choose the right sub-tree.

(3) Repeat step (2) until the leaf node is encountered where the data has to be inserted.

EXAMPLE : The operation of insertion is shown in the figure below :

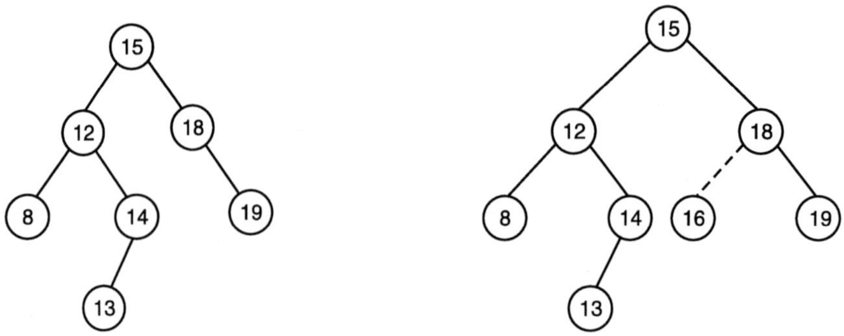

(a) BST before insertion (b) BST after insertion

Fig. 9.8.

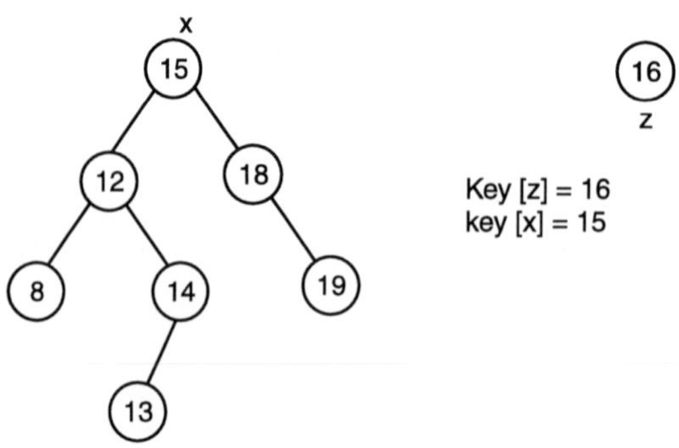

Fig. 9.9

While x ≠ NIL

y ← x

Here key [z] > key [x] i.e., 16 > 15

So x ← right [x]

Now

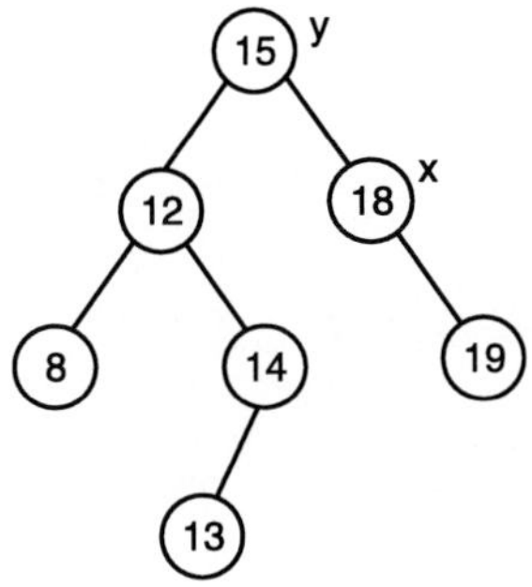

Fig. 9.10.

Now $y \leftarrow x$

key [x] = 18   key [z] = 16

Here key [z] < key [x]

So $x \leftarrow$ left [x]

and here left [x] = NIL.

So $x =$ NIL

and $P[z] \leftarrow y$ *i.e.*

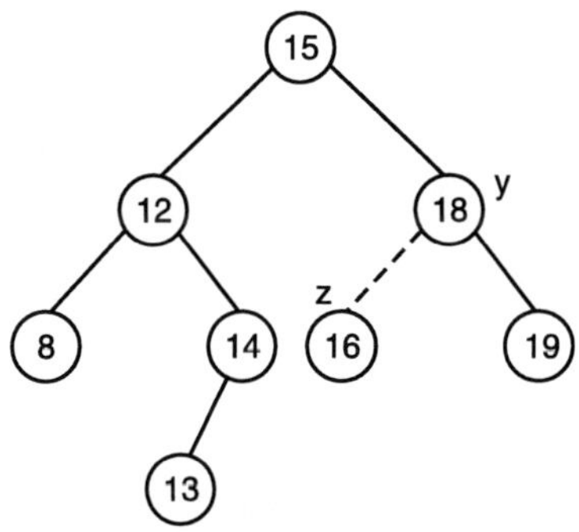

Fig. 9.11.

Here $y \neq$ NIL and key [z] < key [y]

So, z is the left child of y and the final tree after insertion is as follows :

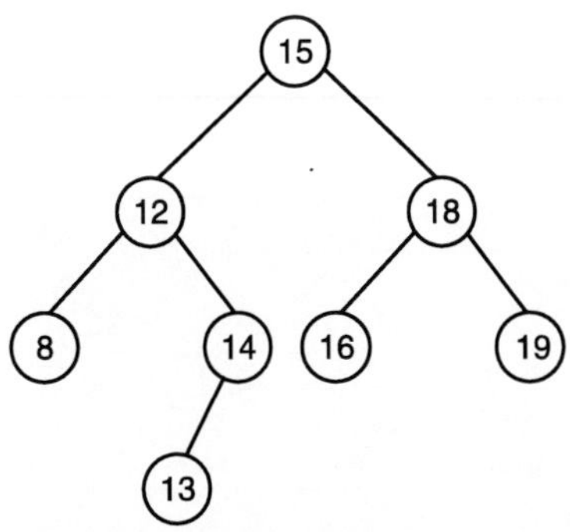

Fig. 9.12.

In our implementation of binary search tree, we will consider following declarations.

```
struct bstree node
    {
```

```
            int info ;
            struct bstree node * left ;
            struct bstree node * right ;
      } ;
   struct bstree node  * bt, * root ;
```

(f) Delete a node : Deletion of a node from a BST depends on the number of its children. Suppose delete a node with key = z from BST T. There are 3 cases that can occur.

Case 1 : z has no child (*i.e.*, leaf node)

Case 2 : z has exactly one child.

Case 3 : z has two child.

Case 1 : If the node to be deleted z is a leaf node, simply make the pointer field of the parent node of z to NULL.

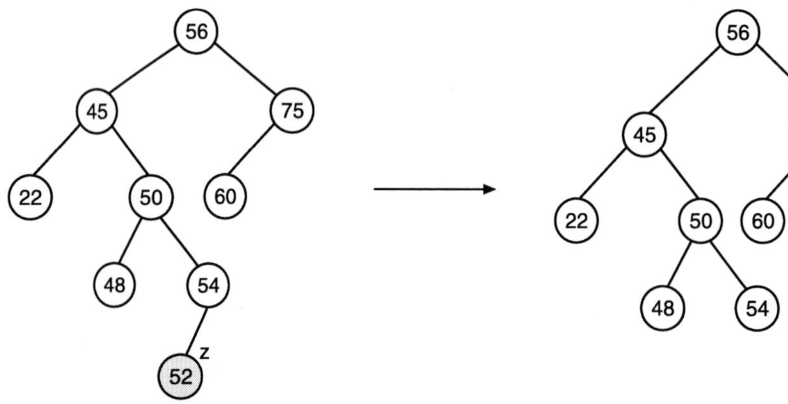

(a) Before deletion of node 52 (b) After deletion of node 52

Fig. 9.13.

Case 2 : If the node to be deleted z has a child then simply set the pointer field of the parent node of z to point to the only child node of z.

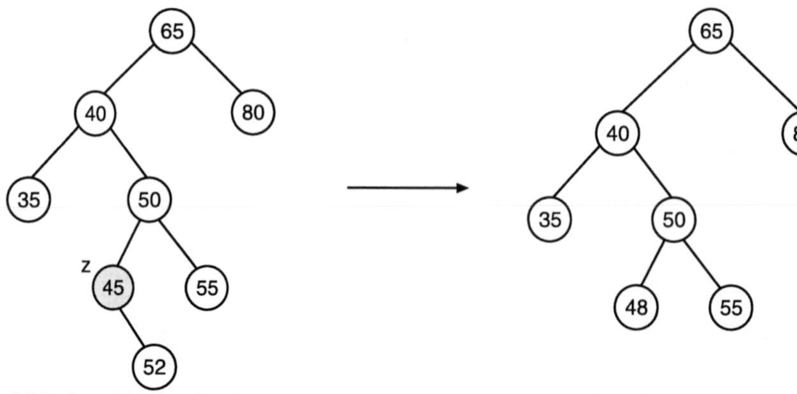

(a) Before deletion of node 45 (b) After deletion of node 45.

Fig. 9.14.

Case 3 : If the node to be deleted z has two children. This case is a little more difficult since two subtrees need to be re-attached and both must maintain the same relation to the parent of the node being deleted and to each other. Here, find the in-order successor of the node to be deleted z.

Place it in the node z. Then delete the in-order successor node similar to case-1 or case-2.

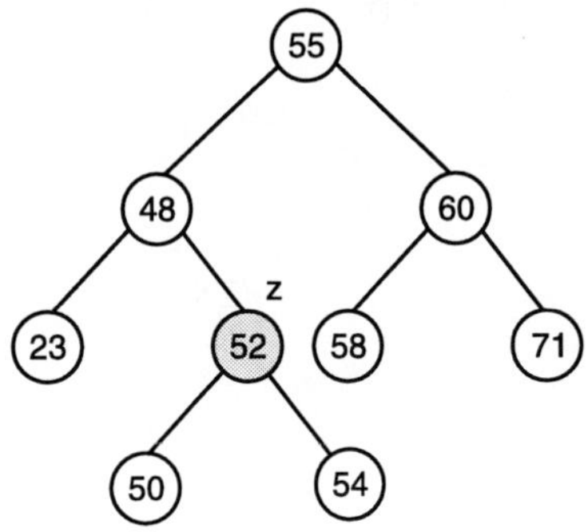

(a) Before deletion of node 52

Fig. 9.15. (a)

Find successor of node 52, we get node 54. Place it in the node 52 and then delete the successor node.

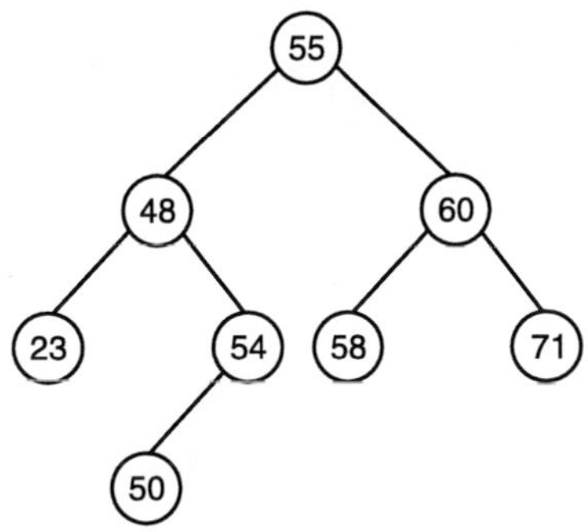

(b) After deletion of node 52

Fig. 9.15. (b)

EXAMPLE 9.1. Suppose the following list of letters is inserted in order into an empty BST

J R D T G E A M H F Q U B

Find the final tree.

Solution: Insert the nodes one after the other maintaining BST property.

1. Insert node J in empty BST 

2. Insert node R

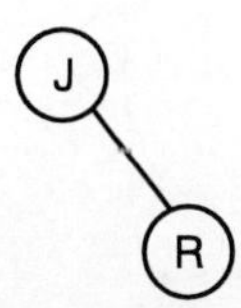

3. Insert node D.

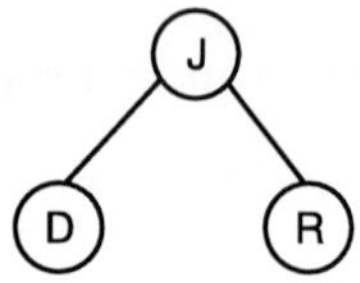

4. Insert node T

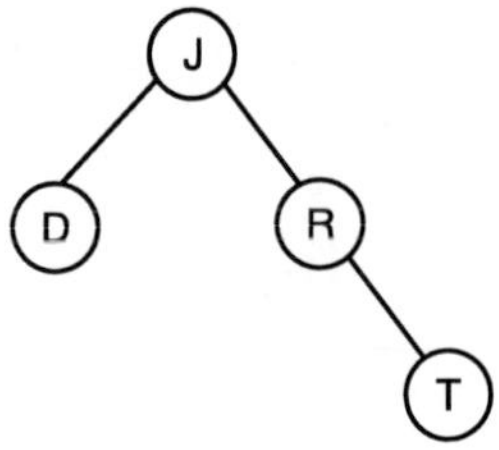

5. Insert node G

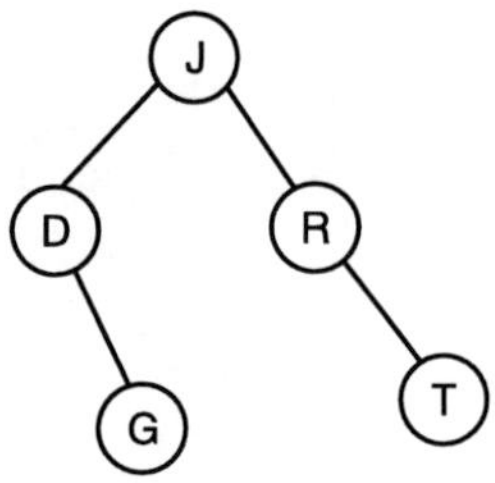

6. Insert node E

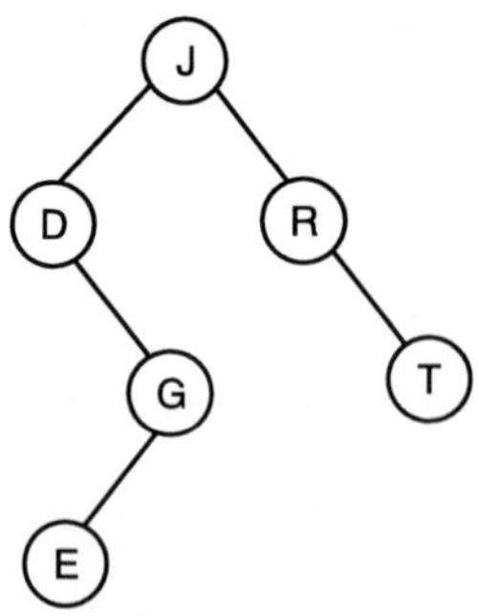

7. Insert node A

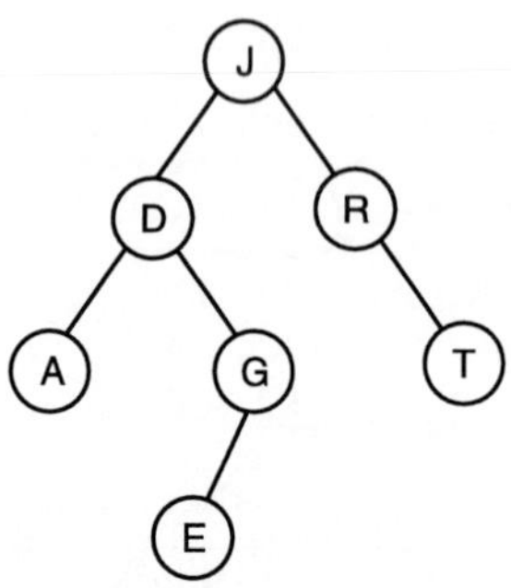

8. Insert node M

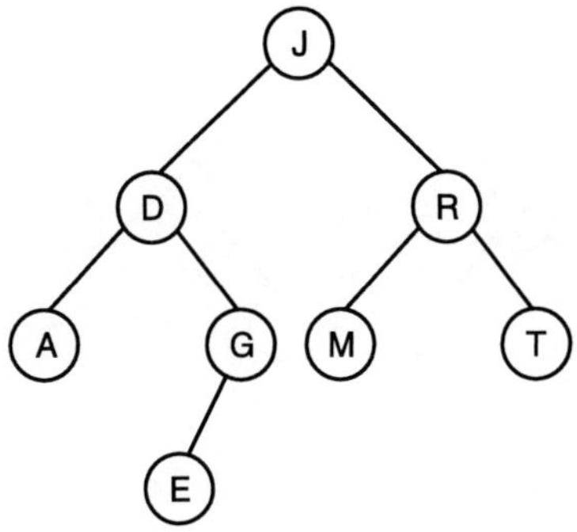

9. Insert node H

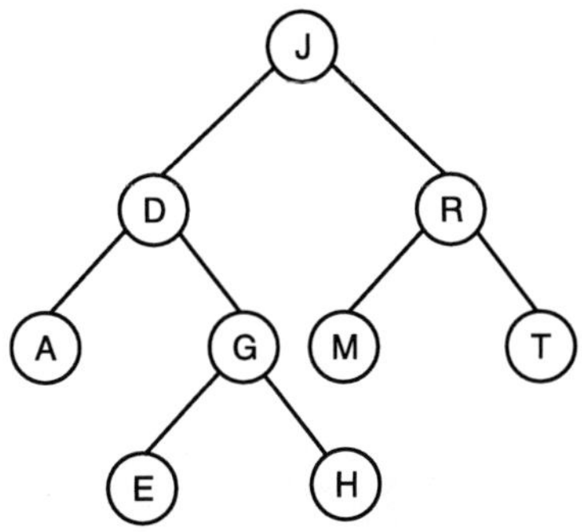

10. Insert node F

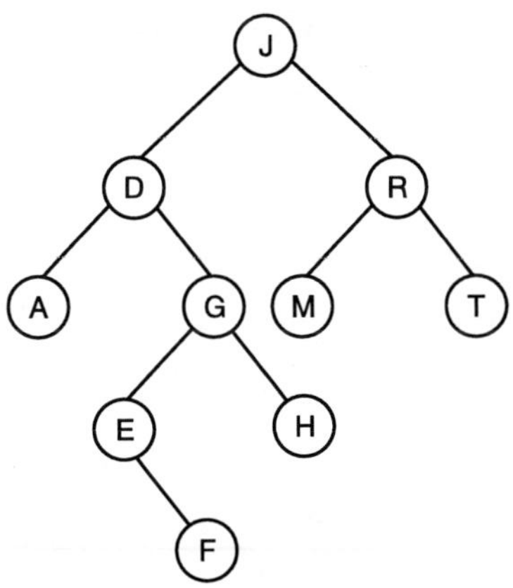

11. Insert node Q

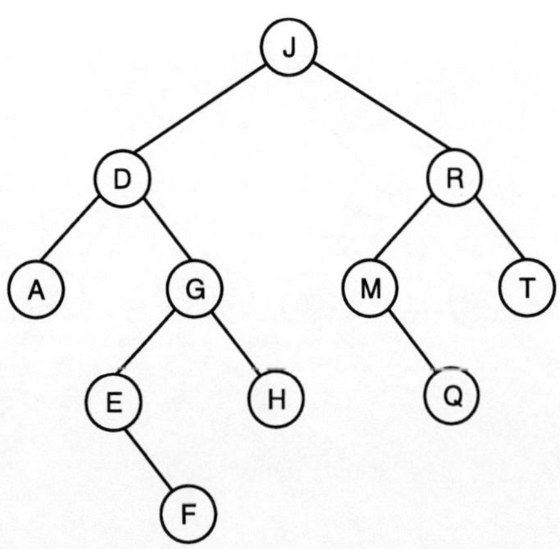

12. Insert node U

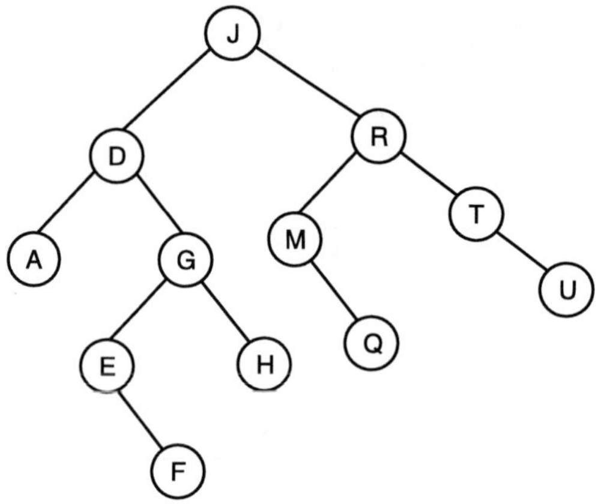

13. Insert node B

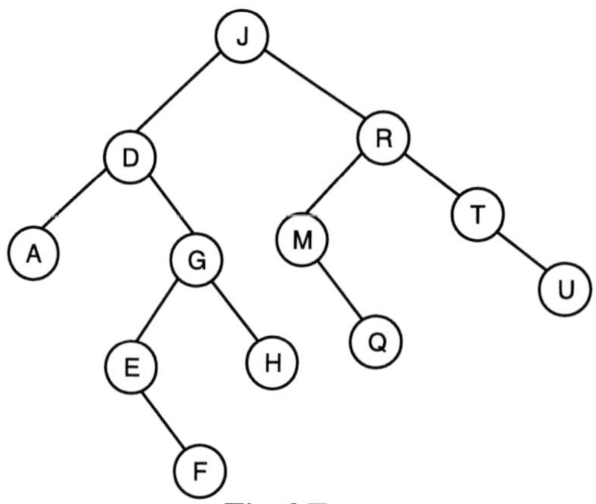

Final Tree

EXAMPLE 9.2. Suppose the following numbers are inserted an empty BST.

40 25 70 22 35 60 80 90 10 30

Draw the tree T.

And applying the following operation an original tree T.

(not successively) (*a*) Delete node 30 (*b*) Delete node 80 (*c*) Delete node 40

Solution:

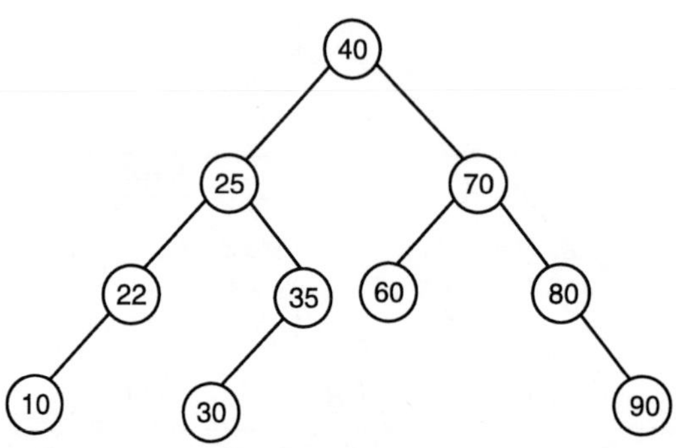

Binary search tree T.

(*a*) Delete node 30

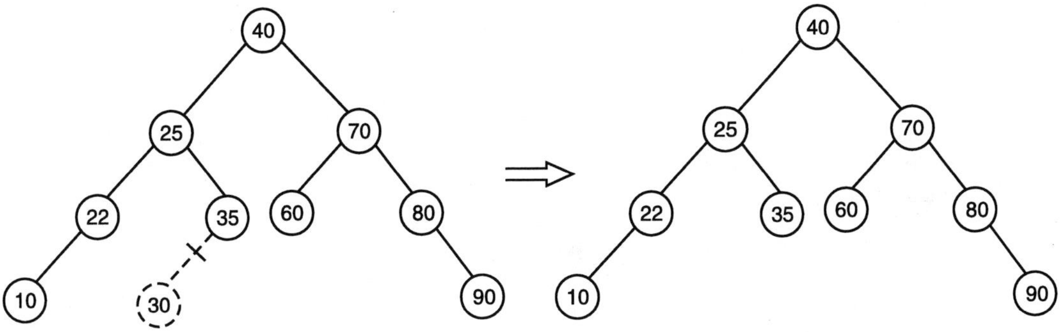

(*b*) Delete node 80

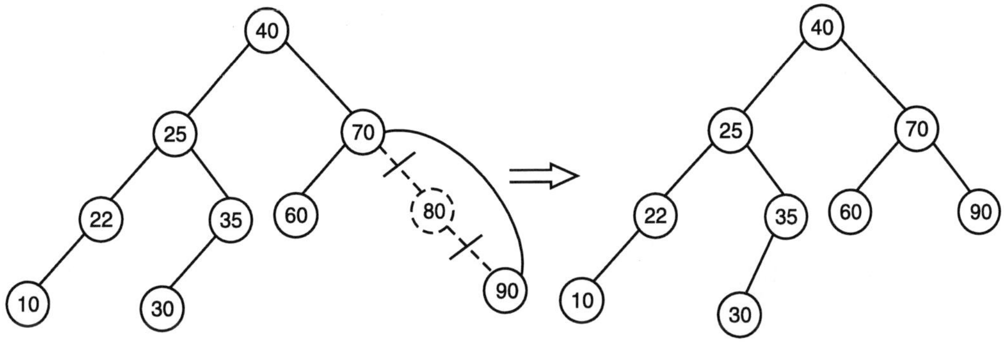

(*c*) Delete node 40

First replace the deleted node, *i.e.*, 40 to its inorder successor, *i.e.*, 60 and then delete the node, *i.e.*,

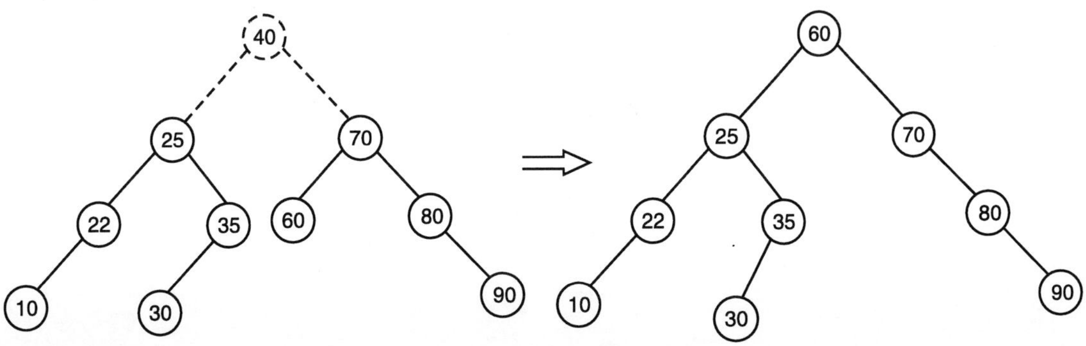

PROGRAM

/* PROGRAM TO CREATE AND TRAVERSE A BINARY SEARCH TREE */

```
#include<stdio.h>
#include<conio.h>
#include<alloc.h>
typedef struct tnode
{
```

```
int info;
struct tnode *l;
struct tnode *r;
} node;
node *root;
void create( )
{
int n, i;
node *newn, *ptr, *pptr;
printf("\n enter no of nodes:");
scanf ("%d", &n);
printf("\n Enter%d values\n", n);
root=(node *) malloc(sizeof(node));
root->l=root->r=NULL;
scanf("%d",&root->info);
for(i=2;i<=n;i++)
{
newn=(node *)malloc(sizeof(node));
newn->l=newn->r=NULL;
scanf("%d",&newn->info);
ptr=root;
while(ptr!=NULL)
{
pptr=ptr;
if(newn->info<ptr->info)
{
ptr=ptr->l;
if(ptr==NULL)
pptr->l=newn;
}
else
{
ptr=ptr->r;
if(ptr==NULL)
pptr->r=newn;
}
}
}
}
void preorder(node *ptr)
```

```
{
if(ptr!=NULL)
{
printf("%d",ptr->info);
preorder(ptr->l);
preorder(ptr->r);
}
}
void inorder(node*ptr)
{
if(ptr!=NULL)
{
inorder(ptr->l);
printf("%d",ptr->info);
inorder(ptr->r);
}
}
void postorder(node*ptr)
{
if(ptr!=NULL)
{
postorder(ptr->l);
postorder(ptr->r);
printf("%d",ptr->info);
}
}
void main()
{
clrscr();
create();
printf("\n preorder traversal:\n");
printf("\n");
preorder(root);
printf("\n inorder traversal:\n");
printf("\n");
inorder(root);
printf("\n postorder traversal:\n");
printf("\n");
postorder(root);
getch();
}
```

PROGRAM :

```
/* Create BST for numerals */
# include <stdio.h>
# include <conio.h>
# include <alloc.h>
struct bst
{
     int data;
     struct bst *lchild;
     struct bst *rchild;
 }
*root=NULL;
void create_bst (struct bst **bstroot,int item)
{
     if (*bstroot==NULL)
     {
          *bstroot=(struct bst*)malloc(sizeof (struct bst));
          (*bstroot)->lchild=NULL;
          (*bstroot)->rchild=NULL;
          (*bstroot)->data=item;
     }
     else
     {
          if (item<(*bstroot)->data)
          create_bst (& ( (*bstroot) ->lchild, item);
          if (item>(*bstroot)->data)
          create_bst (& ( (*bstroot) ->rchild, item);
     }
}
void inorder_traverse (struct bst *bstroot)
{
     if (bstroot !=NULL)
     {
          inorder_traverse (bstroot->lchild);
          printf ("\n%d", bstroot->data);
          inorder_traverse (bstroot->rchild);
     }
}
void preorder_traverse (struct bst *bstroot)
{
     if (bstroot !=NULL)
```

```
    {
        printf ("\n%d", bstroot->data);
        preorder_traverse (bstroot->lchild);
        preorder_traverse (bstroot->rchild);
    }
}
void postorder_traverse (struct bst *bstroot)
{
    if (bstroot !=NULL)
    {
        postorder_traverse (bstroot->lchild);
        postorder_traverse (bstroot->rchild);
        printf ("\n%d", bstroot->data);
    }
}
void main ( )
{
    clrscr ( );
    create_bst (&root, 100);
    create_bst (&root, 110);
    create_bst (&root, 90);
    create_bst (&root, 10);
    create_bst (&root, 101);
    create_bst (&root, 120);
    create_bst (&root, 103);
    create_bst (&root, 105);
    create_bst (&root, 108);
    clrscr ( ) ;
    printf ("\nTraversing bst into inorder");
    inorder_traverse (root);
    printf ("\nTraversing bst into preorder");
    preorder_traverse (root);
    printf ("\nTraversing bst into postorder");
    postorder_traverse (root);
    getch ( );
}
```

PROGRAM :

```
/* Create BST for alphabets */
# include <stdio.h>
# include <conio.h>
# include <alloc.h>
```

```
struct bst
{
     char data;
     struct bst *lchild;
     struct bst *rchild;
 }
*root=NULL;
void create_bst (struct bst **bstroot,char item)
{
     if (*bstroot==NULL)
     {
          *bstroot=(struct bst*)malloc(sizeof (struct bst));
          (*bstroot)->lchild=NULL;
          (*bstroot)->rchild=NULL;
          (*bstroot)->data=item;
     }
     else
     {
          if (item<(*bstroot)->data)
          create_bst (& ( (*bstroot) ->lchild, item);
          if (item>(*bstroot)->data)
          create_bst (& ( (*bstroot) ->rchild, item);
     }
}
void inorder_traverse (struct bst *bstroot)
{
     if (bstroot !=NULL)
     {
          inorder_traverse (bstroot->lchild);
          printf ("\n%c", bstroot->data);
          inorder_traverse (bstroot->rchild);
     }
}
void preorder_traverse (struct bst *bstroot)
{
     if (bstroot !=NULL)
     {
          printf ("\n%c", bstroot->data);
          preorder_traverse (bstroot->lchild);
          preorder_traverse (bstroot->rchild);
     }
```

```
}
void postorder_traverse (struct bst *bstroot)
{
     if (bstroot !=NULL)
     {
          postorder_traverse (bstroot->1child);
          postorder_traverse (bstroot->rchild);
          printf ("\n%c", bstroot->data);
     }
}
void main ( )
{
     clrscr ( );
     create_bst (&root, 'A');
     create_bst (&root, 'B');
     create_bst (&root, 'a');
     create_bst (&root, 'b');
     create_bst (&root, 'C');
     create_bst (&root, 'c');
     create_bst (&root, 'D');
     create_bst (&root, 'd');
     create_bst (&root, 'E');
     clrscr ( ) ;
     printf ("\nTraversing bst into inorder");
     inorder_traverse (root);
     printf ("\nTraversing bst into preorder");
     preorder_traverse (root);
     printf ("\nTraversing bst into postorder");
     postorder_traverse (root);
     getch ( );
}
```

PROGRAM :

```
/* Program to implement BST and traverse in Preorder without using recursion*/
#include <stdio.h>
#include <conio.h>
#include <alloc.h>
#define MAXSIZE 10
int i, top-0, stack [MAXSIZE] ;
struct bst
{
```

```
    int data;
    struct bst *1child;
    struct bst *rchild;
}
*root=NULL;
void insert (struct bst **bstroot, int item)
{
    if (*bstroot ==NULL)
    {
        *bstroot = (struct bst*) malloc (sizeof (struct bst) );
        (*bstroot)->1child=NULL;
        (*bstroot)->rchild=NULL;
        (*bstroot)->data=item;
    }
    else
    {
        if (item<(*bstroot)->data)
        insert (& ( (*bstroot)->1child), item) ;
        if (item>(*bstroot)->data)
        insert (& ( (*bstroot)->rchild), item) ;
    }
}
void preorder (struct bst *node)
{
    int top = 1 ;
    stack [0] = NULL;
    node = root
    while (node ! = NULL)
    {
        printf ("%d", node -> data) ;
        if (node ->rchild !=NULL)
        {
            top++;
            stack [top] = node;
            node = node -> 1child ;
        }
        if (node -> 1child !=NULL)
        node = node ->rchild;
        else
        {
                    node = stack [top] ;
```

```
                    top--;
                }
            }
        }
void main ( )
    {
        clrscr ( );
        insert (&root, 20);
        insert (&root, 60);
        insert (&root, 40);
        insert (&root, 10);
        insert (&root, 30);
        insert (&root, 50);
        insert (&root, 25);
        insert (&root, 28);
        insert (&root, 29);
        clrscr ( ) ;
        printf ("\n Traversing BST into preorder :") ;
        preorder (root) ;
        getch ( ) ;
    }
```

PROGRAM

/* Program to implement BST and traverse in Postorder without using recursion*/

```
#include <stdio.h>
#include <conio.h>
#include <alloc.h>
#define MAXSIZE 10
int i, top=0, stack [MAXSIZE] ;
struct bst
{
    int data;
    struct bst *lchild;
    struct bst *rchild;
}
*root=NULL;
void insert (struct bst **bstroot, int item)
{
    if (*bstroot ==NULL)
    {
        *bstroot = (struct bst*) malloc (sizeof (struct bst) );
```

```
            (*bstroot)->lchild=NULL;
            (*bstroot)->rchild=NULL;
            (*bstroot)->data=item;
      }
      else
      {
            if (item<(*bstroot)->data)
            insert (& ( (*bstroot)->lchild), item) ;
            if (item>(*bstroot)->data)
            insert (& ( (*bstroot)->rchild), item) ;
      }
}
      void postorder (struct bst *node)
      {
            top = 1 ;
            stack [0] = NULL;
            node = root;
            level 1 :
            while (node ! = NULL)
            {
                  top++;
                  stack [top] = node;
                  if (node->rchild !=NULL)
                  {
                        top++;
                        stack [top] = node->rchild;
                        node = node ->lchild ;
                  }
            }
            node =stack [top];
            top--;
            while (node>0)
            {
                  printf ("%d", node -> data) ;
                  node = stack [top];
                  top--;
            }
            if (node<0)
            {
                  goto level 1;
            }
```

```
    }
void main ( )
{
    clrscr ( );
    insert (&root, 100);
    insert (&root, 110);
    insert (&root, 90);
    insert (&root, 10);
    insert (&root, 101);
    insert (&root, 120);
    insert (&root, 103);
    insert (&root, 105);
    insert (&root, 108);
    clrscr ( ) ;
    printf ("\n Traversing bst into postorder") ;
    postorder (root) ;
    getch ( ) ;
    }
```

PROGRAM :

```
/*  Implement Binary Search Tree with different operations */
#include <conio.h>
#include <stdio.h>
#include <alloc.h>
typedef struct bst
{
int data;
struct bst *left, *right;
} node;
void insert(node *, node*);
void inorder(node *);
node *search(node *, int, node **);
void del(node *, int);
void main ( )
{
int ch;
char ans = 'N';
int key;
node *New, *root, *temp, *parent;
node *get_node( );
root=NULL;
clrscr ( );
```

```
printf("\n \t Program for Binary Search Tree");
do
{
printf("\n 1.Create \n 2.Search \n 3. Delete \n 4. Display");
printf("\n\n Enter your choice");
scanf("%d",&ch);
switch(ch)
{
case 1 : do
{
New=get_node( );
printf("\n Enter the Element");
scanf("%d",&New->data);
if(root==NULL)
root=New;
else
insert(root,New);
printf("\n Do u want to conitnue enter more elements(y/n)");
ans=getch( );
} while(ans=='N');
break;
case 2 :
printf("\n Enter the element which u want to search");
scanf("%d",&key);
temp=search(root,key,&parent);
printf("\n Parent of node %d is %d", temp->data,parent->data);
break;
case 3 : printf("\n Enter the Element u wish to delete");
scanf("%d",&key);
del(root,key);
break;
case 4 : if(root==NULL)
printf("Tree is not created");
else
{
printf("\n The tree is :");
inorder(root);
}
break;
}
} while (ch !=5);
```

```
}
node *get_node( )
{
node *temp;
temp=(node*)malloc(sizeof(node));
temp->left=NULL;
temp->right=NULL;
return temp;
}
void insert(node *root,node *New)
{
if(New->data<root->data)
{
if(root->left==NULL)
root->left=New;
else
insert(root->left,New);
}
if(New->data>root->data)
{
if(root->right==NULL)
root->right=New;
else
insert(root->right,New);
}
}
node *search(node *root,int key, node **parent)
{
node *temp;
temp=root;
while(temp !=NULL)
{
if(temp->data==key)
{
printf("\n The %d Element is present",temp->data);
return temp;
}
*parent=temp;
if(temp->data>key)
temp=temp->left;
else
```

```
temp=temp->right;
}
return NULL;
}
void del(node*root,int key)
{
node*temp,*parent,*temp_succ;
temp=search(root,key,&parent);
if(temp->left!=NULL&&temp->right!=NULL)
{
parent=temp;
temp_succ->=temp->right;
while(temp_succ->left!=NULL)
{
parent=temp_succ;
temp_succ=temp_succ->left;
}
temp->data=temp_succ->data;
parent->right=NULL;
printf("\n Now deleted it!");
return;
}
if(temp->left!=NULL&&temp->right==NULL)
{
if(parent->left==temp)
parent->left=temp->left;
else
parent->right=temp->left;
temp=NULL;
free(temp);
printf("\n Now deleted it!");
return;
}
if(temp->left==NULL&&temp->right!=NULL)
{
if(parent->left==temp)
parent->left=temp->right;
else
parent->right=temp->right;
temp=NULL;
free(temp);
```

```
printf("\n Now deleted it!");
return;
}
if(temp->left==NULL &&temp->right==NULL)
{
if(parent->left==temp)
parent->right=NULL;
else
parent->right=NULL;
printf("\n Now Deleted it!");
return;
}
}
void inorder(node *temp)
{
if(temp !=NULL)
{
inorder(temp->left);
printf("%d", temp->data);
inorder(temp->right);
}
}
```

OUTPUT

```
Program for Binary Search Tree
1. Create
2. Search
3. Delete
4. Display
Enter your choice1
Enter the Element 10
Enter your choice 1
Enter the Element 8
Enter your choice 1
Enter the Element 9
Enter your choice 1
Enter the Element 7
Enter your choice 1
Enter the Element 15
Enter your choice 1
Enter the Element 13
Enter your choice 1
Enter the Element 14
Enter your choice 1
```

```
Enter the Element 12
Enter your choice 1
Enter the Element 16
Do u want to continue enter more elements (y/n) y
1. Create
2. Search
3. Delete
4. Display
Enter your choice 4
The tree is : 7 8 9 10 12 13 14 15 16
1. Create
2. Search
3. Delete
4. Display
Enter your choice 2
Enter the element which u want to search 16
The 16 Element is present
Parent of node 16 is 15
1. Create
2. Search
3. Delete
4. Display
Enter your choice 3
Enter the element u wish to delete 13
The 13 Element is present
Now deleted it !
1. Create
2. Search
3. Delete
4. Display
Enter your choice 4
This tree is : 7 8 9 10 12 14 15 16
```

Exercises

1. Compare the contrast between a binary and binary search tree.
2. Name the operations that are specific to binary search trees not to binary trees.
3. Can we insert a new node as an internal node ?
4. Which of the operation on BST is more complex and why ?
5. How the height of a BST effects its performance ?
6. What was the reason behind development of BST ?
7. Suppose the following list of numbers is inserted into an empty BST. Draw the resulting tree.
   20, 10, 18, 4, 8, 5, 13, 16, 17, 1, 27.
8. What is the maximum and minimum number of levels that a binary search tree with 100 nodes can have?
9. Write a recursive "C" function to search for an element in a given binary search tree. Write a non-recursive version for the same.

Chapter 10

AVL Tree

10.1 INTRODUCTION

The first balanced binary search tree was the AVL *tree* (named after its discoveries, **Adelson Velskii** and **Landis**). The AVL tree is a binary search tree that has an additional balance condition. This balance condition must be easy to maintain and ensures that the depth of the tree is O (log N). The simplest idea is to require that the left and right sub-trees have the same height. Recursion dictates that this idea applies to all nodes in the tree, since each node is itself a root of some sub-tree. This balance condition ensures that the depth of the tree is logarithmic, but it is too restrictive because it is too difficult to insert new items while maintaining balance. Thus the AVL trees uses a notion of balance that is somewhat weaker but still strong enough to guarantee logarithmic depth.

Chapter Outline

- Introduction
- Balance Factor
- Representation of an AVL Tree
- Searching an AVL Search Tree
- Insertion in an AVL Search Tree
- Deletion in an AVL Tree

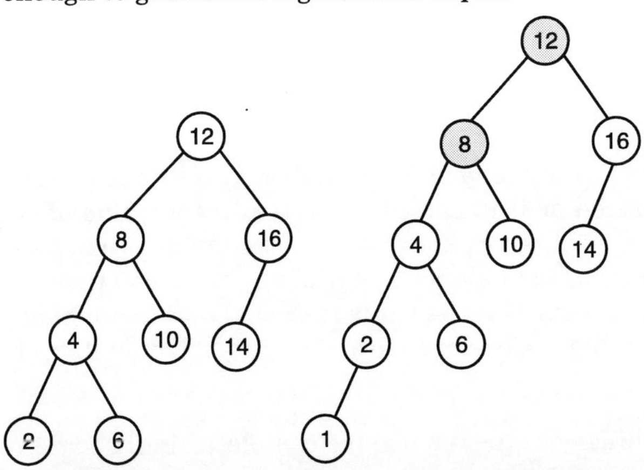

Fig. 10.1. Two binary search trees : the left tree is AVL tree, but the right tree is not (unbalanced nodes are darkened)

DO YOU KNOW ?

The AVL tree was the first balanced binary search tree. An AVL tree was originally called an admissible tree. In 1962, ***Adelson-Velski*** and ***Landis*** introduced a binary tree structure that is balanced with respect to the height of sub-trees. That is why it is called a height-balanced tree. It has historical significance and also illustrates most of the ideas that are used in other schemes.

AVL is a height balanced tree. A *height balanced tree* is one in which the difference in the height of the two sub-trees for any node is less than or equal to some specified amount. In AVL tree, the height difference may be no more than 1. In fact for an AVL tree it will never be greater than **1.44 log (n + 2)**. Because of this, the AVL tree provides a method to ensure that tree searches always stay close to the theoretical minimum of 0 (log n) and never degenerate to 0(n).

Definition

- An empty binary tree B is an AVL tree.
- If B is the non-empty binary tree with B_L and B_R are its left and right sub-trees then B is an AVL tree if and only if:

  (a) B_L and B_R are AVL trees, and

  (b) $|h_L - h_R| \leq 1$

  where h_L and h_R are the heights of B_L and B_R sub-tress, respectively.

10.2 BALANCE FACTOR

To implement an AVL tree each node must contain a balance factor, which indicates its states of balance relative to its sub-trees.

If balance is defined by

Height of left sub-tree – Height of right sub-tree

then the balance factors in a balanced tree can have values of –1, 0 or 1. In figure, numbers placed near the nodes indicate balance factors.

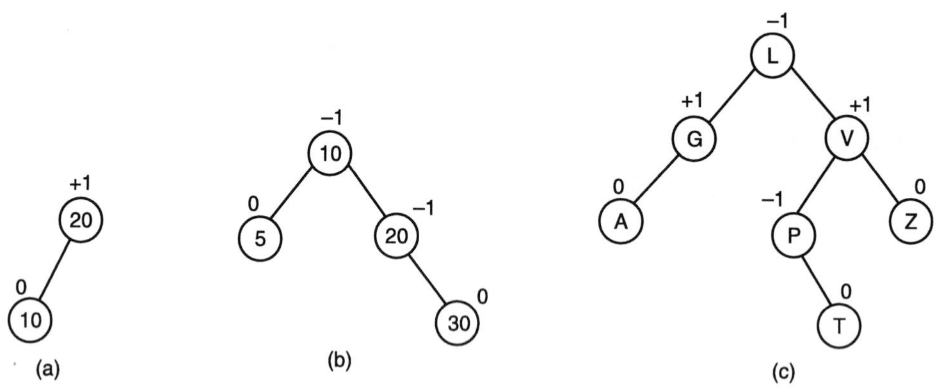

Fig. 10.2.

For an AVL tree, the value of the balance factor of any node is –1, 0, or 1. If it is other than these three values then the tree is not balanced or it is not an AVL tree. If the value of the balance factor of any node is –1, then the height of the right sub-tree of that node is one more than the height of its left sub-tree. If the value of the balance factor of any node is 0 then the height of its left and right sub-tree is exactly the same. If the value of the balance factor of any node is 1 then the height of the left subtree of that node is one more than the height of its right sub-tree.

A height-balanced binary tree is always a binary search tree and a complete binary search tree is always height balanced, but the reverse is not true.

10.3 REPRESENTATION OF AN AVL TREE

AVL search trees like binary search trees are represented using a linked representation. However, every node consists of its balance factor. So, to represent a node of an AVL tree four fields are required-one for data, two for storing the address of the left and right child and an additional field is required to hold the balance factor.

Left	Info	BF	Right

Node

Figure shows the representation of an AVL tree.

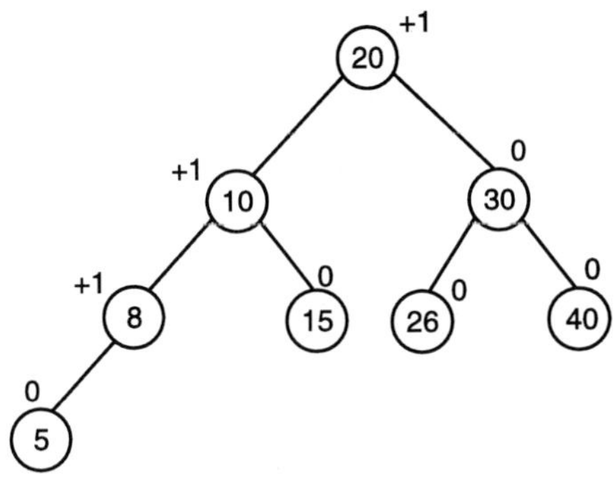

Fig. 10.3.

The number against each node represents its Balance Factor (BF). The linked representation of above tree is as follows :

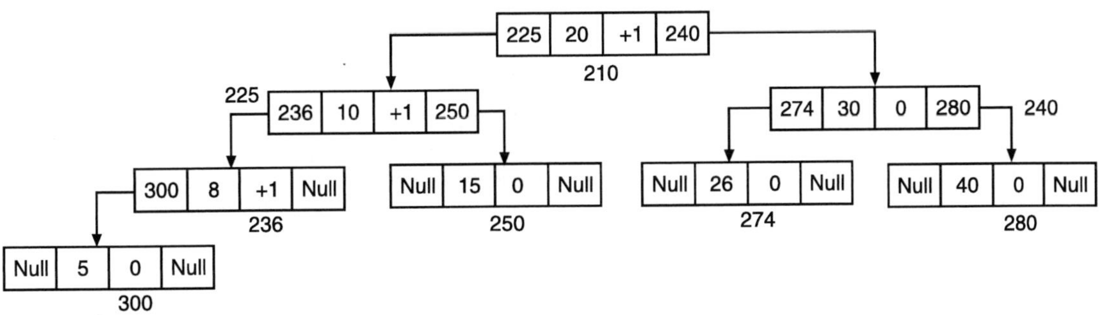

Fig. 10.4.

10.4 SEARCHING AN AVL SEARCH TREE

Searching an AVL search tree for an element is exactly similar to the method used in a BST.

10.5 INSERTION IN AN AVL SEARCH TREE

We can insert a new node in an AVL tree by finding its appropriate position similar to that of the binary search tree. But insertion of a new node involves certain overheads since the tree may become unbalanced due to the increase in its height.

1. If the data item with key 'k' is inserted into an empty AVL tree, then the node with key 'k' is set to be the root node. In this case, the tree is height balanced.

2. If tree contains only single node, *i.e.*, root node, then the insertion of the node with key 'k' depends upon its value. If the value of 'k' is less than the key value of the root then it is inserted to the left of the root otherwise right of the root. In this case the tree is height balanced.
3. If on inserting the node with key 'k' the height of the right sub-tree of the root has increased.

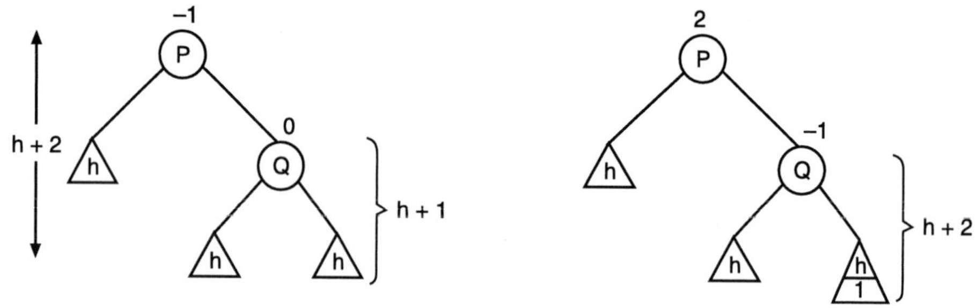

Fig. 10.5.

For example, Inserting node 55

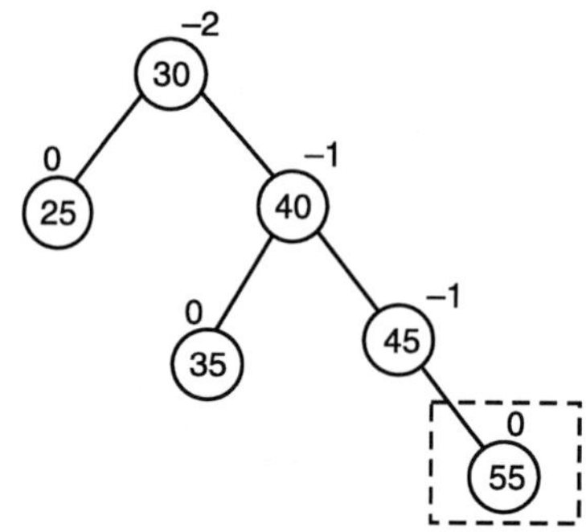

Fig. 10.6.

In this case, the problem can be easily rectified by rotating node Q about its parent P, so that the balance factor of both P and Q becomes zero, *i.e.*,

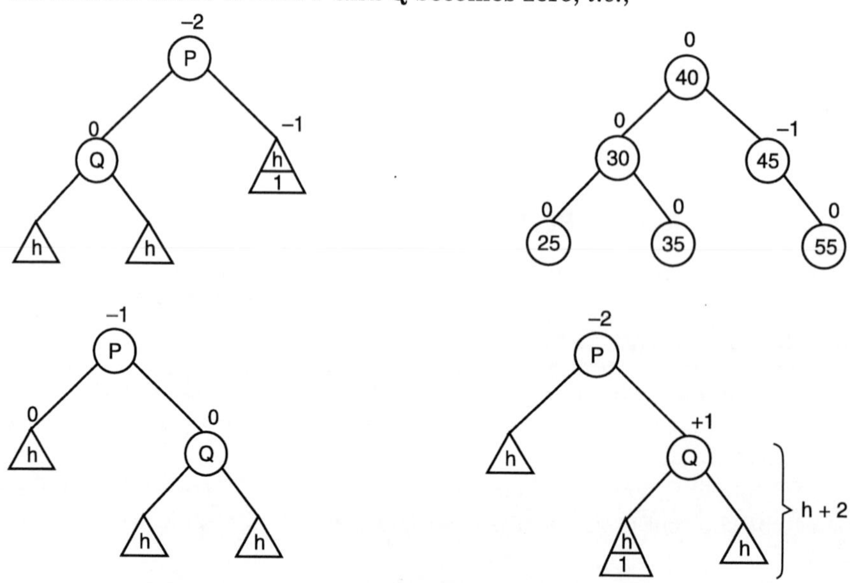

Fig. 10.7.

In this case, insert a node in the left sub-tree of the right child, *i.e.*,

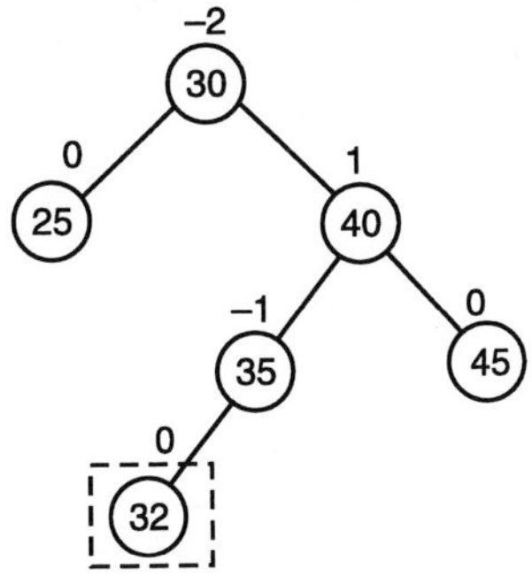

Fig. 10.8.

4. If on inserting a node with key 'k', the height of the left sub-tree of the root is increased *i.e.*,
   (*a*) Height of the right sub-tree of the left sub-tree of the root node is increased, *i.e.*,

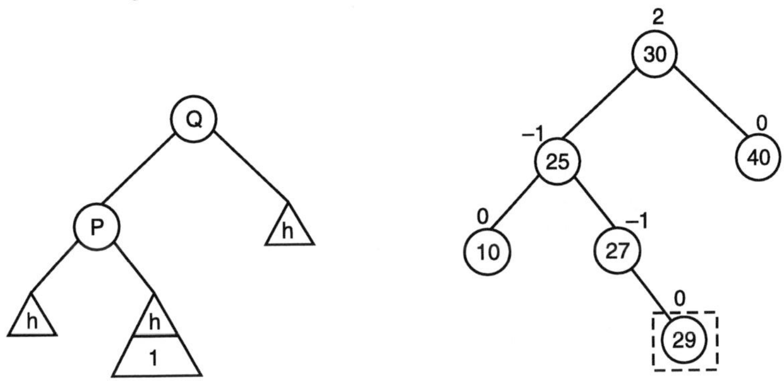

Fig. 10.9.

   (*b*) Height of the left sub-tree of the left sub-tree of the root node is increased, *i.e.*,

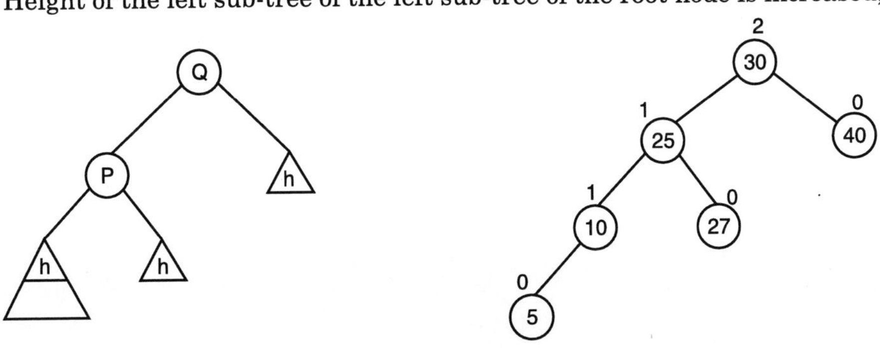

Fig. 10.10.

10.5.1 AVL Rotations

In order to balance a tree, there are four cases of rotations, such as:

LL Rotation	Inserted node is in the left sub-tree of left sub-tree of node.
RR Rotation	Inserted node is in the right sub-tree of right sub-tree of node.
LR Rotation	Inserted node is in the right sub-tree of left sub-tree of node.
RL Rotation	Inserted node is in the left sub-tree of right sub-tree of node.

***(a)* LL Rotation :** The new node x is inserted in the left sub-tree of left sub-tree of A whose balance factor becomes +2 after insertion. To rebalance the search tree, it is rotated so as to allow B to be the root with B_L and A as B_L is the left sub-tree and A is the right sub-tree and B_R and A_R to be the left and right sub-trees of A.

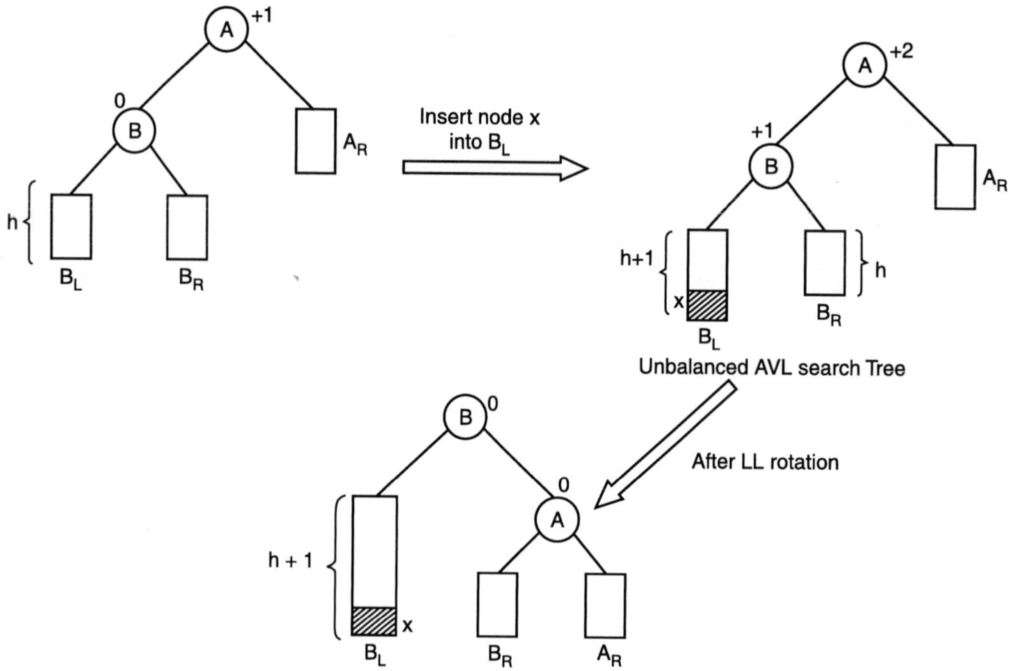

Fig. 10.11. Unbalanced AVL Search Tree.

(Here A_R : Right subtree of A, B_L : Left subtree of B, B_R : Right sub-tree of B)
i.e., in LL Rotation

Left (A) ← Right (B)
Right (B) ← A

Example : Insert node 10, in

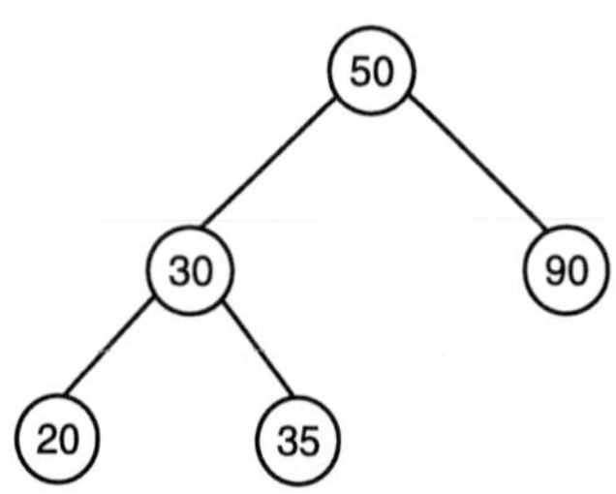

We get

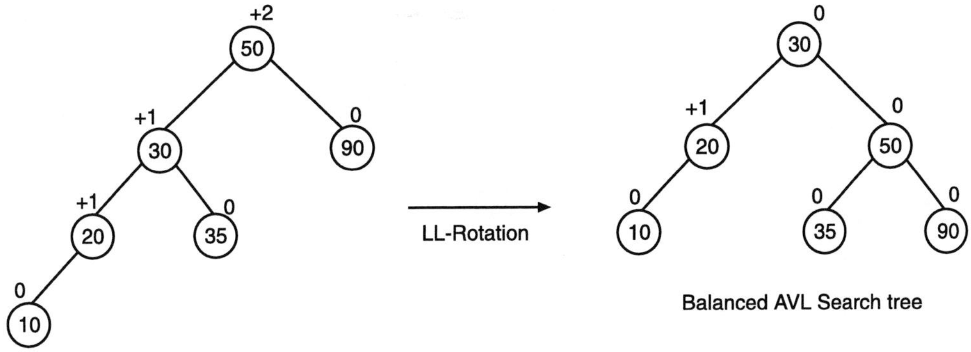

Fig. 10.12.

***(b)* RR Rotation**

Unbalance occurred due to the insertion in the right sub-tree of the right child of the pivot node. So, in this rotation new node is inserted at the right sub-tree of the right sub-tree of root node (pivot node).

Fig. 10.13 illustrates the balancing of an AVL search tree using RR rotation. Here the new node x is in the right sub-tree of A. The re-balancing rotation pushes B upto the root with A as its left child and B_R as its right sub-tree and A_L and B_L as the left and right sub-trees of A.

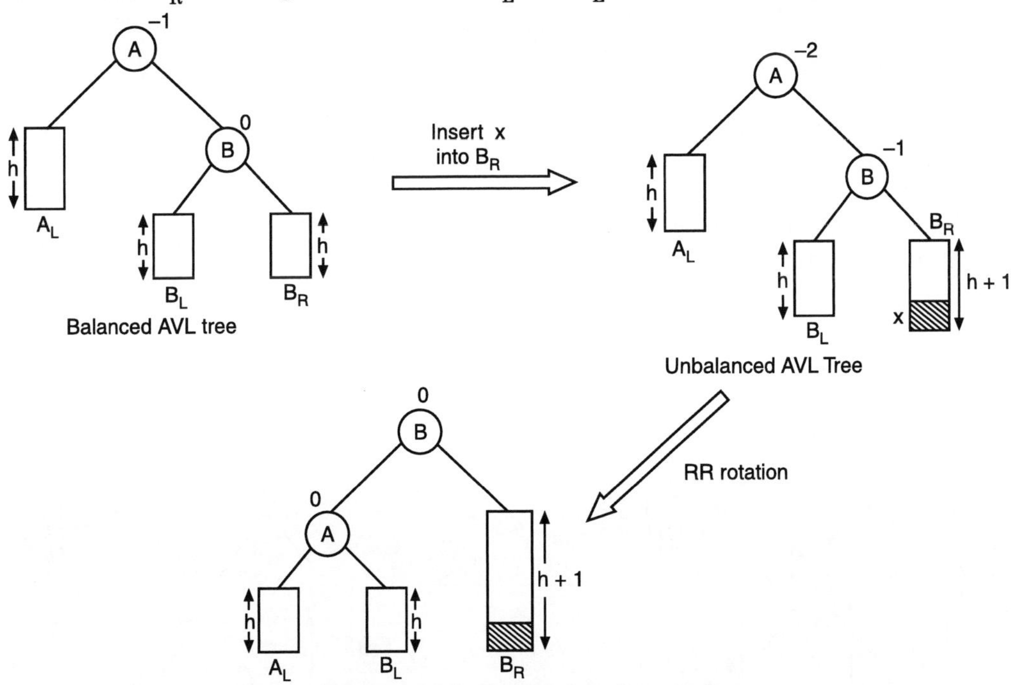

Fig. 10.13.

That is, in RR rotation

Right (A) ← left (B) left (B) ← A

Example : Insert node 85 in.

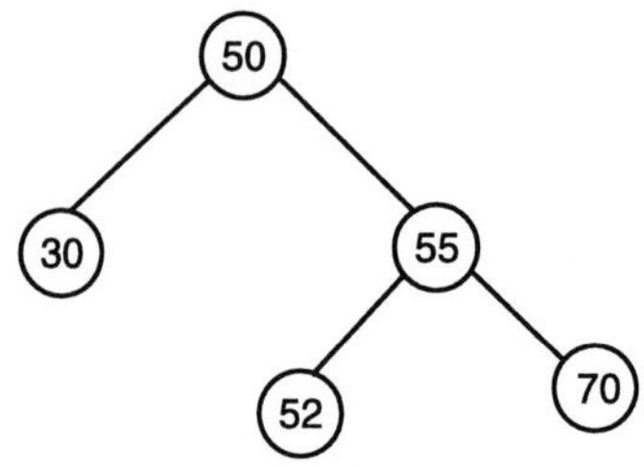

Fig. 10.14.

We get,

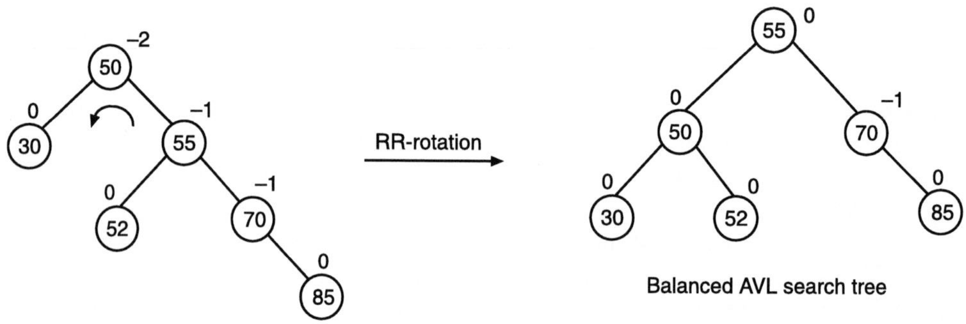

Fig. 10.15.

(c) LR Rotation

In this, unbalance occurred due to the insertion in the right sub-tree of the left child of the root (pivot) node. So, this is known as left to right insertion.

LR rotation involves two rotations for the manipulation in pointers.

> **DO YOU KNOW ?**
>
> Amongst the rotations, LL and RR rotations are called as single rotations and LR and RL are known as double rotations since, LR can be accomplished by RR followed LL rotation and RL can be accomplished by LL followed by RR rotation.

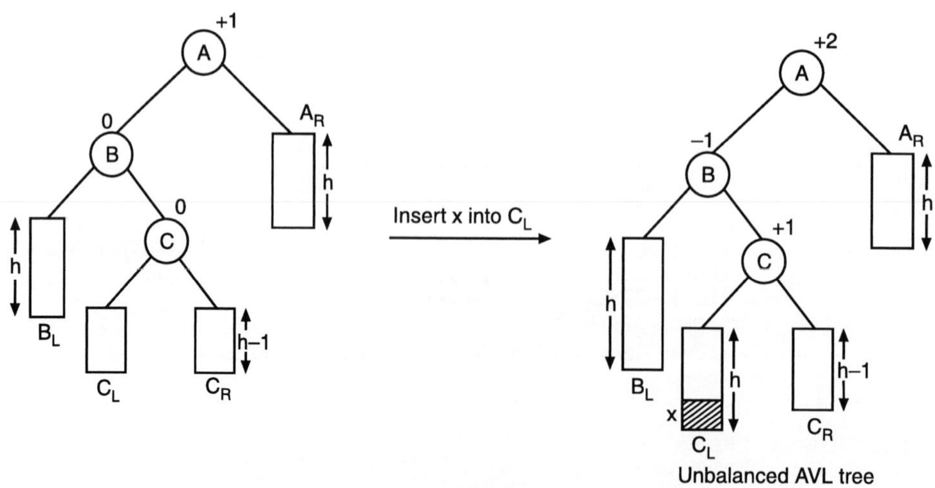

Fig. 10.16.

Rotation 1 : The left sub-tree (C_L) of the right child (C) of the left child (B) of pivot node (A) becomes the right sub-tree of the left child (B). The left child (B) of the pivot node (A) becomes the left child of C (*i.e.*, RR-Rotation).

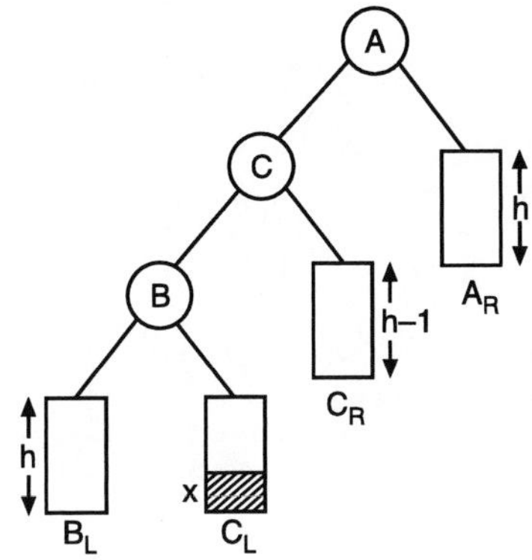

Fig. 10.17.

Rotation 2 : The right sub-tree (C_R) of the left child (C) of the left child (B) of the pivot node (A) becomes the left sub-tree of A and A becomes the right child of C.

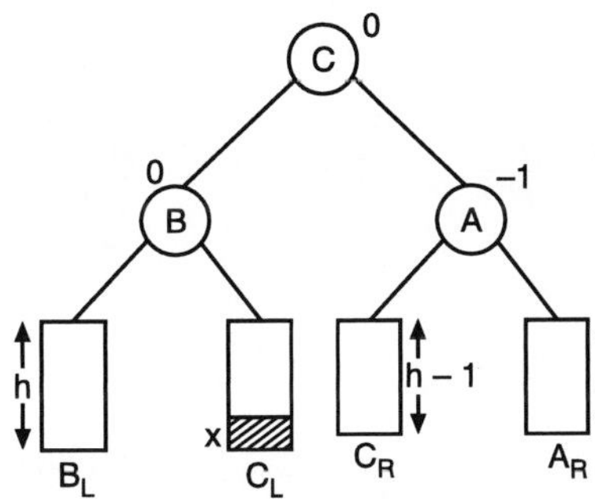

Fig. 10.18. Balanced AVL Search Tree.

The pointer movement is:

Right (B) ← left (C) left (A) ← right (C) left (C) ← B right (C) ← A.

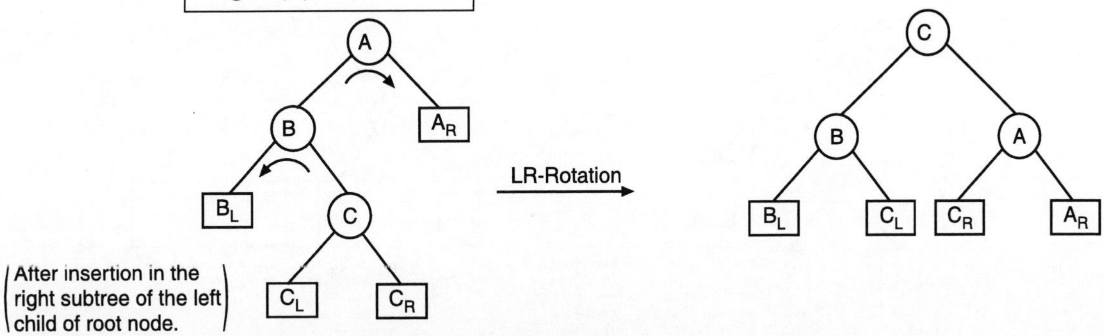

Fig. 10.19.

Example : Consider the AVL tree which is height balanced in Fig. 10.20 below.

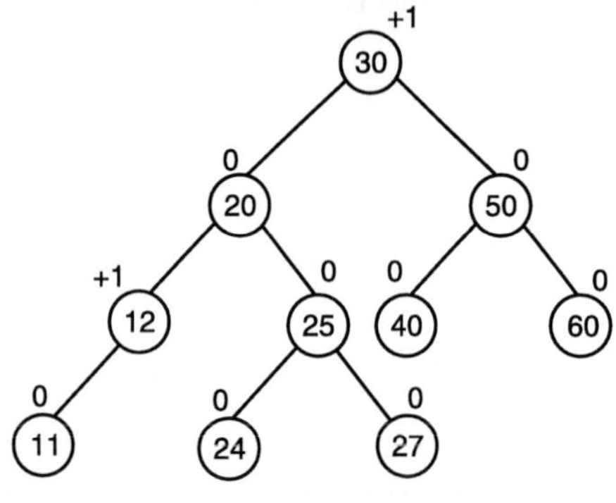

Fig. 10.20.

Insertion of the value 29 makes the tree unbalanced as shown in Fig. 10.21(a). Now AVL rotation has to be performed to make the tree balanced.

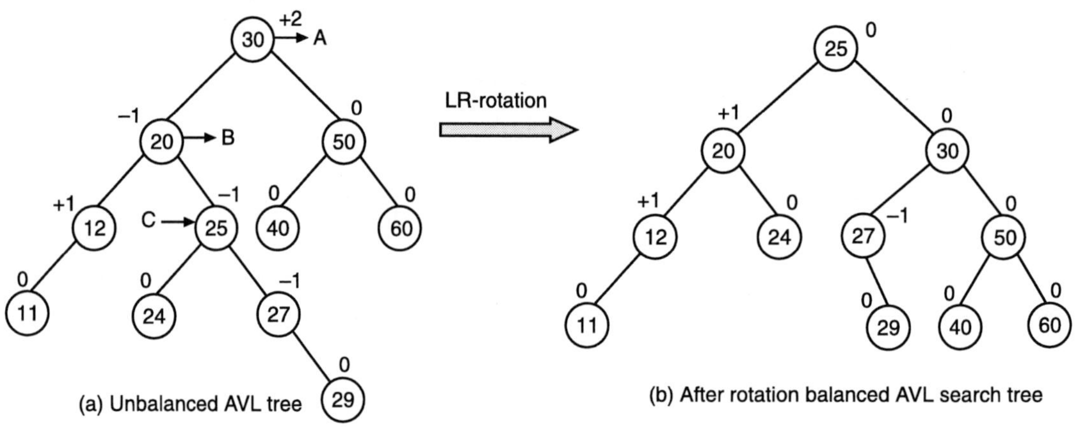

Fig. 10.21.

(*d*) RL Rotation

In this, unbalance occurred due to the insertion in the left sub-tree of the right child of the root (pivot) node. This is known as **right-to-left insertion**.

RL-Rotation is the mirror image of LR-Rotation. Fig. 10.22, illustrates the unbalance appearance and its removal due to the insertion.

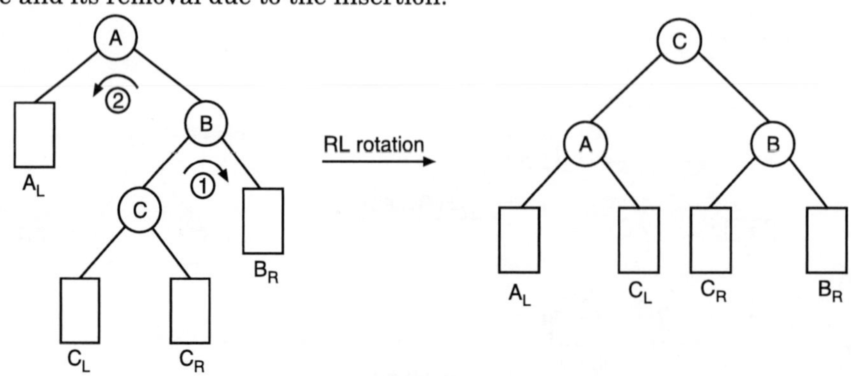

Fig. 10.22.

Here, there are two rotations for the manipulations of pointers. They are the following :

Rotation 1 : The right sub-tree (C_R) of the left child (C) of the right child (B) of root node (A) becomes the left sub-tree of B and the right child (B) of the root node becomes the right child of C.

Rotation 2 : The left sub-tree (C_L) of the left child (C) of the right child (B) of the root node becomes the right sub-tree of A.

Example : Consider the AVL tree as in figure below :

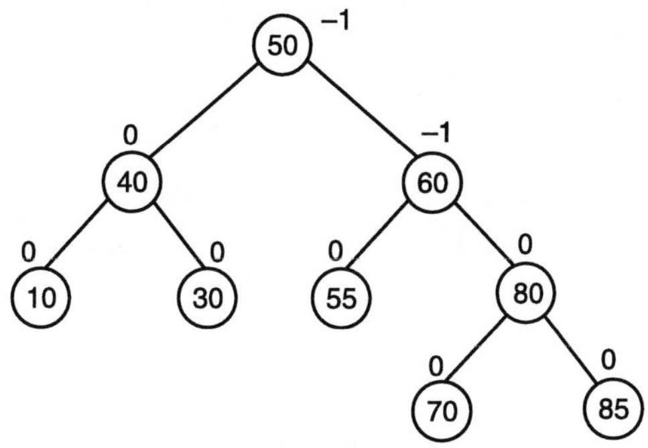

Fig. 10.23.

This is height unbalanced. When 65 is inserted into this tree it becomes

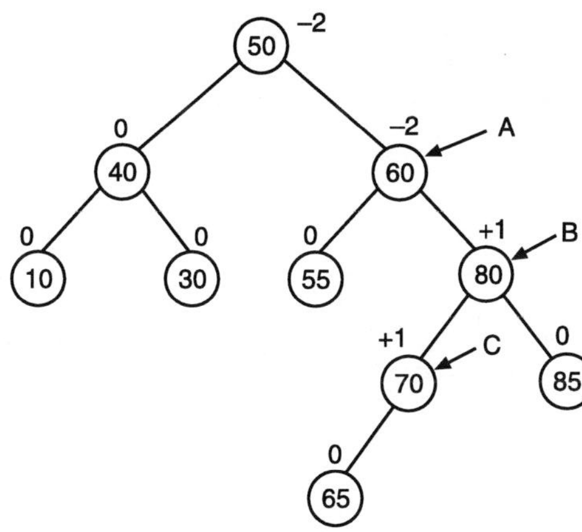

Fig. 10.24.

Now the AVL tree is unbalanced. To make it balanced, do the AVL rotations (RL rotation).

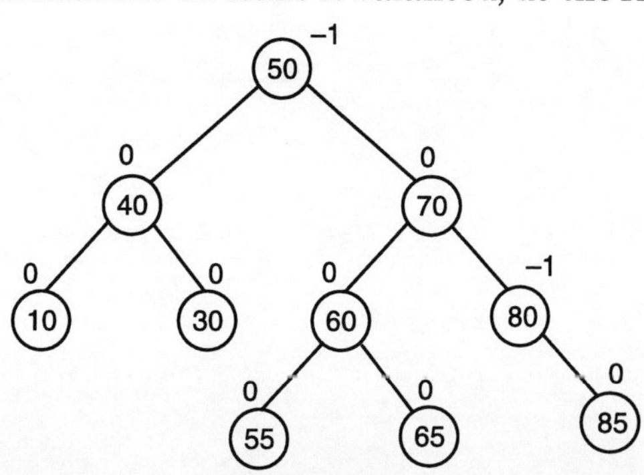

Fig. 10.25.

Now, the tree has become a height-balanced tree.

EXAMPLE 10.1. Create an AVL search tree from the given set of values :
H, I, J, B, A, E, C, F, D, G, K, L

Solution: Insert H

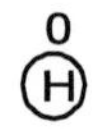

No rebalancing

Insert I

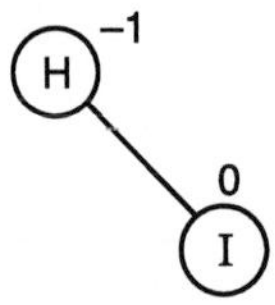

No rebalancing

Insert J

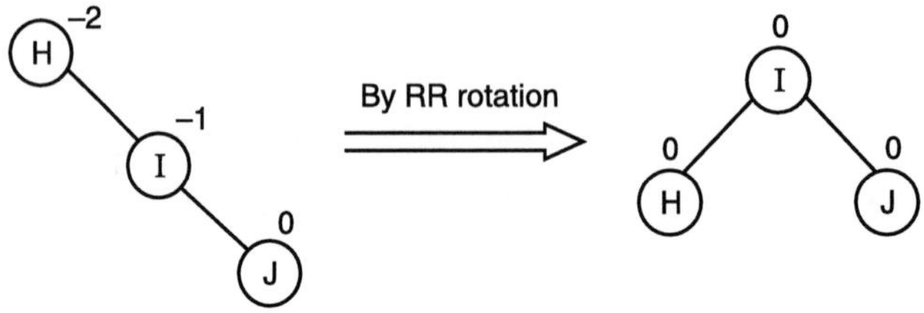

Insert B

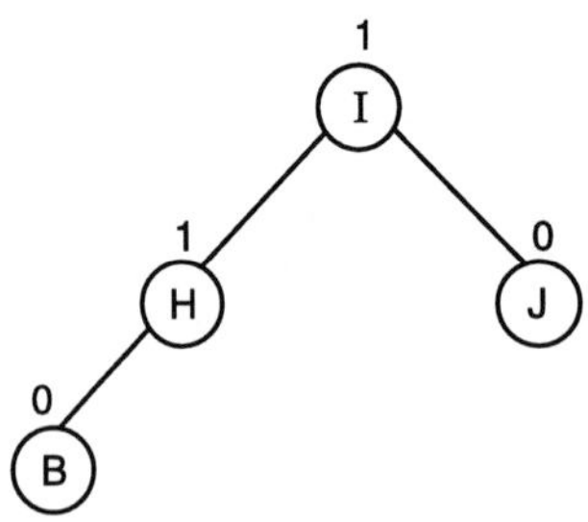

No rebalancing

Insert A

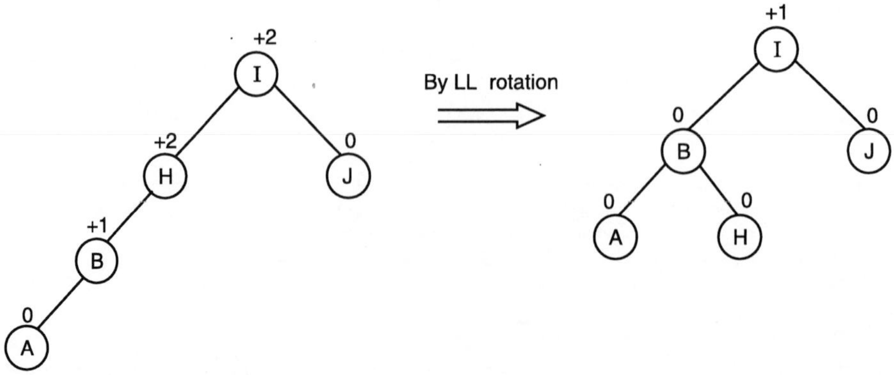

Insert E

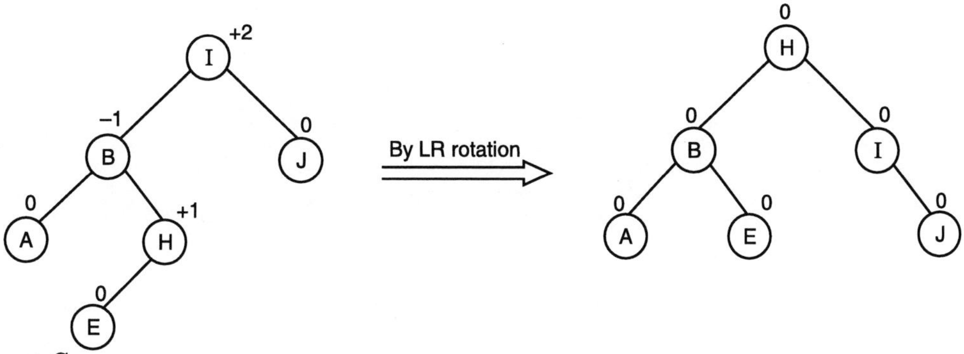

Insert C

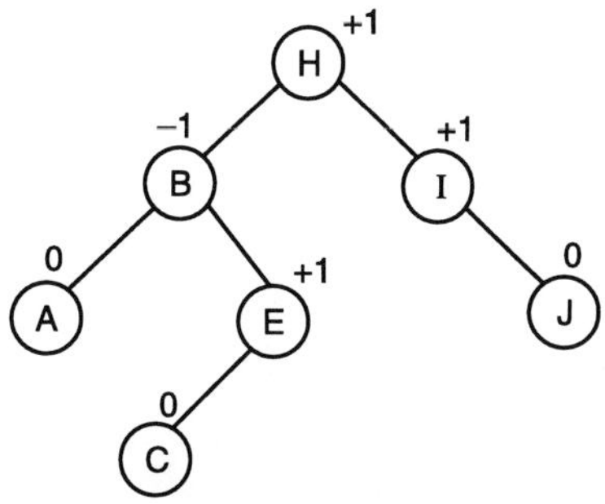

No rebalancing

Insert F

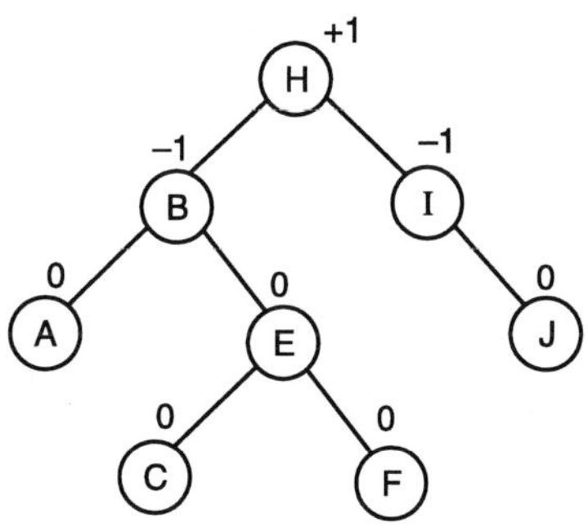

No rebalancing

Insert D

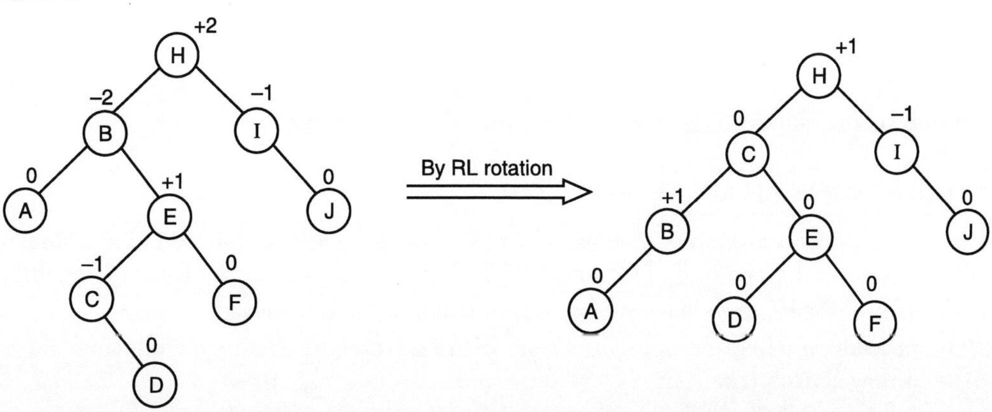

Insert G

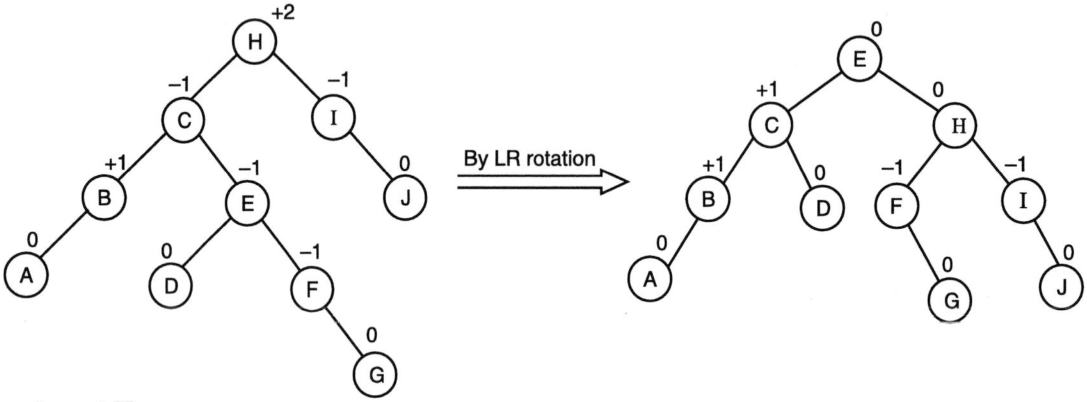

Insert K

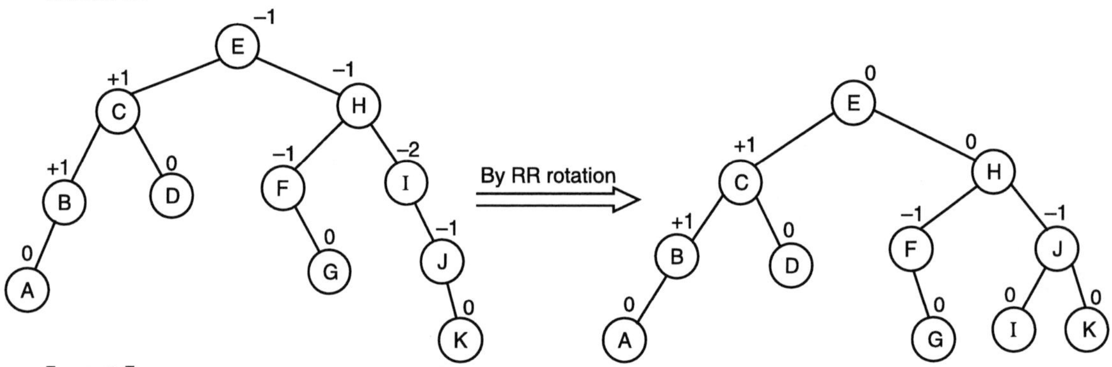

Insert L

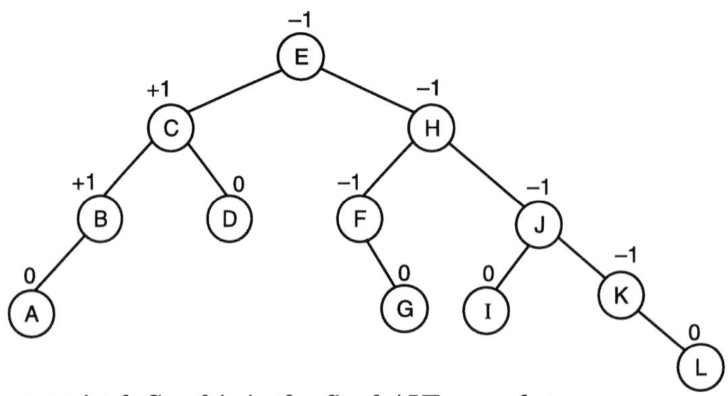

No rebalancing required. So, this is the final AVL search tree.

10.6 DELETION IN AN AVL TREE

The deletion of a node from an AVL tree is exactly the same as the deletion of a node from the BST. The sequence of steps to be followed in deleting a node from an AVL tree is as follows :

1. Initially, the AVL tree is searched to find the node to be deleted.
2. The procedure used to delete a node in an AVL tree is the same as deleting the node in the binary search tree.
3. After deletion of the node, check the balance factor of each node.

4. Rebalance the AVL tree if the tree is unbalanced. For this, AVL rotations are used.

On deletion of a node x from the AVL tree. Let A be the closest ancestor node on the path from x to the root node, with a balance factor of +2 or –2. To restore balance, the rotation is classified as L or R depending on whether the deletion occurred on the left or right sub-tree of A.

Now depending on the value of balance factor of B, where B is the root of the left or right sub-tree of A, the R or L rotation is further classified as R0, R1 and R –1 or L0, L1 and L –1.

RO Rotation and LO Rotation : When balance factor of B is 0. RO rotation is used.

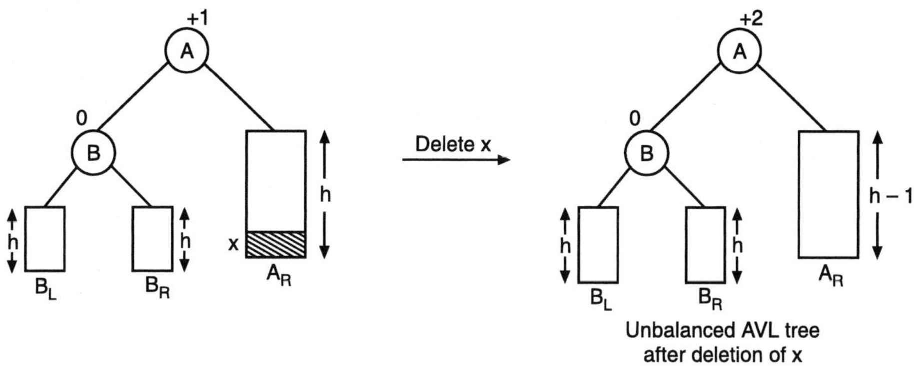

Fig. 10.26.

From RO rotation, we get

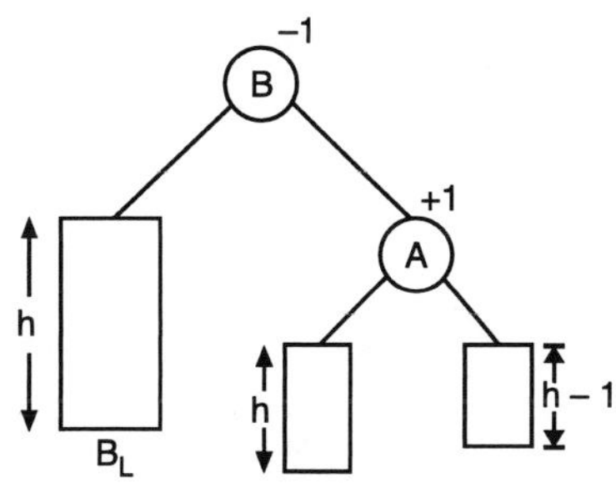

Fig. 10.27.

Example : Delete 90 from the AVL search tree shown in Fig. 10.27.

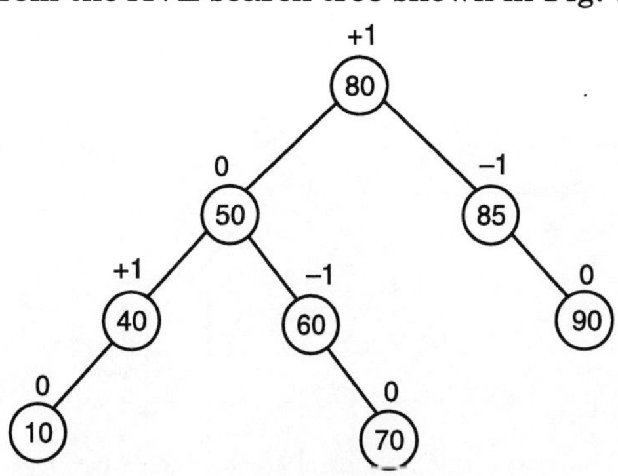

Fig. 10.28.

After deletion of 90, we get unbalanced AVL tree, as

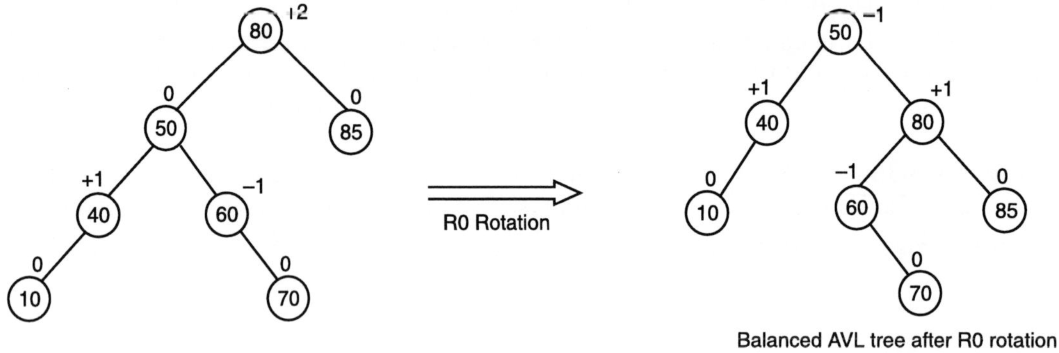

Fig. 10.29.

Similarly LO rotation is executed, when balance factor of B is zero and B is the root of right sub-tree of A. LO rotation is executed as illustrated in Fig. 10.30.

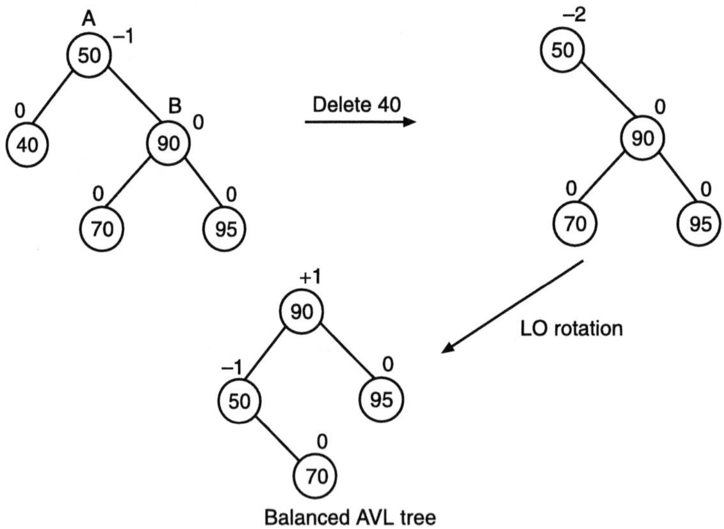

Fig. 10.30.

R1 and L1 Rotation

If balanced factor of B is 1. R1 or L1 rotation is used R1 rotation as illustrated in figure below.

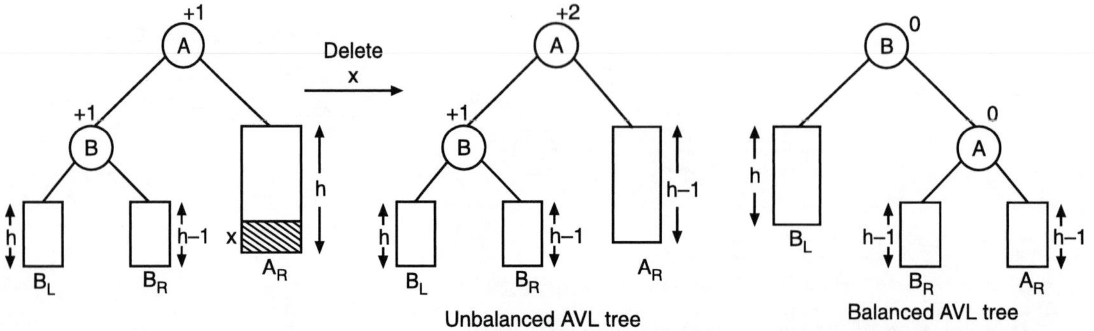

Fig. 10.31.

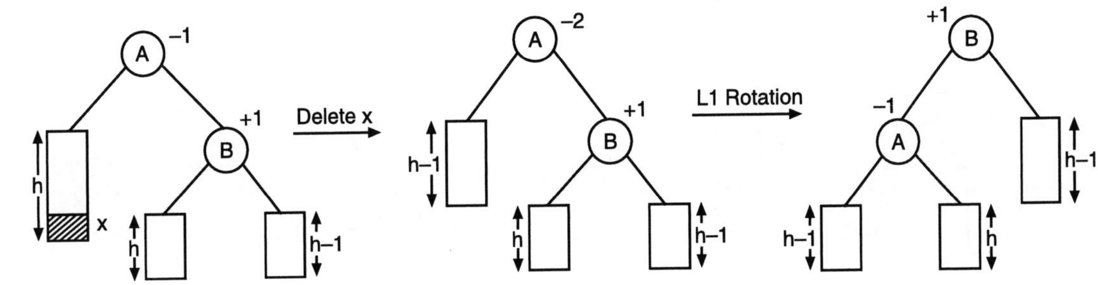

Fig. 10.32.

Example :

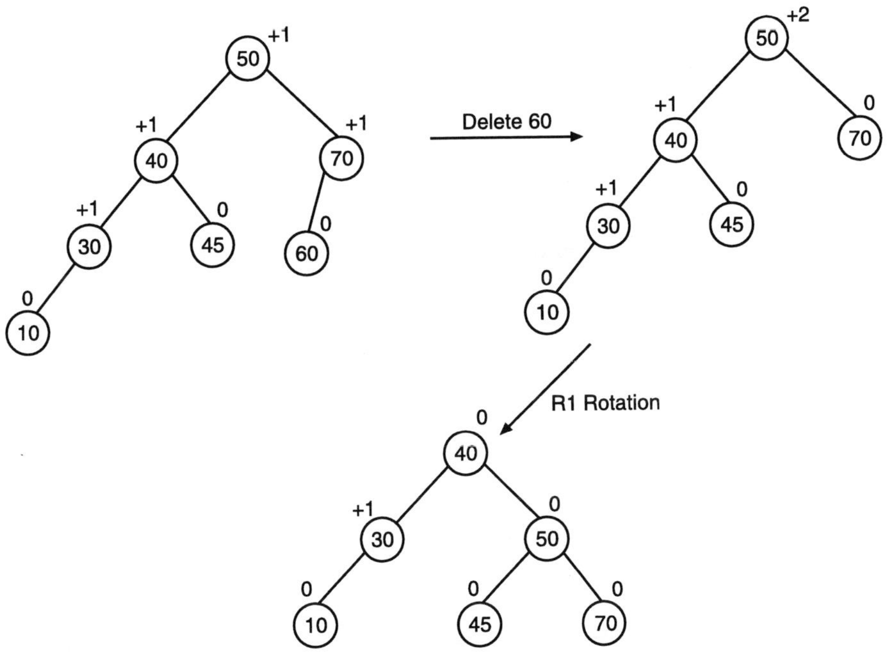

Balanced AVL Tree

Fig. 10.33.

R(–1) and L(–1) Rotation

If balance factor of B is –1, then R(–1) and L(–1) rotation is used. R(–1) rotation is shown in figure.

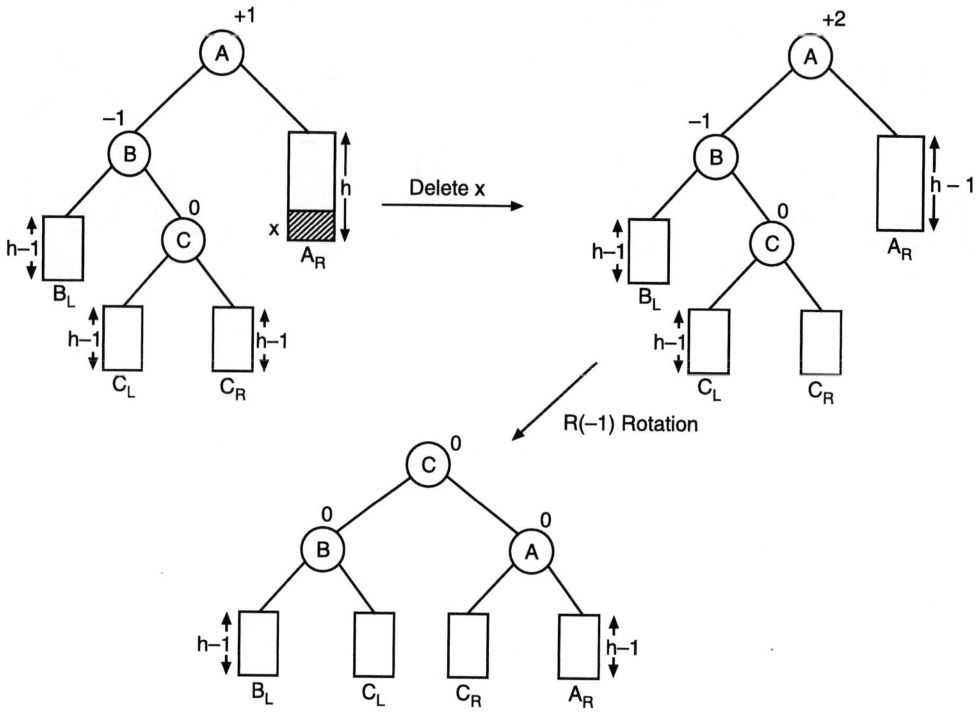

Balance AVL Search Tree

Fig. 10.34.

Example :

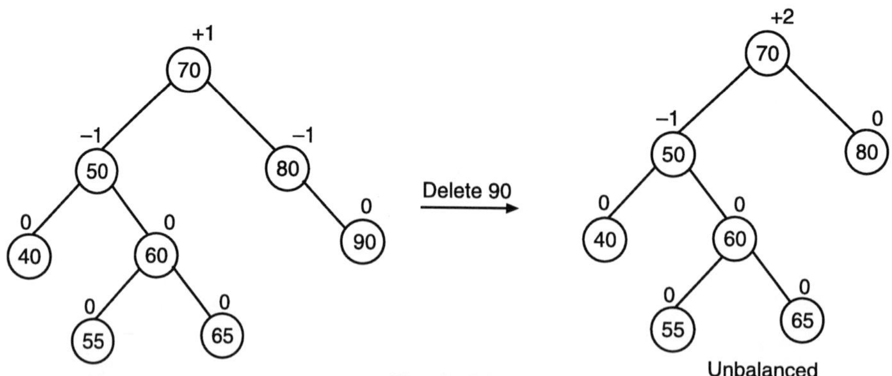

Fig. 10.35.

By R(–1) rotation, we get

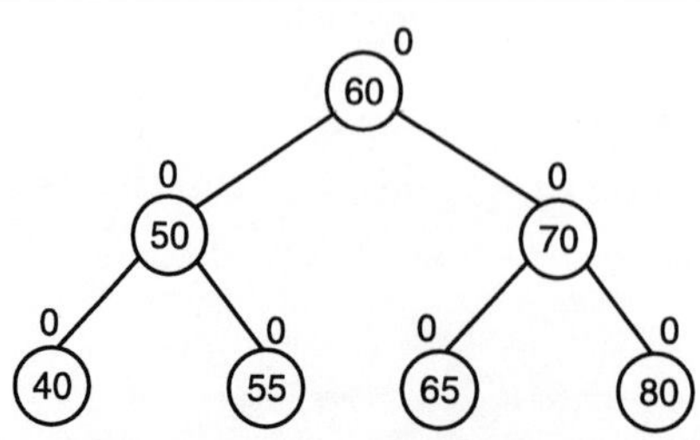

Balance AVL Search Tree

Fig. 10.36.

EXAMPLE 10.2. **Insert the following keys in order shown to construct an AVL tree**

10, 20, 30, 40, 50

Delete the last two keys in the order of LIFO (Last in Fast Out).

Solution: Insert 10, 20

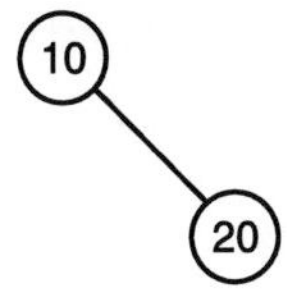

Insert 30

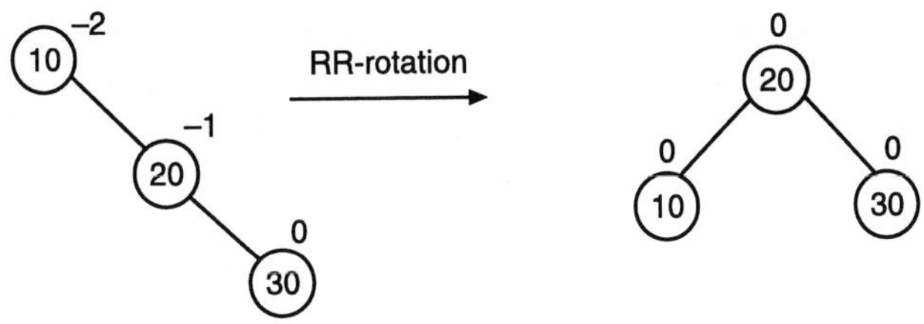

Insert 40

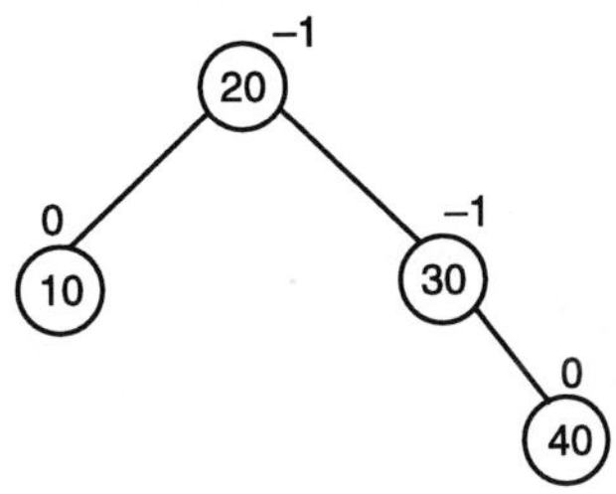

Insert 50

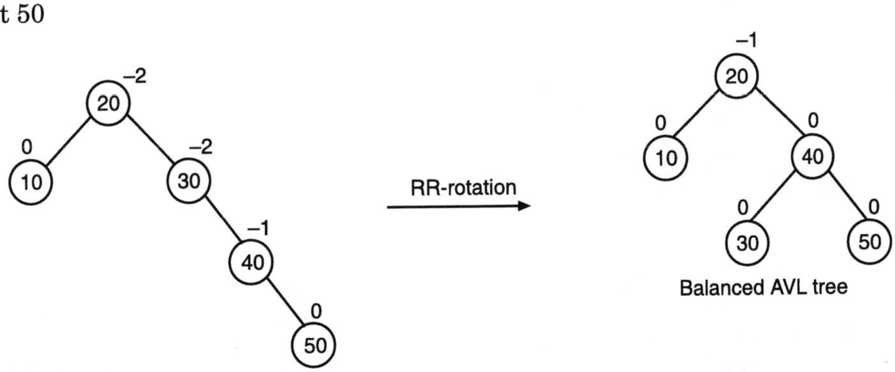

Now Delete 50

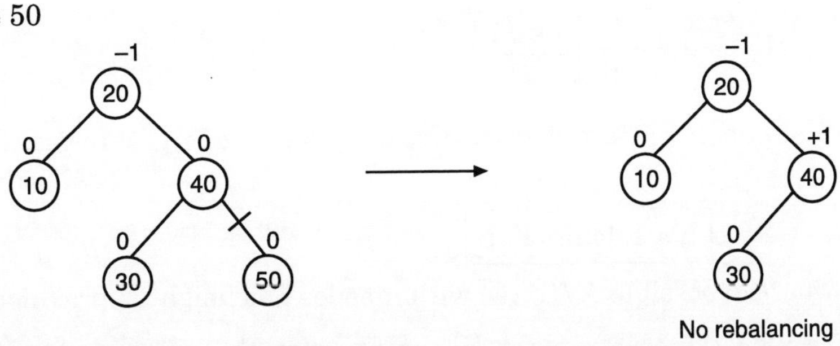

Delete 40

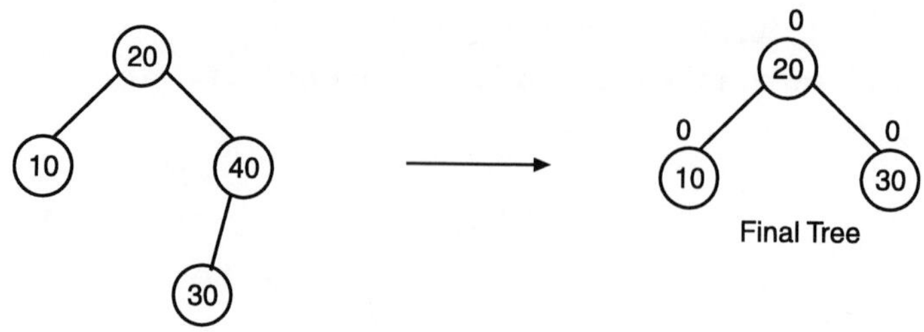

Height of an AVL Tree

To determine the maximum height that an AVL tree with n nodes can have, we can instead ask what is the maximum number of nodes that an AVL tree of height h can have. If F_h is such a tree and left and right sub-trees of its root are F_l and F_r, then one of F_l and F_r must have height h –1, say F_l and the other has height either h–1 or h–2. Since F_h has the minimum number of nodes among AVL trees of height h, it follows that F_l must have the minimum number of nodes among AVL trees of height h–1 and F_r must have height h–2 with minimum number of nodes.

The trees built by the above rule are called ***Fibonacci trees*** such as

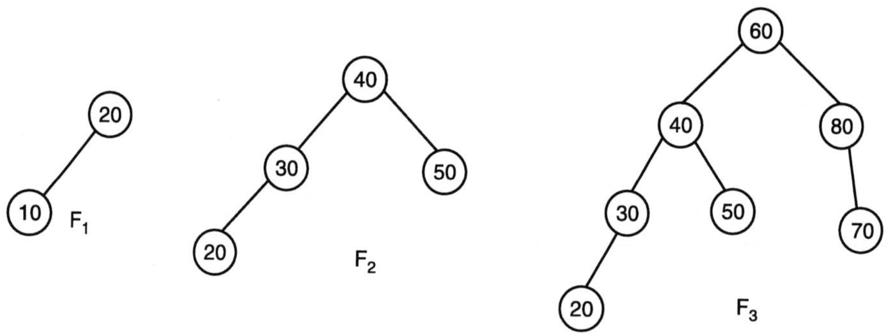

Fig. 10.37.

If we write $|F|$ for the number of nodes in a tree F, we then have (counting the root as well as the sub-trees) the recurrence relation.

$$|F_h| = |F_{h-1}| + |F_{h-2}| + 1$$

where $\quad |F_0| = 1 \quad \text{and} \quad |F_1| = 2$

By adding 1 to both sides, we see that the numbers $|F_h| + 1$ satisfy the definition of the Fibonacci numbers. By evaluation we get

$$|F_h| + 1 \approx \frac{1}{\sqrt{5}} \left[\frac{1+\sqrt{5}}{2}\right]^{h+2}$$

Take logarithms of both sides and discarding all except the largest terms. The approximate result is:

$$\boxed{h \approx 1.44 \lg |F_h|}$$

This means that the possible AVL tree with n nodes has height approximately 1.44 log n.

Theorem : An AVL tree of height H has at least $F_{H+3} - 1$ nodes, where F_i is the *i*th Fibonacci number.

Proof : Let S_H be the size of the smallest AVL tree of height H. Clearly $S_0 = 1$ and $S_1 = 2$. Fig. 10.38 shows that the smallest AVL tree of height H must have sub-trees of height H – 1 and H – 2, because at least one sub-tree has height H – 1 and the balance condition implies that subtree heights can differ by at most one. These sub-trees must themselves have the fewest number of nodes for their height, so $S_H = S_{H-1} + S_{H-2} + 1$. It is then a simple matter to complete the proof by using an induction argument.

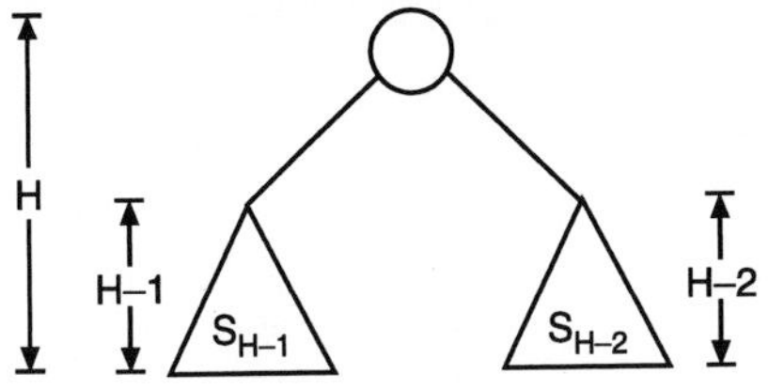

Fig. 10.38.

As we know that, $F_i = \phi^i/\sqrt{5}$, where $\phi = (1 + \sqrt{5})/2 \approx 1.618$. Consequently, an AVL tree of height H has at least (roughly) $\phi^{H+3}/\sqrt{5}$ nodes. This allows us to conclude that its depth is at most logarithmic. A precise calculation allows us to determine that the height of an AVL tree satisfies.

$$H < 1.44 \log (N + 2) - 1.328$$

so the worst-case height is at most 44 percent more than the minimum possible for binary trees.

The AVL tree has height at most roughly 44 percent greater than the minimum.

PROGRAM :

```
/* insertion operation of AVL tree */
#include<stdio.h>
#include<malloc.h>
typedef enum {FALSE, TRUE} bool;
struct node
{
int info;
int balance;
struct node *lchild;
struct node *rchild;
};
struct node *insert(int,struct node *,int*);
struct node *search(struct node *,int);
main()
{
bool ht_inc;
int info;
int choice;
```

```
struct node *root=(struct node *)malloc(sizeof(struct node));
root=NULL;
while(1)
{
printf("1.Insert\n");
printf("2.Display\n");
printf(3.Quit\n");
printf("Enter your choice :");
scanf("%d",&choice);
switch(choice)
{
case 1:
printf("Enter the value to be inserted :");
scanf("%d",&info);
if(search(root,info)==NULL)
root=insert(info,root,&ht_inc);
else
printf("Duplicate value ignored\n");
break;
case 2:
if(root==NULL)
{
printf("Tree is empty\n");
continue;
}
printf("Tree is :\n");
display(root, 1);
printf("\n\n");
printf("Inorder Traversal is: ");
inorder(root);
printf("\n");
break;
case 3:
exit(1);
defalt:
printf("wrong choice\n");
}
}
}
struct node* search(struct node*ptr,int info)
{
```

```
if(ptr!=NULL)
if(info< ptr->info)
ptr=search(ptr->lchild,info);
else if(info > ptr->info)
ptr=search(ptr->rchild,info);
return(ptr);
}
struct node *insert(int info,struct node *pptr,int *ht_inc)
{
struct node *aptr;
struct node *bptr;
if(pptr==NULL)
{
pptr = (struct node *)malloc(sizeof(struct node));
pptr->info = info;
pptr->lchild = NULL;
pptr->rchild = NULL;
pptr->balance = 0;
*ht_inc = TRUE;
return (pptr);
}
if(info < pptr->info)
{
pptr->lchild = insert(info,pptr->lchild,ht_inc);
{
switch(pptr->balance)
{
case -1:
pptr->balance = 0;
*ht_inc = FALSE;
break;
case 0;
pptr->balance = 1;
break;
case 1:
aptr = pptr->lchild;
if( aptr->balance == 1)
{
printf("Left to Left Rotation\n");
pptr->lchild = aptr->rchild;
aptr->rchild = pptr;
```

```
pptr->balance = 0;
aptr->balance = 0;
pptr = aptr;
}
else
{
printf("Left to right Rotation\n");
bptr = aptr->rchild;
aptr->rchild = bptr->lchild;
bptr->lchild = aptr;
pptr->lchild = bptr->rchild;
bptr->rchild = pptr;
if(bptr->balance ==1)
pptr->balance = -1;
else
pptr->balance = 0;
if(bptr->balance == -1)
aptr->balance = 1;
else
aptr->balance = 0;
bptr->balance = 0;
pptr=bptr;
}
*ht_inc = FALSE;
}
}
}
if(info > pptr->info)
{
pptr->rchild = insert(info,pptr->rchild,ht_inc);
if(*ht_inc==TRUE)
{
switch(pptr->balance)
{
case 1:
pptr->balance = 0;
*ht_inc = FALSE;
break;
case 0:
pptr->balance = -1;
break;
```

```
case -1:
aptr = pptr->rchild;
if(aptr->balance == -1)
{
printf("Right to Right Rotation\n");
pptr->rchild= aptr->lchild;
aptr->lchild = pptr;
pptr->balance = 0;
aptr->balance = 0;
pptr = aptr;
}
else
{
printf("Right to Left Rotation\n");
bptr = aptr->lchild;
aptr->lchild = bptr->rchild;
bptr->rchild = aptr;
pptr->rchild = bptr->lchild;
bptr->lchild = pptr;
if(bptr->balance == -1)
pptr->balance = 1;
else
pptr->balance = 0;
if(bptr->balance == -1)
aptr->balance = -1;
else
aptr->balance = 0;
bptr->balance = 0;
pptr = bptr;
}
*ht_inc = FALSE;
}
}
}
return(pptr);
}

display(struct node *ptr,int level)
{
int i;
if (ptr!=NULL)
```

```
{
display(ptr->rchild, level+1);
printf("\n");
for (i = 0;i < level; i++)
printf(" ");
printf("%d",ptr->info);
display(ptr->lchild, level+1);
}
}
inorder(struct node *ptr)
{
if(ptr! = NULL)
(inorder(ptr->lchild);
printf("%d",ptr->info);
inorder(ptr->rchild);
}
}
```

OUTPUT :

```
1. Insert
2. Display
3. Quit
Enter your choice :1
Enter the value to be inserted : 23
1. Insert
2. Display
3. Quit
Enter your choice :1
Enter the value to be inserted : 15
1. Insert
2. Display
3. Quit
Enter your choice :1
Enter the value to be inserted : 13
Left to Left Rotation
1. Insert
2. Display
3. Quit
Enter your choice :2
Tree is :
23
15
13
Inorder Traversal is :13 15 23
```

```
1. Insert
2. Display
3. Quit
Enter your choice :3
```

Advantages of an AVL Tree

Since AVL trees are height balanced trees, operations like insertion and deletion have low time complexity.

Let us consider an example. If we have the following tree with keys 10, 20, 30, 40, 50, 60, 70 then a binary tree would look like this.

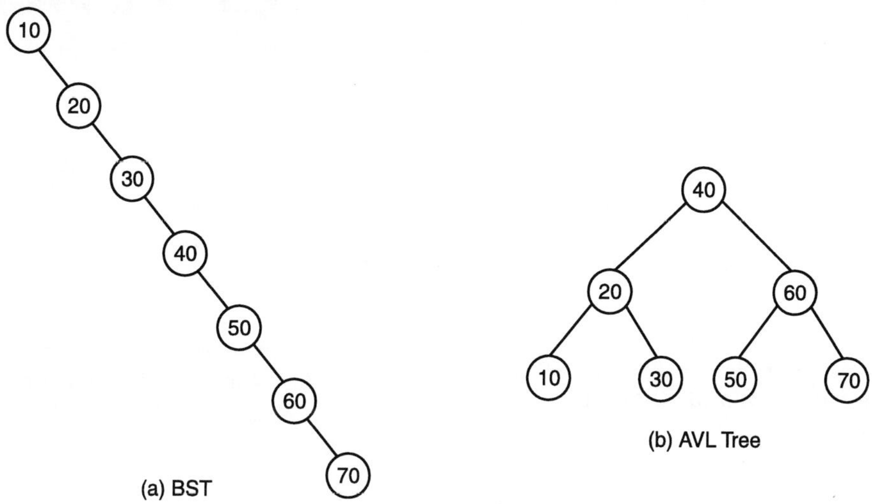

Fig. 10.39.

In order to insert a node with a key 'k' in the binary tree (a), the algorithm requires 7 comparisons, but if we insert the same key in AVL tree (b), the algorithm will just require 3 comparisons, which is less than half of the binary search tree.

Thus, AVL search trees will increase the efficiency of the programs.

Exercises

1. What is height-balanced tree ?
2. What is the purpose for AVL trees ?
3. What conditions defines an AVL tree among all BST trees ?
4. Draw a figure explaining how balance is restored when an insertion into an AVL tree puts a node out of balance.
5. In each of the following, insert the elements in the order shown, to build them into an AVL tree. (*a*) 2, 26, 1, 20, 3, 30 (*b*) 1, 2, 3, 4, 5, 6 (*c*) 1, 26, 5, 24, 8, 10
6. How an AVL tree differs from a BST ? How AVL trees are represented in computer memory?
7. In what way is an AVL tree better than a binary tree ?
8. Describe the AVL trees insertion and deletion method.

Chapter 11

B-Tree and Its Variants

All the data structures discussed so far favour data stored in the internal memory and hence support internal information retrieval. However to favour retrieval and manipulation of data stored in external memory *viz.*, storage devices such as disks etc. there is a need for some special data structures such as m-way search trees B-trees and B^+ trees.

Chapter Outline

11.1 WHY DO WE NEED ANOTHER TREE-STRUCTURE?

- In database programs, the data is too large to fit in memory; therefore, it is stored on secondary storage (disks or tapes).
- Disk access is very expensive; the disk I/O operation takes milliseconds while CPU processes data on the order of nanoseconds, one million times faster.
- When dealing with external storage the disk accesses dominate the running time.
- Balanced binary search trees (AVL & Red-Black) have good performance if the entire data can fit in the main memory.
- These trees are not optimised for external storage and require many disk accesses, thus give poor performance for very large data.
- Data is transfer to and from the disk in block (typically blocks are of 512, 2048, 4096 or 8192 bytes).

- To reduce disk accesses
  - Reduce tree height by increasing the number of children of a node
  - Store multiple records in a block on the disk.
- To achieve above goals we use Multiway (m-way) search tree, which is a generalisation of BST, binary search tree.

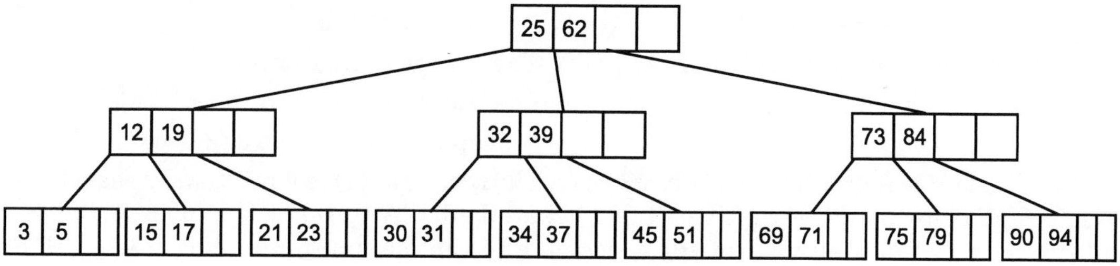

Fig. 11.1. 4- way Search tree

11.2 M-WAY SEARCH TREES

m-way search trees are generalised versions of binary search trees. The goals of m-way search tree is to minimise the accesses while retrieving a key from a file. However, an m-way search tree of height h calls for O (h) number of accesses for an insert/delete/retrieval operation. Hence it pays to ensure that the height h is close to $\log_m (n+1)$, because the number of elements in an m-way search tree of height h ranges from a minimum of h to a maximum of m^h—1. This implies that an m-way search tree of n elements ranges from a minimum height of $\log_m (n+1)$ to a maximum height of n. Therefore there arises the need to maintain balanced m-way search trees. B trees are balanced m-way search trees.

Trees having (m-1) keys and m children are called **m-way search trees**. A binary tree is 2-way tree. It means that it has $m - 1 = 2 - 1 = 1$ key (here m = 2) in every node and it can have maximum of two children. A binary tree is also called an m-way tree of order 2. Similarly an m-way tree of order 3 is a tree in which key values could be either 1 or 2 (*i.e.*, inside every node and it can have maximum of three children). For example, an m-way tree at order (degree) 4 is shown in Fig. 11.2.

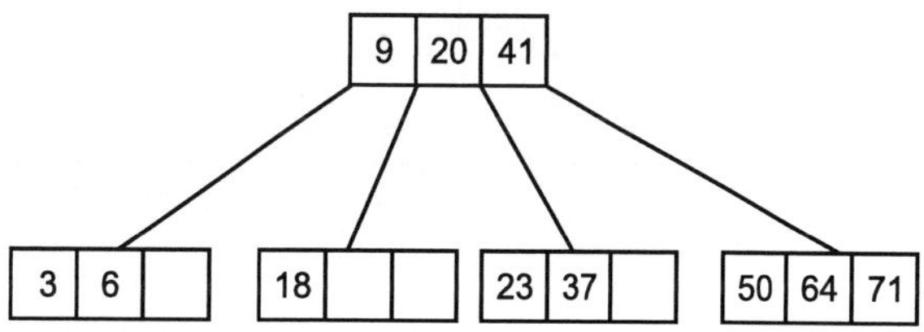

Fig. 11.2.

- In an m-way tree all the nodes have degree ≤ m.
- Each node has the following structure :

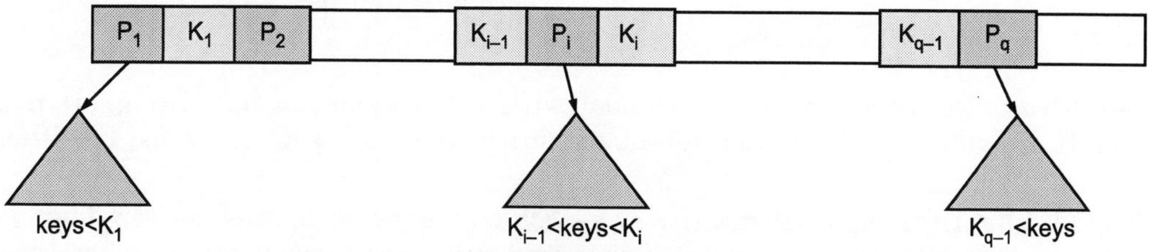

Fig. 11.3.

K_i are the keys, where $1 \le i \le q\text{-}1$, $q <= m$
P_i are pointers to sub-trees, where $1 \le i \le q$, $q <= m$

- The keys in each node are in ascending order $k_1 \le k_2 \le \ldots \le k_i$
- The key K_i is larger than keys in sub-tree pointed by P_i and smaller than keys in subtree pointed by $P_{i+1.}$
- The sub-trees are the m-way trees.
- M-way tree is a generalisation of BST, its working, benefits & issues are same.

Benefits	**Problems**
• Fast information retrieval.	• The tree is not balanced.
• Fast update.	• Leaf nodes are on different levels.
	• Bad space usage, tree can become skew.

Fig. illustrates a number of multi-way search trees. Fig (a) is a multi-way search tree of order 4. The eight nodes of that tree have been labelled A through H. Nodes A,D,E and G contain the maximum number of sub-trees, 4, and the maximum number of keys, 3. Such nodes are called *full nodes*. Nodes B,C,F and H are not full and also contain some empty sub-trees.

The first sub-tree of A contains the keys 5 and 12 but smaller than 15, which is the first key of A. The second sub-tree of A contains 20 and 30, both greater than 15 (the first key of A) and less than 50 (the second key of A). The third sub-tree contains 60,65,66,69,70,80 all of which are between 50 and 90 (the third key of A). Finally the last sub-tree of A contains 95, 98, 100 and 150 all greater than 90, the last key in node A. Similarly each sub-tree of any other node contains only keys between the appropriate two keys of that node and its ancestors.

Fig. 11.4 (*b*) illustrates a **top-down multiway search tree** of order 3 such a tree is characterised by the condition that any non-full node is a leaf.

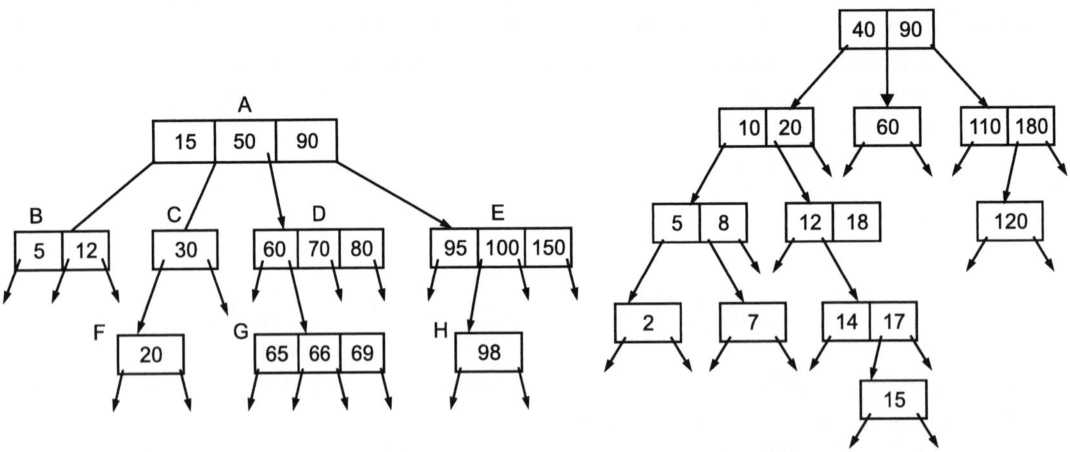

(a) **(b)**

Semi-leaf : Define a **semi-leaf** as a node with at least one sub-tree. In fig. (a), nodes B through H are all semi-leaves. In a top-down multiway tree, a semileaf must be either full or a leaf.

Fig. 11.4. (*c*) is yet another multi-way search tree of order 3. It is not top-down. Since there are 4 nodes with only one key and non-empty sub-trees. However, it does have another property in that it is balanced—That is, all its semileaves are at the same level (3). This implies that all semi-leaves are leaves.

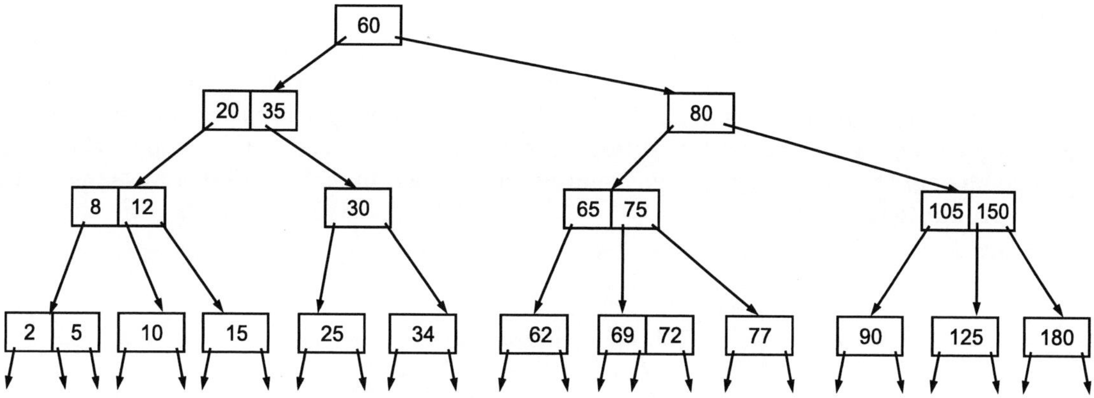

(c)

Fig. 11.4.

11.3 SEARCHING A MULTI-WAY TREE

Searching for a key in a m-way search tree is similar to that of binary search trees. The procedure is as follows.

1. If the key is less than k_1 and k_2 then the search is continued in m-way search tree pointed to by pointer p_1
2. If the key lies between k_1 and k_2, than the search is continued in m-way search tree pointed by p_2
3. If the key lies between k_2 and k_3, then the search is continued in m-way search tree pointed by p_3.

.

.

.

4. If the key is greater than k_{q-1}, then search is continued in m-way search tree pointed to by p_q

These steps are carried on recursively till either the key is found or the m-way search tree is empty.

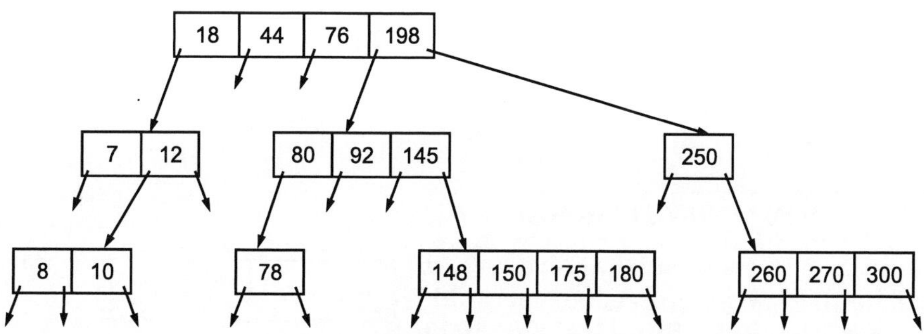

Fig. 11.5 5-way search tree.

To search for 78 in the above 5-way search tree, we begin at the root and as 78>76>44>18 move to the 4th sub-tree of the root node. 78<80 and therefore, we move to the first sub-tree of the node. Since 78 is available in the only node of this subtree, we claim 78 was successfully reached.

11.4 INSERTION IN AN M-WAY SEARCH TREE

To insert a new element into an m-way search tree, we proceed in the same way as one would in order to search an element. To insert 5 into the 5-way search tree shown in above figure, we proceed to search for 5 and find that we fall off the tree at the node [7,12]. Since the node has only two keys and a 5-way search tree can accommodate upto 4 keys in a node. 5 is inserted into the node.

Insert 5

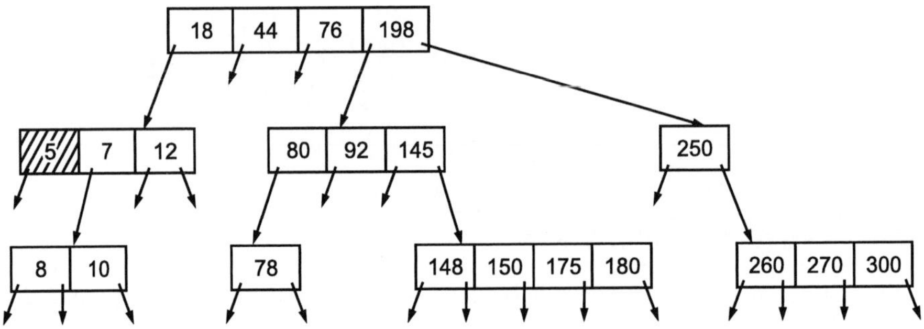

Fig. 11.6.

But to insert 160, the node [148,150,175,180] is already full hence we open a new child and insert 160 into it.

Insert 160

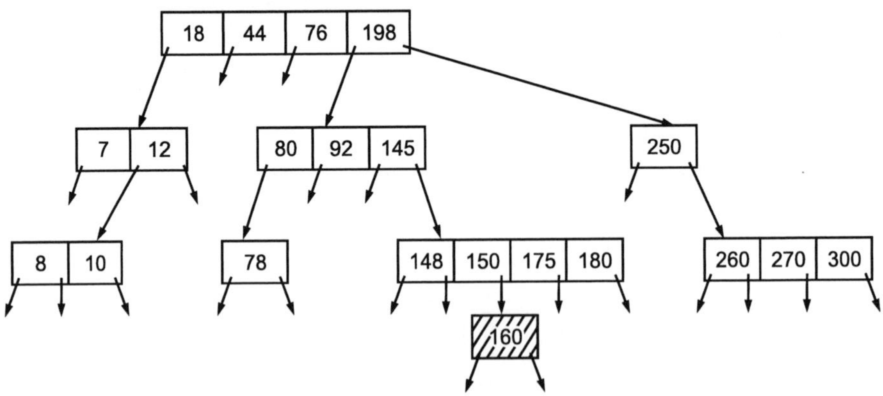

Fig. 11.7.

11.5 DELETION IN M-WAY SEARCH TREE

Let K be a key to be deleted from the m-way search tree. To delete the key we proceed as one would to search for the key. Suppose the node accommodating the key as follows.

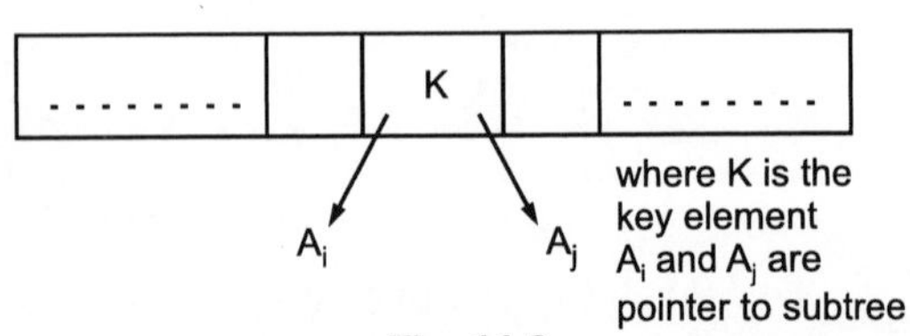

Fig. 11.8.

- If ($A_i = A_j = NULL$) then delete K.
- If ($A_i \neq NULL$, $A_j = NULL$) then choose the largest of the key elements K′ in the child node pointed to by A_i delete the key K′ and replace K by K′. Obviously deletion of K′

may call for subsequent replacements and therefore deletions in a similar manner, to enable the key K′ move up the tree.

- If (A_i=*NULL*, $A_j \neq$ *NULL*) then choose the smallest of the key elements K″ from the sub-tree pointed to by A_j delete K″ and replace K by K″. Again deletion of K″ may trigger subsequent replacements and deletions to enable K″ move up the tree.
- If ($A_i \neq$ *NULL*, $A_j \neq$ *NULL*) then choose either the largest of the key elements K′ in the sub-tree pointed to by A_i or the smallest of the key elements K″ from the sub-tree pointed to by A_j to replace K. As mentioned before to move K′ or K″ up the tree it may call for subsequent replacements and deletions.

To delete **150**, we search for 150 and we get that in the leaf node [148, 150, 175,180] where it is present. Here both A_i = *NULL*, A_j = *NULL*. So we therefore delete 150 and the node becomes [148, 175, 180].

To delete **250**, we find its left and right sub-tree pointers, A_i and A_j respectively and we get A_i = *NULL* and $A_j \neq$ *NULL*. Now we choose the smallest element from the right child node [260, 270,300] i.e. node 260. Delete 260 and replace 250 with 260.

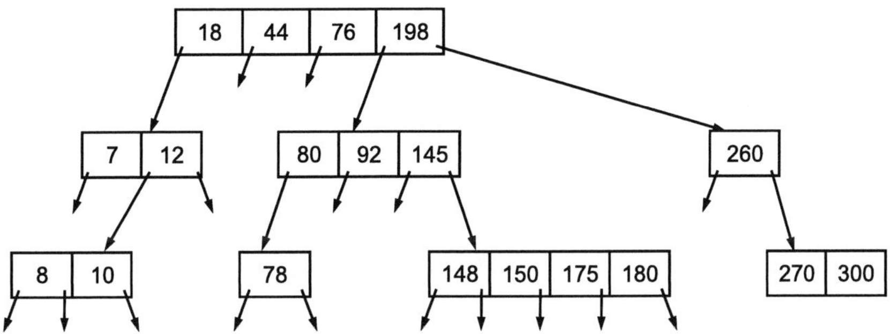

Fig. 11.9. Delete node 250.

EXAMPLE 11.1. Construct a 3-way search tree for the list of keys in the order shown below, what are your observations?

List A: 10,15,20,25,30,35,40,45

List B: 20,35,40,10,15,25,30,45

Solution: 3-way search tree for list A.

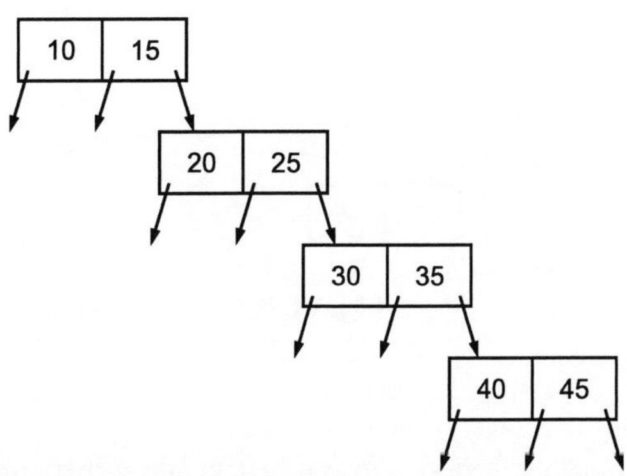

Fig. 11.10.

3-way search tree for list *B*.

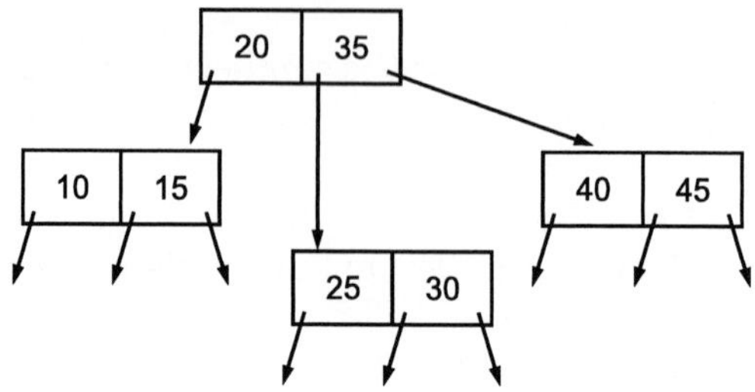

Fig. 11.11.

The observation is that for the same set of keys, various m-way search trees of varying heights can be constructed, In the problem, while the 3-way search tree corresponding to list *B* is of balanced height with a value equal to $\log_m (n + 1)$, the same corresponding to list A shows an imbalance in height.

The height of an m-way search tree varies from a low of $\log_m (n + 1)$ to a high of n. For minimum number of accesses related to the efficient retrieval of keys, it is essential that the height of an m-way search tree is close to $\log_m (n + 1)$.

EXAMPLE 11.2. In the following 4-way search tree trace the tree after deletion of (*i*) *U* and (*ii*) *M*.

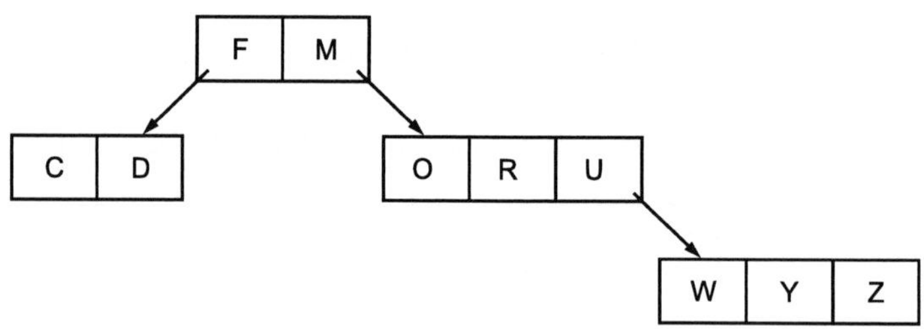

Fig. 11.12.

Solution: The 4-way search tree after deletion of U and M is as follows:

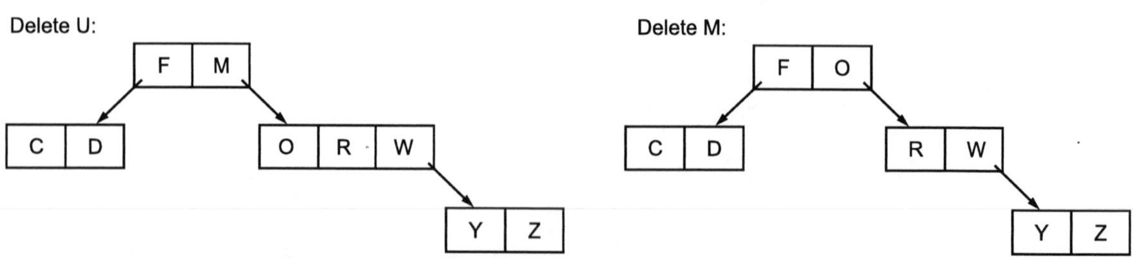

Fig. 11.13.

11.6 B TREE

In multi-way search tree, so many nodes have left subtree but no right sub-tree. Similarly they have right sub-tree but no left sub-tree. Insertion of key also increases the height of tree. As we know access time in tree is totally dependent on level of tree. So our aim is to minimise the access time which can be through balanced tree only. So we have a need to take all the leaf nodes at same level and non leaf nodes should not contain the empty sub-tree. So for a order n tree, maximum number of children should be n and each node can contain k keys where $k <= n - 1$. For balancing the tree each node should contain n/2 keys (except root). So the B-tree of order n can be defined as :

1. Each node has at least $n/2$ and maximum n non-empty children.
2. All leaf nodes will be at same level.
3. All leaf nodes can contain maximum $n - 1$ keys.
4. All non-leaf nodes can contain $m - 1$ keys where m is the number of children for that node.
5. Keys in non-leaf node will divide the left and right sub-tree where value of left sub-tree keys will be less and value of right sub-tree keys will be more than that particular key.

Let us take a B-tree of order 5:

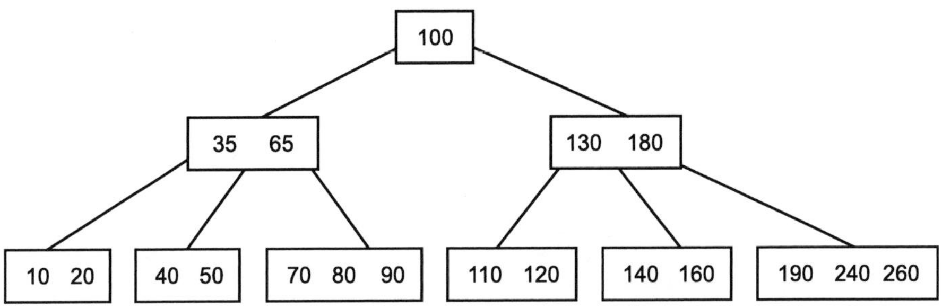

Fig. 11.14.

Here we can see all leaf nodes are at same level. All non-leaf nodes have no empty sub-tree and they have keys 1 less than number of their children.

A balanced **order-n,** multi-way search free in which each non root node contains at least $\lfloor (n-1)/2 \rfloor$ keys is called a B-tree of order n. A B-tree of order n is also called an $n - (n - 1)$ tree or an $(n - 1) - n$ tree. That is each node in the tree has a maximum of $n - 1$ keys and n children Thus, a $4 - 5$ tree is a B-tree of order 5, as is a $5 - 4$ tree. In particular, a $2 - 3$ (or $3 - 2$) tree is the most elementary nontrivial (that is, non-binary) B-tree, with one or two keys per node and two or three sons per node.

In discussing B-trees, the word "order" is used differently by different authors. It is common to find the **order** of a B-tree defined as the minimum number of keys in a non-root node. *i.e.,* $(n - 1)/2$ and the **degree** of a B-tree to mean the maximum number of sons *i.e.,* n. Still other authors USE "order" to mean the maximum number of keys in a node *i.e.,* $n - 1$.

A B-tree is a rooted tree (whose root is root [T]) having the following properties:

1. Every node x has the following fields:
   (*a*) $n[x]$ the number of keys currently stored in node x
   (*b*) the $n[x]$ keys themselves stored in non decreasing order, so that
   $key_1[x] \le key_2[x] \le \ldots \le key_{n[x]}[x]$
   (*c*) leaf $[x]$, a boolean value is TRUE if x is a leaf and FALSE if x is an internal node.
2. Each internal node x also contains $n[x] + 1$ pointers $c_1(x), c_2[x] \ldots c_{n[x]+1}[x]$ to its children Leaf nodes have no children, so their c_i fields are underfined.
3. The keys key $_i[x]$ separate the range of keys stored in each subtree: if k_i is any key stored in the subtree with root $c_i[x]$then.
   $k_1 \le key_1[x] \le k_2 \le key_2[x] \le \ldots \le key_{n[x]}[x] \le key_{n[x]+1}$

4. All leaves have the same depth, which is the tree's height h.
5. There are lower and upper bounds on the number of keys a node can contain. These bounds are expressed in terms of a fixed integer $t \geq 2$ called the **minimum degree** of the B-tree:
   (*a*) Every node other than the root must have **at least $t - 1$ key**. Every internal node other than the root has at least t children. If the tree is non-empty, the root must have at least one key.
   (*b*) Every node can contain **at most $2t - 1$key**. Therefore, an internal node can have **atmost $2t$ children**. We say that a node is **full** if it contains **exactly $2t - 1$keys**.

The simplest B-tree occurs when $t = 2$. Every internal node then has either 2, 3, or 4 children, and we have a $2 - 3 - 4$ tree. In practice, however much larger values of t are used typically adjusted to block size of hard disks for efficient *I/O*.

Theorem : If $n \geq 1$,then for any n-keys B-tree T of height h and minimum degree $t \geq 2$, then

$$h \leq \log_t \frac{n+1}{2}$$

Proof. The root contains at least one key. All other nodes contain at least $t - 1$ keys. There are at least 2 nodes at depth 1, at least $2t$ nodes at depth 2, at least $2t^{i-1}$ nodes at depth i and $2t^{h-1}$ nodes at depth h.

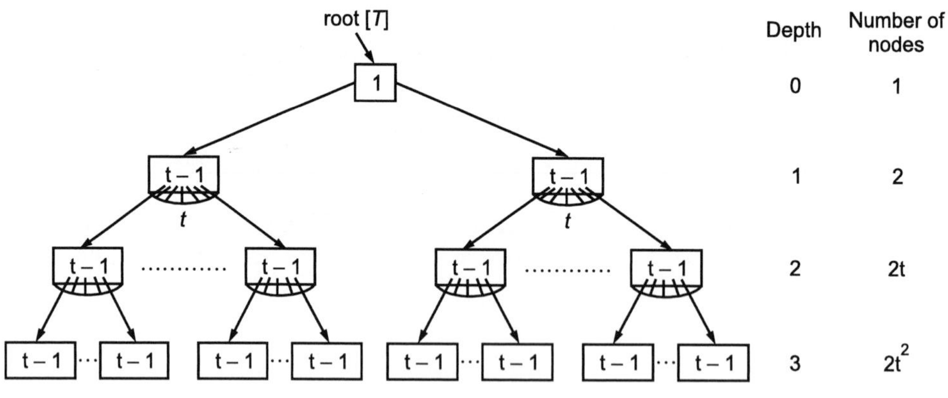

Fig. 11.15.

So we write, $n \geq 1 + (t - 1)\ (2 + 2t + 2t^2 + 2t^3 + ... + 2t^{h-1})$

$$= 1 + 2\ (t - 1)\ (1 + t + t^2 + t^3 + ... + t^{h-1})$$

$$= 1 + 2\ (t - 1) \left(\frac{(t^h - 1)}{(t - 1)} \right)$$

$$= 2t^{h-1}$$

we get, $t^h \leq (n + 1)/2$

Taking base-t logarithms of both sides, we get $h \leq \log_t (n + 1)/2$.

11.7 INSERTION IN B-TREE

The insertion of a key in a B-tree requires first traversal in B-Tree. Through traversal it will find that key to be inserted is already existing or not. Suppose key does not exist in tree then through traversal it will reach leaf node. Now we have two cases for inserting the key.

1. Node is not full.
2. Node is already full

If the leaf node in which the key is to be inserted is not full, then the insertion is done in the node. A node is said to be full if it contains a maximum of $(m - 1)$ keys, given the order of the B-tree to be m.

If the node were to be full, then insert the key in order into the existing set of keys in the node, split the node at its median into two nodes at the same level, pushing the median element up by one level. Note that the split nodes are only half full. Accommodate the median element in the parent node if it is not full. Otherwise repeat the same procedure and this may even call for rearrangement of the keys in the root node or the formation of a new root itself.

Thus a major observation pertaining to insertion in a B-tree is that, since the leaf nodes are all at the same level, unlike m-way search trees, the tree grows upwards.

EXAMPLE 11.3. Let us take a list of keys and create a B-tree of order 5.

10, 20, 50, 60, 40, 80, 100, 70, 130, 90, 30, 120, 140, 25, 35, 160, 180

Solution:

1. Insert 10 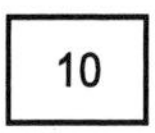

2. Insert 20 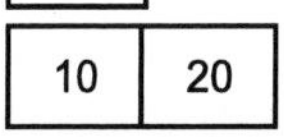

3. Insert 50 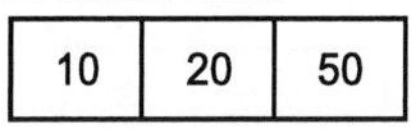

4. Insert 60 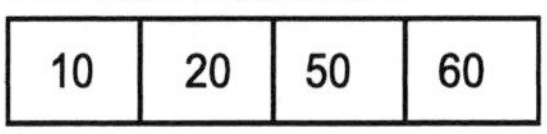

5. Insert 40

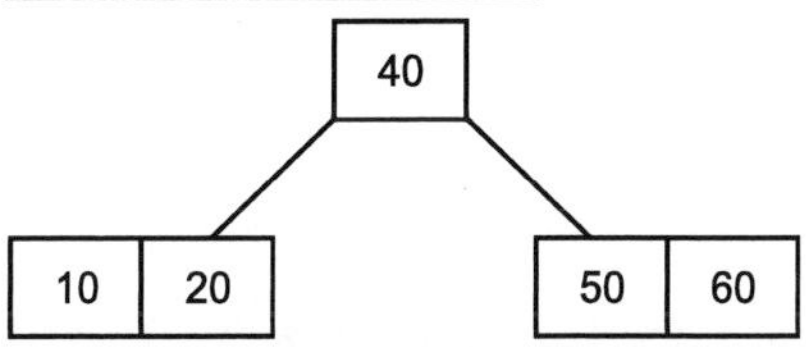

Here node was already full, so after insertion of 40, it is

splitted into two nodes, 40 is the median key so it will go into parent node and it will become root.

Insert 80

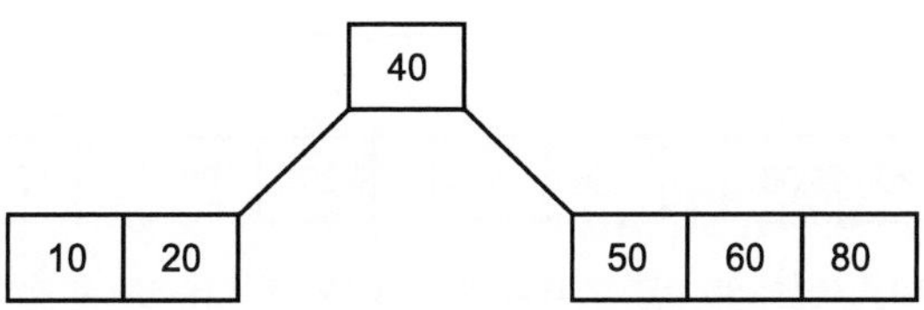

Insert 100

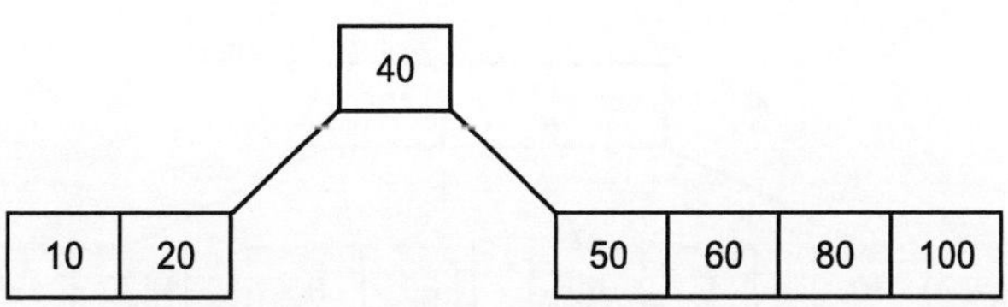

Insert 70

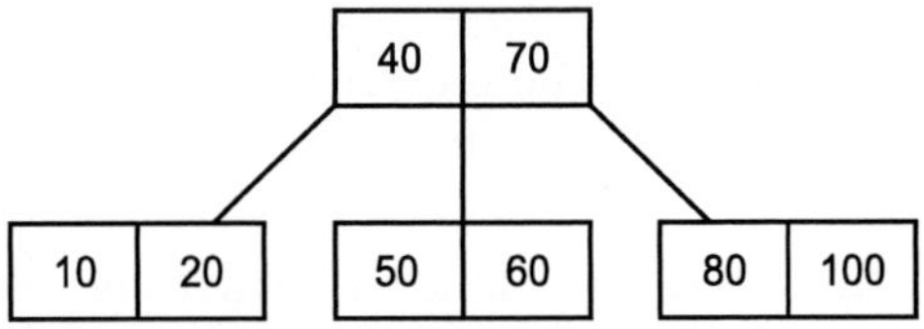

Here node was already full, so after insertion of 70 it splitted into two nodes, 70 is the median key, so it will go to the parent node.

Insert 130

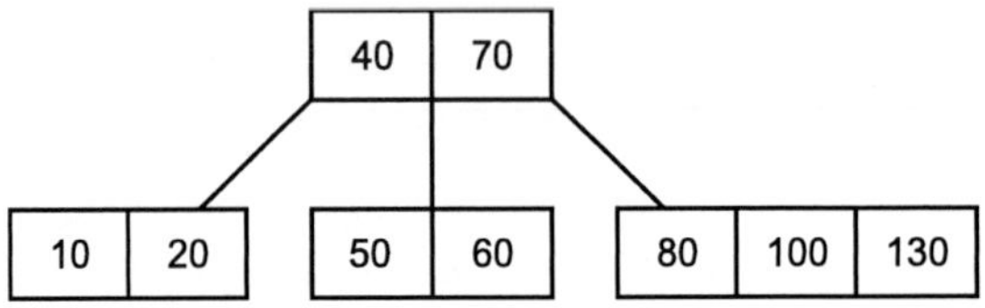

Insert 90

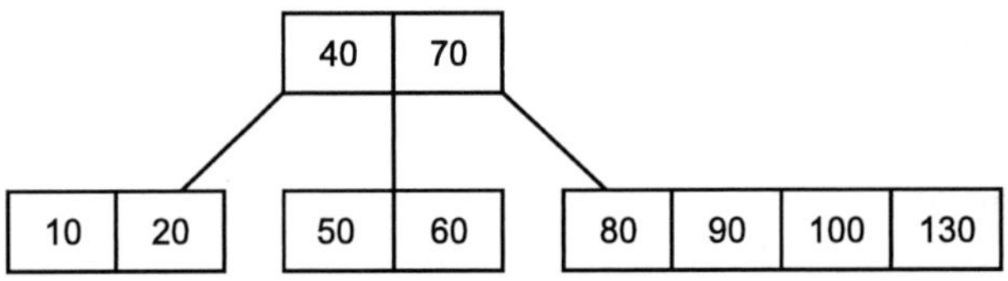

Here after insertion of 90, the keys in node will be sorted.

Insert 30

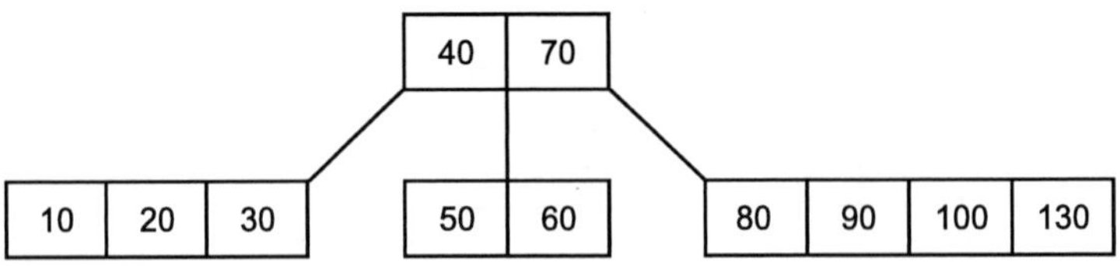

Insert 120

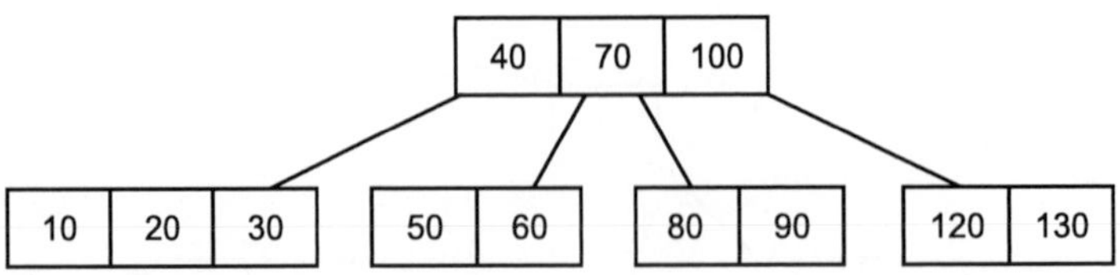

Here node was already full, so after insertion of 120 it splitted in two nodes, 100 is the median key so it will go into parent.

Insert 140

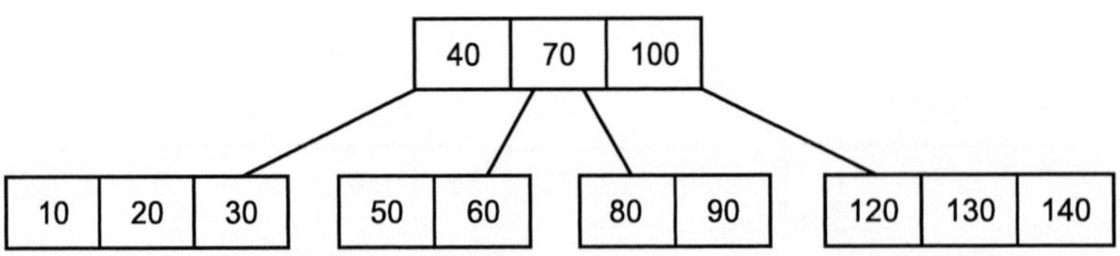

Insert 25

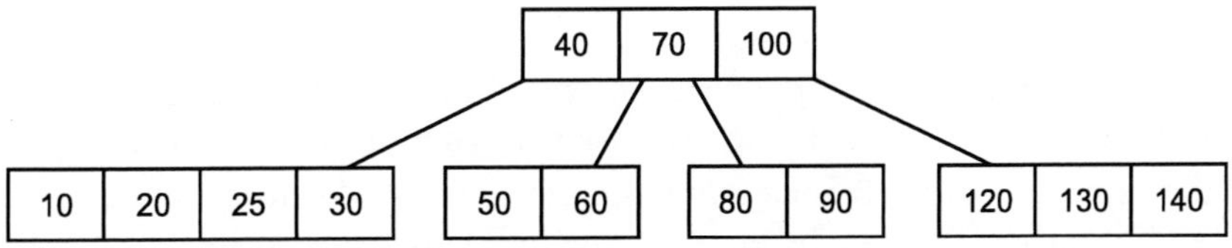

Insert 35

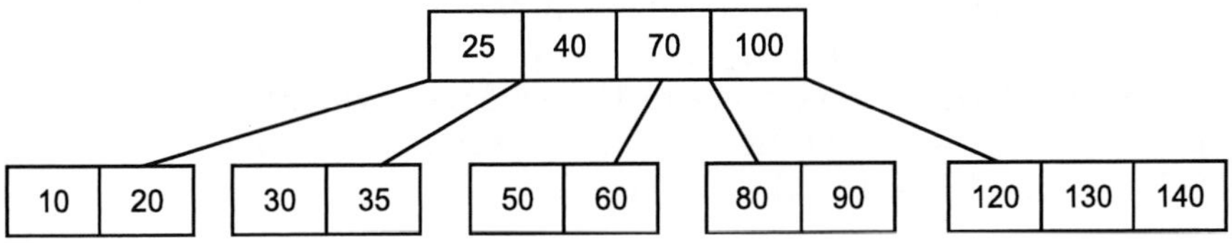

Insert 160

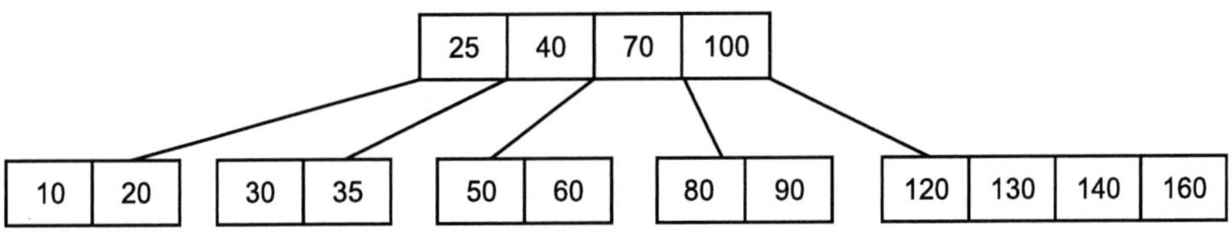

Insert 180

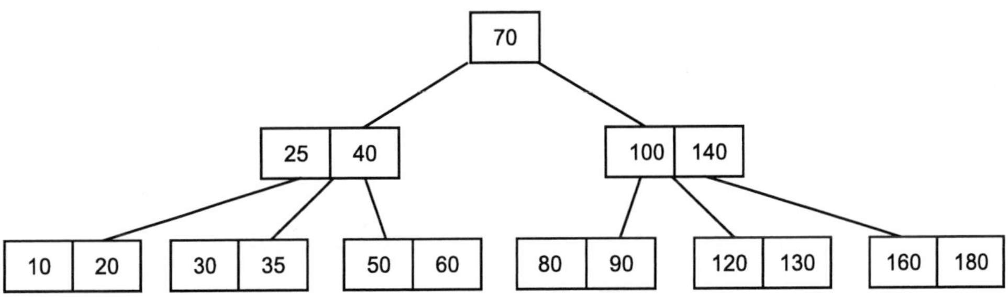

Here node was already full, so after insertion 180 it splitted in two nodes. 140 is the median key so it will go into parent node. Here parent is already full, so it splitted in two nodes and 70 is the median key so it will become the new root.

11.8 DELETION IN B-TREE

Deletion of key also requires first traversal in B-tree, After reaching on particular node, two cases may occur:

1. Node is leaf node
2. Node is non-leaf node

For the **first case** suppose node has more than minimum number of keys then it can be easily deleted. But suppose it has only minimum number of keys then first we will see the number of keys in adjacent leaf node if it has more than minimum number of keys then first key of the adjacent node will go to the parent node and key in parent node which is partitioning will be combined together in one node. Suppose now parent has also less than the minimum number of keys then the same thing will be repeated until it will get the node which has more than the minimum number of keys.

For the **second case** key will be deleted and it's predecessor or successor key will come on its place. Suppose both nodes of predecessor and successor key have minimum number of keys then the nodes of predecessor and successor keys will be combined.

Thus, while removing a key from a leaf node, if the node contains more than the minimum number of elements, then key can be easily removed. However, if the leaf node contains just the minimum number of elements, then scout for an element from either the left sibling node or right sibling node to fill the vacancy. If the left sibling node has more than the minimum number of keys, pull the largest key up into the parent node and move down the intervening entry from the parent node to the leaf node where the key is to be deleted. Otherwise, pull the smallest key of the right sibling node to the parent node and move down the intervening parent element to the leaf node.

If both the sibling nodes have only minimum number of entries, then create a new leaf node out of the two leaf nodes and the intervening element of the parent node, ensuring that the total number does not exceed the maximum limit for a node. If while borrowing the intervening element from the parent node, it leaves the number of keys in the parent node to be below the minimum number, then we propagate the process upwards ultimately resulting in a reduction of height of the B-tree.

EXAMPLE : Let us take a B-tree of order 5.

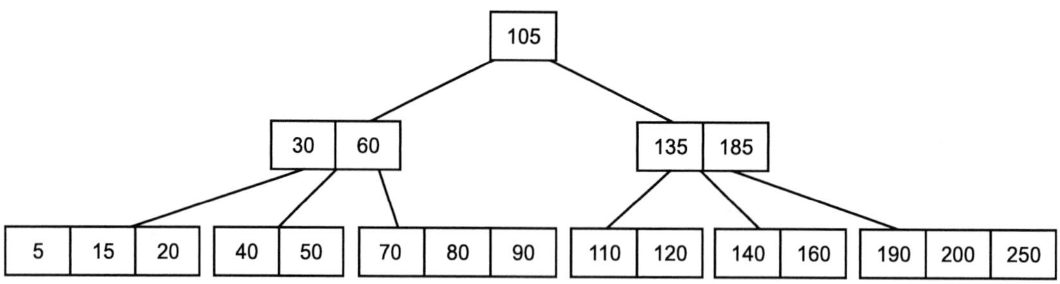

Delete 190.

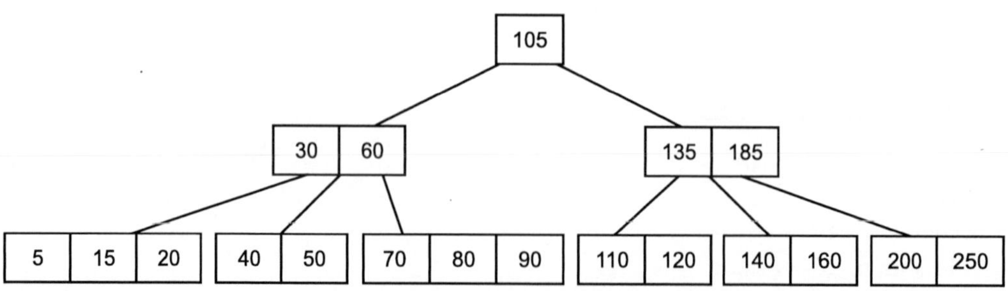

Here 190 is in leaf node, so delete it from only leaf node.

Delete 60.

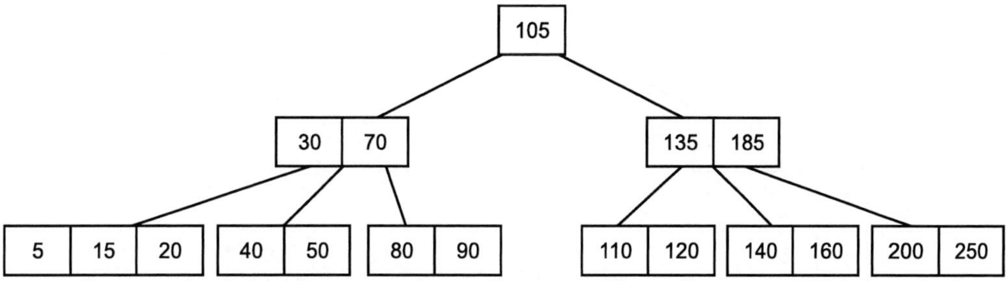

Here 60 is in non leaf node. So first it will be deleted from the node and then the element of the right child will come in that node.

Delete 40.

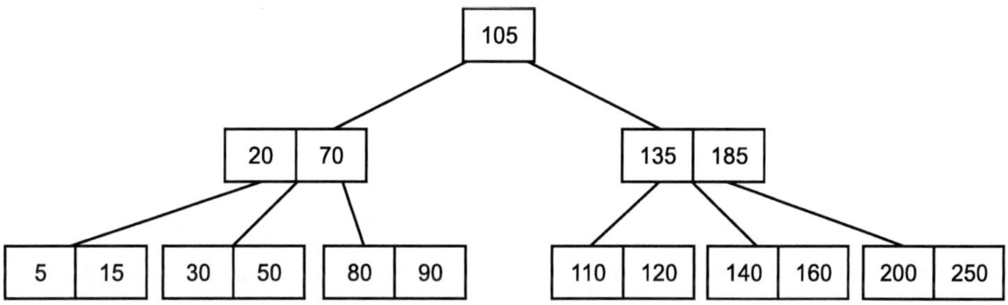

Here first 40 will be deleted from leaf node then left side element in the parent node will come in leaf node and then last element of the left side node of the parent node will come in parent node.

Delete 140.

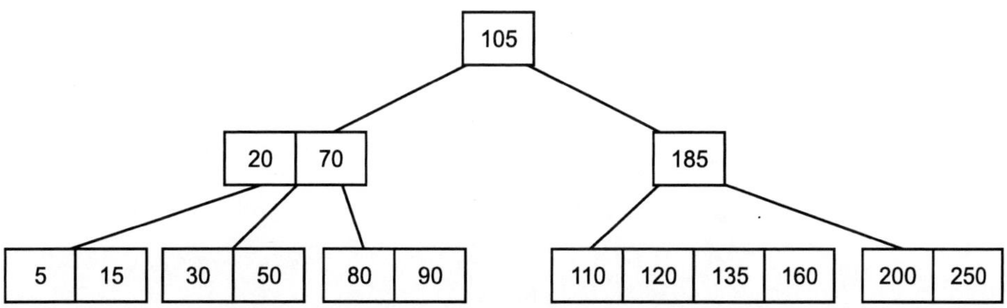

But at least two elements should be in child node so it will go in root node.

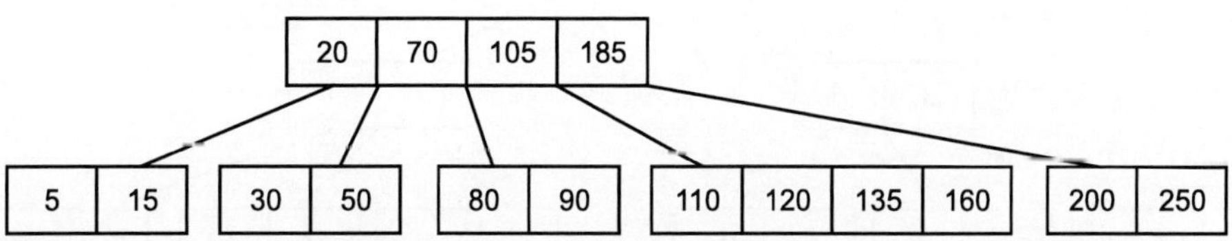

EXAMPLE : Suppose a B-tree of order-5 is shown in the Fig.

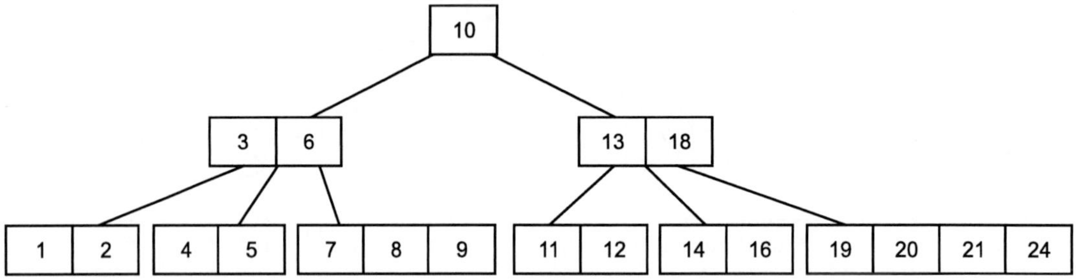

Delete 8.

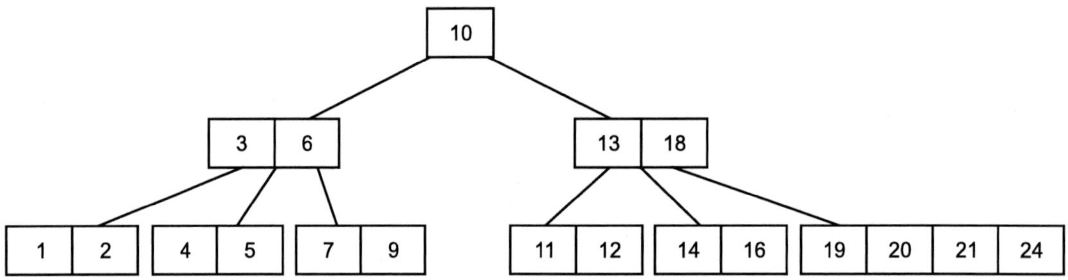

Deletion of 8 is from a leaf node with more than the minimum number of elements and hence leaves no problem.

Delete-18.

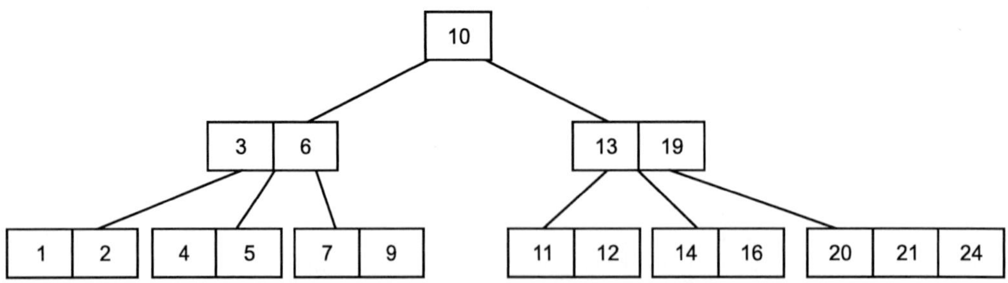

Deletion of 18 is from non-leaf node and therefore the immediate successor is placed at 18, *i.e.*, 19.

Delete-16.

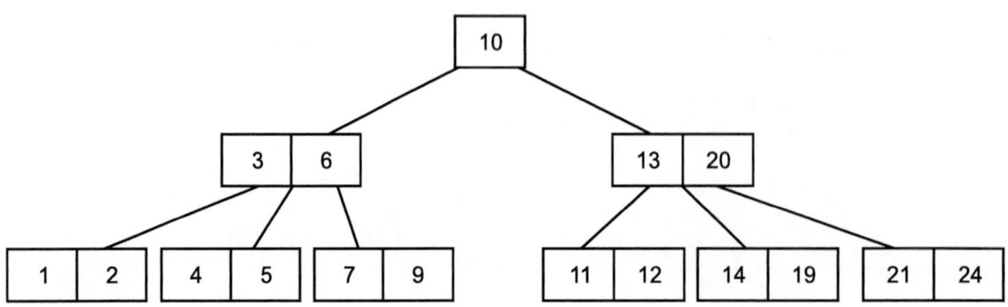

Deletion of 16 leaves its nodes with too few elements. The element 19 from parent node is therefore brought down and replaced by the element 20.

Delete-4.

The deletion of 4 leaves the node with less than minimum number of elements and neither of its sibling nodes can spare an element. The node therefore is combined with one of the siblings and with the median element from the parent node.

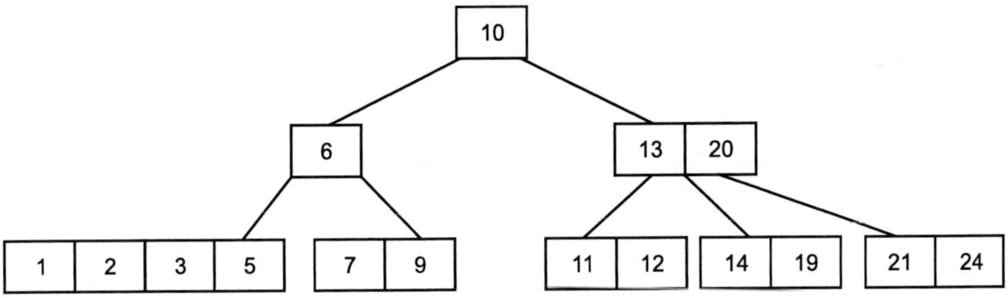

But at least two elements should be in the child of root so it will go in root node.

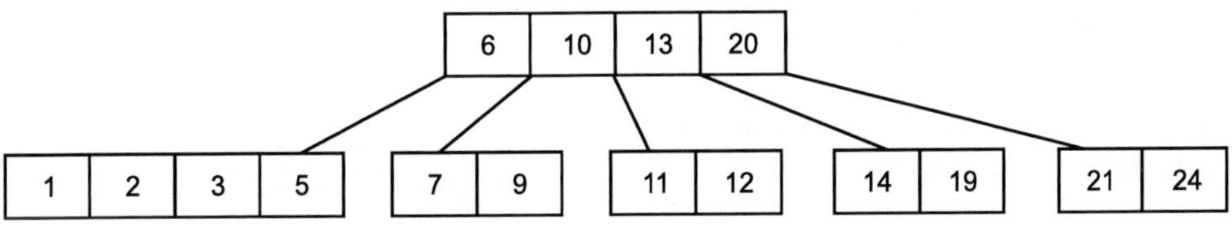

It reduces the height of tree also.

11.9 APPLICATION OF A B-TREE

The main application of a B-tree is the organisation of a huge collection of records into a file structure. The organisation should be in such a way that any record in it can be searched very efficiently *i.e.,* insertion, deletion and modification operations can be carried out perfectly and efficiently.

11.10 B$^+$- TREES

The B-tree structure is the standard organisation for indexes in a database system. There are several variations of the B-tree, most well known being the B*-tree and the B$^+$-tree. The B-tree guarantees at least 50% storage utilisation, that is, at any given time, the tree has each of its nodes at least 50% full. The B$^+$-tree is a slightly different data structure, which in addition to indexed access, also allows sequential data processing and stores all data in the lowest level of the tree.

One of the major drawbacks of the B-tree is the difficulty of traversing the keys sequentially. B$^+$ tree retains the rapid random access property of the B-tree, while also allowing rapid sequential access. In the B$^+$ tree, all keys are maintained in leaves and keys are replicated in non-leaf nodes to define paths for locating individual records. The leaves are linked together to provide a sequential path for traversing the keys in the tree.

The B$^+$-Tree is called a *balanced tree* because every path from the root node to a leaf node is the same length. A balanced tree means that all searches for individual values require the same number of nodes to be read from the disk.

B^+ tree

- is a structure of nodes linked by pointers
- is anchored by a special node called the root and bounded by leaves
- has a unique path to each leaf, and all paths are equal length
- stores keys only at leaves, and stores reference values in other, internal nodes
- guides key search, via the reference values, from the root to the leaves

A B + tree of order M (M>3) is an M-ary tree with the following properties:

1. The data items are stored at leaves.
2. The root is either a leaf or has between two and M children.
3. Node:
   - The (internal) node (non-leaf) stores up to M – 1 keys (redundant) to guide the searching; key i represents the smallest key in subtree $i + 1$.
   - All nodes (except the root) have between $\lfloor M/2 \rfloor$ and M children
4. Leaf :
   - A leaf has between $\lfloor L/2 \rfloor$ and L data items, for some L (usually L < < M, but we will assume M=L in most examples).
   - All leaves are at the same depth.
5. Less disk accesses due to fewer levels in the tree.
6. B^+tree provides faster sequential access of data.

Fig. 11.16 illustrates a B^+ Tree

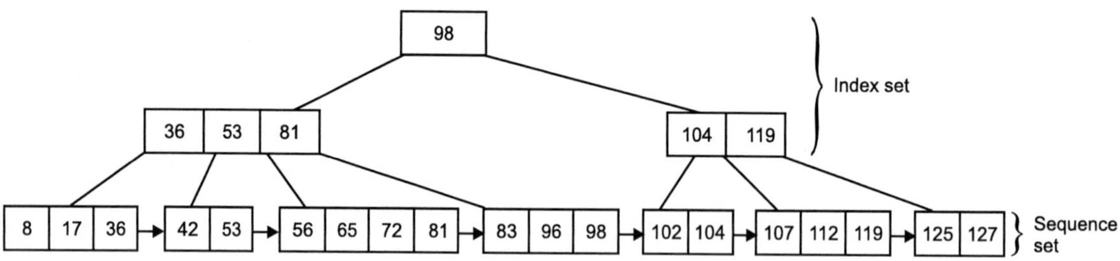

Fig. 11.16.

As we can see in the figure, the leaves have been connected to form a linked list of keys in sequential order. The B^+ tree has two parts the first part is the **index set** that constitutes interior nodes and the second part is the **sequence set** that constitutes leaves.

B^+-Tree Structure

B^+-tree consists of two parts :

Index Set

- Provides indexes for fast access of data.
- Consist of internal nodes that store only key & sub tree pointers.

Sequence Set

- Consists of leaf nodes that contain data pointers.
- Provides efficient sequential access of data (using doubly linked list).

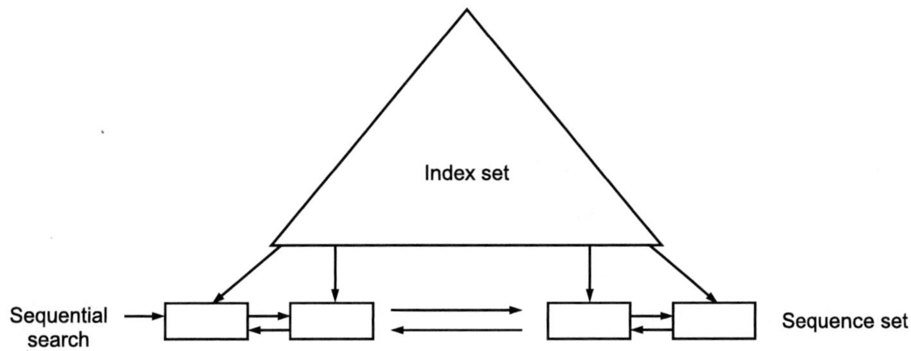

Fig. 11.17.

B⁺-Tree: Index Node Structure

The basic structure of the B^+-tree index node of order m is same as that of B-tree node of order m.

The root has at least two sub-trees unless it is a leaf.

Each non-root index node holds $q - 1$ keys and q subtree pointers, where $\lceil m/2 \rceil \le q \le m$

Only **difference** is that index node does not contain data pointers.

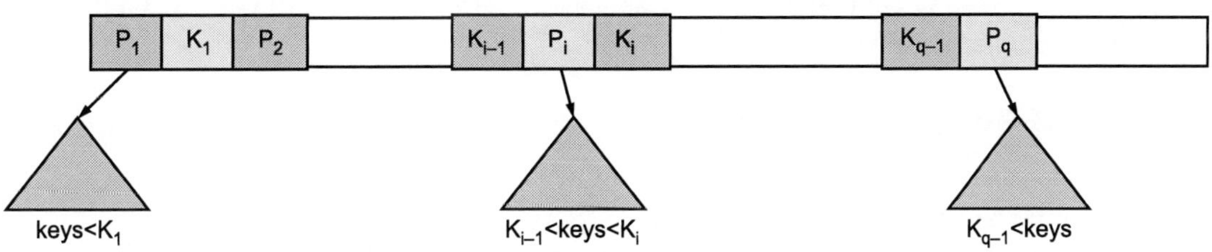

Fig. 11.18.

B⁺-Tree: Sequence Node Structure

The structure of the B^+-tree sequence node is as follows:

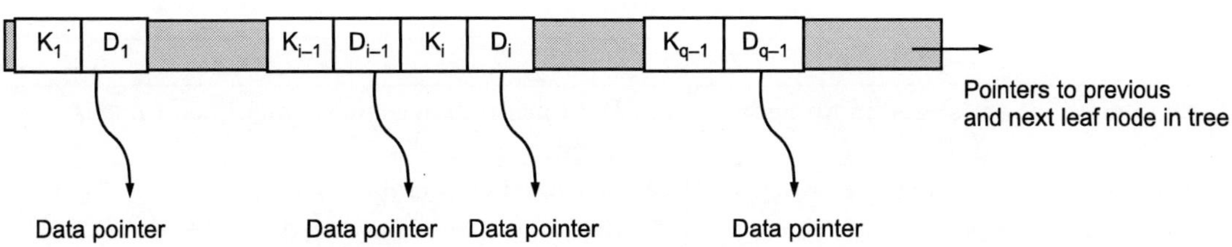

Fig. 11.19.

11.11 SEARCHING KEY IN A B⁺ TREE

- Start from the root

- If an internal node is reached:
  - search KEY among the keys in that node
    - Linear search or binary search
  - If KEY ≤ smallest key, follow the leftmost child pointer down
  - If KEY > largest key, follow the rightmost child pointer down
  - If $K_i \leq$ KEY $< K_j$, follow the child pointer between K_i and K_j
- If a leaf is reached:
  - Search KEY among the keys stored in that leaf
    - linear search or binary search
  - If found, return the corresponding record; otherwise report not found.

11.12 INSERTION INTO A B$^+$ TREE

In order to insert a key into a B$^+$ Tree, first a B$^+$ Tree search is performed to find the correct location for the key. Actually, the procedure to insert a new key value into a B$^+$–Tree is almost as for B–Tree.

The sequence of steps required to insert a node in a B$^+$ Tree are as follows:

1. The search operation is used to find the leaf node in which the key value of the node has to be inserted.
2. If the key value already exits in a leaf node, no more insertion is needed. Else if the said key value does not exist, insert the value in the leaf node in an ordered fashion.
3. When a node is split, the middle key is retained in the left half-node as well as being promoted to the father.

11.13 DELETION FROM A B$^+$ TREE

The sequence of steps required to delete a key value from the B$^+$ Tree is as follows:

1. Search the B$^+$ Tree for the key value.
2. If the key value is in the B$^+$ Tree, remove it from the tree as that of B Tree.
3. When a key value is deleted from a leaf there is no need to delete that key from the index set of the key. That key value still can direct searches to proper leaves, since it is still a valid separator between the keys in the nodes below.

11.14 COMPARISON BETWEEN B-TREE AND B$^+$ TREE

B-Tree	*B$^+$ Tree*
1. Data pointers are stored in all nodes.	1. Data printers are stored only in leaf nodes (sequence set).
2. Search can end at any node.	2. Search always ends at leaf node.
3. No redundant keys.	3. Redundant keys may exist.
4. Slow sequential access.	4. Efficient sequential access.
5. Higher trees.	5. Flatter trees. No data pointers in index set nodes.

Exercises

1. What is an M-way search tree?
2. What is a B^+tree? Write down its applications in file organisation.
3. Differentiate between the B^+tree index files and B-tree index files.
4. Write an algorithm and a C routine to delete a record from a B-tree of order n.
5. Are B- trees of order 2 full binary trees? If so, how?
6. What is the difference between a B-tree and B^+tree? When you might prefer to use B^+ tree instead of a B-tree?

Chapter

12 Graphs

12.1 INTRODUCTION

A graph is another important non-linear data structure. The only non-linear data structure that we have seen so far is tree. In a tree data structure, there is a hierarchical relationship between parent and children, *i.e.*, one parent and many children. A tree in fact is a special type of graph. In a graph, the relationship is less restricted, *i.e.*, the relationship is from many parents to many children.

Graphs are data structures which have wide-ranging applications in real-life like. Airlines, analysis of electrical circuits, source, destination networks, finding shortest routes, flow chart of a program, statistical analysis etc.

Figure 12.1. shows the non-linear data structure called a graph.

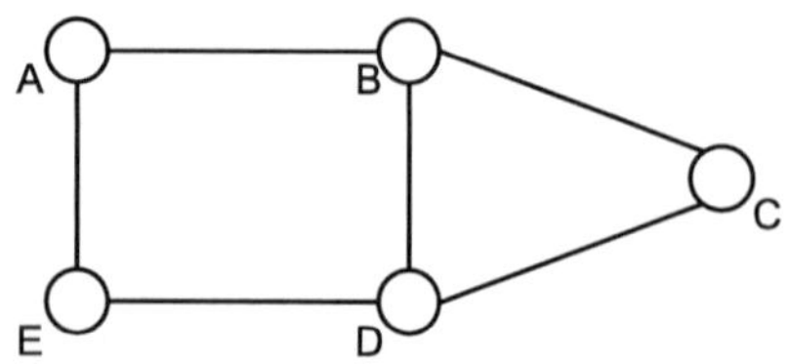

Fig. 12.1.

Chapter Outline

- Introduction
- Definition
- Terminologies Used
- Representation of Graph
- Implementation of a Graph
- Operations on Graph
- Traversing a Graph
- Breadth First Search
- Depth First Search
- Topological Sort
- Strongly Connected Components
- Spanning Tree
- Minimum Spanning Trees
- Kruskal's Algorithm
- Prim's Algorithm
- Shortest Paths
- Dijkstra's Algorithm
- Warshall's Algorithm
- Transitive Closure

12.2 DEFINITION

A graph G consists of a set V, where members are called the **vertices** of G and the set E, whose members are called the **edges**. The set of vertices is non-empty and the set of edges contains pair of vertices. If $e = (u, v)$ is an edge with vertices u and v, then u and v are said to lie on e and e is said to be incident with u and v.

Thus, a graph G is a collection of two sets V and E, where V is the vertices v_0, v_1, ... v_{n-1} and E is the collection of edges e_1, e_2, ... e_n. This can be represented as $G = (V, E)$ where,

$$V(G) = (v_0, v_1, \ldots.. v_n) \text{ or set of vertices}$$
$$E(G) = (e_1, e_2, \ldots.. e_n) \text{ or set of edges}$$

A graph can be of two types :

1. Undirected Graph
2. Directed Graph.

Undirected Graph : If the pair of vertices are unordered then Graph G is called an **undirected graph**. That means if there is an edge between v_1 and v_2 then it can be represented as (v_1, v_2) or (v_2, v_1) also.

If the Fig. 12.2.

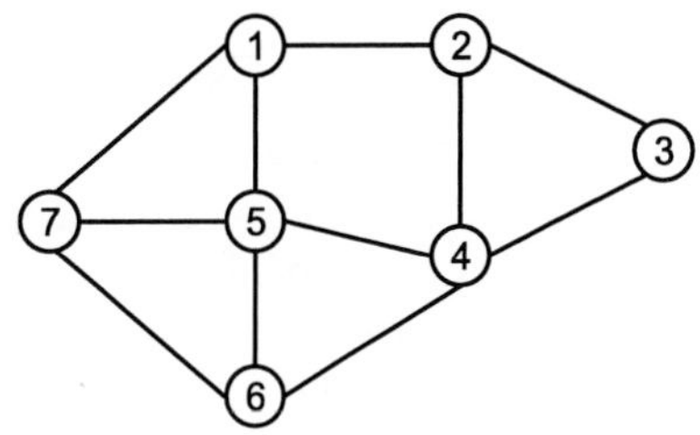

Fig. 12.2.

$$V(G) = \{1, 2, 3, 4, 5, 6, 7\}$$
$$E(G) = \{(1, 2), (2, 3), (3, 4), (2, 4), (4, 5), (5, 6), (7, 5), (1, 5)\}$$

That is, this graph 7 nodes and 8 edges.

Directed Graph : If the pair of vertices are ordered then Graph G is called **directed graph**. That is, a directed graph or digraph is a graph which has ordered pair of vertices (v_1, v_2) where v_1 is the tail and v_2 is the head of the edge. If the graph is directed then the line segments or arcs have arrow heads indicating the direction.

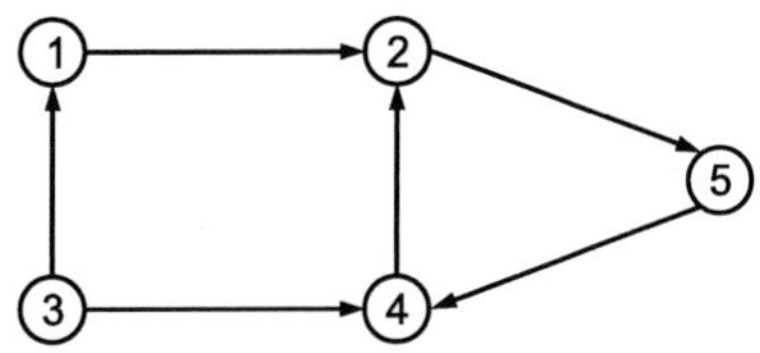

Fig. 12.3.

Directed graph is often shortend to diagraph.

$$V(G) = \{1, 2, 3, 4, 5\}$$
$$E(G) = \{(1, 2), (2, 5), (5, 4), (4, 2), (3, 4), (3, 1)\}$$

This graph has 5 nodes and 6 nodes.

12.3 TERMINOLOGIES USED

1. **Weighted Graph :** A graph is said to be a weighted graph if all the edges in it are labelled with some numbers. It is shown in the Fig. 12.4 below.

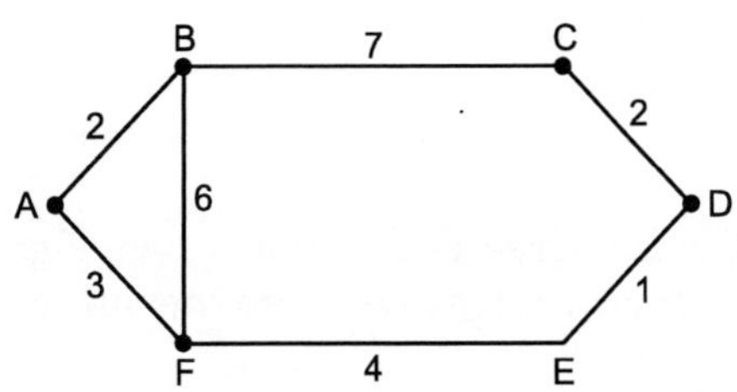

Fig. 12.4.

2. **Self Loop :** If there is an edge whose starting and end vertices are same that is (v_2, v_2) is an edge then it is called a self loop or simply a loop. It is shown in Fig. 12.5.
3. **Parallel Edges :** If there are more than one edge between the same pair of vertices then they are known as parallel edges.

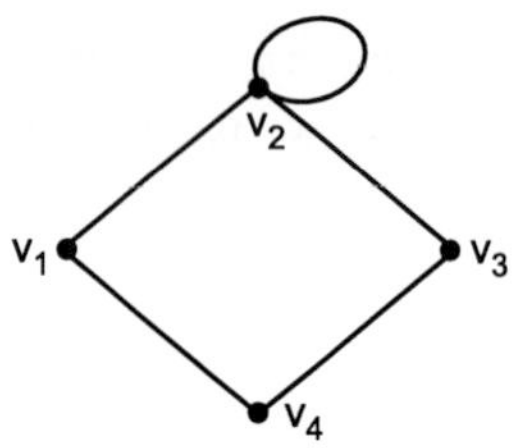

Fig. 12.5.

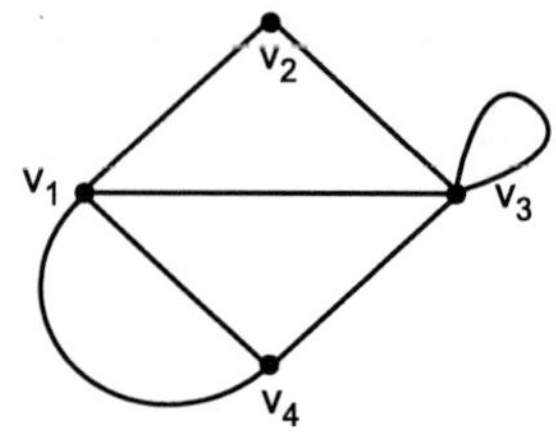

Fig. 12.6.

In Fig. 12.6, there are two edges between the same pair of vertices v_1 and v_4.

4. **Adjacent vertices :** A vertex u is adjacent to (or the neighbour of) other vertex v if there is an edge from u to v. In undirected graph if (v_1, v_2) is an edge then v_1 is adjacent to v_2 and v_2 is adjacent to v_1. In a directed graph if (v_1, v_2) is an edge then v_1 is adjacent to v_2 and v_2 is adjacent from v_1.
5. **Incidence :** In an undirected graph the edge (u, v) is incident on vertices u and v. In a digraph the edge (u, v) is incident from node u and is incident to node v.
6. **Degree of vertex :** The degree of a vertex is the number of edges incident to that vertex. In an undirected graph, the number of edges connected to a node is called the degree of that node.

Consider the graph, as shown in Fig. 12.7.

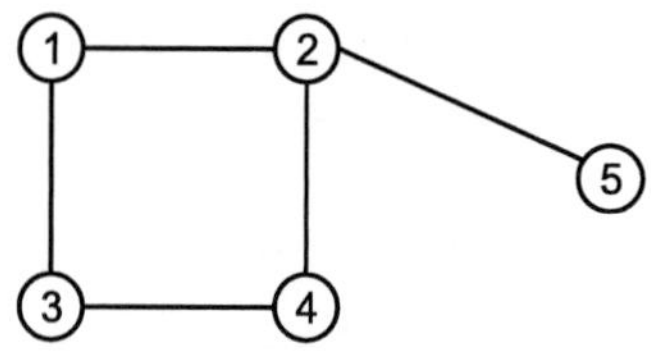

Fig. 12.7.

Here, for the vertex 2, the degree is 3 because 3 edges are incident to the vertex 2.

In a digraph, there are two degrees for every node.

1. In degree.
2. Out degree.

In degree of a vertex : The indegree of a vertex is the number of edges coming to that vertex or in other words, edges incident to it.

Outdegree of a vertex : The outdegree of a vertex is the number of edges going outside from that node or in other words, the edges incident from it.

Here, for the vertex 2, the indegree is 2 because two edges reach 2 and for the vertex 4, three are two outgoing edges so the outdegree of the vertex 4 is 2.

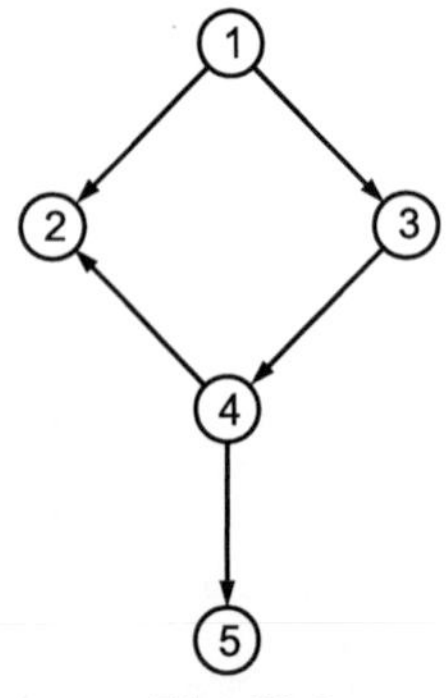

Fig. 12.8.

7. **Simple graph :** A graph or directed graph which does not have only self-loop or parallel edges is called a simple graph.
8. **Multi-graph :** A graph which has either a self-loop or parallel edges or both is called a multi-graph.
9. **Complete graph :** A graph is complete graph if each vertex is adjacent to every other vertex in graph or we can say that there is an edge between any pair of nodes in the graph. An undirected complete graph will contain $n(n - 1)/2$ edges.

10. **Regular graph :** A graph is regular if every node is adjacent to the same number of nodes.
    Here every node is adjacent to 3 nodes.

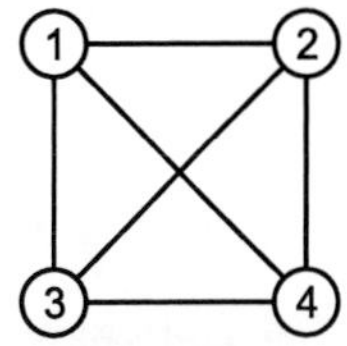

Fig. 12.9. Regular graph

11. **Planer graph :** A graph is planar if it can be drawn in a plane without any two edges intersecting.
12. **Connected graph :** In a graph *G*, two vertices v_1 and v_2 are said to be connected if there is a path in G from v_1 to v_2 or v_2 to v_1. A graph is said to be connected if there is a path from any node of graph to any other node, *i.e.*, for every pair of distinct vertices in *G*, there is a path.

(a) Connected graph (b) Not connected graph

Fig. 12.10.

(*a*) ***Strongly Connected Graph :*** A directed graph is said to be strongly connected graph if for every pair of distinct vertices in *G*, there is a path.

(*b*) ***Weakly Connected Graph :*** A diagram is called weakly connected or unilaterally connected if for any pair of nodes *u* and *v*, there is a path from *u* to *v* or a path from *v* to *u*. If from the diagraph we remove the directions and the resulting undirected graph is connected then that diagraph is weakly connected. Figure 12.11 shows a weakly connected graph.

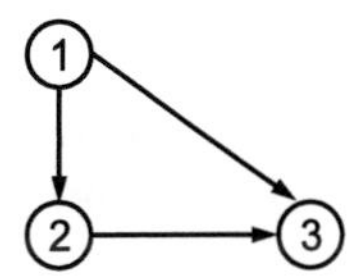

Fig. 12.11.

13. **Cycle :** If there is a path containing one or more edges which starts from a vertex and terminates into the same vertex then the path is called as a cycle.
14. **Acyclic graph :** If a graph (digraph) does not have any cycle then it is called as acyclic graph.
15. **Cyclic graph :** A graph that has cycles is called a cyclic graph.
16. **Maximum edges in graph :** In an undirected graph there can be $\frac{n(n-1)}{2}$ maximum edges and in a directed graph there can be $n(n-1)$ maximum edges. Where n is the total number of vertices in the graph.
17. **Articulation point :** If on removing a node from the graph, the graph becomes disconnected then that node is called the articulation point.
18. **Bridge :** If on removing an edge from the graph, the graph becomes disconnected then that edge is called the bridge.
19. **Biconnected graph :** A graph with no articulation points is called a biconnected graph.

12.4 REPRESENTATION OF GRAPH

Usually, a graph can be represented in many ways. Some of these representations are:

1. Set representation.

2. Sequential Representation
   (*i*) Adjacency
   (*ii*) Incidence
3. Linked representation.

Set Representation

In this representation, two sets are maintained. They are :

1. Set of vertices v, and
2. Set of edges E, which is the subset of $V \times V$.

For the Fig.

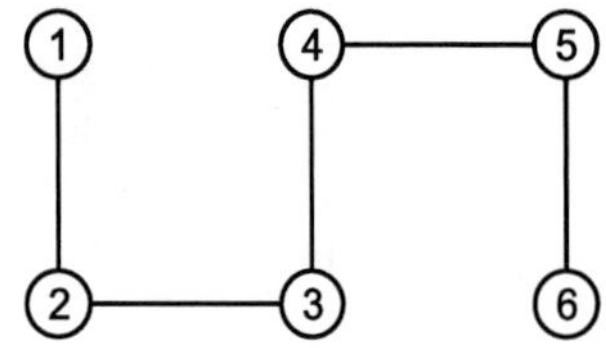

Fig. 12.12.

$$V(G) = \{1, 2, 3, 4, 5, 6\}$$
$$E(G) = \{(1, 2), (2, 3), (3, 4), (4, 5), (5, 6)\}$$

Advantage

1. From the memory point of view it is most efficient method.

Disadvantages

1. This representation does not allow the storing of parallel edges in a multigraph.
2. Multiplication of the graph is difficult.

Sequential Representation

The graphs can be represented a matrices in sequential representation. There are two most common matrices.

1. Adjacency matrix.
2. Incidence matrix.

Adjacency Matrix Representation

Adjacency matrix is the matrix, which keeps the information of adjacent nodes. In other words, we can say that this matrix keeps the information that whether this vertex is adjacent to any other vertex or not.

The general representation of the adjacency matrix shown in the Fig. 12.13.

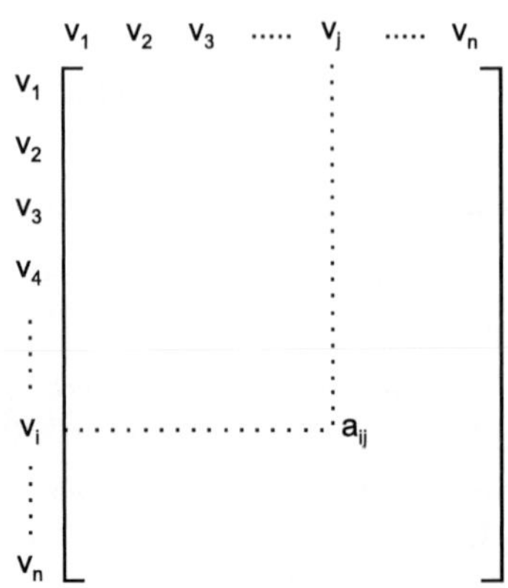

Fig. 12.13.

The entries in this matrix are placed according to the following rule :

$$a_{ij} = \begin{cases} 1 & \text{if there is an edge from } v_i \text{ to } v_j \\ 0 & \text{otherwise} \end{cases}$$

This adjacency matrix is also called a **bit matrix** or **Boolean matrix** because the entries are either 0 or 1.

Let us take a graph.

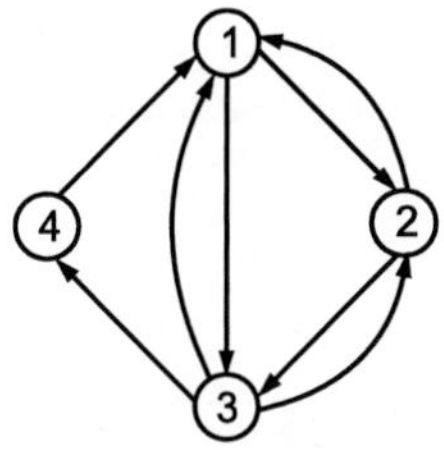

Fig. 12.14.

The corresponding adjacency matrix for this graph will be:

$$\begin{array}{c} \\ 1 \\ 2 \\ 3 \\ 4 \end{array}\begin{array}{c} \begin{array}{cccc} 1 & 2 & 3 & 4 \end{array} \\ \begin{bmatrix} 0 & 1 & 1 & 0 \\ 1 & 0 & 1 & 0 \\ 1 & 1 & 0 & 1 \\ 1 & 0 & 0 & 0 \end{bmatrix} \end{array}$$

Now let us take an undirected graph

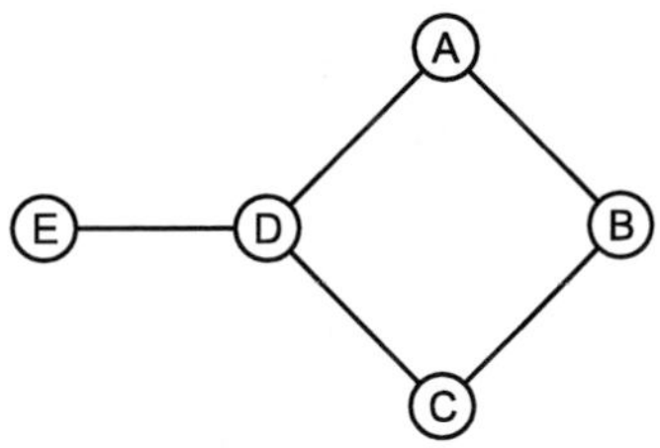

Fig. 12.15.

The corresponding adjacency matrix for this graph will be:

$$\begin{array}{c} \\ A \\ B \\ C \\ D \\ E \end{array}\begin{array}{c} \begin{array}{ccccc} A & B & C & D & E \end{array} \\ \begin{bmatrix} 0 & 1 & 0 & 1 & 0 \\ 1 & 0 & 1 & 0 & 0 \\ 0 & 1 & 0 & 1 & 0 \\ 1 & 0 & 1 & 0 & 1 \\ 0 & 0 & 0 & 1 & 0 \end{bmatrix} \end{array}$$

Note that, the adjacency matrix for an undirected graph is symmetric as the edge (AB) in $E(G)$ iff the edge (BA) is also in $E(G)$.

The adjacency matrix for a directed graph need not be symmetric.

Note : The space needed to represent a graph using its adjacency matrix is n^2 bits.

- From the adjacency matrix of an undirected graph, the degree of any vertex i is its row sum.
- From the adjacency matrix of a directed graph the row sum is the out-degree while the column sum is the in-degree.

- If A is an adjacency matrix of the graph G and if $A = A^T$ where A^T is the transpose of A then graph G is a simple undirected graph.
- An adjacency matrix is also useful to store multi-graphs as well as weighted graphs. In case of a multi-graph, instead of the entry 1, the entry will be the number of edges between two vertices and in the case of a weighted graph, the entries in the adjacency matrix are weights of the edges between the vertices instead of 0 or 1.

Let us take the graph,

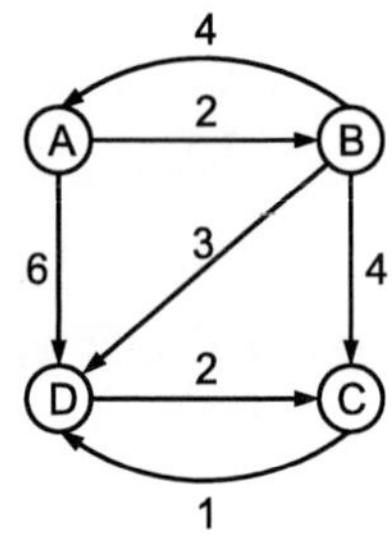

Fig. 12.16.

The corresponding adjacency matrix is:

$$\begin{array}{c} \\ A \\ B \\ C \\ D \end{array}\begin{array}{c} \begin{array}{cccc} A & B & C & D \end{array} \\ \begin{bmatrix} 0 & 2 & 0 & 6 \\ 4 & 0 & 4 & 3 \\ 0 & 0 & 0 & 1 \\ 0 & 0 & 2 & 0 \end{bmatrix} \end{array}$$

EXAMPLE 12.1. Draw the graphs corresponding to the adjacency matrices.

$$A(G) = \begin{bmatrix} 0 & 1 & 0 & 1 & 0 \\ 1 & 0 & 0 & 1 & 1 \\ 0 & 0 & 0 & 1 & 1 \\ 1 & 1 & 1 & 0 & 1 \\ 0 & 1 & 1 & 1 & 0 \end{bmatrix} \qquad A(G') = \begin{bmatrix} 0 & 0 & 1 & 1 \\ 0 & 0 & 1 & 0 \\ 1 & 1 & 0 & 1 \\ 1 & 0 & 1 & 0 \end{bmatrix}$$

Solution: (*a*) Since $A(G)$ is a 5×5 matrix, hence G will have 5 vertices say v_1, v_2, v_3, v_4 and v_5.

Draw an edge from v_i to v_j where $a_{ij} = 1$.

The required graph is shown in figure 12.17.

(*b*) Since $A(G')$ is a 4×4 square matrix, the graph G' has 4 vertices v_1, v_2, v_3, v_4 and the required graph is shown in figure 12.18.

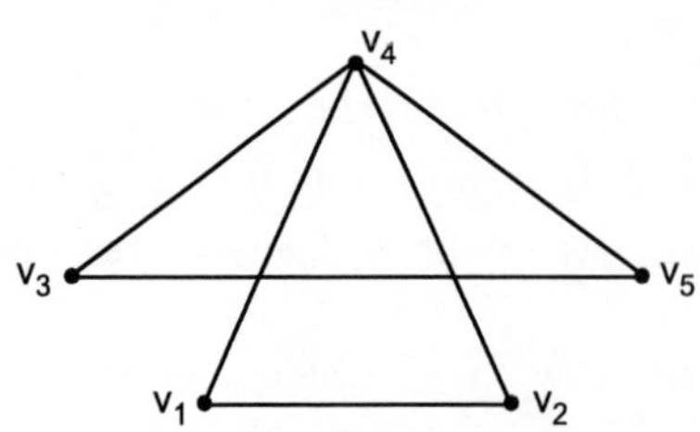

Fig. 12.17.

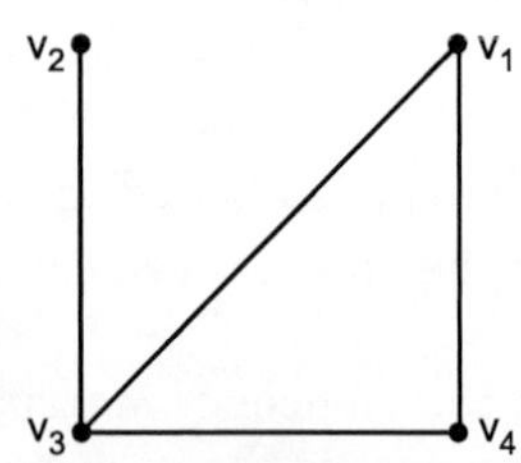

Fig. 12.18.

Powers of Adjacency Matrix

If A be an adjacency matrix of a diagraph then $A^n = [a_{ij}^n]$ gives the number of paths of length n from v_i to v_j where A^n is the nth power matrix of A.

Consider the graph

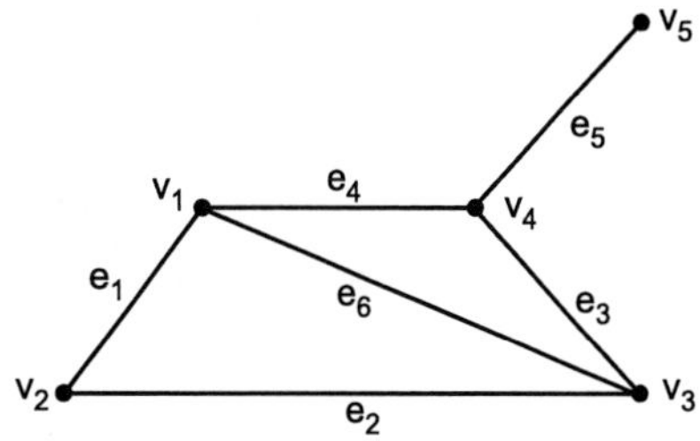

Fig. 12.19.

Whose adjacency matrix is

$$\begin{array}{c} \\ v_1 \\ v_2 \\ v_3 \\ v_4 \\ v_5 \end{array} \begin{array}{c} \begin{array}{ccccc} v_1 & v_2 & v_3 & v_4 & v_5 \end{array} \\ \begin{bmatrix} 0 & 1 & 1 & 1 & 0 \\ 1 & 0 & 1 & 0 & 0 \\ 1 & 1 & 0 & 1 & 0 \\ 1 & 0 & 1 & 0 & 1 \\ 0 & 0 & 0 & 1 & 0 \end{bmatrix} \end{array}$$

The matrices A^2, A^3 and A^4 for the graph as follows:

$$A^2 = \begin{bmatrix} 3 & 1 & 2 & 1 & 1 \\ 1 & 2 & 1 & 2 & 0 \\ 2 & 1 & 3 & 1 & 1 \\ 1 & 2 & 1 & 3 & 0 \\ 1 & 0 & 1 & 0 & 1 \end{bmatrix} \quad A^3 = \begin{bmatrix} 4 & 5 & 4 & 5 & 1 \\ 5 & 2 & 5 & 2 & 2 \\ 5 & 5 & 4 & 6 & 1 \\ 6 & 2 & 6 & 2 & 3 \\ 1 & 2 & 1 & 2 & 0 \end{bmatrix}$$

$$A^4 = \begin{bmatrix} 14 & 8 & 14 & 9 & 5 \\ 9 & 10 & 9 & 12 & 2 \\ 15 & 9 & 16 & 10 & 6 \\ 12 & 12 & 10 & 15 & 2 \\ 5 & 2 & 5 & 2 & 2 \end{bmatrix}$$

Note that $a_{2,4}^2 = 2$. So there are two paths of length 2 from v_2 to v_4 *i.e.*, $\{e_2, e_3\}$ and $\{e_1, e_4\}$.

Also $a_{4,5}^3 = 3$, So there are 3 paths of length 3 from v_4 to v_5.

Incidence Matrix

Let G be a graph with n vertices and e edges then the incidence matrix is a matrix of order $n \times e$, whose n rows correspond to the n vertices and e columns correspond to the e edges as follows :

$$a_{ij} = \begin{cases} 1 & \text{if the } j^{\text{th}} \text{ edge } e_j \text{ is incident on } i^{\text{th}} \text{ vertex } v_i \\ 0 & \text{otherwise} \end{cases}$$

For example, suppose the graph G is as follows

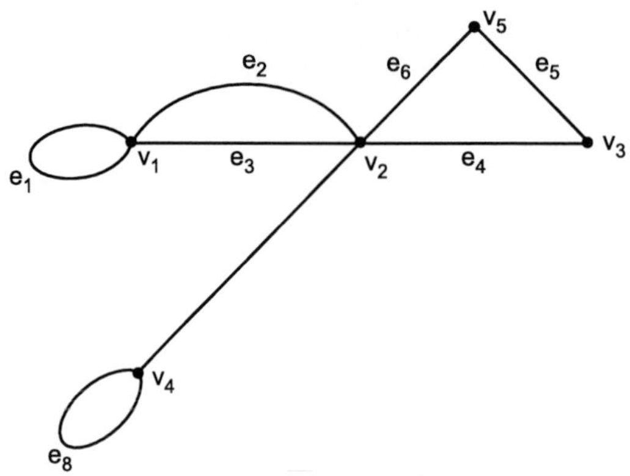

Fig. 12.20.

The incidence matrix for this graph is

$$\begin{array}{c} \\ v_1 \\ v_2 \\ v_3 \\ v_4 \\ v_5 \end{array} \begin{array}{c} \begin{array}{cccccccc} e_1 & e_2 & e_3 & e_4 & e_5 & e_6 & e_7 & e_8 \end{array} \\ \begin{bmatrix} 1 & 1 & 1 & 0 & 0 & 0 & 0 & 0 \\ 0 & 1 & 1 & 1 & 0 & 1 & 1 & 0 \\ 0 & 0 & 0 & 1 & 1 & 0 & 0 & 0 \\ 0 & 0 & 0 & 0 & 0 & 0 & 1 & 1 \\ 0 & 0 & 0 & 0 & 1 & 1 & 0 & 0 \end{bmatrix} \end{array}$$

Path Matrix

Let G be a simple directed graph with n vertices $v_1, v_2, \ldots, v_n$. An $n \times n$ matrix $P = [P_{ij}]$ defined as follows

$$P_{ij} = \begin{cases} 1 & \text{if there is a path from } v_i \text{ to } v_j \\ 0 & \text{otherwise} \end{cases}$$

Suppose there is a path from v_i to v_j then there must be simple path from v_i to v_j when $v_i \neq v_j$ or there must be a cycle from v_i to v_j when $v_i = v_j$.

Matrix P is called the path matrix or **reachability matrix** of the graph G.

The path matrix P of a given graph G can be obtained from its adjacency matrix by following steps :

1. From the adjacency matrix of A, we can determine whether there exists an edge from one vertex to another.
2. Find A^n for some possible integer n.
3. Add the matrices $A, A^2, A^3, \ldots, A^n$.
   i.e., $B^n = A + A^2 + \ldots + A^n$
4. Now path matrix P can be obtained from B^n as follows :
   $P_{ij} = 1$ if and only if there is a non-zero element in the i, j entry of the matrix B^n.
   otherwise $P_{ij} = 0$.

Linked Representation

Let G be a directed graph with n vertices. The sequential representation of G in memory has a number of major drawbacks. First of all, it may be difficult to insert and delete nodes in G. This is because the size of matrix may need to be changed and the nodes may need to be reordered, so there may be many changes in the matrix. Furthermore, if the number of edges is $O(n)$ or $O(n \log_2 n)$ then the matrix will be sparse (will contain many zeros). So, G is represented in memory by a linked representation.

In linked representation, two node structures are used.

1. For non-weighted graph

INFO	Adj- list

2. For weighted graph

Weight	INFO	Adj- list

where Adj-list is the adjacency list *i.e.,* the list of vertices which are adjacent for the corresponding node.

In the linked representation of graphs, the number of list depends on the number of vertices in the graph. The header node in each list maintains a list of all adjacent vertices of that node for which the header node is meant. Consider a graph adjacency list representation of this undirected graph is:

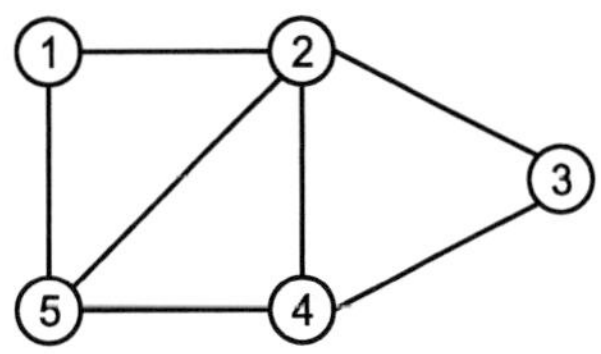

Fig. 12.21.

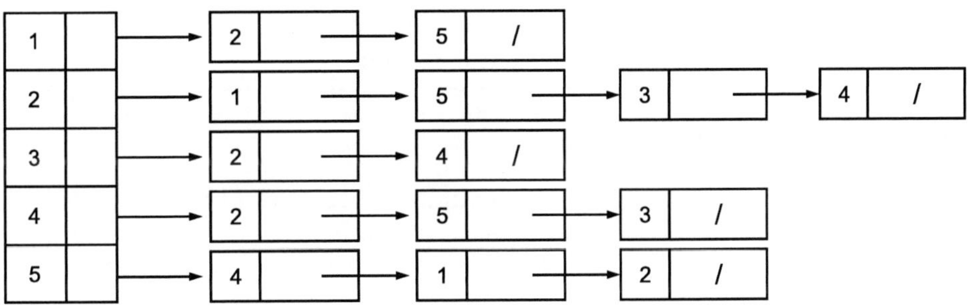

Suppose a directed graph.

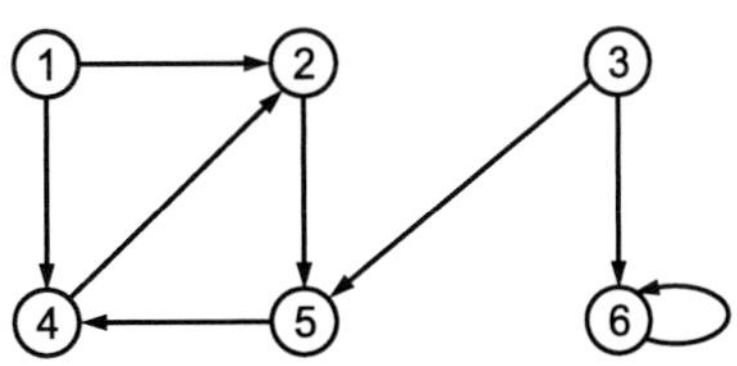

Fig. 12.22.

the adjacency list representation of this graph.

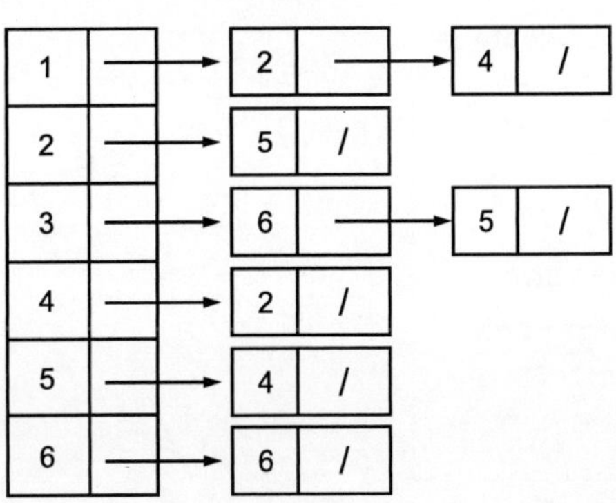

Comparison of various representations :

Representation	*Advantages*	*Disadvantages*
1. Set representation	1. It is the most efficient method from the memory point of view.	1. It does not allow storing of parallel edges. 2. Manipulation has a lot of difficulities.
2. Matrix representation	1. Manipulation of the graph with this representation is very easy. 2. It allows storing of parallel edges.	1. Space is wasted. 2. Insertion and deletion operations are not that much easy when compared to the linked representation
3. Linked representation	1. It is a space-saving method. 2. It is dynamic in nature. 3. It allows storing of parallel edges. 4. Insertion, deletion and other operations arts easy to perform.	1. Pointer fields are used as extra fields in addition to the data fields. 2. There is no faster way than to search the entire list for the presence of the edge.

12.5 IMPLEMENTATION OF A GRAPH

Let us first consider non-weighted graph with number of vertices MAX, where MAX is defined equal to 20, then the basic node type which store the information of **non-weighted graph** is defined as:

```
#define MAX 30
typedef struct node
     {
          int vertex;
          struct node *next;
     }node1;
```

Now, we can declare the adjacency list on the basis of number of vertices of graph as:

```
node1 *adj [MAX];
```

and for **weighted graph,** it becomes

```
typedef struct node
     {
          int vertex;
               int weight ;
          struct node *next;
     }node 2;
node2 *adj [MAX];
```

12.6 OPERATIONS ON GRAPH

Most of the algorithms developed assumed that the input graph is represented in adjacency-list representation.

(a) Creating an empty Graph :

```
void createGraph (node1 *adj[], int num)
    {
        int i;
        for (i=1, i<=num; i++)
            adj[i]=NULL;
    }
```

To create an empty graph, the entire adjacency list contains the NULL value.

(b) Inserting values in a Graph

(i) Non-weighted graph

```
void input (node1 *adj[], int num)
    {
        node1 *ptr, *last;
        int i, j, k, value;
        for (i=1; i<=num; i++)
            {
                last = NULL;
                printf("\n Number of Nodes in the adjacency list of
                    node %d:", i);
                scanf ("%d", &k);
                for (j=1; j <=k; j++)
                    {
                        printf("Enter the node %d:", j);
                        scanf ("%d", &value);
                        ptr = (node1*) malloc (sizeof (node1));
                        ptr → vertex = value ;
                        ptr → next = NULL;
                        If (adj [i] == NULL)
                            adj [i] = last = ptr;
                        else
                            {
                                last → next = ptr;
                                last = ptr;
                            }
                    }
            }
    }
```

(ii) Weighted Graph

```
void inputweighted (node2 * adj[], int n)
    {
        node2 *ptr, *last;
        int i, j, k, value, w;
        for (i = 1; i <=n; i++)
            {
                last=NULL;
                printf("\n No. of nodes in the adjacency list of node
                    %d:", i);
                scanf ("%d",&k);
                for (j=1; j <=k; j++)
                    {
                        printf ("Enter node %d:", j);
                        scanf ("%d", &value);
                        printf ("Enter weight for edge (%d %d)",i,value");
                        scanf ("%d", &w);
                        ptr=(node2*) malloc (sizeof (node2));
                        ptr → vertex = value;
                        ptr → weight = w;
                        ptr → next = NULL;
                        if (adj[i] == NULL)
                            adj [i] = last = ptr;
                        else
                            {
                                last → next = ptr;
                                last = ptr;
                            }
                    }
            }
    }
```

(c) Printing a Graph

This shows the way graph information can be outputted

```
void printgraph (node1*adj[],int n)
    {
        node1 * ptr;
        int i;
        for (i=1, i<=n; i++)
            {
                ptr = adj[i];
```

```
                    printf("%d",i);
                    while(ptr!=NULL)
                        {
                            printf("→ %d", ptr → vertex);
                            ptr = ptr → next;
                        }
                    printf("\n");
                }
        }
```

(d) Deleting a Graph :

This shows a graph removed from memory

```
void delegraph (node1 *adj[], int n)
    {
        int i;
        node1 *temp, *ptr;
        for (i=1; i<=n; i++)
            {
                ptr=adj[i];
                while(ptr!=NULL)
                    {
                        temp=ptr;
                        ptr=ptr → next;
                        free (temp);
                    }
                adj[i] = NULL;
            }
    }
```

12.7 TRAVERSING A GRAPH

As we know that traversing is nothing but visiting each node in some systematic approach. Graph is represented by its nodes and edges, so traversal of each node is the traversing in graph. There are two graph traversal methods :

1. Breadth First Search [BFS]
2. Depth First Search (DFS)

In BFS, we use **queue** for keeping nodes, which will be used for next processing and in DFS, we use **stack,** which keeps the node for next processing.

12.7.1 Difference Between Traversal in Graph and Tree

Traversal in graph is different from traversal in tree or list because of the following reasons :

(*a*) There is no first node or root node in a graph, hence the traversal can start from any node.

(*b*) In tree or list when we start traversing from the first node, all the nodes are traversed but in graph only those nodes will be traversed which are reachable from the starting node. So if we want to traverse all the reachable nodes we again have to select another starting node for traversing the remaining nodes.

(*c*) In tree or list while traversing we never encounter a node more than once but while traversing graph, there may be a possibility that we reach a node more than once. So to ensure that each node is visited only once we have to keep the status of each node whether it has been visited or not.

(*d*) In tree or list we have unique traversals. For example, if we are traversing the tree in inorder there can be only one sequence in which nodes are visited. But in graph, for the same technique of traversal there can be different sequences in which nodes can be visited.

12.8. BREADTH FIRST SEARCH

This graph traversal technique uses *queue* for traversing all the nodes of the graph. In this, first we take any node as a starting node then we take all the nodes adjacent to that starting node. Similar approach we take for all other adjacent nodes, which are adjacent to the starting node and so on. We maintain the status of visited node in one array so that no node can be traversed again.

Suppose V_0 is our starting node and V_1, V_2, V_3 are nodes adjacent to it. V_{11}, V_{12}, V_{13} are nodes adjacent to V_1, V_{21}, V_{22} are nodes adjacent to V_2 and V_{31} is adjacent to V_3. So we will traverse V_0 first and then all nodes adjacent to V_0, *i.e.*, V_1, V_2, V_3. Then we will traverse all nodes adjacent to V_1, *i.e.*, V_{11}, V_{12}, V_{13} and then we will traverse nodes adjacent to V_2, *i.e.*, V_{21}, V_{22} and then nodes adjacent to V_3, *i.e.*, V_{31}. Traversal will be in following order ;

$V_0\ V_1\ V_2\ V_3\ V_{11}\ V_{12}\ V_{13}\ V_{21}\ V_{22}\ V_{31}$

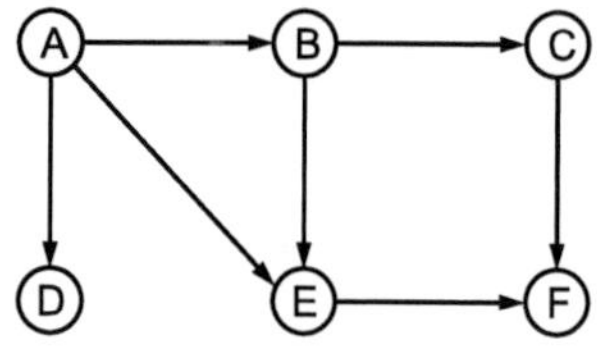

Fig. 12.23.

Let us take a graph and apply BFS to it

Take the node *A* as the starting node and start the traversal of the given graph.

First we traverse node *A*, then we traverse all adjacent nodes to node *A, i.e., B, D* and *E*. We can traverse these nodes in any order. Suppose we traverse in *B, D, E* order, so now traversal is:

A, B, D, E

Now we traverse all nodes adjacent to *B*, then all the nodes adjacent to *D* then all the nodes adjacent to *E, i.e., C, F* we can see that node adjacent to node *B* is *E* and *C*, but *E* is already traversed so we will ignore it and now the traversal is:

A, B, D, E, C, F

This was the traversal when we take node *A* as the starting node.

BFS through Queue

Take an array queue which will be used to keep the unvisited neighbours of the node. Take a boolean array *visited* which will have value *true* if the node has been *visited* and will have value *false* if the node has *not been visited.*

Initially queue is empty and front = – 1 and rear = – 1.

Initially visited [*i*] = false where *i* = 1 to *n*, *n* is total number of nodes.

Procedure

1. Insert starting node into the queue.

> 2. Delete front element from the queue and insert all its unvisited neighbours into the queue at the end, and traverse them. Also make the value of visited array true for these nodes.
> 3. Repeat step 2 until the queue is empty.

Let us take node 1 as the starting node for traversal.

Consider the graph G in Fig. 12.24. below:

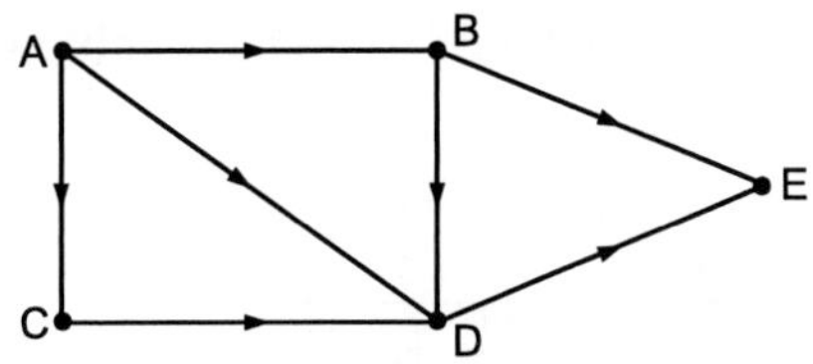

Fig. 12.24.

The linked list or adjacency list representation of the graph is follows :

Vertex	*Adjacency list*
A	*B, C, D*
B	*D, E*
C	*D*
D	*E*
E	–

Suppose the source vertex is *A*. Then following steps will illustrate the BFS.

Step 1 : Initially push *A* (the source vertex) to the queue.

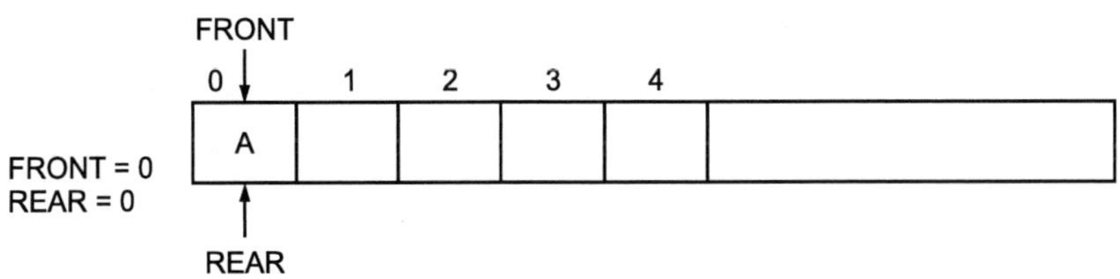

Step 2 : Remove the front element *A* from the queue (by incrementing FRONT = FRONT + 1) and display it. Then push all the neighbouring vertices of *A* to the queue (by incrementing REAR = REAR + 1) if it is not in queue.

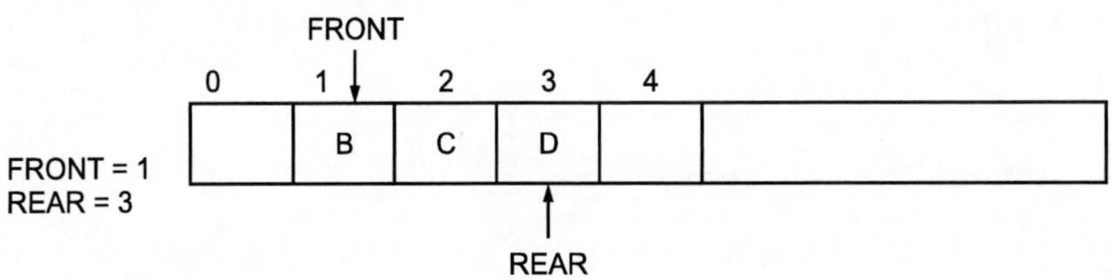

Traversed nodes = *A*.

Step 3 : Remove the front element *B* from the queue and display it then add the neighbouring vertices of *B* to the queue, if it is not in queue.

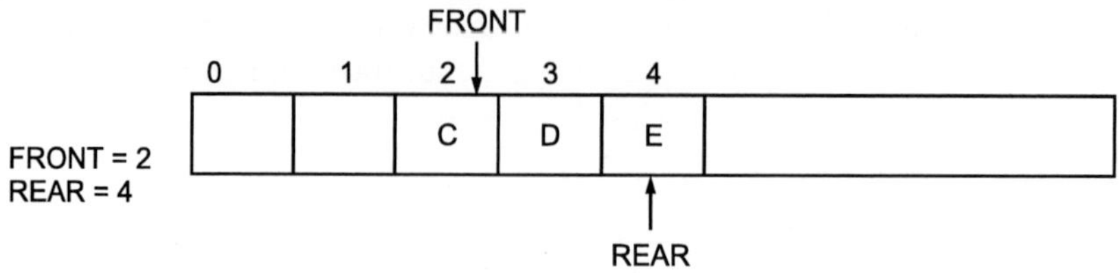

Traversed nodes = *A*, *B*.

Step 4 : Remove the front element *C* and add the neighbouring vertex if it is not present in the queue, *i.e.*,

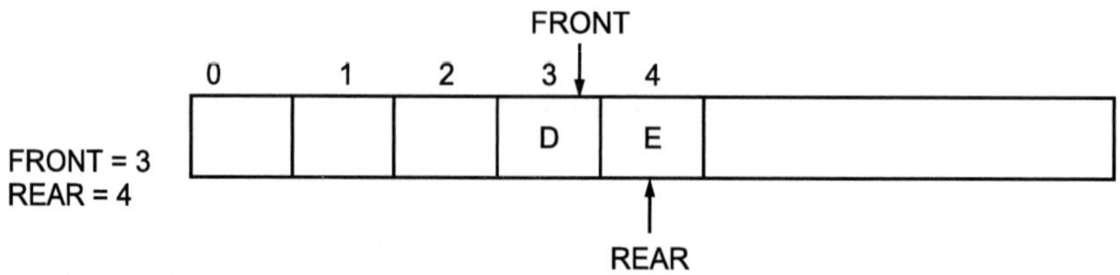

Traversed nodes = *A*, *B*, *C*.

Step 5 : Again the process is repeated (until FRONT > REAR), *i.e.*, remove the front element *D* of the queue and add the neighbouring vertex if it is not present in the queue.

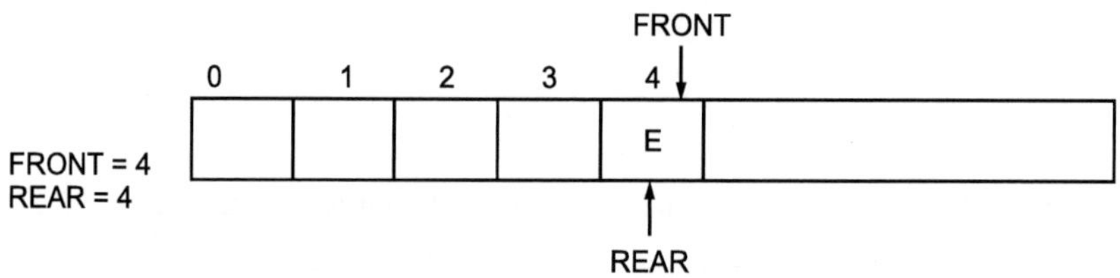

Traversed nodes = *A*, *B*, *C*, *D*.

then,

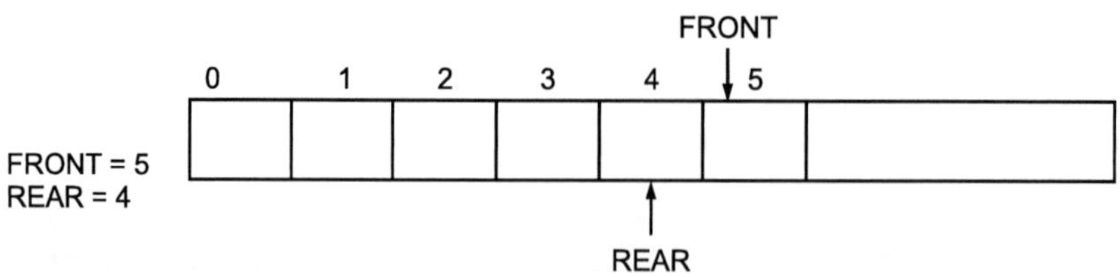

Traversed nodes = *A*, *B*, *C*, *D*, *E*.

So, *A*, *B*, *C*, *D*, *E* is the BFS traversal of the graph.

C Implementation

```
BFS (int v)
    {
        int i, front, rear;
        int que[20];
        front=rear=-1;
        printf("%d",V);
        visited[v]=true;
        rear++;
```

```
        front++;
        que[rear]=v;
        whle(front<=rear)
        {
            v=que[front]; /* delete fron queue */
            front++;
            for(i=1;i<=n;i++)
            {
                /* Check or adjacent unvisited nodes */
                if(adj[v][i]==1 && visited[i]==false)
                {
                    printf("%d",i);
                    visited[i]=true;
                    rear++;
                    que[rear]=i;
                }
            }
        }
    }
```

BFS by Colouring Scheme

To keep track of the progress, BFS colours each vertex WHITE, GRAY or BLACK, where vertex WHITE colour indicates that the vertex is undiscovered, vertex with GRAY colour indicates that the vertex is discovered but not yet processed and the vertex with BLACK colour indicates that the vertex has been processed. So all the vertices are initially WHITE and later on become GRAY and then BLACK.

Thus if $(u, v) \in E$ and vertex u is BLACK, then vertex v is either GRAY or BLACK, *i.e.*, all vertices adjacent to BLACK vertices have been discovered. Gray vertices may have some adjacent WHITE vertices.

```
BFS (G, s)
    for each vertex u ∈ V[G] - {s}
        do color[u] ← WHITE
            d[u] ← ∞/* distance from vertex u */
            π[u] ← NIL
    color[s] ← GRAY
    d[s] ← 0
    π[s] ← NIL
    Q ←φ
    ENQUEUE (Q, s)
    while Q ≠φ
        do u ← DEQUEUE (Q)
```

```
for each v ∈ Adj [u]
    do if color[v] = WHITE
        then color[v] ← Gray
            d[v] ← d[u] + 1
            π[v ] ← u
            ENQUEUE (Q, v)
    color[u] ← Black
```

EXAMPLE 12.2. Consider the graph *G* in Fig. 12.25. Describe the whole process of breadth first search. Using vertex 3 as the source.

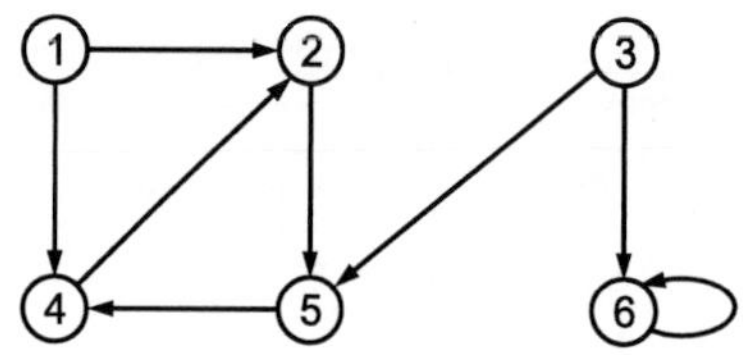

Fig. 12.25.

Solution: First, we create adjacency-list representation.

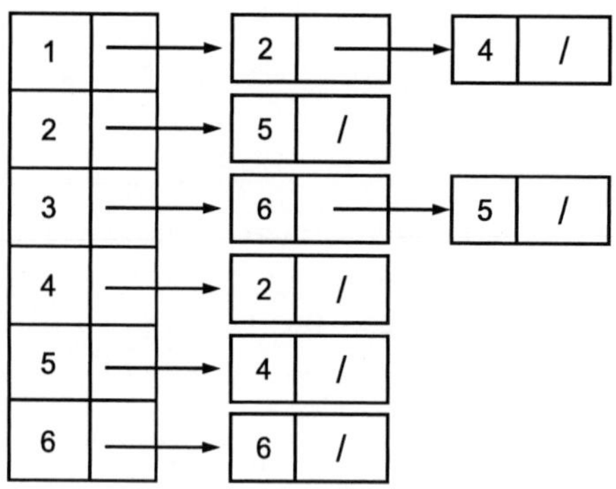

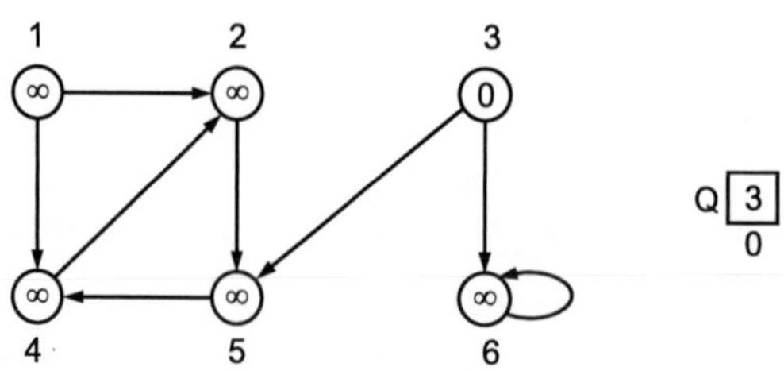

Fig. 12.26.

So, adj[3] = [6, 5] so color [6] = GRAY

color [5] = GRAY

and $d[6] = 1 \quad d[5] = 1$

$\pi[6] \leftarrow 3 \quad \pi[5] \leftarrow 3$

Now,

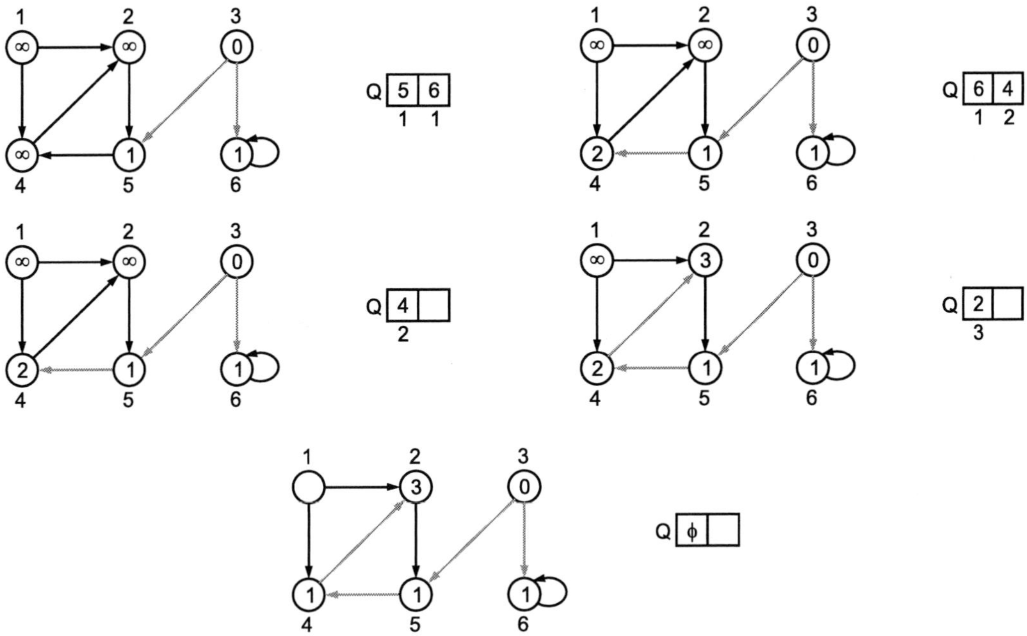

Fig. 12.27.

Thus vertex 1 cannot be reachable from source.

12.9 DEPTH FIRST SEARCH

The Depth First Search (DFS), as its name suggest, is to search deeper in the graph, whenever possible. Given as input graph $G = (V, E)$ and a source vertex S, from where the searching starts. First we visit the starting node, then we travel through each node along a path, which begins at S. That is, we visit a neighbour vertex of S and again a neighbour of a neighbour of S and so on. DFS also works on both directed as well as on undirected graphs.

DFS uses **stack**.

Let us take the graph (see Fig. 12.28).

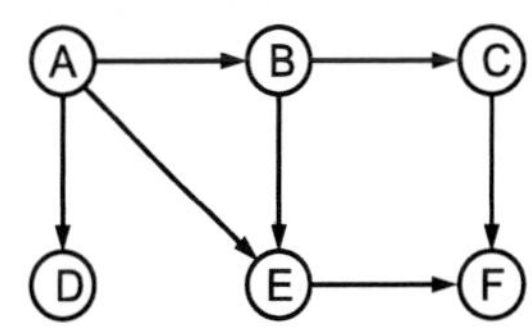

Fig. 12.28.

Suppose node A as the starting node. First, we will traverse node A. Then, we will traverse any node adjacent to node A. Suppose we traverse node B, then traverse node E, which is adjacent to node B, then traverse adjacent node of E, *i.e.,* node F. Now there is no node adjacent to node F, *i.e.,* dead end. So we will move backward. Till now the traversal is:

$$A, B, E, F$$

Now, we reach node E, see if there is any node adjacent to it. There is no such node. So we will move backward. Now, we reach node B, see if there is any node adjacent to it and not traversed yet. Here node is C. So we traverse it. Now there is no untraversed node adjacent to node C. So, we will move backward. On reaching node A we see that node D is adjacent to it and has not been traversed. So we traverse it. Now there is no untraversed node adjacent to node D. We can't move forward and we can't move backward also, so we will stop. So, the traversal is:

A, B, E, F, C, D.

Depth First Search Through Stack

Depth Search technique uses stack. Take an array **STACK**, which will be used to keep the unvisited neighbours of the node. Take a another Boolean array **VISITED,** which will have value **TRUE** if the node has been *visited* and will have value **FALSE** if the node has *not been visited.*

Initially stack is empty and TOP = – 1.

Initially VISITED [i] = FALSE where $i = 1$ to n, n is total number of nodes.

Procedure

1. Push starting node into the stack.
2. Pop an element from the stack, if it has not been traversed then traverse it, if it has already been traversed then just ignore it. After traversing make the value of visited array true for this node.
3. Now push all the unvisited adjacent nodes of the popped element on stack. Push the element even if it is already on the stack.
4. Repeat steps 3 and 4 until stack is empty.

Let us consider a graph.

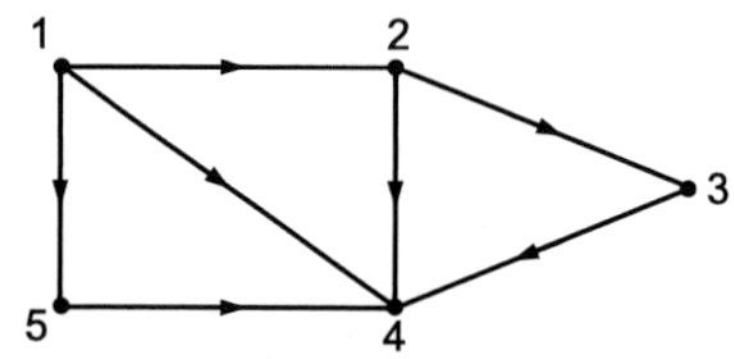

Fig. 12.29.

Suppose the starting node for traversal is 1.

Step 1 : Push node 1 into stack.

Now, TOP = 0 and STACK = 1.

Step 2 : POP node 1 from stack and traverse it

So, Traversed node = 1 and VISITED [1] = TRUE

Now push all the unvisited adjacent nodes 2, 5, 4 of the popped element on the stack.

Now TOP = 2 and STACK = 2, 5, 4

Traversal = 1.

Step 3 : POP the element node 4 from the stack, traversed it and push all its unvisited adjacent nodes. There is no adjacent nodes. So,

Traversed node = 4

VISITED [4] = TRUE

TOP = 1 STACK = 2, 5

Traversal = 1, 4

Step 4 : POP the element 5 from the stack, traverse it and push all its unvisited adjacent nodes. Here node 4 is adjacent node of 5 but it is visited node. So,

Traversed node = 5

VISITED [5] = TRUE

TOP = 0 STACK = 2

Traversal = 1, 4, 5

Step 5 : Pop the element 2 from the stack, traverse it and push all its unvisited adjacent nodes, *i.e.*, node 3.

Now, Traversed node = 2

VISITED[2] = TRUE

TOP = 0 STACK = 3

Traversal = 1, 4, 5, 2

Step 6 : POP the element 3 from the stack, traverse it and push all its unvisited adjacent nodes. No node is here.

Now, Traversed node = 3

VISITED[3] = TRUE

TOP = – 1 STACK = NULL

Traversal = 1, 4, 5, 2, 3

Since the stack is empty, so we will stop our process. The function for DFS is as follows :

C Implementation

```
DFS (int v)
    {
        int i, stack[MAX], TOP = - 1, POP, j, k;
        int ch;
        TOP ++;
        stack[TOP] = v;
        while(TOP>=0)
            {
                POP = stack[TOP];
                TOP--;
                if(visited[POP] == false)
                    {
                        printf("%d",pop);
                        visited[pop] = true;
                    }
                else
                    continue
                for(i=n; i>=1; i--)
                    {
                        if(adj[pop][i] == 1 && visited[i] == false)
                          {
                              top++;
                              stack[top]=i;
                          }
                    }
            }
    }
```

DFS by Colouring Scheme

To keep track of the progress, the DFS also uses the same colouring scheme as described for BFS. In addition, the DFS makes uses a stack structure to maintain the order in which the vertices are to be processed, instead of queue.

In depth-first search, edges are explored out of the most recently discovered vertex v that still has unexplored edges leaving it. When all of v's edges have been explored, the search backtracks to explore edges leaving the vertex from which v was discovered. This process continues until we have discovered all the vertices that are reachable from the original source vertex. If any undiscovered vertices remain, then one of them is selected as a new source and the search is repeated from that source. This entire process is repeated until all vertices are discovered.

As in the BFS, whenever a vertex v is discovered during a scan of the adjacency list of an already discovered vertex u, DFS records this by setting v's predecessor field $\pi[v]$ to u. In DFS predecessor sub-graph is composed of several trees, because the search may be repeated from multiple sources.

In DFS, each vertex v has two time-stamps: the first time-stamp $d[v]$, *i.e.,* discovery time records when v is first discovered, *i.e.,* grayed, and the second timestamp $f[v]$, *i.e.,* finishing time records when the search finishes examining v's adjacency list, *i.e.,* blacked. For every vertex $d[u] < f[u]$.

```
DFS(G)
    for each vertex u ∈ V[G]
        do color[u] ← WHITE
            π[u] ← NIL
    time ← 0
    for each vertex u ∈ V[G]
        do if color[u] = WHITE
            then DFS-VISIT(u)
```

```
DFS-VISIT(u)
    color[u] ← GRAY              /*White vertex u has just been
                                   discovered*/
    time ← time + 1
    d[u] ← time
    for each v ∈ Adj[u]          /*Explore edge(u, v)*/
        do if color[v] = WHITE
            then π[v] ← u
                DFS-VISIT(v)
    color[u] ← BLACK             /*Blacken, u it is finished*/
    f[u] ← time ← time + 1
```

EXAMPLE 12.3. Show the progress of DFS algorithm on a directed graph.

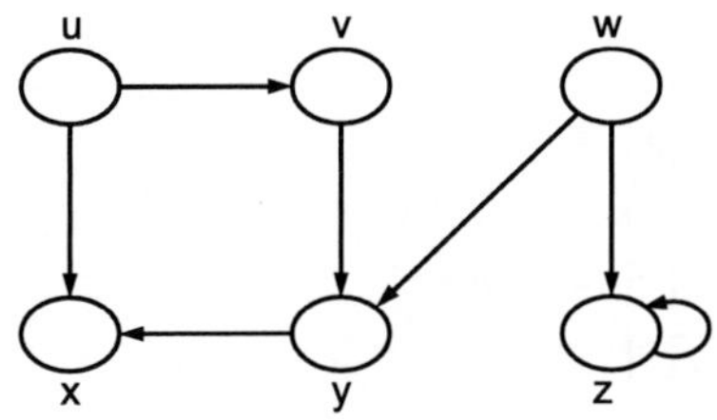

Fig. 12.30.

Solution: Initially for each vertex color is white and time = 0.

Now call DFS-VISIT(u) since color [u] = white.

Make color[u] = GRAY.

and time = 0 + 1 = 1

$d[u]$ = 1.

adj[u] = v and x.

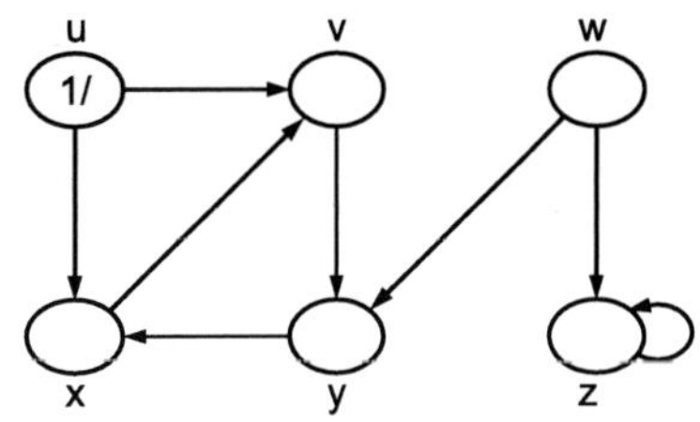

Fig. 12.31.

Here, color[v] = white, So call DFS-VISIT (v)

Make color[v] = Gray

time = 1 + 1 = 2.

$d[v]$ = 2.

adj[v] = y

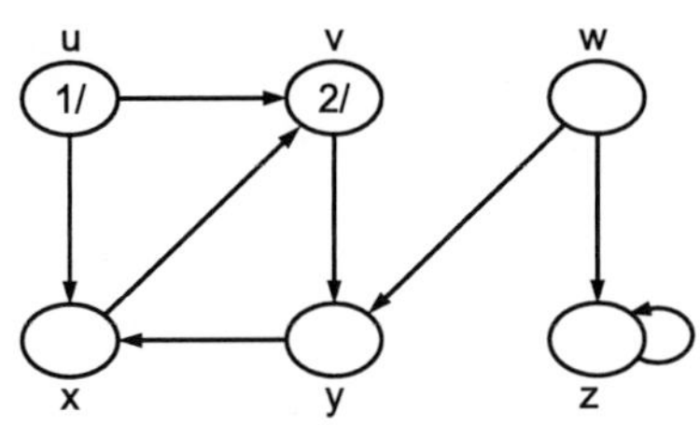

Fig. 12.32.

and color [y] = white, so call DFS-VISIT [y]

Make color[y] = GRAY

time = 2 + 1 = 3

$d[y]$ = 3

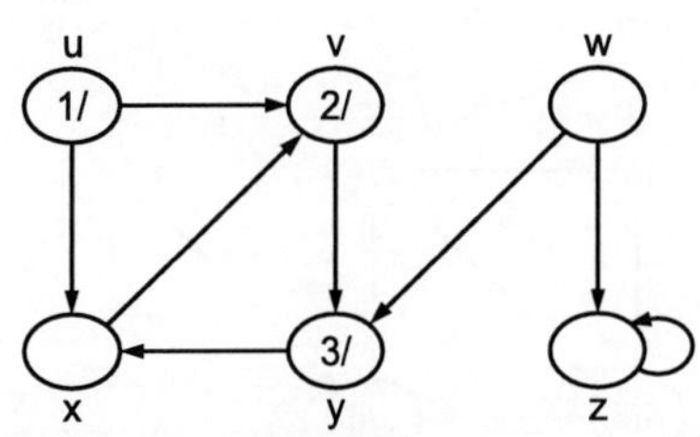

Fig. 12.33.

Now, adj[y] = x

Here, color[x] = white, so call DFS-VISIT [x]

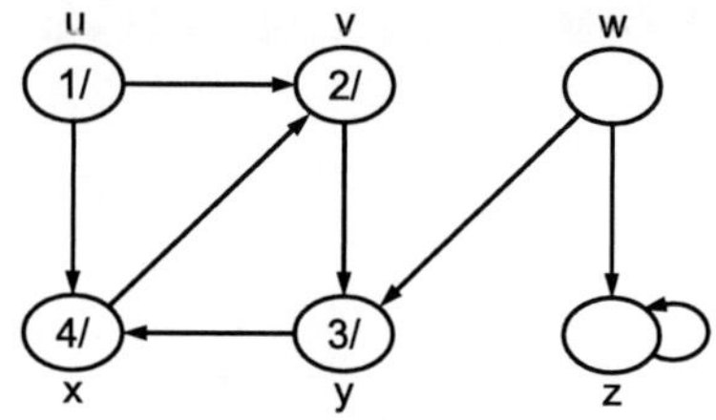

Fig. 12.34.

Make, $\text{color}[x] = \text{GRAY}$

$\text{time} = 3 + 1 = 4$

$d[x] = 4$

$\text{adj}[x] = v$

Here, $\text{color}[v] \neq \text{white}$

So, $\text{color}[x] = \text{Black}$

and $f[x] = 4 + 1 = 5$

Now,

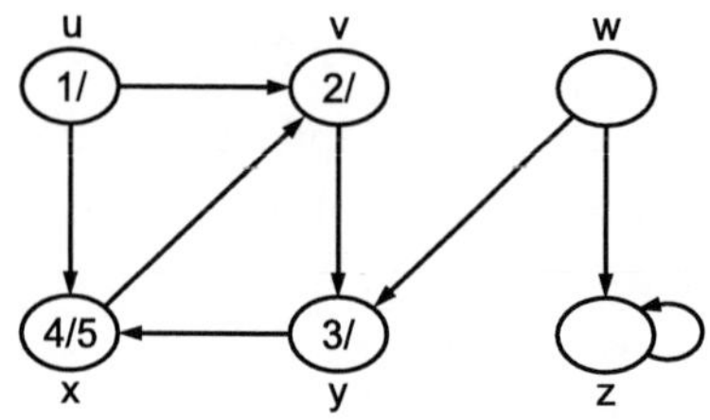

Fig. 12.35.

Now, make color $[y] = \text{Black}$

and $f[y] = 5 + 1 = 6$

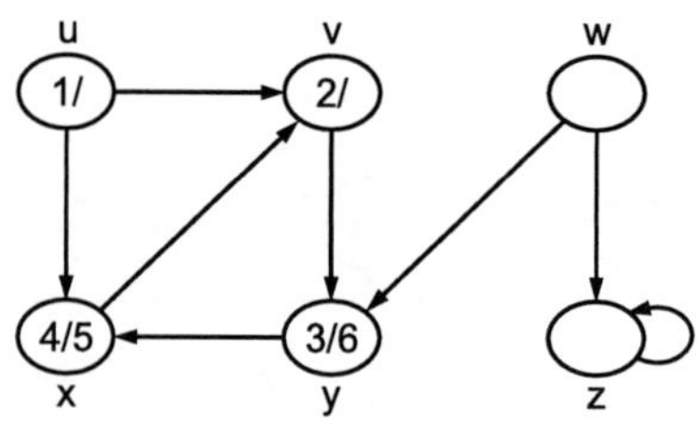

Fig. 12.36.

Similarly, make color $[v] = \text{Black}$

and $f[v] = 6 + 1 = 7$

then, we take other adjacency vertex of u, *i.e.*, x.

Now, color $[x] \neq \text{white}$

So, make $\text{color}[u] = \text{Black}$

and $f[u] = 7 + 1 = 8$

Now,

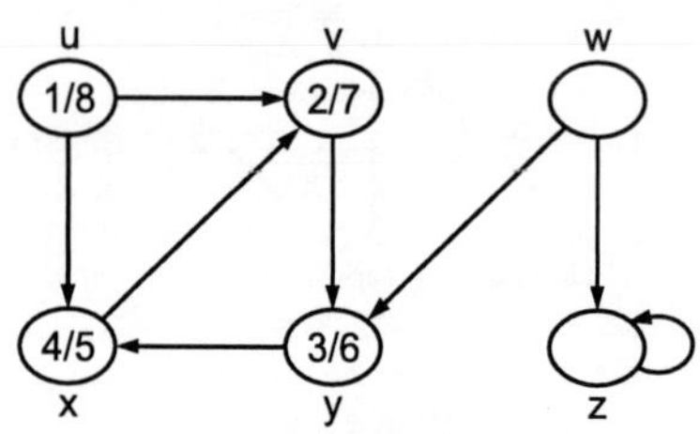

Fig. 12.37.

Now, we take other vertex, *i.e.*, w.

Here	color[w]	= white. So call DFS-VISIT[w]
	color[w]	= Gray and $d[w] = 8 + 1 = 9$
and	adj[w]	= z and y
Now,	color[z]	= white
So, call DFS-VISIT[z]		
Make	color[z]	= GRAY
and	$d[z]$	$= 9 + 1 = 10$
Here,	adj[z]	= z and color[z] $\neq$ white
So,	make color[z]	= Black
and	$f[z]$	$= 10 + 1 = 11$

i.e.,

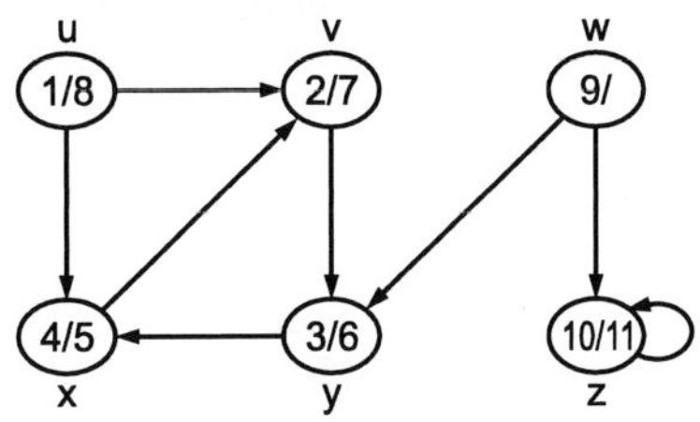

Fig. 12.38.

Now, take other adjacency vertex of w, *i.e.*, y.

	color[y]	$\neq$ white
So,	make color[w]	= Black
and	$f[w]$	$= 11 + 1 = 12$

i.e.,

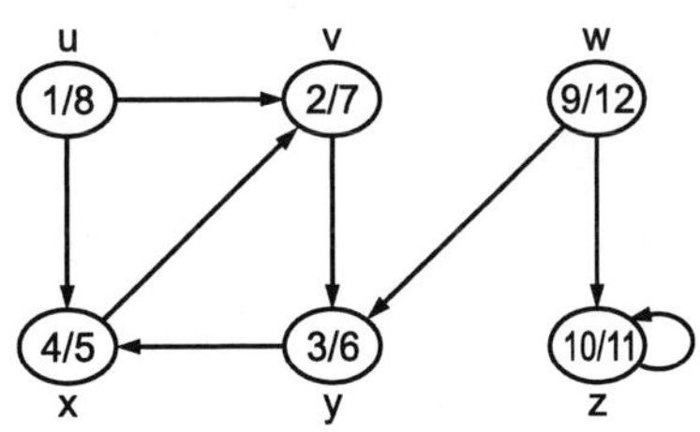

Fig. 12.39.

12.10 TOPOLOGICAL SORT

If G is a directed graph with no directed cycles, *i.e.*, G is a directed a cyclic graph, then a topological order T is a sequential listing of all the vertices in G, such that for all vertices u, $v \in G$ if there is an edge from u to v, then u precedes v in the topological ordering.

> Thus, a topological sort of a directed acyclic graph $G = (V, E)$ is a linear ordering of u, $v \in V$ such that if $(u, v) \in E$ then u appears before v in this ordering. If G is a cyclic graph, then no such ordering exists.

The main idea behind our algorithm to find a topological order of a graph without cycles is that any vertex v with zero in-degree (*i.e.*, without any predecessors) may be selected as the first element in the order.

Algorithm

1. Find the in-degree of each vertex *v* of *G*.
2. Insert all vertices with zero degree in a queue.
3. Repeat steps (4) and (5) while queue is not NULL.
4. Remove front vertex *v* of the queue.
5. For each neighbour *w* of the vertex *v*.
   (*a*) Decrease the in-degree of *w* by 1, *i.e.*, remove the edge from *v* to *w*.
   (*b*) If in-degree of *w* is 0 (zero), then add *w* to the rear of the queue.
6. STOP.

Topological-Sort (G)

```
for each vertex u ∈ V
     do indegree[u] ← 0
for each vertex u ∈ V
     do for each v ∈ Adj[u]
          indegree[v] ← indegree + 1
Q ←φ
for each vertex u ∈ V
     if indegree [u] = 0
          then ENQUEUE (Q, u)
while Q ≠φ
     do u ← DEQUEUE (Q)
          print u
          for each v ∈ Adj[u]
               indegree[v] ← indegree[v] - 1
               if indegree[v] = 0
                    then ENQUEUE (Q, v)
```

EXAMPLE 12.4. Let us take a directed graph G.

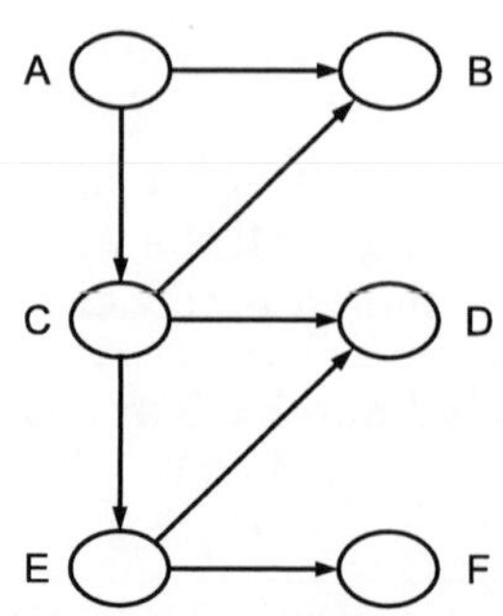

Fig. 12.40.

Solution: Adjacency list of G.

$$A \to B, C$$
$$B \to$$
$$C \to B, D, E$$
$$D \to$$
$$E \to D, F$$
$$F \to$$

Step 1 : Indegree of nodes are

$$A = 0, B = 2, C = 1, D = 2, E = 1, F = 1$$

Taking all nodes which have zero indegree *i.e.,* node A.

Step 2 : (*i*) Add zero indegree node to queue.

∴ Queue : A Front = 1 Rear = 1

(*ii*) Delete node A and edges going from A.

Now Queue : ϕ Front = 0 Rear = 0 Sort : A

and the indegree of nodes will be

$$B = 1, C = 0, D = 2, E = 1, F = 1$$

Step 3 : Add C to queue

Queue : C Front = 1 Rear = 1 Sort : A

delete node C and edges going from C

So, Queue : ϕ Front = 0 Rear = 0 Sort : A, C

and the indegree of nodes will be

$$B = 0, D = 1, E = 0, F = 1$$

Step 4 : Add B, E to queue

Queue : B, E Front = 1 Rear = 2 Sort : A, C

Delete the node B and edges from B

No change in the indegree of nodes.

Queue : E Front = 1 Rear = 1 Sort : A, C, B

Now Delete the node E and edges going from E.

Queue : ϕ Front = 0 Rear = 0 Sort : A, C, B, E

and the indegree of nodes will be

$$D = 0 \quad F = 0$$

Step 5 : Add D, F to queue

Queue : D, F Front = 0 Rear = 0 Sort : A, C, B, E

Delete the node D and edges from D.

Queue : F Front = 1 Rear = 1 Sort : A, C, B, E, D

Now Delete the node F and edges from F.

Queue : ϕ Front = 0 Rear = 0 Sort : A, C, B, E, D, F

Now we have no more nodes in the graph, so the topological sorting of graph will be

$$A, C, B, E, D, F$$

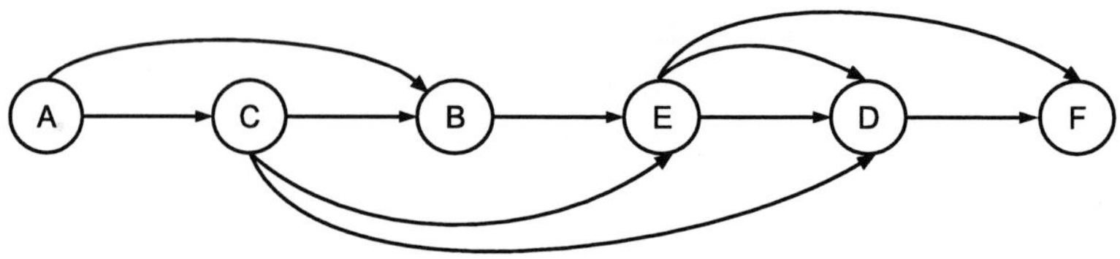

Fig. 12.41.

Another way to perform topological sorting

Topological-Sort (G)

1. Call DFS (G) to compute finishing times $f(v)$ for each vertex $v \in G$.
2. As each vertex is finished, insert it onto the front of a linked list.
3. Return the linked list of vertices.

12.11 STRONGLY CONNECTED COMPONENTS

A directed graph $G = (V, E)$ is called **strongly connected,** if for every pair of vertices u and v there is a path from u to v and a path from v to u. The strongly connected components of a directed graph are its maximal strongly connected sub-graphs.

For example,

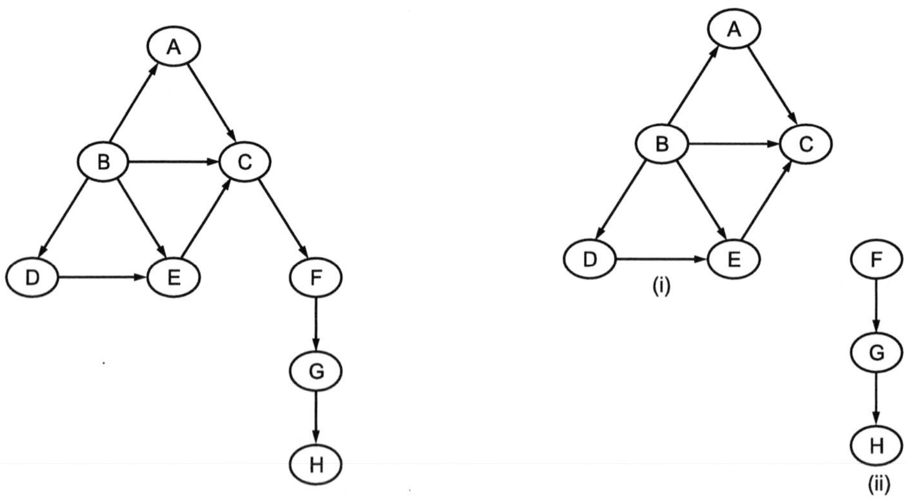

(*a*) Directed graph (*b*) Strongly Components

Fig. 12.42.

Kosaraju's algorithm efficiently computes the strongly connected components of a directed graph.

Kosaraju's algorithm

1. Call DFS (G) to compute finishing times $f[u]$ for each vertex u.
2. Compute transpose of G, *i.e.*, G^T.
3. Call DFS (G^T) and consider the vertices in order of decreasing $f[u]$.
4. Produce as output the vertices of each tree in the DFS forest formed in line (3) as a separate strongly connected component.

EXAMPLE 12.5. Find the strongly connected components in the Fig. 12.43.

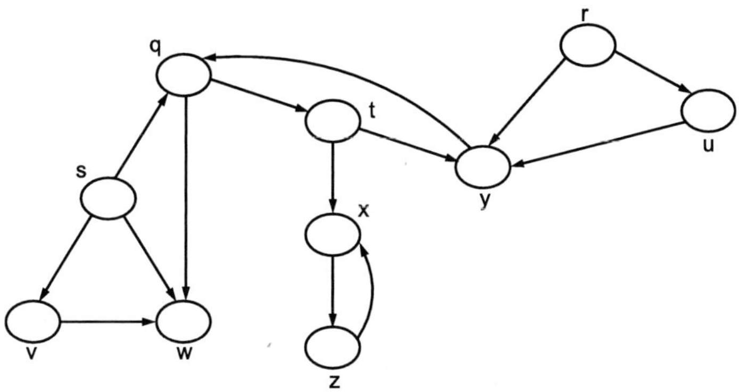

Fig. 12.43.

Solution: Call DFS, we get,

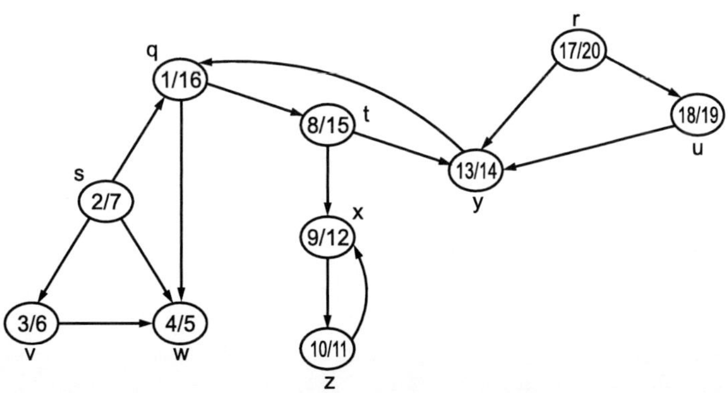

Fig. 12.44.

DFS gives the list,

$$w, v, s, z, x, y, t, q, u, r$$

then the reverse order of turning black is

$$r, u, q, t, y, x, z, s, v, w$$

Now, G^T as follows

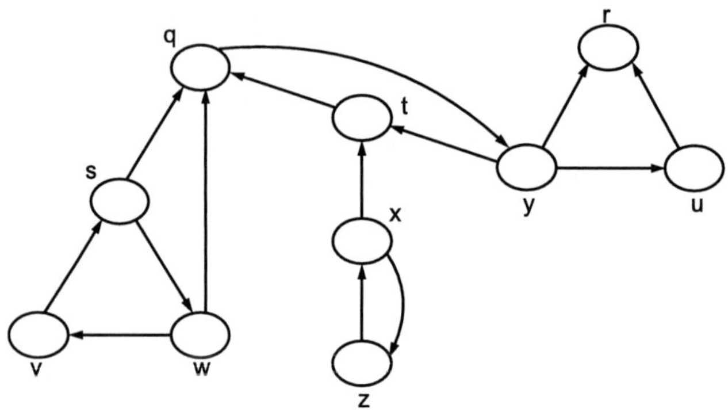

Fig. 12.45.

Call DFS (G^T).

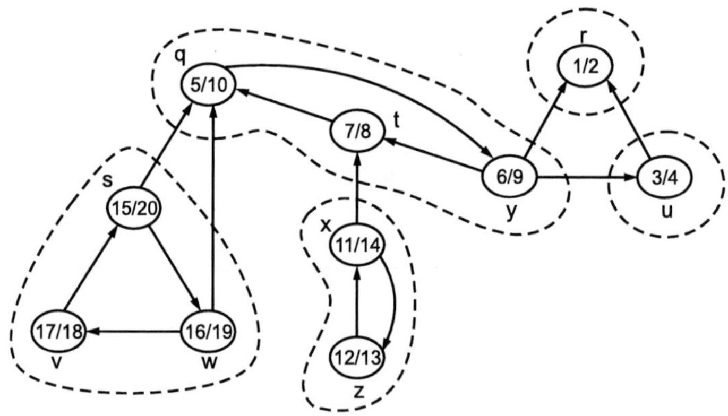

Fig. 12.46.

Thus, strongly connected components are

$$\{r\}, \{u\}, \{q, t, y\}, \{x, z\}, \{s, v, w\}$$

12.12 SPANNING TREE

A spanning tree of a graph is just a sub-graph that contains all the vertices and is a tree. That is, a spanning tree of a connected graph G contains all the vertices and has the edges which connects all the vertices. So, the number of edges will be 1 less than the number of nodes.

Let us take a connected graph.

Here all these trees are spanning tree and the number of edges is one less than the number of nodes.

If graph is not connected, *i.e.*, a graph with n vertices has edges less than $n - 1$ then no spanning tree is possible. A graph may have many spanning trees. **DFS and BFS spanning trees** are the spanning trees which are obtained by DFS and BFS traversal respectively.

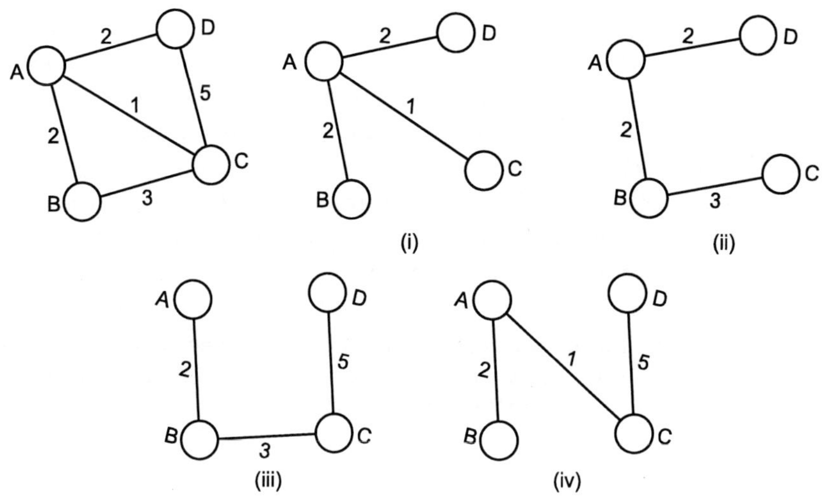

Fig. 12.47.

12.13 MINIMUM SPANNING TREES

Given a connected weighted graph G, it is often desired to create a spanning tree T for G such that the sum of the weights of the tree edges in T is as small as possible. Such a tree is called a **minimum spanning tree** and represents the "cheapest" way of connecting all the nodes in G.

There are number of techniques for creating a minimum spanning tree for a weighted graph but the most famous methods are **Prim's** and **Kruskal algorithm**.

12.14 KRUSKAL'S ALGORITHM

This algorithm creates a *forest of trees*. Initially, the forest consists of n single node trees and no edges. That is, in the method initially we take n distinct trees for all n nodes of the graph. At each step, we add one (the cheapest one) edge, so that it joins two trees together. If it forms a cycle, it does simply links two nodes that were already part of a single connected tree, so that this edge does *not included.*

The steps are as follows :

1. Initially construct a separate tree for each node in a graph.
2. Edges are placed in a *priority queue, i.e.,* we take edges in ascending order. We can use a heap for the priority queue.
3. Until we have added $n - 1$ edges.
   (*a*) Extract the cheapest edge from the queue.
   (*b*) If it forms a cycle, reject it
   else
   add it to the forest.
4. Whenever we insert an edge, two trees will be joined, *i.e.,* every step will have joined two trees in the forest together so that at the end, there will be only one tree.

In this method, first we examine all the edges one-by-one starting from the smallest edge. To decide whether the selected edge should be included in the spanning tree or not, we will examine the two nodes connecting the edge. If the two nodes belong to the same tree then we will not include the edge in the spanning tree, since the two nodes are in the same tree, they are already connected and adding this edges would result in a cycle. So we will insert an edge in the spanning tree only if it's nodes are in different trees.

Now, we will see how to decide whether two nodes are in the same tree or not. For this, we need a **UNION-FIND** structure.

To understand the UNION-FIND structure, we need to look at a **partition of a set.**

(*a*) Every element of the set belongs to one of the sets in the partition.

(*b*) No element of the set belongs to more than one of the sub-sets.

(*c*) Every element of a set belongs to one and only one of the sets of a partition.

A partition of a set may be thought of as a set of equivalence classes. Each sub-set of the partition contains a set of equivalent elements. For each subset, we denote one element as the representative of that subset. Each element in the sub-set is, some how, equivalent and represented by the representative. When we add elements to the sub-set, we arrange that all the elements point to their representative. Initially, each node is its own representative. As the initial pairs of nodes are joined to form a tree, the representative pointer of one of the nodes is made to point to the other, which becomes the representative of the tree. As trees are joined, the representative pointer of the representative of one of them is set to point to any element of the other. Let x denote an object, we wish to support the following operations.

1. **MAKE -SET (*x*)** creates a new set whose only member (and thus representative) is pointed to by x.
2. **FIND-SET (*x*)** returns a pointer to the representative of the set containing x.
3. **UNION (*x*, *y*)** unites the dynamic sets that contain x and y, say S_x and S_y into a new set that is the union of these two sets. The representative of the resulting set is some member of $S_x \cup S_y$. Since we require the sets in the collection to be disjoint, we "destroy" sets S_x and S_y removing them from the collection.

MST-KRUSKAL (*G*, *W*)

```
1.   A ←ϕ.
2.   for each vertex v ∈ V[G]
          do MAKE-SET (v)
3.   sort the edges of E into increasing order by weight w.
4.   for each edge (u, v) ∈ E taken in increasing order.
          do if FIND-SET (u) ≠ FIND-SET (v)
               then A ← A U {u, v}
                    UNION (u, v)
5.   return A.
```

EXAMPLE 12.6. Find the minimum spanning tree of the following graph using Kruskals algorithm.

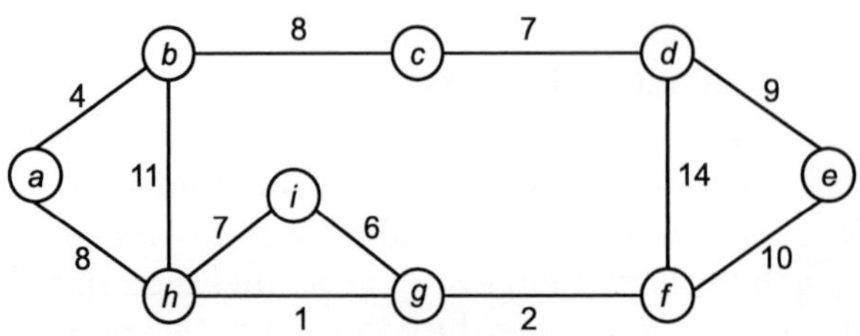

Fig. 12.48.

Solution: First we initialise the set A to the empty set and create $|v|$ trees, one containing each vertex with MAKE-SET procedure. Then sort the edges in E into order by non-decreasing weight, *i.e.*,

Edge	*Weight*
(h, g)	1
(g, f)	2
(a, b)	4
(i, g)	6
(h, i)	7
(c, d)	7
(b, c)	8
(a, h)	8
(d, e)	9
(e, f)	10
(b, h)	11
(d, f)	14

Now, check for each edge (u, v), whether the end points u and v belong to the same tree. If they do, then the edge (u, v) cannot be added. Otherwise, the two vertices belong to different trees and the edge (u, v) is added to A and the vertices in the two trees are merged in by UNION procedure.

So, first take (h, g) edge

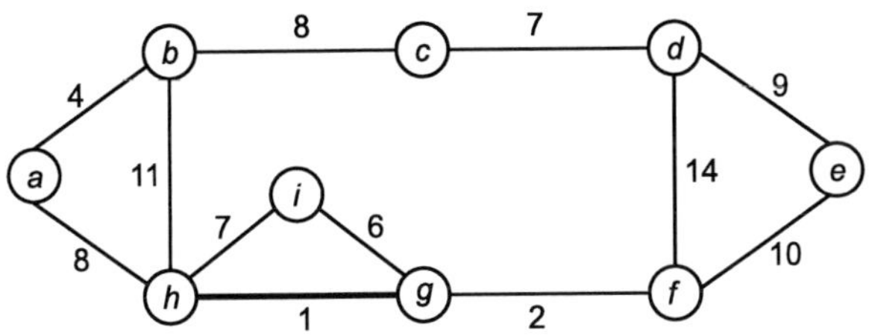

Fig. 12.49.

the (g, f) edge.

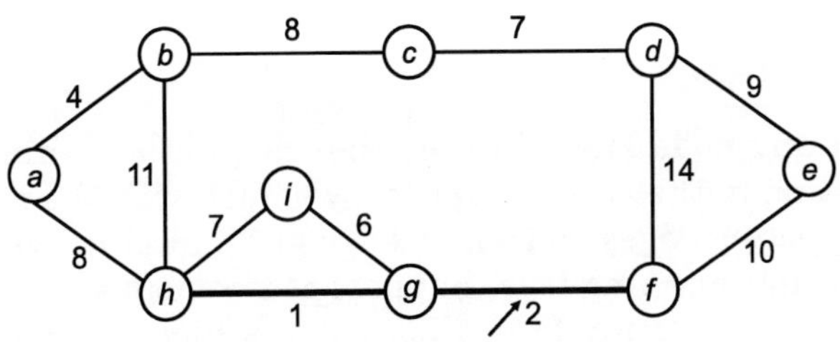

Fig. 12.50.

then (a, b) and (i, g) edges are considered and forest becomes.

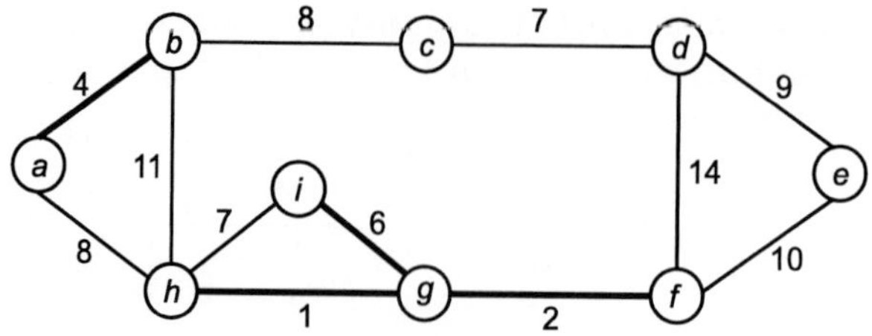

Fig. 12.51.

Now, edge (h, i). Both h and i vertices are in same set, thus it creates a cycle. So this edge is *discarded.*

Then edge (c, d), (b, c), (a, h), (d, e), (e, f) are connected and forest becomes.

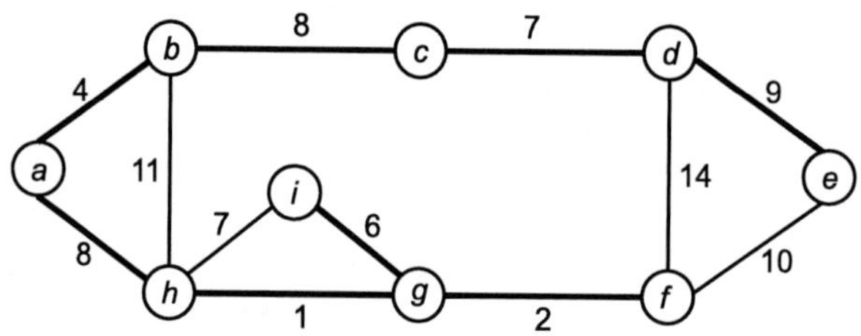

Fig. 12.52.

In (e, f) edge both end point e and f exist in same tree so *discarded* this edge. Then (b, h) edge, it also creates a cycle.

After that edge (d, f) and the final spanning tree is shown as in dark lines.

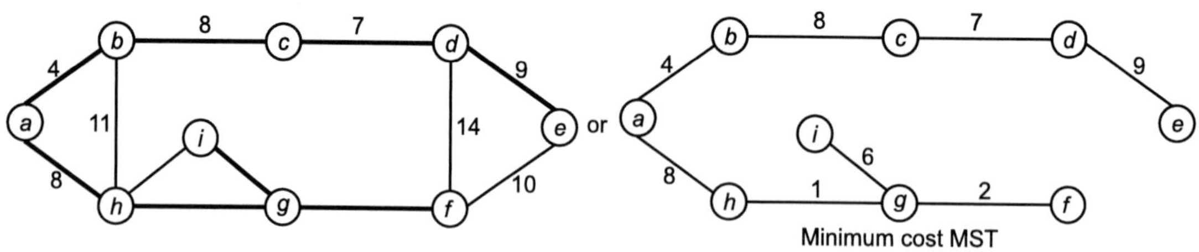

Fig. 12.53.

12.15 PRIM'S ALGORITHM

In this, we start with any node and add the other node in spanning tree on the basis of weight of edge connecting to that node. Suppose we start with the node 'u' then we have a need to see all the connecting edges and which edge has minimum weight. Then we will add that edge and node to the spanning tree. Suppose if two nodes u_1 and u_2 are in spanning tree and both have edge connecting to v_3 then edge. Which has minimum weight will be added in spanning tree.

The main idea of Prim's algorithm is similar to that of **Dijkstra's algorithm** (discussed later) for finding shortest path in a given graph. Prim's algorithm has the property that the edges in the set A *always from a single tree.* We begin with some vertex v in a given graph $G = (V, E)$, defining the initial set of vertices A. Then, in each iteration, we choose a minimum-weight edge (u, v) connecting a vertex v in the set A to the vertex u outside to set A. Then vertex u is brought in to A. This process is repeated until a spanning tree is formed. Like Kruskal's algorithm, here too, the important fact about MSTs is we always choose the smallest-weight edge joining a vertex inside set A to the one outside the set A. The implication of this fact is that it adds only edges that are safe for A; therefore when the algorithm terminates, the edge in set A form a MST.

MST-Prim (G, w, r)

```
for each u ∈ V[G]
     do key[u] ← ∞
          π[u] ← NIL
key[r] ← 0
Q ← V[G]
while Q ≠ϕ
     do u ← EXTRACT-MIN (Q)
          for each v ∈ Adj[u]
               do if v ∈ Q and w (u, v) < key [v]
                    then π[u] ← u
                         key[v] ← w(u, v)
```

EXAMPLE 12.7. Find the minimum spanning tree vising Prim's algorithm for the given graph.

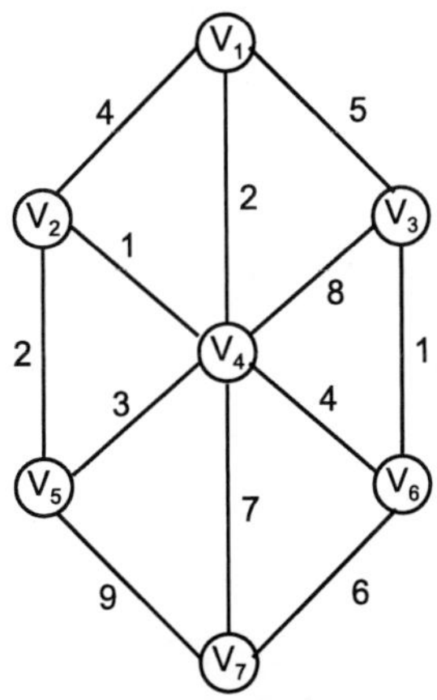

Fig. 12.54.

Solution: First we intialize the priority queue Q and key of each vertex to ∞ except for the root whose key is o. Suppose V_1 vertex is the root.

$$\text{Adj }[V_1] = \{V_2, V_3, V_4\}$$

Removing V_1 from the set Q and adds it in the tree.

Now, update the key and π fields of every vertex adjacent to V_1 but not in the tree.

$$\text{key }[V_2] = \infty$$

$$\omega(V_1, V_2) = 4 \quad i.e., \quad \omega(V_1, V_2) < \text{key }[V_2]$$

So, $\pi[V_2] = V_1$ and $\text{key }[V_2] = 4$

Similarly, $\text{key }[V_2] = \infty$

$$\omega(V_1, V_3) = 5 \quad i.e., \quad \omega(V_2, V_3) < \text{key}[V_3]$$

So, $\pi[V_3] = V_1$ and key $[V_3] = 5$

and $\pi[V_4] = V_1$ and key $[V_4] = 2$

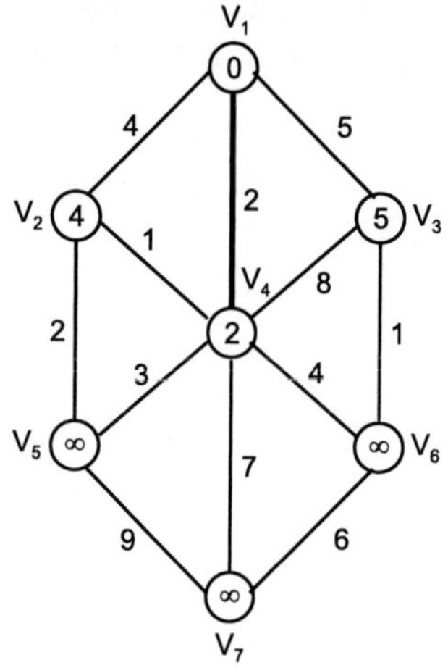

Fig. 12.55.

Now, by EXTRACT_MIN (Q). Remove V_4 node becuase key $[V_4] = 2$, which is minimum.

$$\text{Adj}[V_4] = \{V_1, V_2, V_3, V_5, V_6, V_7\}$$

Removing V_4 from the set Q and adds it to the set $V - Q$ of vertices in the tree where V_1 is already there.

$$\text{key}[V_2] = 4$$

$$\omega(V_2, V_4) = 1 \quad \omega(V_2, V_4) < \text{key}[V_2]$$

So, $\pi[V_2] = V_4$ and key $[V_2] = 1$

Similarly, $\pi[V_5] = V_4$ key $[V_5] = 3$

$\pi[V_7] = V_4$ key $[V_7] = 7$

$\pi[V_6] = V_4$ key $[V_6] = 4$

But $\omega(V_3, V_4) = 8$ and key $[V_3] = 5$

i.e., $\omega(V_3, V_4) > \text{key}[V_3]$

So, no change and V_1 is not in the Q.

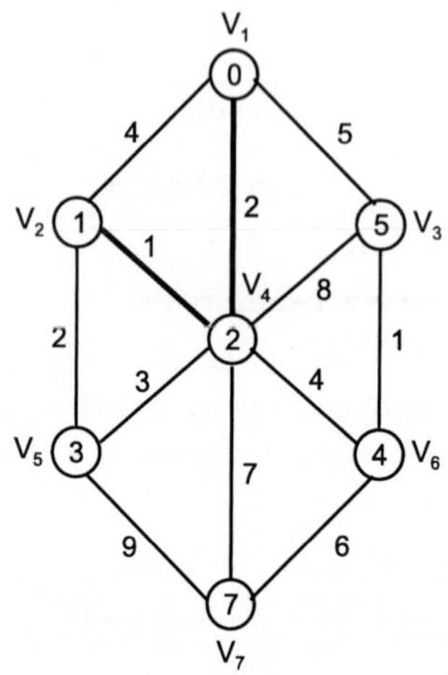

Fig. 12.56.

Now, By EXTRACT_MIN(Q). Remove V_2 because key $[V_2] = 1$ which is minimum and adds it to the set $V\text{-}Q$ of vertices in the tree. The elements in tree now $\{V_1, V_4, V_2\}$

$$\text{Adj } [V_2] = \{V_1, V_4, V_5\}$$

V_1, V_4 are not in the Q so

$$\omega (V_2, V_5) = 2$$
$$\text{key } [V_5] = 3$$
$$\omega (V_2, V_5) < \text{key } [V_5]$$

So, $\pi (V_5) = V_2$ and key $[V_5] = 2$

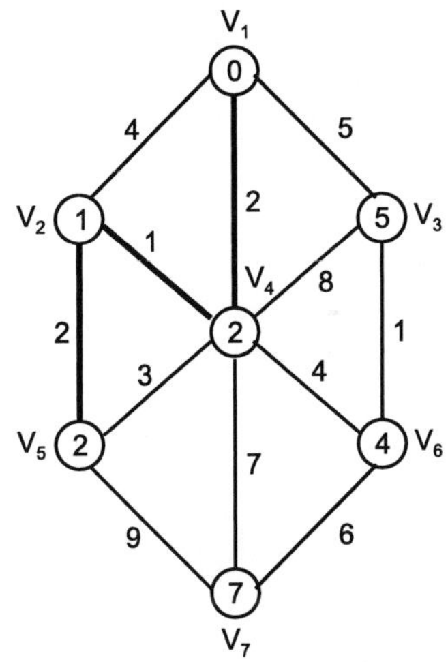

Fig. 12.57.

Now, by EXTRACT_MIN (Q). Remove V_5 because key $[V_5] = 2$ which is minimum. Now,

$$\text{Adj } [V_5] = \{V_2, V_4, V_7\}$$

where V_2, V_4 not in Q. So,

$$\text{key } [V_7] = 7 \quad \text{and} \quad \omega (V_5, V_7) = 9$$
$$\omega (V_5, V_7) > \text{key } [V_7]. \text{ No change.}$$

Now, by EXTRACT_MIN (Q). Remove V_6 because key $[V_6] = 4$

$$\text{Adj } [V_6] = \{V_3, V_4, V_7\}$$
$$\omega (V_3, V_6) = 1, \quad \text{key } [V_3] = 5$$
$$\omega (V_3, V_6) < \text{key } [V_3]$$

So, key $[V_3] = 1$ and $\pi [V_3] = V_6$

Similarly, $\omega (V_6, V_7) = 6$ key $[V_7] = 7$

$$\omega (V_6, V_7) < \text{key } [V_7]$$

So, key $[V_7] = 6$ and $\pi [V_7] = V_6$

Now, Remove V_3 because key $[V_3] = 1$

$$\text{Adj } [V_3] = \{V_1, V_4, V_6\}$$

No node in Q.

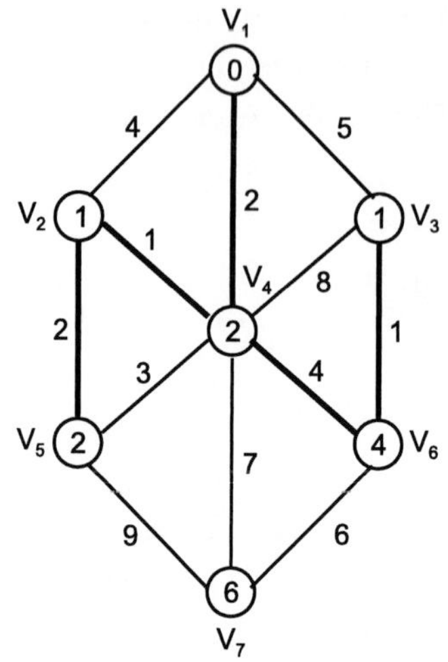

Fig. 12.58.

Now, by EXTRACT_MIN (Q) remove V_7 which is the last node in Q. So, final spanning tree is

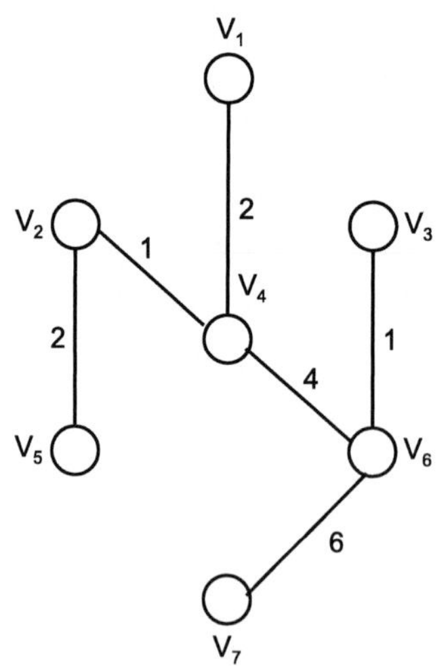

Fig. 12.59.

The edges that belong to MST are

$(V_1, V_4), (V_4, V_2), (V_2, V_5), (V_4, V_6), (V_6, V_3), (V_6, V_7)$

Weight of MST will be

$$2 + 1 + 2 + 4 + 1 + 6 = 16.$$

EXAMPLE 12.8. Generate minimum cost spanning tree for the following graph using Prim's algorithm.

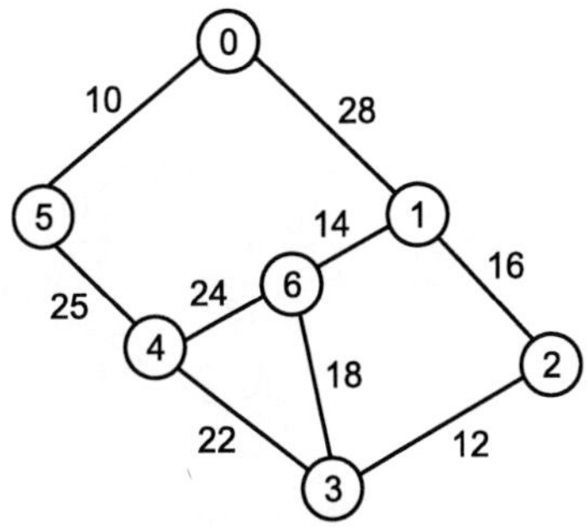

Fig. 12.60.

Solution: In Prim's algorithm, first we initialise the priority queue Q to contain all the vertices and the key of each vertex to ∞ except for the root, whose key is set to 0. Suppose 0 vertex is the root, *i.e.*, r. By EXTRACT-MIN (Q) produre, now $u = r$ and Adj[u] = [5, 1].

Removing u from the set Q and adds it to the set $V - Q$ of vertices in the tree. Now, update the key and π fields of every vertex v adjacent to u but not in the tree.

$$\text{key}[5] = \infty$$

$$w(0, 5) = 10 \quad i.e., \quad w(u, v) < key[5]$$

so, $$\pi[5] = 0 \quad \text{and} \quad key[5] = 10.$$

and $$key[1] = \infty$$

$$w(0, 1) = 28 \quad i.e., \quad w(u, v) < key[5]$$

so $$\pi\,[1] = 0 \quad \text{and} \quad key[1] = 28.$$

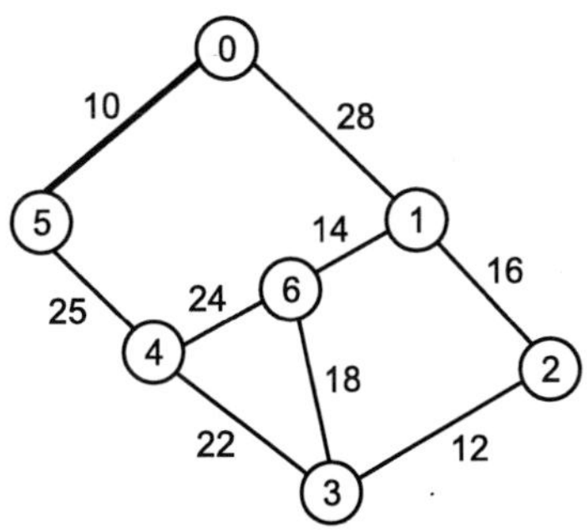

Fig. 12.61.

Now, by EXTRACT_MIN (Q) Remove 5 because key [5] = 10 which is minimum so $u = 5$

$$\text{Adj}[5] = \{4\}$$

$$w(u, v) < key\,[v] \text{ then } key\,[4] = 25$$

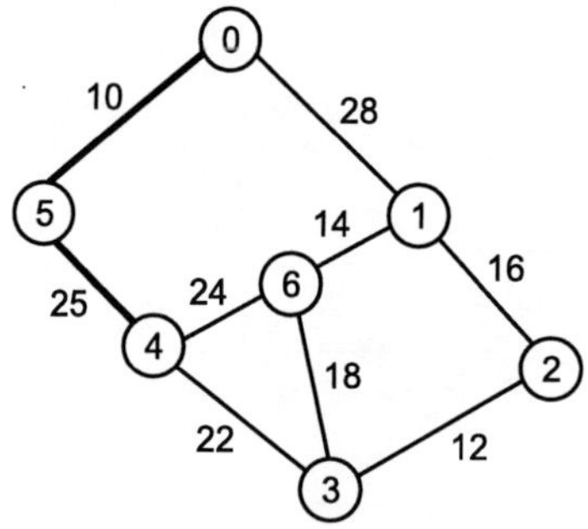

Fig. 12.62.

Now remove 4 because *key*[4] = 25 which is minimum so $u = 4$

Adj[4] = {6, 3}

$w(u, v) < key\ [v]$ then $key\ [6] = 24$

$key\ [3] = 22$

Now remove 3 because *key* [3] = 22 is minimum so $u = 3$

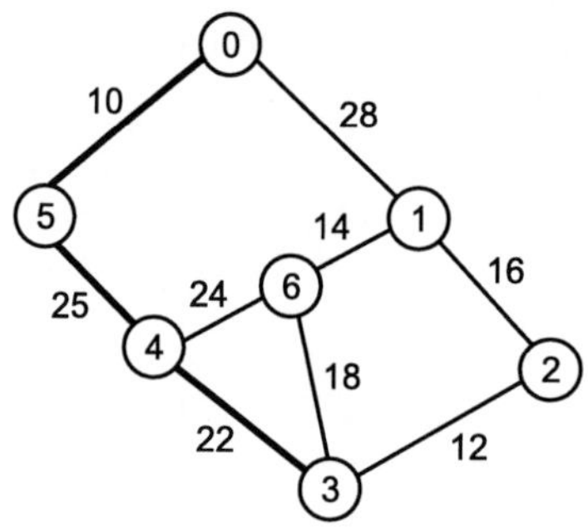

Fig. 12.63.

Adj[3] = {4, 6, 2}

$4 \notin Q$ *key* [6] = 24 now becomes *key* [6] = 18

key [2] = 12

Now, in Q *key* [2] = 12, key [6] = 18, key [1] = 28

By EXTRACT_MIN (Q) Remove 2, because key [2] = 12 is minimum

Adj[2] = {3, 1}

$3 \notin Q$

Now, *key* [1] = 16.

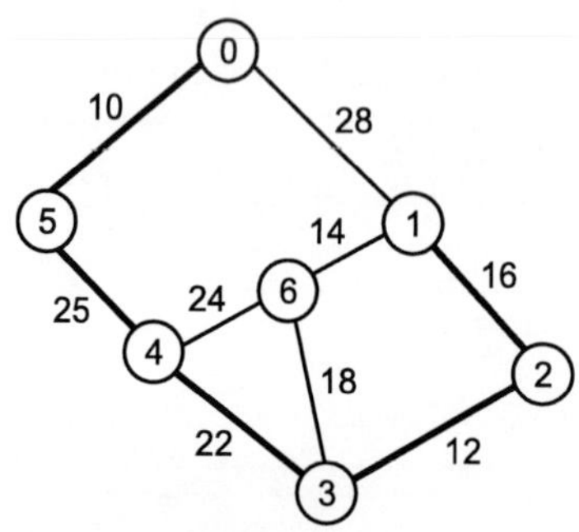

Fig. 12.64.

By EXTRACT_MIN (Q) Remove 1 because key [1] = 16 is minimum

$$\text{Adj}[1] = \{0, 6, 2\}$$

$$\{0, 2\} \notin Q, \textit{key}\ [6] = 14.$$

Now because Q contains only one vertex 6. By EXTRACT_MIN remove it

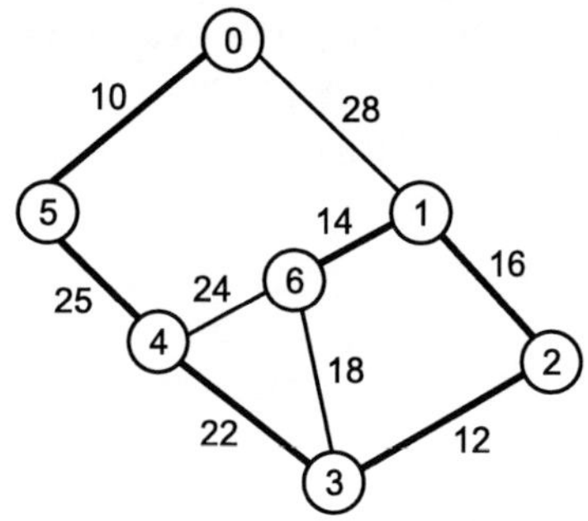

Fig. 12.65.

Thus, the final spanning tree is

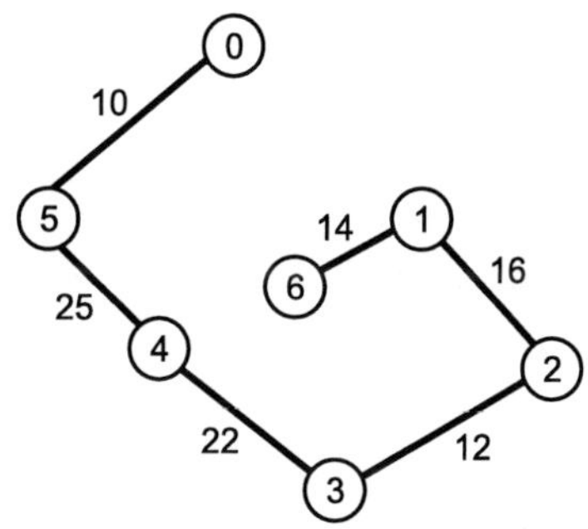

Fig. 12.66.

12.16 SHORTEST PATHS

A path from source vertex s to t is shortest path from s to t if there is no path from s to t with lower weights.

In a **shortest-paths problem,** we are given a weighted directed graph $G = (V, E)$, with weight function $w : E \rightarrow R$ mapping edges to real-valued weights. The weight of path $p = < v_0, v_1, ... v_k>$ is the sum of the weights of its constituent edges.

$$w(p) = \sum_{i=1}^{k} w(v_{i-1}, v_i)$$

we define the shortest-path weight from u to v by

$$\delta(u, v) = \min (w(p) : u \rightarrow v),$$

If there is a path from u to v. The shortest paths are not necessarily unique.

(*a*) Shortest Path for Given Source and Destination :

In order to find a shortest path from a given source to given destination, we discuss **Dijkstra's shortest-path algorithm.**

Relaxation

The single-source shortest-paths algorithms are based on a technique known as *relaxation*, a method that repeatedly decreases an upper bound on the actual shortest-path weight of each vertex until the upper bound equals the shortest-path weight. For each vertex $v \in V$, we maintain an attribute $d[v]$, which is an upper bound on the weight of a shortest path from source s to v. We call $d[v]$ a **shortest-path estimate**. We initialize the shortest-path estimates and predecessors by the following procedure.

```
INITIALIZE-SINGLE-SOURCE (G, s)
    for each vertex v ∈ V[G]
          do d[v] ←∞
          π[v] ← NIL
    d[s] ← 0
```

d[v] : represents the shortest distance of v from the source and

π[v] : represents the predecessor of node v, *i.e.,* the node which precedes the given node in the shortest-path from source.

After initialisation, $\pi[v]$ = NIL for all $v \in V$, $d[v] = 0$ for $v = s$, and $d[v] = \infty$ for $v \in V - \{s\}$.

The **process of relaxing** an edge (u, v) consists of testing whether we can improve the shortest path to v found so far by going through u and, if so, updating $d[v]$ and $\pi[v]$. A relaxation step may decrease the value of the shortest-path-estimate $d[v]$ and update v's predecessor field $\pi[v]$. The following code performs a relaxation step on edge (u, v).

```
RELAX (u, v, w)
    if d[v] > d[u] + w(u, v)
          then d[v] ← d[u] + w(u, v)
               π[v] ← u
```

For example,

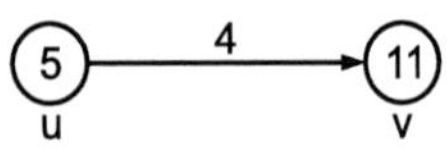

Here, $d[u] = 5$

$d[v] = 11$

$w(u, v) = 4$

i.e., $d[v] > d[u] + w(u, v)$

$11 > 5 + 4$

$11 > 9$

By apply RELAX, we get

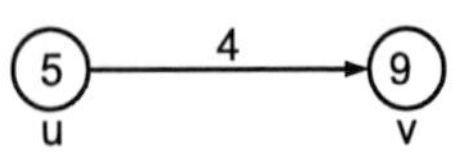

i.e., $d[v] = 9$

$d[u] = 5$ and $w(u, v) = 4$

i.e.,

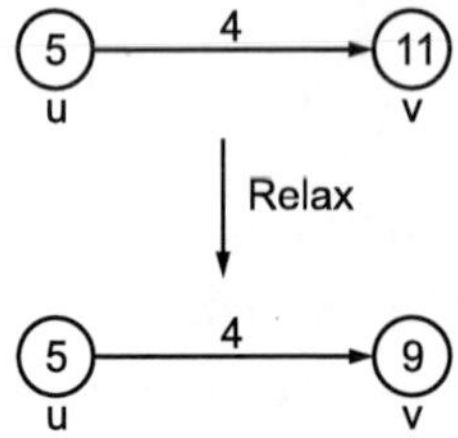

12.17 DIJKSTRA'S ALGORITHM

Dijkstra's algorithm, named after its discoverer, Dutch computer scientist **Edsger Dijkstra,** is a greedy algorithm that solves the single-source shortest path problem for a directed graph $G = (V, E)$ with non-negative edge weights, *i.e.*, we assume that $w(u, v) \geq 0$ each edge $(u, v) \in E$.

Dijkstra's algorithm maintains a set S *of* vertices whose final shortest-path weights from the source s have already been determined. That is, for all vertices $v \in S$, we have $d[v] = \delta(s, v)$. The algorithm repeatedly selects the vertex $u \in V - S$ with the minimum shortest-path estimate, inserts u into S, and relaxes all edges leaving u. We maintain a priority queue Q that contains all the vertices in $v - s$, keyed by their d values. Graph G is represented by adjacency lists.

```
DIJKSTRA (G, w, s)
INITIALIZE-SINGLE-SOURCE (G, s)
S ← φ
Q ← V[G]
while Q ≠φ
     do u ← EXTRACT-MIN (Q)
          S ← S ∪ {u}
          for each vertex v ∈ Adj[u]
          do RELAX (u, v, w)
```

Because Dijkstra's always chooses the "lightest" or "closest" vertex in $V - S$ to insert into set S, we say that it uses a **greedy strategy.**

Dijkstra's algorithm bears some similarity to both **breadth-first search** and **Prim's algorithm** for computing minimum spanning trees. It is like breadth-first search in that set S correspond to the set of black vertices in a breadth-first search; just as vertices in S have their final shortest- path weights, so black vertices in a breadth-first search have their correct breadth-first distances. Dijkstra's algorithm is like Prim's algorithm in that both algorithms use a priority queue to find the "lightest" vertex outside a given set (the set S in Dijkstra's algorithm and the tree being grown in Prim's algorithm), insert this vertex into the set, and adjust the weights of the remaining vertices outside the set accordingly.

EXAMPLE : Consider *A* as a source vertex.

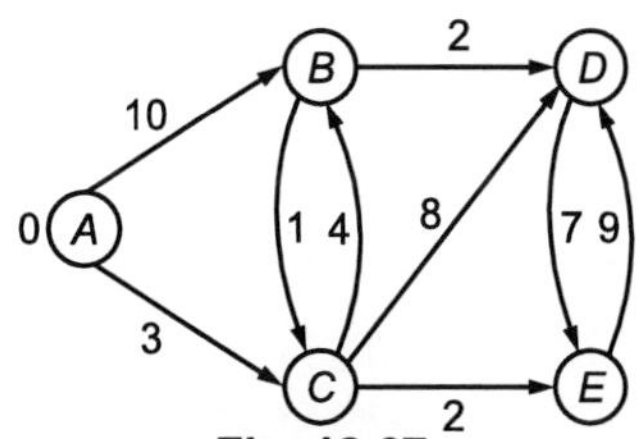

Fig. 12.67.

Initialise :

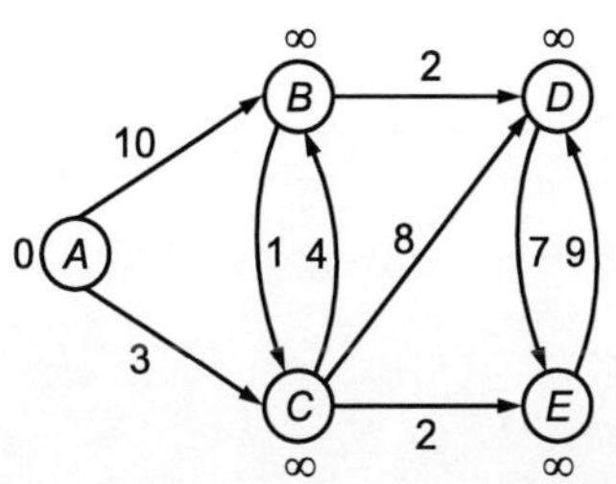

Q:	A	B	C	D	E
	0	∞	∞	∞	∞

S : {}

"A" ← EXTRACT-MIN (Q) :

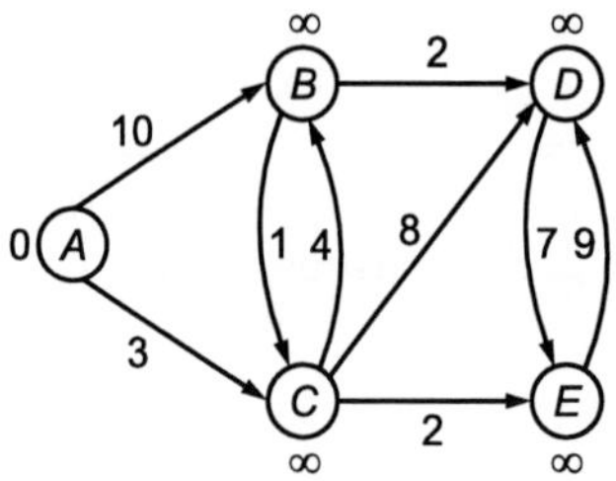

Q:	A	B	C	D	E
	0	∞	∞	∞	∞

S : {A}

Relax all edges leaving A :

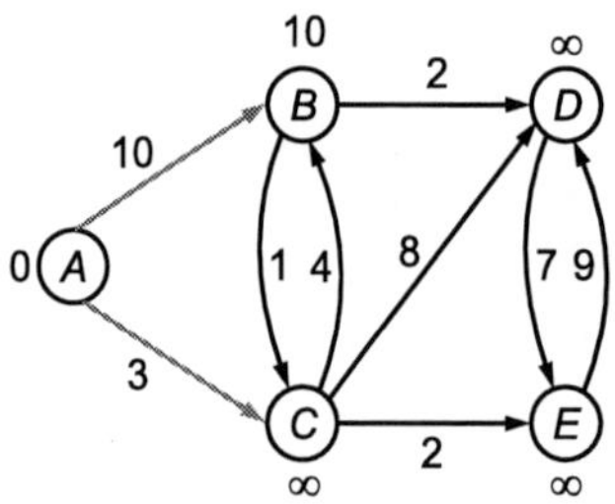

Q:	A	B	C	D	E
	0	∞	∞	∞	∞
		10	3	–	–

S : {A}

"C" ← EXTRACT-MIN (Q) :

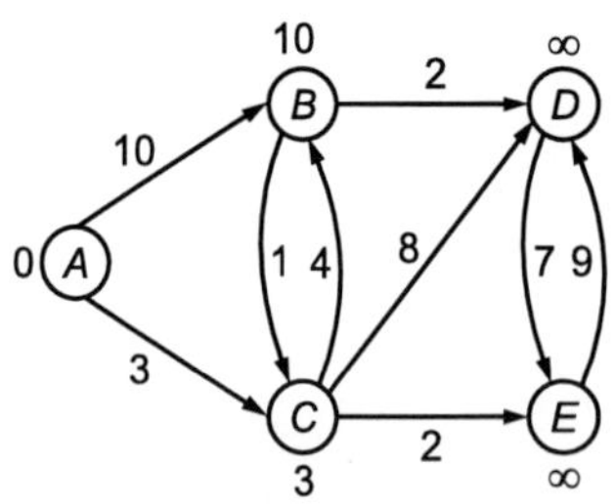

Q:	A	B	C	D	E
	0	∞	∞	∞	∞
		10	3	–	–

S : {A, C}

Relax all edges leaving C :

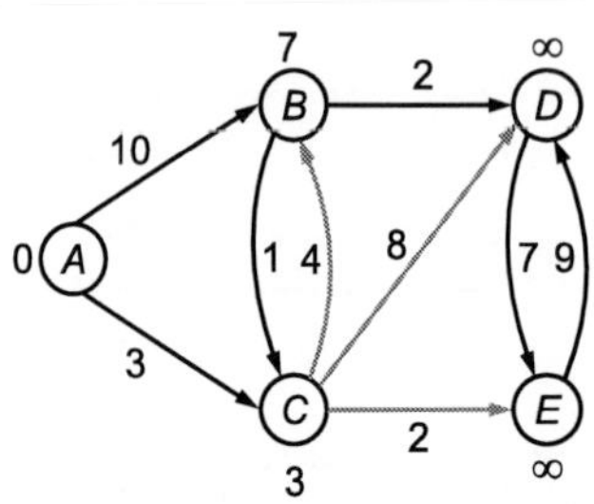

Q:	A	B	C	D	E
	0	∞	∞	∞	∞
		10	3	–	–
		7		11	5

S : {A, C}

"E" ← EXTRACT-MIN (Q) :

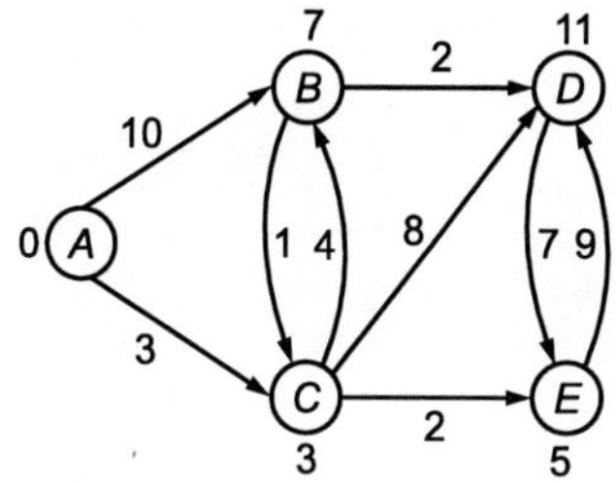

Q:	A	B	C	D	E
	0	∞	∞	∞	∞
		10	3	–	–
		7		11	5

S : {A, C, E}

Relax all edges leaving E :

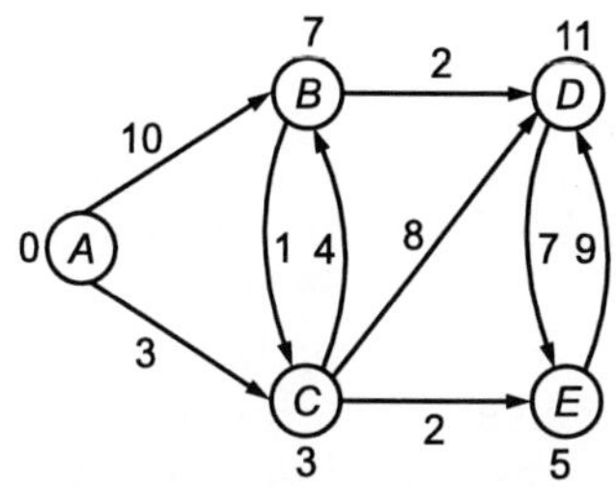

Q:	A	B	C	D	E
	0	∞	∞	∞	∞
		10	3	∞	∞
		7		11	5
		7		11	

S : {A, C, E}

"B" ← EXTRACT-MIN (Q) :

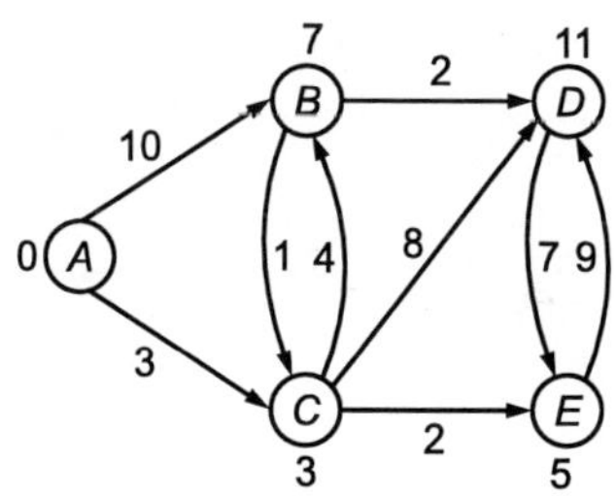

Q:	A	B	C	D	E
	0	∞	∞	∞	∞
		10	3	∞	∞
		7		11	5
		7		11	

S : {A, C, E, B}

Relax all edges leaving B :

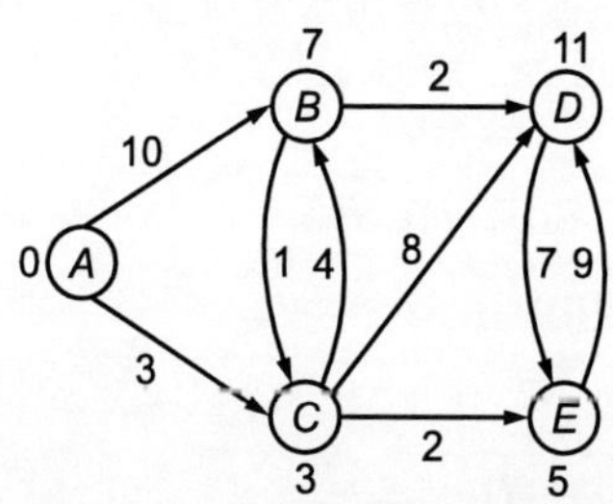

Q:	A	B	C	D	E
	0	∞	∞	∞	∞
		10	3	∞	∞
		7		11	5
		7		11	
				9	

S : {A, C, E, B}

"D" ← EXTRACT-MIN (Q) :

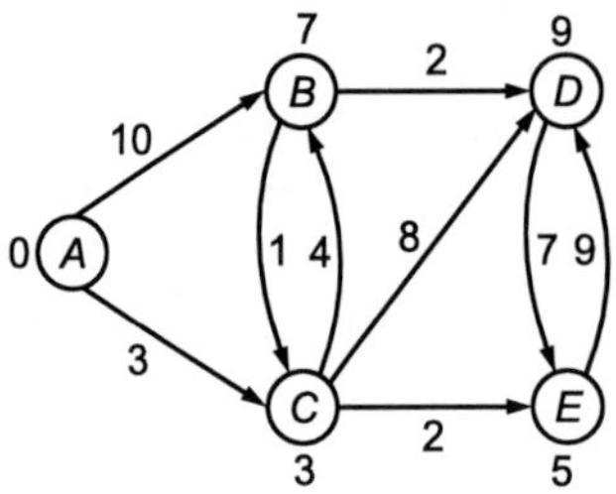

Q:	A	B	C	D	E
	0	∞	∞	∞	∞
		10	3	∞	∞
		7		11	5
		7		11	
				9	

S : {*A*, *C*, *E*, *B*, *D*}

12.18 WARSHALL'S ALGORITHM

The algorithm considers the "*intermediate*" vertices of a shortest path, where an ***intermediate*** vertex of a simple path $p = \langle v_1, v_2, \ldots\ldots v_m \rangle$ is any vertex of p other than v_1 or v_m, that is, any vertex in the set $\{v_2, v_3, \ldots\ldots v_{m-1}\}$.

The Floyd-Warshall algorithm is based on the following observation. Let the vertices of G be $V = \{1, 2, \ldots\ldots, n\}$, and consider a sub-set $\{1, 2, \ldots\ldots, k\}$ of vertices for some k. For any pair of vertices $i, j \in V$, consider all paths from i to j whose intermediate vertices are all drawn from $\{1, 2, \ldots\ldots, k\}$, and let p be a minimum-weight path from among them. (Path p is simple, since we assume that G contains no negative-weight cycles). The Floyd-Warshall algorithm exploits a relationship between path p and shortest paths from i to j with all intermediate vertices in the set $\{1, 2, \ldots\ldots, k-1\}$. The relationship depends on whether or not k is an intermediate vertex of path p.

If k is not an intermediate vertex of path p, then all intermediate vertices of path p are in the set $\{1, 2, \ldots\ldots, k-1\}$. Thus, a shortest path from vertex i to vertex j with all intermediate vertices in the set $\{1, 2, \ldots\ldots, k-1\}$ is also a shortest path from i to j with all intermediate vertices in the set $(1, 2, \ldots\ldots, k)$.

If k is intermediate vertex of path p, then we break p down into $i \xrightarrow{p_1} k \xrightarrow{p_2} j$.

Let $d_{ij}^{(k)}$ be the weight of a shortest path from vertex i to vertex j with all intermediate vertices in the set $\{1, 2, \ldots\ldots, k\}$.

A recursive definition is given by

$$d_{ij}^{(k)} = \begin{cases} w_{ij} & if\, k = 0 \\ \min\left(d_{ij}^{(k-1)}, d_{ik}^{(k-1)} + d_{kj}^{(k-1)}\; if\, k \geq 1\right) \end{cases}$$

and

$$w_{ij} = \begin{cases} 0 & \text{if}\, i = k \\ w(i,j) & \text{if}\, i \neq j\, \text{and}\, (i,j) \in E \\ \infty & \text{if}\, i \neq j\, \text{and}\, (i,j) \notin E \end{cases}$$

Thus,

$$d_{ij}^{0} = \begin{cases} 0 & \text{if}\, i = j \\ w(i,j) & \text{if}\, i \neq j\, \text{and}\, (i,j) \in E \\ \infty & \text{otherwise} \end{cases}$$

Let us define the $V \times V$ matrix

$$D^{(m)} = (d_{ij}^{(m)})$$

$$d_{ij}^{(m)} = \text{the length of the shortest path from } i \text{ to } j \text{ with} \leq \text{m edges.}$$

When $m = 0$, there is a shortest path from i to j with no edges if and only if $i = j$.

$$d_{ij}^{(0)} = \begin{cases} 0 & \text{if } i = j \\ w_{ij} & \text{if } i \neq j, (i, j) \in E \\ \infty & \text{otherwise} \end{cases}$$

```
FLOYD-WARSHELL (W)
    n ← rows [W]
    D(0) ← W
    for k ← 1 to n
        do for i ← 1 to n
            do for j ← 1 to n
                do d(k)ij ← min (d(k - 1)ij, d(k - 1)ik + d(k - 1)kj)
    return D(n)
```

Constructing a Shortest Path

There are a variety of different methods for constructing shortest paths in the Floyd-Warshell algorithm. One way is to compute the matrix D of shortest-path weights and then construct the predecessor matrix Π from the D matrix. This method can be implemented to run in $O(n^3)$ time. Given predecessor matrix Π, the PRINT-ALL-PAIRS-SHORTEST-PATH procedure can be used to print the vertices on a given shortest path.

We can compute the predecessor matrix Π just as the Floyd-Warshall algorithm computes the matrices $D(k)$. Specifically, we compute a sequence of matrices $\Pi^{(0)}, \Pi^{(1)}, \ldots.., \Pi^{(n)}$, where $\Pi = \Pi^{(n)}$ and is defined to be the predecessor of vertex j on a shortest path from vertex i with all intermediate vertices in the set (1, 2, ..., k).

The recursive formulation of $\pi_{ij}^{(k)}$. When $k = 0$, a shortest path from i to j has no intermediate vertices at all. Thus,

$$\pi_{ij}^{(0)} = \begin{cases} \text{NIL} & \text{if } i = j \text{ or } w_{ij} = \infty \\ i & \text{if } i = j \text{ and } w_{ij} < \infty \end{cases}$$

For $k \geq 1$, if we take the path $i \rightarrow k \rightarrow j$, where $k \neq j$ then the predecessor of j we choose is the same as the predecessor of j we chose on a shortest path from k with all intermediate vertices in the set (1, 2, ..., $k - 1$). Otherwise , we choose the same predecessor of j that we chose on a shortest path from i with all intermediate vertices in the set (1, 2, ..., $k - 1$). Formally, for $k \geq 1$,

$$\pi_{ij}^{(k)} = \begin{cases} \pi_{ij}^{(k-1)} & \text{if } d_{ij}^{(k-1)} \leq d_{ik}^{(k-1)} + d_{kj}^{(k-1)} \\ \pi_{kj}^{(k-1)} & \text{if } d_{ij}^{(k-1)} > d_{ik}^{(k-1)} + d_{kj}^{(k-1)} \end{cases}$$

EXAMPLE 12.9. Apply Floyd-Warshall algorithm for constructing shortest path. Show the matrix $D^{(k)}$ that results each iteration.

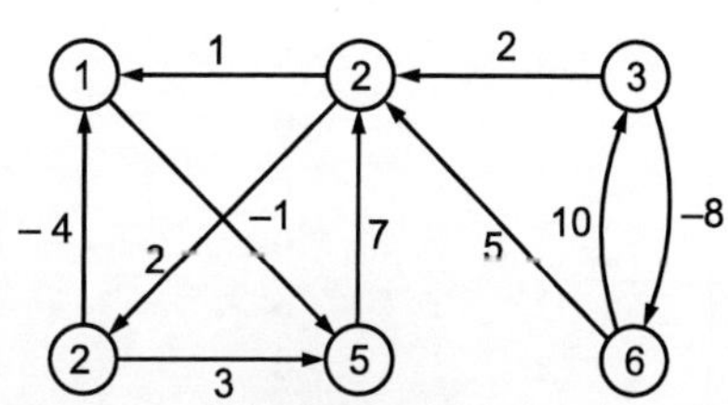

Fig. 12.68.

Solution:

$$D^{(0)} = \begin{bmatrix} 0 & \infty & \infty & \infty & -1 & \infty \\ 1 & 0 & \infty & 2 & \infty & \infty \\ \infty & 2 & 0 & \infty & \infty & -8 \\ -4 & \infty & \infty & 0 & 3 & \infty \\ \infty & 7 & \infty & \infty & 0 & \infty \\ \infty & 5 & 10 & \infty & \infty & 0 \end{bmatrix}$$

We know,

$$d_{ij}^{(k)} = \min [d_{ij}^{(k-1)}, d_{ik}^{(k-1)} + d_{kj}^{(k-1)}]$$

k	i	j	$d_{ij}^{(k-1)}$	$d_{ik}^{(k-1)}$	$d_{kj}^{(k-1)}$	d_{ij}^{k}
1	1	1	0	0	0	0
1	1	2	∞	0	∞	∞
1	1	3	∞	0	∞	∞
1	1	4	∞	0	∞	∞
1	1	5	– 1	0	– 1	– 1
1	1	6	∞	0	∞	∞
1	2	1	1	1	0	1
1	2	2	0	1	∞	0
1	2	3	∞	1	∞	∞
1	2	4	2	1	∞	2
1	2	5	∞	1	– 1	0
1	2	6	∞	1	∞	∞
1	3	1	∞	∞	0	∞
1	3	2	2	∞	∞	2
1	3	3	0	∞	∞	0
1	3	4	∞	∞	∞	∞
1	3	5	∞	∞	– 1	∞
1	3	6	– 8	∞	∞	– 8
1	4	1	– 4	– 4	0	– 4
1	4	2	∞	– 4	∞	∞
1	4	3	∞	– 4	∞	∞
1	4	4	0	– 4	∞	0
1	4	5	3	– 4	– 1	– 5
1	4	6	∞	– 4	∞	∞
1	5	1	∞	∞	0	∞
1	5	2	7	∞	∞	7
1	5	3	∞	∞	∞	∞
1	5	4	∞	∞	∞	∞
1	5	5	0	∞	– 1	0
1	5	6	∞	∞	∞	∞
1	6	1	∞	∞	0	∞
1	6	2	5	∞	∞	5
1	6	3	10	∞	∞	10
1	6	4	∞	∞	∞	∞
1	6	5	∞	∞	– 1	∞
1	6	6	0	∞	∞	0

So, the $D^{(1)}$ matrix is

$D^{(1)}$	1	2	3	4	5	6
1	0	∞	∞	∞	–1	∞
2	1	0	∞	2	0	∞
3	∞	2	0	∞	∞	–8
4	–4	∞	∞	0	–5	∞
5	∞	7	∞	∞	0	∞
6	∞	5	10	∞	∞	0

Similarly, we get

$D^{(2)}$	1	2	3	4	5	6
1	0	∞	∞	∞	–1	∞
2	1	0	∞	2	0	∞
3	3	2	0	4	2	–8
4	–4	∞	∞	0	–5	∞
5	8	7	∞	9	0	∞
6	6	5	10	7	5	0

$D^{(3)}$	1	2	3	4	5	6
1	0	∞	∞	∞	–1	∞
2	1	0	∞	2	0	∞
3	3	2	0	4	2	–8
4	–4	∞	∞	0	–5	∞
5	8	7	∞	9	0	∞
6	6	5	10	7	5	0

$D^{(4)}$	1	2	3	4	5	6
1	0	∞	∞	∞	–1	∞
2	–2	0	∞	2	–3	∞
3	0	2	0	4	–1	–8
4	–4	∞	∞	0	–5	∞
5	5	7	∞	9	0	∞
6	3	5	10	7	2	0

$D^{(5)}$	1	2	3	4	5	6
1	0	6	∞	8	–1	∞
2	–2	0	∞	2	–3	∞
3	0	2	0	4	–1	–8
4	–4	2	∞	0	–5	∞
5	5	7	∞	9	0	∞
6	3	5	10	7	2	0

and

$D^{(6)}$	1	2	3	4	5	6
1	0	6	∞	8	– 1	∞
2	– 2	0	∞	2	– 3	∞
3	(– 5)	– 3	0	– 1	– 6	– 8
4	– 4	2	∞	0	(– 5)	∞
5	5	7	∞	9	0	∞
6	3	(5)	10	7	2	0

The shortest path from 3 to 1 is

3 →6 → 2 → 4 → 1 and is – 5 units.

Similarly, if we want to find the shortest path from 6 node to 2 node then it is 5 units [6 → 2] and the shortest path from 4 to 5 is – 5, *i.e.*, 4 → 1 → 5.

12.19 TRANSITIVE CLOSURE

The transitive closure of a graph G is defined to be the graph G' such that G' has the same nodes as G and there is an edge (v_i, v_j) in G' whenever there is a path from v_i to v_j in G. Accordingly the path matrix P of the graph G is precisely the adjacency matrix of its transitive closure G′.

The transitive closure of a graph G is defined as G^* or $G' = (V, E^*)$.

where,

$$E^* = \{(i, j) \text{ there is a path from veretx } i \text{ to vertex } j \text{ in } G\}$$

For $i, j, k = 1, 2, \ldots, n$ we define $t_{ij}^{(k)}$ to be 1 if there exists a path in graph G from vertex i to vertex j with all intermediate vertices in the set $\{1, 2, \ldots, k\}$ and 0 otherwise.

We construct the transitive closure $G^* = (V, E^*)$ by putting edge (i, j) into E^* if and only if $t_{ij}^{(n)} = 1$.

The recursive definition of $t_{ij}^{(k)}$ is

$$t_{ij}^{(0)} = \begin{cases} 0 & \text{if } i \neq j \text{ and } (i, j) \notin E \\ 1 & \text{if } i = j \text{ or } (i, j) \in E \end{cases}$$

and for $k \geq 1$

$$t_{ij}^{(k)} = t_{ij}^{(k-1)} \vee (t_{ik}^{(k-1)} \wedge t_{kj}^{(k-1)})$$

```
Transitive-closure (G)
    n ← v[G]
    for i ← 1 to n.
        do for j ← 1 to n
            do if i = j or (i, j) ∈ E[G]
                then t_ij^(0) ← 1
            else
                t_ij^(0) ← 0
    for k ← 1 to n
    do for i ← 1 to n
        do for j ← 1 to n
            do t_ij^(k) ← t_ij^(k - 1) ∨ (t_ik^(k - 1) ∧ t_kj^(k - 1))
```

Exercises

1. Which of the multigraph in Fig. 12.69 are (*a*) connected (*b*) loop-free and (*c*) graphs ?

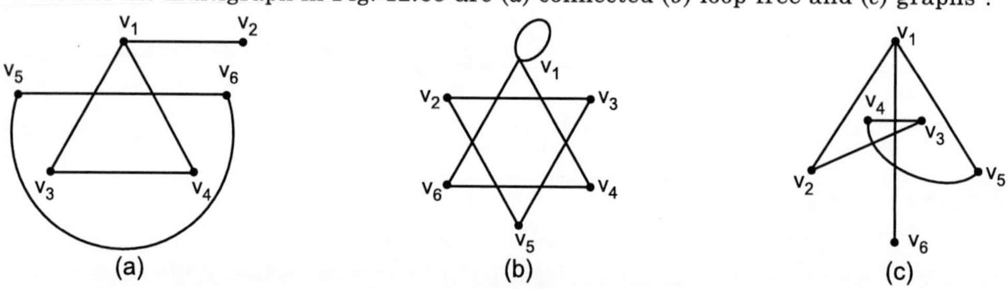

Fig. 12.69.

2. Consider the graph G in Fig. 12.70. Suppose the nodes are stored in memory in an array DATA as follows :

$$\text{Data} : X, Y, Z, S, T$$

(a) Find the adjacency matrix A of G.

(b) Find the incidence matrix of G.

(c) Find the path matrix of G.

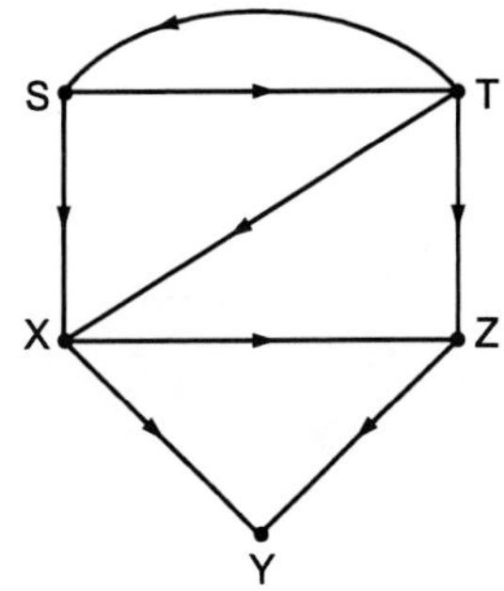

Fig. 12.70.

3. The following is the adjacency matrix A of an undirected graph G :

$$A = \begin{bmatrix} 0 & 1 & 1 & 0 \\ 1 & 0 & 0 & 1 \\ 0 & 0 & 0 & 1 \\ 0 & 1 & 1 & 0 \end{bmatrix}$$

Draw G and find its incidence matrix M.

4. Write a procedure which determines whether or not G is an undirected graph.
5. List the different types of representations of graphs.
6. What is the difference between DFS and BFS ?
7. What is meant by minimum spanning tree ?
8. List the different algorithms used for implementing the MST ?
9. What is meant by topological ordering of vertices ?
10. Prove that in finite graph the number of vertices of odd degree is always even.
11. In Dijkstra's shortest path algorithm, what technique is used to choose the next vertex to process ?
12. Write a procedure to compute length of shortest path of a directed graph.
13. (a) Execute BFS algorithm on the above graph with 3 as the starting node.

(b) Explain Kruskal's algorithm for finding minimum spanning tree in a graph.

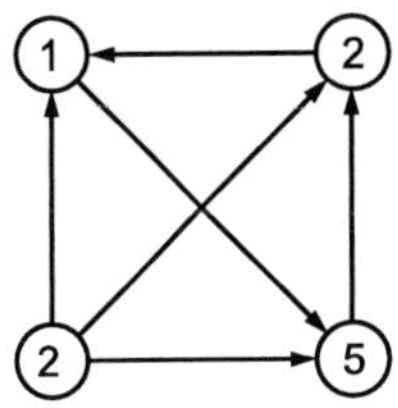

Fig. 12.71.

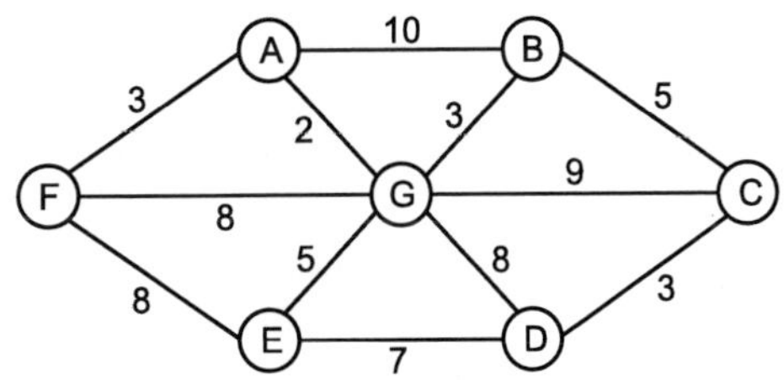

Fig. 12.72.

14. Using Dijkastra's shortest path algorithm, find out a least cost path to all other nodes 1 through 6 for the following graphs :

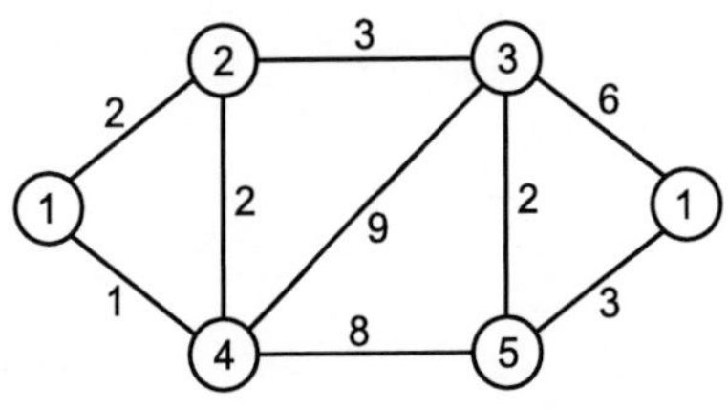

(a)

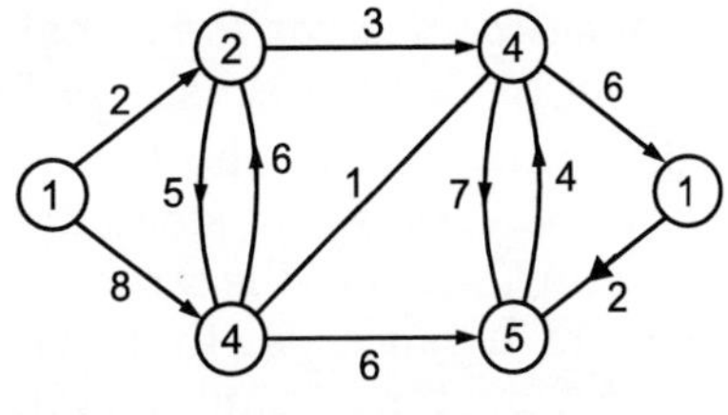

(b)

Fig. 12.73.

Chapter

13 Sorting Techniques

13.1 INTRODUCTION

The efficiency of data handling can often be substantially increased if the data are sorted according to some criteria of order. For example, it would be practically impossible to find a name in the telephone directory if the name were not alphabetically ordered. The same can be said about dictionaries, book indexes, payrolls, bank accounts, student lists, and other alphabetically organised materials. The convenience of using sorted data is unquestionable and must be addressed in computer science as well.

So, retrieval of information is made easier when it is stored in some predefined order. Sorting is, therefore, a very important computer application activity. Many sorting algorithms are available. Different environments require different sorting methods. Sorting algorithms can be characterised in the following two ways :

1. *Simple algorithms*, which require the order of n^2 (written as $O(n^2)$) comparisons to sort n items.
2. *Sophisticated algorithms* that require the $O(n \log_2 n)$ comparisons to sort n items.

The difference lies in the fact that the first method moves data only over small distances in the process of sorting, whereas the second method moves data over large distances, so that items settle into the proper order sooner, thus resulting in fewer comparisons. Performance of a sorting algorithm can also depend on the degree of order already present in the data.

Chapter Outline

The two main **properties** of sorting techniques are :

- **Stable :** If the sorting algorithm preserves the relative order of any two equal elements, then the sorting algorithm is stable.
- **In Place :** If an algorithm does not require an extra memory space except for few memory units, then the algorithm is said to be in place.

There are two basic **categories** of sorting methods :

1. **Internal Sorting :** (sorting of data items in the main memory)
2. **External Sorting :** (when the data to be sorted is so large that some of the data is present in the memory and some kept in *auxiliary memory*).

Internal sorting is applied when the entire collection of data to be sorted is small enough so that the entire sorting can take place within the main memory. The time required to read or write is not considered to be significant in evaluating the performance of internal sorting methods. External sorting methods are applied to larger collection of data which reside on secondary devices. Read and write access times are a major concern in determining sorting performance of such methods.

So, *internal sorting* takes place in the main memory of a computer, where we can use the random access capability of the main memory to advantage in various ways and *external sorting* is necessary when the number of objects to be sorted is too large to fit in main memory. There, the bottleneck is usually the movement of data between main and secondary storage, and we must move data in large blocks to be efficient. The fact that physically contiguous data is most conveniently moved in one block constrains the kinds of external sorting algorithms we can use.

13.2 BASIC TERMINOLOGY

We now present some basic terminology. A **file** of size n is a sequence of n items r[0], r[1], r[3]... r[n-1]. Each item in the file is called a **record**. A key, k[i], is associated with each record r[i]. The key is usually (but not always) a subfield of the entire record. The file is said to be **sorted on the key** if i < j implies that k[i] precedes k[j] in some ordering on the keys.

It is possible for two records in a file to have the same key. A sorting technique is called **stable** if for all records i and j such that k[i] equals k[j], if r[i] precedes r[j] in the original file, r[i] precedes r[j] in the sorted file. That is, stable sort keeps records with the same key in the same relative order that they were in before the sort.

13.3 INTERNAL SORTING

By internal sorting means we are arranging the numbers within the array only which is in computer primary memory. The various internal sorting methods are :

1. Bubble sort
2. Selection sort
3. Insertion sort
4. Quick sort
5. Merge sort
6. Heap sort
7. Radix sort

The *sorting* problem is to arrange a sequence of records so that the values of their key fields form a non-decreasing sequence. That is, given records $r_1, r_2, ..., r_n$, with key values $k_1, k_2, ..., k_n$, respectively, we must produce the same records in an order r_{i1}, r_{i2}, r_{in}, such that $k_{i1} \leq k_{i2} \leq ... \leq k_{in}$.

The records all need not have distinct values, nor do we require that records with the same key value appear in any particular order.

The time required to read and write is not considered to be significant in evaluating the performance of internal sorting methods. The three important efficiently criteria are :

- Use of storage space
- Use of computer time
- Programming effort

13.4 BUBBLE SORT

Bubble sort, also known as exchange sort, is a simple sorting algorithm. It works by repeatedly stepping through the list to be sorted. Comparing two items at a time and swapping them if they are in the wrong order. The pass through the list is repeated until no swaps are needed, which means the list is sorted. The algorithm gets its name from the way smaller elements "bubble" to the top (*i.e.*, the beginning) of the list via the swaps. In this method, we assume that the lower value items from the set of items are light and bubble up to the top. Because it only uses comparisons to operate on elements, it is a *comparison sort*. This is the easiest comparison sort to implement. The idea applied for the bubble sort is as follows :

(a) Suppose if the array contains n elements, then (n-1) iterations are required to sort this array.

(b) The set of items in the array are scanned again and again and if any two adjacent items are found to be out of order, they are reversed.

(c) At the end of the first iteration, the lowest value is placed in the first position.

(d) At the end of the second iteration, the next lowest value is placed in the second position and so on.

(e) It is very efficient in large sorting jobs. For n data items, this method requires n(n-1)/2 comparisons.

The bubble sort is generally considered to be the most inefficient sorting algorithm in common usage. Under best-case conditions (the list is already sorted), the bubble sort can approach a constant $O(n)$ level of complexity. General-case is an extremely bad $O(n^2)$.

Let us take the following values.

43	50	10	90	40

First Iteration :

43	50	10	90	40

Here 90 is greater than 40 so interchange.

43	50	10	40	90

Here 10 is less than 40 so interchange.

43	50	10	40	90

Here 50 is greater than 10 so interchange.

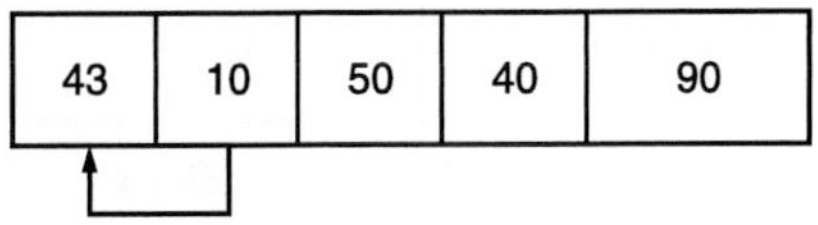

Here 43 is greater than 10 so interchange.

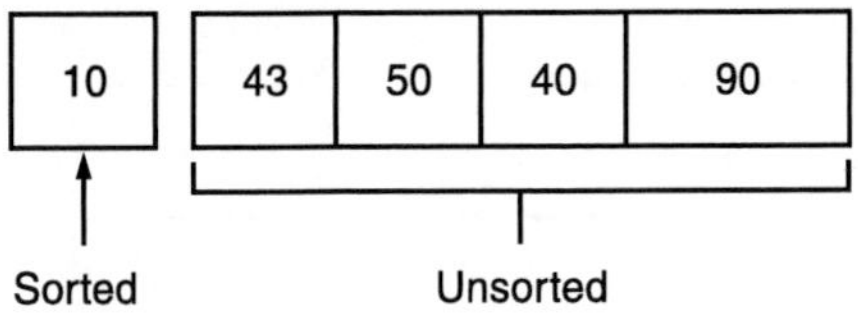

Second Iteration :

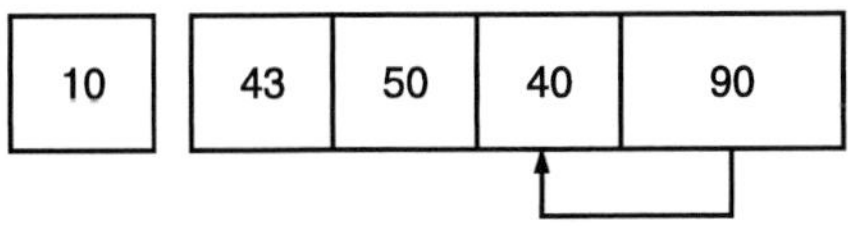

Here 40 is less than 90, no interchange.

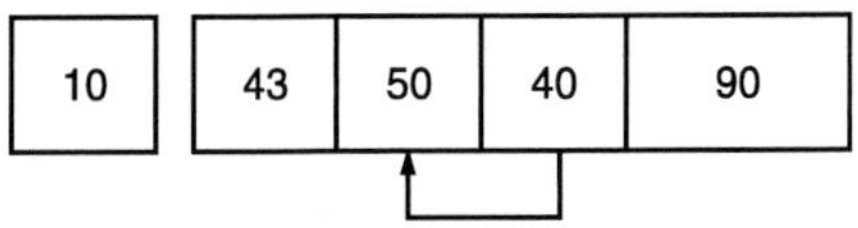

Here 50 is greater than 40, so interchange.

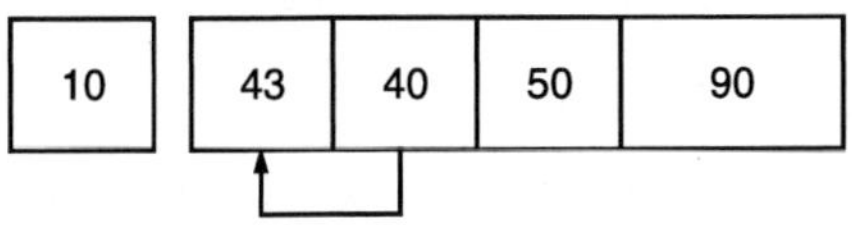

Here 43 is greater than 40 so interchange.

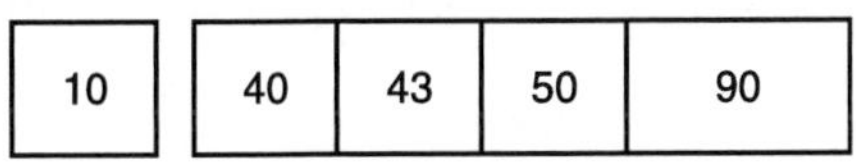

Now

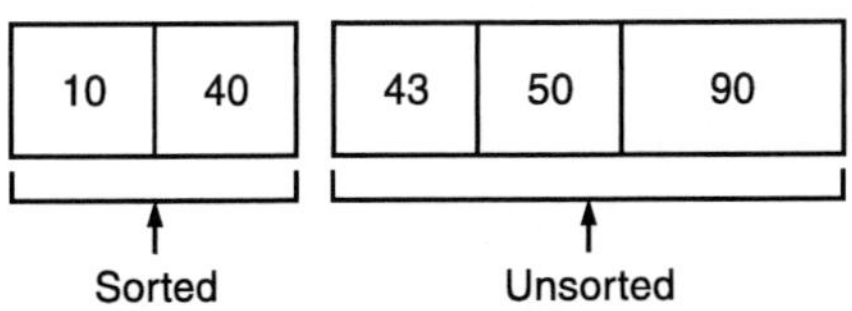

Third Iteration :

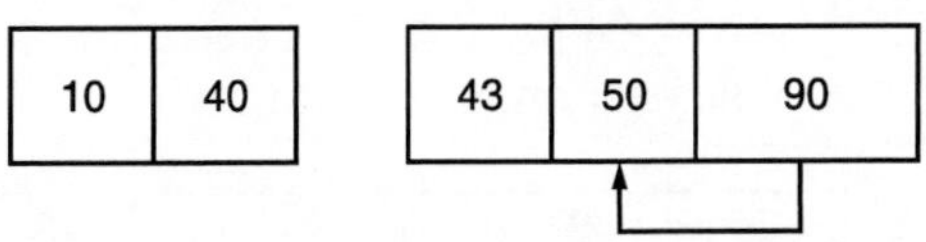

Here 50 is less than 90, so no interchange.

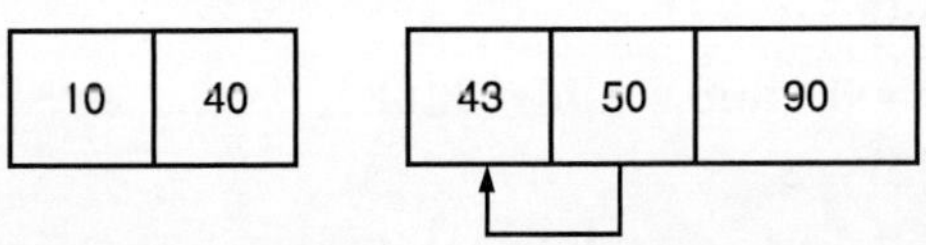

Here 43 is less than 50, so no interchange.

Now

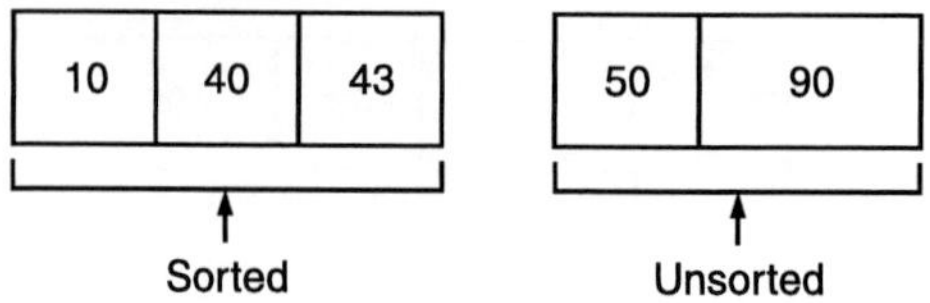

Fourth Iteration :

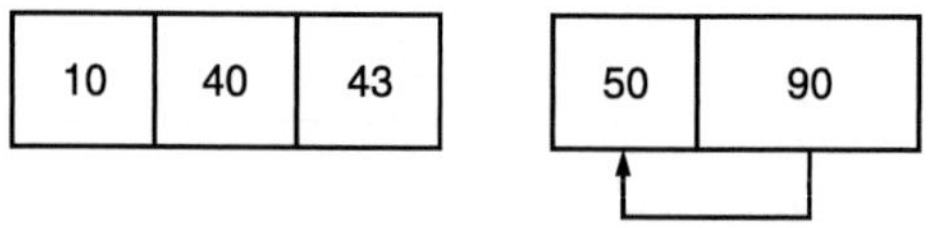

Here 50 is less than 90, so no interchange.

Final sorted array is:

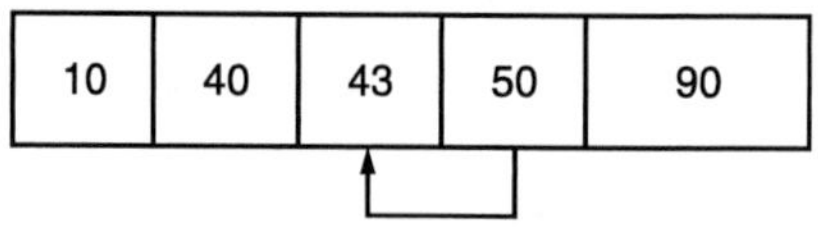

BUBBLE SORT (A)

```
for i ← 1 to length[A]
    for j ← length[A] down to i+1
        if A[j]<A[j-1]
            exchange (A[j], A[j-1])
```

The outer loop is executed n-1 times. Each time the outer loop is executed, the inner loop is executed. Inner loop executes n-1 times at first, linearly dropping to just once. On average, inner loop executes about n/2 times for each execution of the outer loop. In the inner loop, the comparison is always done (constant time), the swap might be done (also constant time). Thus result is n*n/2*k, that is, $O(n^2/2 + k) = O(n^2)$.

EXAMPLE 13.1. Illustrate the operation of Bubble sort on the array

A = <5, 2, 1, 4, 3, 7, 6>

Solution: Here A = <5, 2, 1, 4, 3, 7, 6>

So length[A] = 7

i = 1 to 7 and j = 7 to 2

Take i = 1 and j = 7

A[7] = 6 and A[6] = 7

Here A[7] < A[6] so exchange A[7] and A[6]

i.e., now A[ ] =

5	2	1	4	3	6	7

Now, i = 1, j = 6 then A[6] = 6

A[5] = 3 and A[5] < A[6]

Now, i = 1, j = 5 then A[5] = 3

A[4] = 4 and A[5] < A[4]

So, exchange (A[5], A[4])

and A[ ] =

5	2	1	3	4	6	7

Now, i = 1, j = 4 then A[4] = 3
A[3] = 1 and A[4] > A[3]
Now, i = 1, j = 3 then A[3] = 1
A[2] = 2 and A[3] < A[2]
So, exchange (A[3], A[2])

then A[ ] =

5	1	2	3	4	6	7

Now, i = 2, j = 2 then A[2] = 1
A[1] = 5 and A[2] < A[1]
So, exchange (A[2], A[1])

then A[ ] =

1	5	2	3	4	6	7

Now, i = 1, j = 7 then A[7] = 7
A[6] = 6 and A[7] > A[6] No exchange
Similarly, i = 2, j = 6, 5, 4 No exchange
then i = 2, j = 3
A[3] = 2
A[2] = 5 and A[3] < A[2]
So exchange, (A[3], A[2])

and A[ ] =

1	2	5	3	4	6	7

Now, i = 3, j = 7, 6, 5 No change
then i = 3, j = 4
A[4] = 3
A[3] = 5 and A[4] < A[3]
So exchange, (A[4], A[3])

then A[ ] =

1	2	3	5	4	6	7

Now, i = 4, j = 7 No change
then i = 4, j = 6 No change
Now, i = 4, j = 5
then A[5] = 4 and A[4] = 5
A[5] < A[4]
So exchange, (A[5] and A[4]), we get

A[ ] =

1	2	3	4	5	6	7

This is the final sorted array.

Program /* Bubble Sort */

```
#include <stdio.h>
#include <conio.h>
#define MAXSIZE 50
void bubble(int x[], int n)
{
int i,j,tmp;
for(i=0;i<n;i++)
{
    for(j=0;j<n-1;j++)
    {
        if(x[j]>x[j+1])
        {
            tmp = x[j]
            x[j] = x[j+1]
            x[j+1]=tmp;
        }
    }
}
}
void main( )
{
int x[MAXSIZE], n, i;
clrscr();
printf("Enter the number of elements :");
scanf("%d", &n);
printf("\nEnter the elements: \n");
for(i=0; i<n;i++)
scanf("%d", &x[i]);
bubble(x,n);
printf("\nThe sorted output: \n");
for(i=0; i<n;i++)
printf("\n%d", x[i]);
getch( );
}
```

13.5 SELECTION SORT

The idea of selection sort is rather simple: we repeatedly find the next largest (or smallest) element in the array and move it to its final position in the sorted array. Assume that we wish to sort the array in increasing order, *i.e.*, the smallest element at the beginning of the array and the largest element at the end. We begin by selecting the largest element and moving it to the highest index position. We can do this by swapping the element at the highest index and the largest element. We then reduce the effective size of the array by one element and repeat the process on the smaller sub-array. The process stops when the effective size of the array becomes 1 (an array of 1 element is already sorted). Thus, the selection sort works by selecting the smallest unsorted item remaining in the list, and then swapping it with the item in the next position to be filled.

This is the simple method of sorting. In this method, if n elements have to be sorted then (n-1) iterations are required. The selection sort has a complexity of $O(n^2)$.

EXAMPLE : Let us take the values of array A[ ] as:

<40, 50, 10, 90, 30>

Here there are 5 values, so 4 iterations are required to sort these elements

First Iteration :

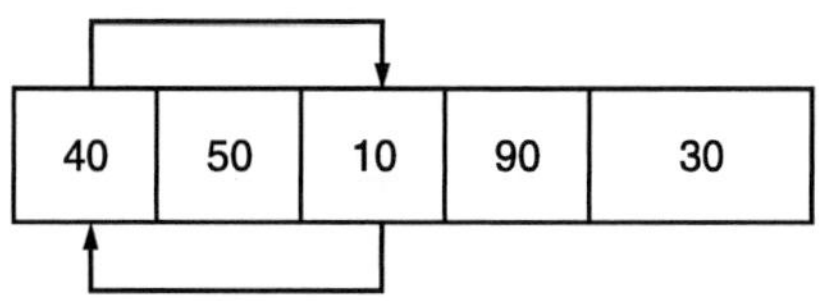

Here 10 is smaller than 40, so exchange it.

10	50	40	90	30

Second Iteration :

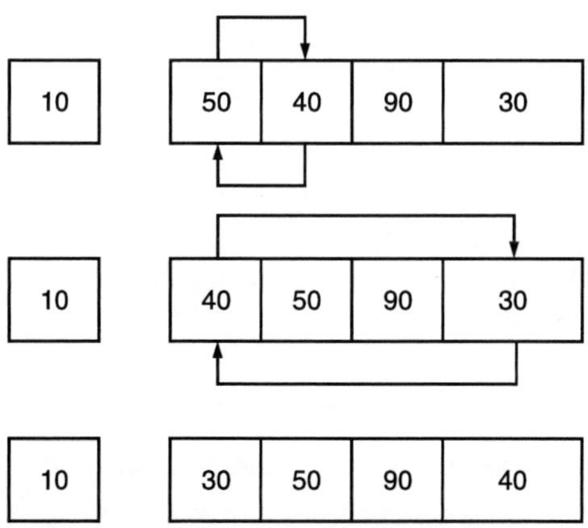

So

10	30

50	90	40

Third Iteration :

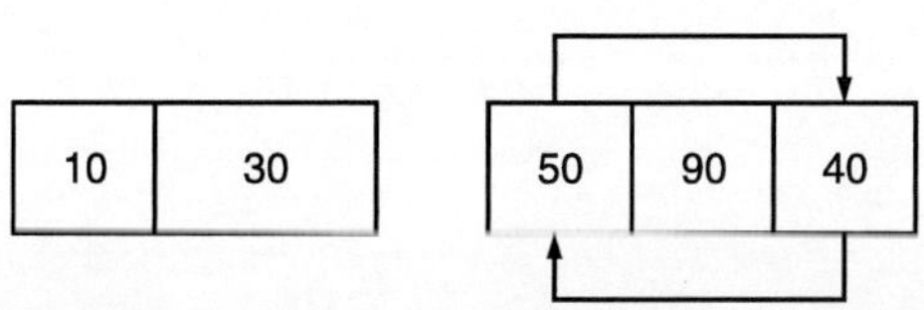

10	30

40	90	50

So,

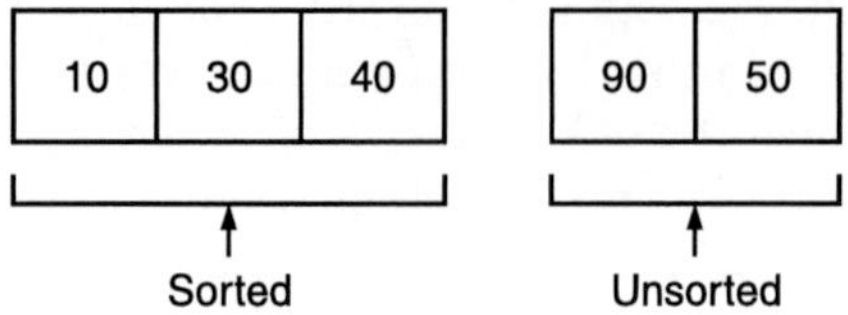

Fourth Iteration :

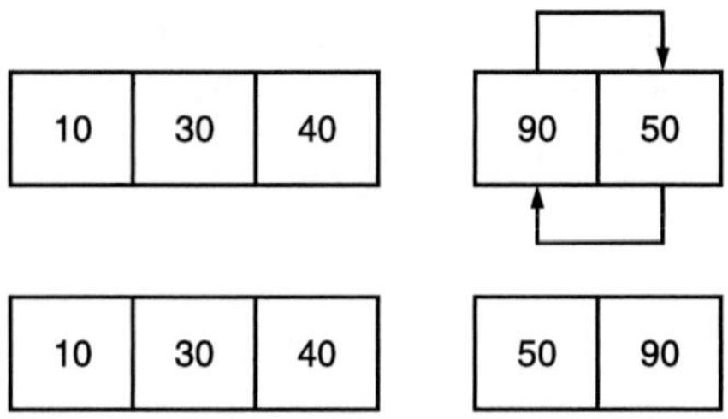

So, final sorted array is

A[ ] =

10	30	40	50	90

SELECTION-SORT (A)

```
n ← length[A]
for j ← 1 to n-1
    smallest ← j
    for i ← j+1 to n
    if A[i]<A[smallest]
        then smallest ← i
            exchange(A[j], A[smallest])
```

Selection sort is very easy to analyse since none of the loops depend on the data in the array. Selecting the lowest element requires scanning all n elements (this takes n-1 comparison) and then swapping it into the first position. Finding the next lowest element requires scanning the remaining n-1 elements and so on, for a total of (n-1) + (n-2) + ... + 2 + 1 = $\theta(n^2)$ comparisons. Each of these scans requires one swap for a total of n-1 swaps (the final element is already in place). Thus, the comparisons dominate the running time, which is $\theta(n^2)$.

EXAMPLE 13.2. Sort the following array using selection sort : A[ ] = (5, 2, 1, 4, 3).

Solution:

A[ ] =

1	2	3	4	5
5	2	1	4	3

Here n = 5

For j = 1 to 4.

j = 1, smallest = 1

For i = 2 to 5.

i = 2, smallest = 1

A[2] = 2   A[1] = 5   A[2] < A[1]

then smallest = 2

Now   i = 3, smallest = 2

A[3] = 1   A[2] = 2   A[3] < A[2]

then smallest = 3

Now   i = 4, smallest = 3

A[4] = 4

A[3] = 1   A[4] > A[3]   No change

Now   i = 5, smallest = 3

A[5] = 3

A[3] = 1   A[5] > A[3]   So, No change

then

exchange(A[1], A[smallest])

i.e., exchange(5, 1)

Now

A[ ] =

1	2	5	4	3

Now j = 2, smallest = 2

For i = 3 to 5.

Now   i = 3, smallest = 2

A[3] = 5

A[2] = 2   A[3] > A[2]   No change

Now   i = 4, smallest = 2   No change

i = 5, smallest = 2   No change

Now   j = 3, smallest = 3

For i = 4 to 5.

Now   i = 4, smallest = 3

A[4] = 4

A[3] = 5   A[4] < A[3]   then smallest = 4

Now   i = 5, smallest = 4

A[5] = 3

A[4] = 4   A[5] < A[4]   then smallest = 5

Now exchange [A[3], A[5]]

then

A[ ] =

1	2	3	4	5

Now j = 4, smallest = 4,   i = 5

A[5] = 5

A[4] = 4   A[5] > A[4]   No change

Hence sorted array is

1	2	3	4	5

Program /* Selection Sort */

```
#include <stdio.h>
#include <conio.h>
#define MAXSIZE 50
void selection(int x[], int n)
{
int i, j, larg, pos, c=0;
for (i = n-1; i>0; i--){
        larg = x[0];
        pos = 0;
        for (j = 1; j<=i; j++)
        {
        c++;
            if (x[j] > larg){
                larg = x[j];
                pos = j;
            }
        }
        x[pos] = x[i];
        x[i] = larg;
        }
        printf("\n Number of comparisons required = %d", c);
}
void main( )
{
        int x[MAXSIZE], n, i;
        clrscr ( );
        printf("\n Enter number of elements");
        scanf("%d", &n);
        printf("\n Unsorted list is as follows\n");
        for (i=0; i<n; i++)
        {
            printf("\n Enter element %d", i+1);
            scanf("%d", &x[i]);
        }
        selection(x, n);
        printf("\n Sorted list is :\n\n");
        for (i=0; i<n; i++)
```

```
            printf("%d ", x[i]);
  getch( );
  }
```

13.6 INSERTION SORT

The insertion sort works just like its name suggest - it inserts each item its proper place in the final list. The simplest implementation of this requires two list structures - the source list and the list into which sorted items are inserted. To save memory, most implementations use an in-place sort that works by moving the current item past the already sorted items and repeatedly swapping it with the preceding item until it is in place.

During the first iteration, the element position 2nd is compared with the element at the 1st position. During the second iteration, the element at the position 3rd is compared with the elements at the 2nd and 1st positions. This process is repeated for all the elements in the array up to (n-1) iterations.

This method is widely used by card players.

Example : Let us take the following values

40	50	12	30	90	18	06	60

Here, there are 8 values, so 7 iterations are required to sort these values.

First Iteration :

40	50	12	30	90	18	06	60

Second Iteration :

40	50	12	30	90	18	06	60

12	40	50	30	90	18	06	60

Third Iteration :

12	40	50	30	90	18	06	60

Fourth Iteration :

12	30	40	50	90	18	06	60

Here 90 is greater than all the elements before it so on exchange occurs.

Fifth Iteration :

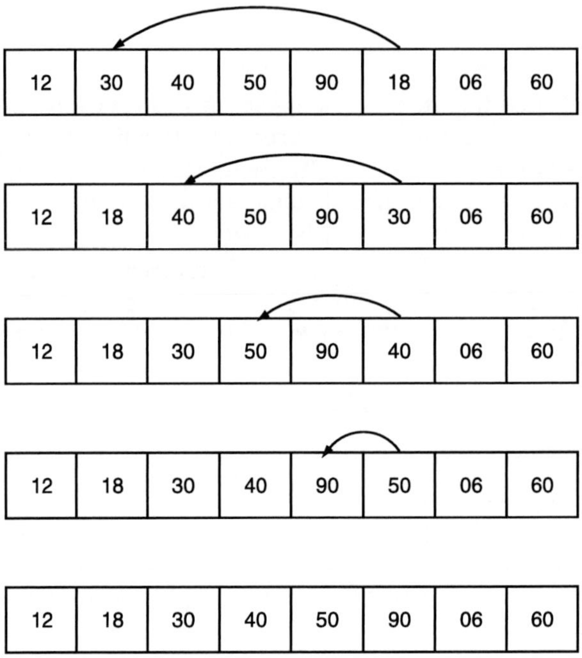

Six Iteration :

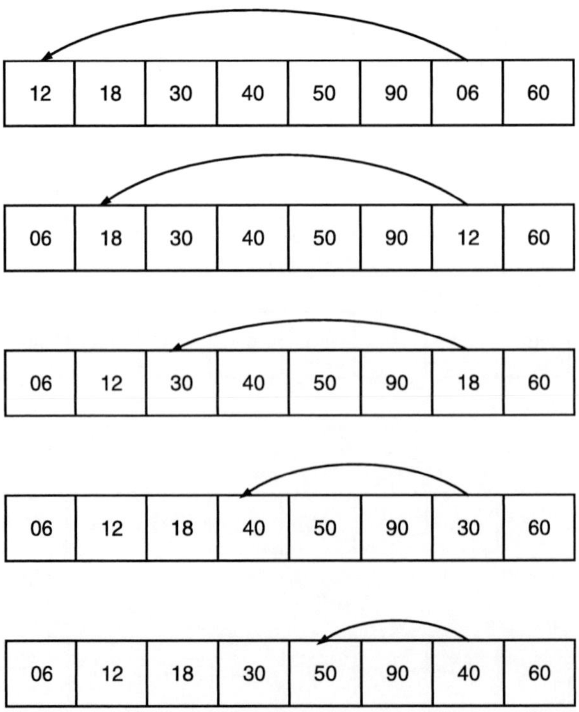

06	12	18	30	40	90	50	60

06	12	18	30	40	50	90	60

Seventh Iteration :

06	12	18	30	40	50	90	60

Here 60 is greater than all the values placed before 90 so

06	12	18	30	40	50	60	90

This is the sorted array.

INSERTION-SORT (A)

```
for j ← 2 to length[A]
     do key ← A[j]
/*Insert A[j] into the sorted sequence A[1..j-1].*/
     i ← j-1
     while i > 0 and A[i] > key
          do A[i + 1] ← A[i]
               i ← i - 1
     A[i+1] ← key
```

EXAMPLE 13.3. Illustrate the operation of INSERTION SORT on the array

A = <2, 13, 5, 18, 14>

Solution:

	1	2	3	4	5
A[ ] =	2	13	5	18	14

For j = 2 to 5.

Now j = 2, key = A[2] i.e. key = 13

i = 2 – 1

i.e. i = 1

while i > 0 and A[1] > 13.

condition false. So no change

Now j = 3, key = A[3] = 5

i = 3 – 1 = 2

i = 2, key = 5

while i > 0 and A[2] > key

condition is **true**

So A[2+1] ← A[2]

A[3] ← A[2]

i.e.,

2		13	18	14

and i = 2-1 = 1, i = 1

while 1>0 andA[1]>key

condition **false**. So no change

then A[1+1] ← key

A[2] ← 5

That is:

2	5	13	18	14

For j = 4

key = A[4]

key = 18, i = 3

Now, while 3>0 and A[3]>18

condition is **false**. No change

Similarly, j = 5

key = A[5]

So key = 14, i = 4

Now, while 4>0 and A[4]>14

condition is **true**.

So A[5] = 18 and i = 4-1 = 3

Now, while 3>0 and A[3]>14

condition is **false**.

So A[3+1] = A[4] = 14

and the sorted array is

A[ ] =

2	5	13	14	18

Program /* Insertion Sort */

```
#include <stdio.h>
#include <conio.h>
void insertion(int x[], int n)
{
int i,j,tmp;
for(i=0;i<n;i++)
{
    temp = x[i];
```

```
        for(j=i-1;j>=0;j--)
        {
            if(temp<x[j])
                x[j+1]=x[j];
            else
                break;
        }
        x[j+1]=temp;
    }
}
void main( )
{
int x[10], n, i,;
clrscr( );
printf("Enter the number of elements :");
scanf("%d", &n);
printf("\nEnter the elements: \n");
for(i=0; i<n;i++)
scanf("%d", &x[i]);
insertion(x,n);
printf("\nThe sorted output: \n");
for(i=0; i<n;i++)
printf("\n%d", x[i]);
getch( );
}
```

Program. Write a C program to sort a list of N strings using insertion sort algorithm. The necessary data must be read from a sequential file.

```
#include <stdio.h>
#include <conio.h>
main( )
{
int i=0,j,k,m;
char fname[10], name[200][30],s[30];
FILE *fp;
clrscr( );
printf("Enter the file name\n");
scanf("%s", fname);
fp=fopen(fname, "r");
if(fp==NULL)
```

```
{ printf("Can't open ..........\n");
exit(0);
}
printf("The names in %s file are\n", fname);
for(;;)
{
fscanf(fp,"%s",s);
if(feof(fp))/* reads the names from the file */
break;
strcpy(name[i++],s);/*store the name taken from the file int array*/
printf("%s",s);
}
m=i;
printf("\nThe sorted order is \n");
for(j=0;j<m;j++)
{ strcpy(s,name[j]);
for(k=j-1; (k>=0&&strcmp(name[k],s)>0);k--)
{
strcpy(name[k+1],name[k]);
}
strcpy(name[k+1],s);
}
for(i=0;i<m;i++)
printf("%s\n", name[i]);
getch( );
}
```

13.7 SHELL SORT

Shell sort, developed by **Donald L. Shell**, is a non-stable in-place sort. Shell sort is the modification of the insertion sort. Shell sort improves on the efficiency of insertion sort by *quickly* shifting values to their destination. Average sort time is $\mathbf{O(n^{1.25})}$, while worst-case time is $\mathbf{O(n^{1.5})}$. In this method, instead of sorting the entire array at once, the array is first divided into smaller segments, *e.g.*, k segments. Then these segments are sorted separately using the insertion sort. The shell sort is also called diminishing increment sort because the value of k continuously decreases, where k is preferably a prime number.

- Shell sort is a sorting algorithm which requires $O(n^2)$ comparisons and exchanges in the worst case.
- Shell sort is a generalisation of insertion sort, with two important observations :
  - Insertion sort is efficient if the input is "almost sorted".
  - Insertion sort is inefficient, on average, because it moves values just one position at a time.
- Shell sort improves insertion sort by comparing elements separated by a gap of several positions.

- This lets an element take "bigger steps" toward its expected position. Multiple passes over the data are taken with smaller and smaller gap sizes.
- The last step of Shell sort is a plain insertion sort, but by then, the array of data is guaranteed to be almost sorted.

EXAMPLE :

45	36	75	20	05	90	80	65	30	50	10	75	85

This distance between the elements to be compared = 3

The sub-files generated with the distance of 3 are as follows :

Sub-file 1	a[0]	a[3]	a[6]	a[9]	a[12]
Sub-file 2	a[1]	a[4]	a[7]	a[10]	
Sub-file 3	a[2]	a[5]	a[8]	a[11]	

3

Input to Pass 1 with distance = 3

45	36	75	20	05	90	80	65	30	50	10	75	85

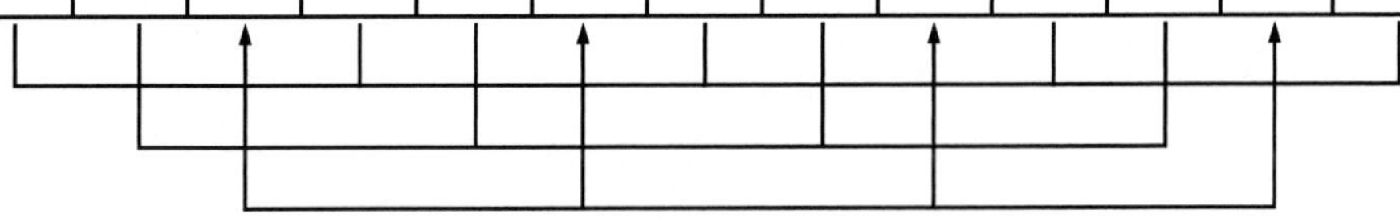

Output of Pass 1 is input to Pass 2 and distance = 2

20	05	30	45	10	75	50	36	75	80	65	90	85

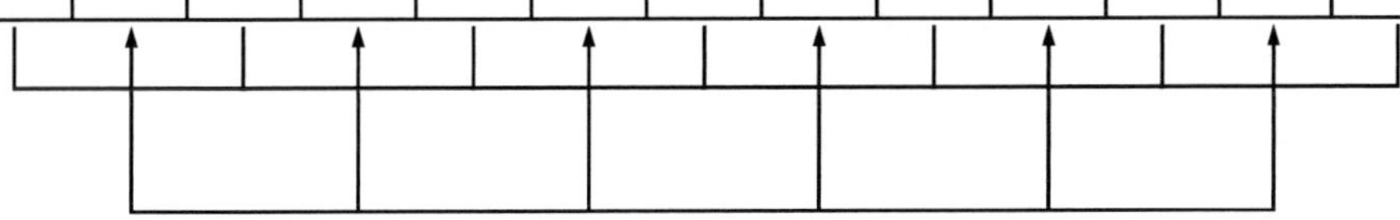

Output of Pass 2 is input to Pass 3 and distance = 1

10	05	20	36	30	45	50	75	65	80	75	90	85

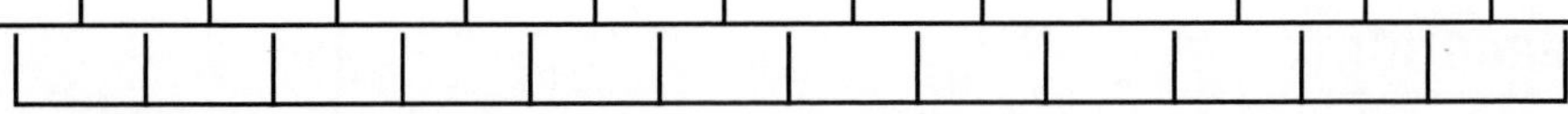

Output of Pass 3

05	10	20	30	36	45	50	65	75	75	80	85	90

Program /* Shell Sort */

```
#include <stdio.h>
#include <conio.h>
#define MAX 15
```

```
void main( )
{
int arr[MAX], i,j,k,n,incr;
clrscr( );
printf("\nEnter the number of elements :");
scanf ("%d",&n);
for (i=0;i < n;i++)
{
printf ("\nEnter element %d :", i+1);
scanf("%d",&arr[i]);
}
printf ("\nUnsorted list is :\n");
for (i = 0; i < n; i++)
printf("%d",arr[i]);
printf ("\nEnter maximum increment (odd value) :");
scanf("%d", &incr);
/*Shell sort algorithm is applied */
while(incr >=1)
{
for (j=incr;j<n;j++)
{
k=arr[j];
for (i=j-incr; i >= 0 && k < arr[i]; i = i-incr)
arr[i+incr] = arr[i];
arr[i+incr] = k;
}
incr=incr-2; /*Decrease the increment */
}
printf("\nSorted list is : \n");
for (i = 0; i < n; i++)
printf("%d", arr[i]);
getch( );
}
```

13.8 MERGE SORT

Merge sort is a sorting algorithm that uses the idea of divide and conquer approach.

MERGE_SORT (A, p, r)

1. if p < r
2. then q ← $\lfloor (p + r)/2 \rfloor$
3. MERGE-SORT (A, p, q)
4. MERGE-SORT (A, q + 1, r)
5. MERGE (A, p, q, r)

MERGE (A, p, q, r)

```
n1 ← q - p + 1
n2 ← r - q
create arrays L[1..n1+1] and R[1..n2+1]
for i ← 1 to n1
      do L[i] ← A[p+i-1]
for j ← 1 to n2
      do R[j] ← A[q+j]
L[n1 + 1] ← ∞
R[n2 + 1] ← ∞
i ← 1
j ← 1
for k ← p to r
      do if L[i] ≤ R[j]
            then A[k] ← L[i]
                  i ← i + 1
            else A[k] ← R[j]
                  j ← j + 1
```

The procedure MERGE-SORT (A, p, r) sorts the elements in the sub-array A[p..r]. If $p \le r$, the sub array has at most one element and is therefore already sorted. Otherwise, the divide step simply computes an index q that partitions A[p..r] into two sub-arrays : A[p..q], containing [n/2] elements, and A[q+1..r], containing [n/2] elements.

To sort the entire sequence A = (A[1], A[2], ... A[n]) we call MERGE-SORT (A, 1, length[A]) where once again *length*[A] = n. If we look at the operation of the procedure bottom-up when n is a power of two, the algorithm consists of merging pairs of 1-item sequences to form sorted sequences of length 2, merging pairs of sequences of length 2 to form sorted sequences of length 4, and so on, until two sequences of length n/2 are merged to form the final sorted sequence of length n.

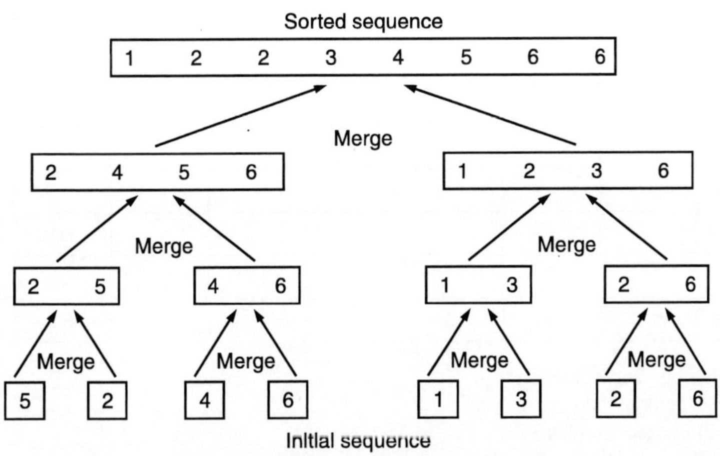

Fig. 13.1. Initial sequences.

Example of Merge Sort

0 MS	85	24	63	45	17	31	96	50
1 Div	85	24	63	45	17	31	96	50
1 MS	85	24	63	45				
2 Div	85	24	63	45				
2 MS	85	24						
3 Div	85	24						
3 MS	85							
3 MS		24						
3 Merge	24	85						
2 MS			63	45				
3 Div			63	45				
3 MS			63					
3 MS				45				
3 Merge			45	63				
2 Merge	24	45	63	85				
2 Merge	24	45	63	85				
1 MS					17	31	96	50
2 Div					17	31	96	50
2 MS					17	31		
3 Div					17	31		
3 MS					17			
3 MS						31		
3 Merge					17	31		
2 MS							96	50
3 Div							96	50
3 MS							96	
3 MS								50
3 Merge							50	96
2 Merge					17	31	50	96
1Merge	17	24	31	45	50	63	85	96

* MS → Merge_Sort operation
* Dir → Divide operation
* Merge → Merge operation

EXAMPLE 13.4. **Use merge-sort algorithm to sort the following elements**

15, 10, 5, 20, 25, 30, 40, 35

Solution: Suppose A[ ] :

1	2	3	4	5	6	7	8
15	10	5	20	25	30	40	35
p			↑ q				r

Here $p = 1$

$r = 8$

$1 < 8$ then $q = \lfloor(1+8)/2\rfloor = \lfloor 4.5\rfloor = 4$

i.e., $q = 4$

Call MERGE-SORT (A, 1, 4)

MERGE-SORT (A, 5, 8)

MERGE (A, 1, 4, 8)

First call **MERGE-SORT (A, 1, 4)**

Here $p = 1$

$r = 4$

$1 < 4$, so $q = \left\lfloor \frac{1+4}{2} \right\rfloor = \lfloor 2.5\rfloor = 2$

$\therefore$ $q = 2$

then call MERGE-SORT (A, 1, 2)

MERGE-SORT (A, 3, 4)

MERGE (A, 1, 2, 4)

Now call **MERGE-SORT (A, 1, 2)** firstly

Here $p = 1$

$r = 2$

$1 < 2$, then $q = \left\lfloor \frac{1+2}{2} \right\rfloor = 1$

So, call MERGE-SORT (A, 1, 1)

MERGE-SORT (A, 2, 2)

MERGE (A, 1, 1, 2)

Call MERGE-SORT (A, 1, 1)

Here $1 < 1$ No change

then call MERGE-SORT (A, 2, 2)

Here $2 < 1$ so, no change

then call MERGE-SORT (A, 1, 1, 2)

Here $p = 1, q = 1, r = 2$

So, $n_1 = q - p + 1 = 1 - 1 + 1$

$n_1 = 1$

$n_2 = 2 - 1$

$n_2 = 1$

Create arrays L[1..2] and R[1..2]

A[ ]	15	10

	1
L[ ]	15

	1
R[ ]	10

For $i = 1$ to 1
$L[1] = A[1 + 1 - 1]$
$L[1] = A[1]$
For $j = 1$ to 1
$R[1] = A[1 + 1]$
$R[1] = A[2]$
Now $L[2] = \infty$ and $i = 1$
$R[2] = \infty$ $j = 1$
For $k = 1$ to 2
$k = 1$
if $L[1] \le R[1]$
$15 \le 10$ (False)
So, $A[1] = R[1]$ *i.e.*, $A[1] = 10$
and $j = 1 + 1 = 2$
Now $k = 2$ and $j = 2$
$L[1] \le R[2]$
$15 \le \infty$ (True)
So $i = i + 1 = 2$ and $A[2] = 15$

Now A[ ] :

	1	2	
A[ ] :	10	15	

Now MERGE-SORT (A, 1, 2) is completed.

then call **MERGE-SORT (A, 3, 4)**

Here $p = 3$
$r = 4$

$3 < 4$, then $q = \left\lfloor \frac{p + r}{2} \right\rfloor = \left\lfloor \frac{3 + 4}{2} \right\rfloor = 3$

Call
- MERGE-SORT (A, 3, 3) (No change)
- MERGE-SORT (A, 4, 4) (No change)
- MERGE-SORT (A, 3, 3, 4)

So, call **MERGE (A, 3, 3, 4)**

Here $p = 3$
$q = 3$
$r = 4$
So, $n_1 = 1$
$n_2 = 1$

So create array L and R
For $i = 1$ to 1
$L[1] = A[3 + 1 - 1]$
$L[1] = A[3]$ i.e $L[1] = 5$

For $j = 1$ to 1

$R[1] = A[3 + 1]$

i.e., $R[1] = 20$

$L[2] = \infty$

and $R[2] = \infty$

and $i = 1$ and $j = 1$

Now for $k = 3$ to 4

$k = 3, \quad i = 1, \quad j = 1$

$L[1] = 5 \qquad$ Here $5 \leq 20$ (True)

$R[1] = 20 \quad$ then $A[3] = 5$ and $i = i + 1 = 2$.

$k = 4, \quad i = 2, \quad j = 1$

$L[2] = \infty$

$R[1] = 20$

$L[2] \nless R[1]$

So $A[4] = 20$

i.e., Now

1	2	3	4	
10	15	5	20	

Now call **MERGE (A, 1, 2, 4)**

Here $p = 1$

$q = 2$

$r = 4$

So, $n_1 = 2 - 1 + 1$

$n_1 = 2 \qquad$ i.e. $n_1 = 2$

$n_2 = 2$

$n_2 = 4 - 2 = 2$

Create array L[1...3] and R[1...3]

For $i = 1$ to 2

$i = 1 \quad L[1] = A[1 + 1 - 1]$

$L[1] = A[1]$

i.e., $L[1] = 10$

For $i = 2 \quad L[2] = A[1 + 2 - 1]$

$= A[2]$

$L[2] = 15$

i.e., L[ ] =

1	2	3
10	15	∞

For $j = 1$ to 2

$R[1] = A[2 + 1] = 5$

$R[2] = A[2 + 2] = 20$

i.e., R[ ] =

1	2	3
5	20	∞

Now i = 1 and j = 1

For k = 1 to 4

k = 1 L[1] ≤ R[1] *i.e.,* 10 ≤ 5 (False)

So A[1] = R[1] *i.e.,* A[1] = 5 and j = 1 + 1 = 2

k = 2 L[1] ≤ R[2] *i.e.,* 10 ≤ 20 (True)

So A[2] = L[1] *i.e.,* A[2] = 10 and i = 1 + 1 = 2

k = 3 L[2] ≤ R[2] *i.e.,* 15 ≤ 20 (True)

So A[3] = L[2] *i.e.,* A[3] = 15 and i = 2 + 1 = 3

k = 4 L[3] ≤ R[2] *i.e.,* ∞ ≤ 20 (False)

So A[4] = R[2] *i.e.,.* A[4] = 20 and j = 2 + 1 = 3

i.e., Now array is

A[ ]

1	2	3	4	
5	10	15	20	

Now call **MERGE-SORT (A, 5, 8)**

Here p = 5

r = 8

$5 < 8$, so $q = \left\lfloor \frac{5+8}{2} \right\rfloor = 6$ *i.e.,* $q = 6$.

then call

MERGE-SORT (A, 5, 6) MERGE-SORT (A, 7, 8) MERGE-SORT (A, 5, 6, 8)

First we call **MERGE-SORT (A, 5, 6)**

Here p = 5

r = 6

$5 < 6$, then $q = \left\lfloor \frac{5+6}{2} \right\rfloor = 5$

then call

MERGE-SORT (A, 5, 5) MERGE-SORT (A, 6, 6) MERGE (A, 5, 5, 6)

Call **MERGE-SORT (A, 5, 5)**

Here 5 < 5 (False) So no change.

then call **MERGE-SORT (A, 6, 6)**

Here 6 < 6 (False) So no change.

Call **MERGE (A, 5, 5, 6)**

Here $p = 5 \quad q = 5 \quad r = 6.$

$n_1 = 1$

$n_2 = 1$

Create arrays L[1..2] and R[1..2]

L[1] = A[5] *i.e.*, L[1] = 25

R[1] = A[6] *i.e.*, R[1] = 30

L[2] = ∞ *i.e.*, L[ ]

1	2
25	∞

R[2] = ∞ R[ ]

1	2
30	∞

i = 1 and j = 1

For k = 5 to 6

k = 5 L[1] ≤ R[1] i.e. 25 ≤ 30 (True)

So A[5] = 25 and i = 1 + 1 = 2.

k = 6 L[2] ≤ R[1] i.e. ∞ ≤ 30 (False)

So A[6] = 30 and j = j + 1 = 2.

Thus now array is

A[ ] =

1	2	3	4	5	6	
5	10	15	20	25	30	

Now call **MERGE-SORT (A, 7, 8)**

We get

A[ ] =

1	2	3	4	5	6	7	8
5	10	15	20	25	30	35	40

We call **MERGE (A, 5, 6, 8)**

We get the array

A[ ] =

1	2	3	4	5	6	7	8	
5	10	15	20	25	30	35	40	

Now call **MERGE (A, 1, 4, 8)**

We get the sorted array as:

A[ ] =

1	2	3	4	5	6	7	8
5	10	15	20	25	30	35	40

This is final sorted array.

Program /* Implement Merge Sort Through Recursion */

```
#include <stdio.h>
#include <conio.h>
#define MAX 20
int array [MAX];
void merge (int low, int mid, int high)
{
int temp[MAX];
int i = low;
int j = mid + 1;
int k = low;
while((i <= mid) && (j <= high))
{
if (array[i] <= array[j])
temp[k++] = array [i++];
else
temp[k++] = array [j++];
}
while (i <= mid)
temp[k++] = array [i++];
while (j <= high)
temp[k++] = array [j++];
for (i = low; i <= high; i++)
array[i]=temp[i];
}
void merge_sort(int low, int high)
{
int mid;
if (low != high)
{
mid = (low + high)/2;
merge_sort(low, mid);
merge_sort(mid + 1, high);
merge(low, mid, high);
}
}
void main( )
{
int i, n;
```

```
clrscr( );
printf ("\nEnter the number of elements :");
scanf ("%d", &n)
for (i = 0; i < n; i++)
{
printf ("Enter element %d :", i + 1);
scanf ("%d", &array[i]);
}
printf ("\nUnsorted list is :\n");
for (i = 0; i < n; i++)
printf ("%d", array[i]);
merge_sort(0, n-1);
printf ("\nSorted list is :\n");
for (i = 0; i < n; i++)
printf ("%d", array[i]);
getch( );
}
```

Program /* Implement Merge Sort Without Recursion */

```
#include <stdio.h>
#include <conio.h>
#define MAX 20
void main( )
{
int a[MAX], temp[MAX], i, j, k, n, size, l1, h1, l2, h2;
clrscr( );
printf("\nEnter the number of elements :");
scanf("%d", &n);
for (i=0; i<n; i++)
{
printf("\nEnter elements %d :" i + 1);
scanf("%d", &a[i]);
}
printf("\nUnsorted list is :");
for (i=0; i<n; i++)
printf("%d", a[i]);
/*l1 lower bound of first pair and so on */
for (size = 1; size < n; size = size*2)
{
l1 = 0;
```

```
k = 0; /*Index for temp array */
while (l1+size < n)
{
h1=l1+size-1;
l2=h1+1;
h2=l2+size-1;
if(h2 >=n)
h2=n-1;
/*Merge the two pairs with lower limits l1 and l2 */
i=l1;
j=l2;
while(i<=h1 && j<=h2)
{
if (a[i] <= a[j])
temp[k++] = a[i++];
else
temp[k++]=a[j++];
}
while(i<=h1)
temp[k++] = a[i++];
while(j<=h2)
temp[k++] = a[j++];
l1 = h2 + 1; /*Take the next two pairs for merging */
}
for (i=l1; k<n; i++) /* any pair left */
temp[k++] = a[i];
for(i=0; i<n; i++)
a[i]=temp[i];
printf("\nSize=%d\nElements are :", size);
for (i = 0; i<n; i++)
printf("%d", a[i]);
}
printf ("\nSorted list is :\n");
for (i = 0; i<n; i++)
printf("%d", a[i]);
getch( );
}
```

13.9 QUICK SORT

C.A.R. Hoare implements quick sort by *divide and conquer method*. That means divide the big problem into two small problems and then those two small problems into two small ones and so on. In Quick sort, we divide the original list into two sub-lists. We choose the item from list called *key* or *pivot* from which all the left side of elements are smaller and all the right side of elements are greater than that element. So we can create two lists, one list on the left side of pivot element and the second list is on light side of the pivot.

Thus, Quick sort works by partitioning a given array A[p..r] into two non-empty sub- arrays A[p..q] and A[q+1..r] such that every key in A[p..q] is less than or equal to every key in A[q+1..r]. Then the two sub-arrays are sorted by recursive calls to Quick sort. The exact position of the partition depends on the given array and index q is computed as a part of the partitioning procedure.

```
QUICK-SORT (A, p, r)
    If p < r then
    q ← PARTITION (A, p, r)
    QUICK-SORT (A, p, q-1)
    QUICK-SORT (A, q+1, r)
```

Note that to sort entire array, the initial call is Quick Sort (A, 1, length [A])

As a first step, Quick Sort chooses as pivot one of the items in the array to be sorted. Then array is partitioned on either side of the pivot. Elements that are less than or equal to pivot will move towards the left and elements that are greater than or equal to pivot will move toward the right.

Partitioning the Array

Partitioning procedure rearranges the sub arrays in-place.

```
PARTITION (A, p, r)
  x ← A[r]
  i ← p-1
  for j ← p to r-1
      do if A[j] ≤ x
         then i ← i+1
              exchange A[i] ↔ A[j]
  exchange A[i+1] ↔ A[r]
  return i+1
```

Partition selects the first key, A[p] as a pivoted key about which the array will be partitioned :

- Keys ≤ A[p] will be moved towards the left.
- Keys ≥ A[p] will be moved towards the right.

EXAMPLE 13.5. Use quicksort algorithm to sort

36, 15, 40, 1, 60, 20, 55, 25, 50, 20

is a stable sorting algorithm ?

Solution: Stable Sorting : We say that a sorting algorithm is stable if, when two records have the same key, they stay in their original order.

	1	2	3	4	5	6	7	8	9	10
Let A[ ] =	35	15	40	1	60	20	55	25	50	20

Here p = 1 r = 10

$x = A[10]$ *i.e.,* $x = 20$

$i = p\text{-}1$ *i.e.,* $i = 0$

$j = 1$ to 9

Now, $j = 1$ and $i = 0$

$A[j] = A[1] = 36$ and $36 \nleq 20$

So $j = 2$ and $i = 0$

$A[2] = 15 \leq 20$ (True)

then $i = 0 + 1 = 1$

and $A[1] \leftrightarrow A[2]$

i.e.,

1	2	3	4	5	6	7	8	9	10
15	36	40	1	60	20	55	25	50	20

Now, $j = 3$ and $i = 1$

$A[3] = 40 \nleq 20$

$j = 4$ and $i = 1$

$A[4] = 1 \leq 20$ (True)

then $i = 1 + 1 = 2$

and $A[2] \leftrightarrow A[4]$

i.e.,

1	2	3	4	5	6	7	8	9	10
15	1	40	36	60	20	55	25	50	20

Now, $j = 5$ and $i = 2$

$A[5] = 60 \nleq 20$

$j = 6$ and $i = 2$

$A[j] = 20 \leq 20$ (True)

then $i = i + 1 = 2$ i.e. $i = 2 + 1 = 3$

and $A[3] \leftrightarrow A[6]$

i.e.,

1	2	3	4	5	6	7	8	9	10
15	1	20	36	60	40	55	25	50	20

Now, $j = 7$ and $i = 3$

$A[7] = 55 \nleq 20$

$j = 8$ and $i = 3$

$A[8] = 25 \nleq 20$

$j = 9, i = 3$

$A[9] = 50 \nleq 20$

Now $A[i+1]$ i.e. $A[4] \leftrightarrow A[10]$

i.e.,

1	2	3	4	5	6	7	8	9	10
15	1	20	20	60	40	50	25	50	36

and two sub-arrays are

15	1	20

and

60	40	50	25	50	36

Thus, this is a stable algorithm.

Program /* Quick Sort */

```
#include <stdio.h>
#include <conio.h>
void quicksort (int a[], int low, int high);
int partition (int a[], int low, int high);
int a[5] = {55, 1, 78, 13, 45};
void main( )
{
clrscr( );
int i, n;
printf("\nOriginal array");
for(i=0; i < 5; i++)
printf("%4d", a[i]);
quicksort(a, 0, 4);
printf("\nThe sorted array is \n");
for(i=0; i < 5; i++)
printf("%4d", a[i]);
getch( );
}
void quicksort (int a[], int low, int high)
{
int j;
if (low < high)
{
j = partition(a,low, high);
quicksort (a, low, j-1);
quicksort(a, j+1, high);
}
}
int partition(int a[], int low, int high)
{
int i, j, temp, key;
key = a[low];
i = low + 1;
```

```
j = high;
while (1)
{
while (i < high && key >= a[i])
i++;
while (key < a[j])
j--;
if (i < j)
{
temp = a[i];
a[i] = a[j];
a[j] = temp;
}
else
{
temp = a[low];
a[low] = a[j];
a[j] = temp;
}
return j;
}
}
```

Program /* Quick Sort */

```
#include <stdio.h>
#include <conio.h>
void quicksort (int *, int, int);
int split(int *, int, int);
int a[10] = {11, 2, 9, 13, 57, 25, 17, 1, 90, 3};
void main( )
{
int i;
clrscr( );
printf("\nQuick Sort\n");
printf("\nArray before sorting :");
for(i=0; i < 10; i++)
printf("%4d", a[i]);
quicksort(a, 0, 9);
printf("\nArray after sorting :");
for(i=0; i < 10; i++)
printf("%4d", a[i]);
```

```
getch( );
}
void quicksort (int a[], int lower, int upper)
{
int i;
if (upper > lower)
{
i = split(a, lower, upper);
quicksort(a, lower, i-1);
quicksort(a, i+1, upper);
}
}
int split(int a[], int lower, int upper)
{
int i, p, q, t;
p = lower + 1;
q = upper;
i = a[lower];
while (q >= p)
{
while(a[p] < i)
p++;
while(a[q] > i)
q--;
if (q > p)
{
t = a[p];
a[p] = a[q];
a[q] = t;
}
}
t = a[lower];
a[lower] = a[q];
a[q] = t;
return q;
}
```

13.10 HEAP SORT

Heap sort was invented by **John Williams** and uses the approach just opposite to selection sort. The selection sort finds the smallest element among n elements, then the smallest element among n-1 elements and so on, until the end of the array whereas heap sort finds the largest element and puts it at the end of the array, then the second largest item is found and this process is repeated for all other elements.

Before discussing heap sort, we will begin by defining a new structure, the heap. We can define minimum heap and maximum heap.

A **max (or min) heap** is a complete binary tree with the property that the value at each node is at least as large as (or as small as) the values of its children (if they exist). This is called **Heap property.**

A heap is a complete balanced binary tree in which each node satisfies the heap property and every leaf is of depth d or d-1.

The definition of a **max heap** implies that one of the largest elements is at the root of the heap. If the elements are district, then the root contains the largest element.

Binary Heap

The binary heap data structure is an array that can be viewed as a complete binary tree. Each node of the binary tree corresponds to an element of the array. The array is completely filled on all levels except possibly lowest.

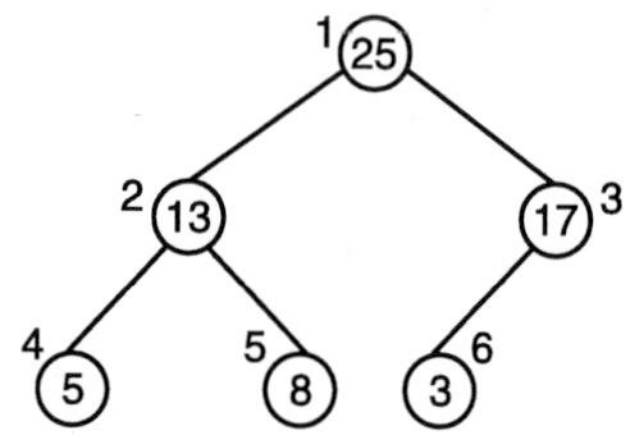

Fig. 13.2.

We represent heaps in level order, going from left to right. The array corresponding to the heap in Fig. 13.2 above is:

	1	2	3	4	5	6
A	25	13	17	5	8	3

If an array A contains key values of nodes in a heap, length [A] is the total number of elements.

heat-size [A] = length [A] = number of elements.

The root of the tree A[1] and given index i of a node the indices of its parent, left child and right child can be computed.

```
PARENT (i)
        return floor (i/2)
LEFT (i)
        return 2i
RIGHT (i)
        return 2i + 1
```

Let us try these out on a heap to make sure we believe they are correct, take this heap.

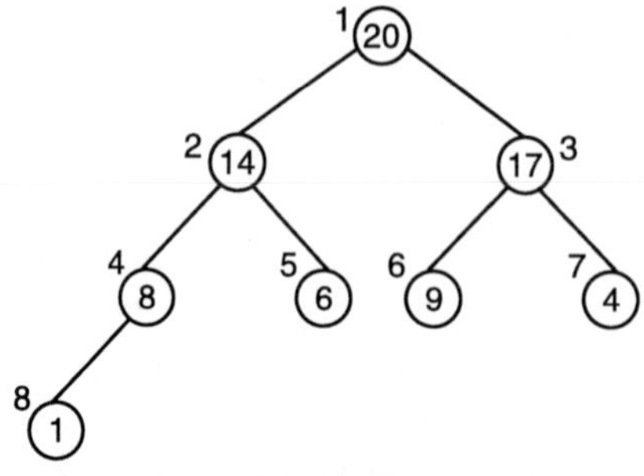

Fig. 13.3.

which is represented by the array.

1	2	3	4	5	6	7	8	9
20	14	17	8	6	9	4	1	

The index of the 20 is 1

To find the index of the left child, we calculate 1*2=2

This takes us (correctly) to the 14.

Now, we go right, so we calculate 2*2+1 = 5

This takes us (again, correctly) to the 6.

Now, 4's index is 7, we want to go to the parent, so we calculate 7/2 = 3 which takes us to the 17.

Heap Property

In a Max-heap, for every node i other than the root, the value of a node is greater than or equal (at most) to the value of its parent

$$A[\text{ PARENT } (i)] \geq A[i]$$

Thus, the largest element in a heap is stored at the root. Following is an example of MAX-HEAP (see Fig. 13.4).

and a Min-Heap is organized in the opposite way *i.e.*, the min-heap property is that for every node i other than the root is:

$$A[\text{ PARENT } (i)] \leq A[i] \text{ (see Fig. 13.5)}$$

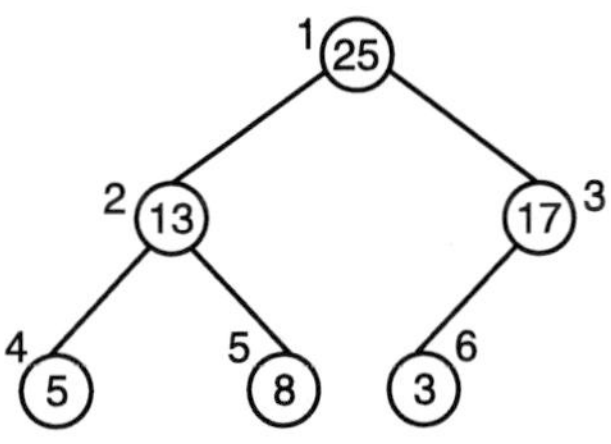

Fig. 13.4.

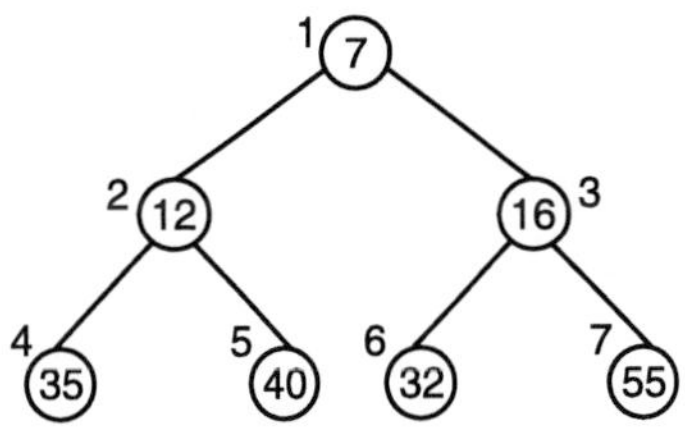

Fig. 13.5.

Heapify

Maintaining the Heap Property

Heapify is a procedure for manipulating heap data structures. It is given an array A and index i into the array. The sub-tree rooted at the children of A[i] are heap but node A[i] itself may possibly violate the heap property, *i.e.*, A[i] < A[2i] or A[i] < A[2i+1]. The procedure 'Heapify' manipulates the tree rooted at A[i] so it becomes a heap.

Outline of Procedure Heapify

Heapify picks the largest child key and compares it to the parent key. If parent key is larger then heapify quits, otherwise it swaps the parent key with the largest child key. So that the parent now becomes larger than its children. It is important to note that swap may destroy the heap property of the sub-tree rooted at the largest child node. If this is the case, heapify calls itself again using largest child node as the new root.

```
MAX-HEAPIFY (A, i)
i ← left [i]
r ← right [i]
if l ≤ heap-size [A] and A[l] > A[i]
      then largest ← l
```

```
      else largest ← i
if r ≤ heap-size [A] and A[r] > A [largest]
      then largest ← r
   if largest ≠ i
      then exchange A[i] ↔ A[largest]
             MAX-HEAPIFY [A, largest].
```

Example : Suppose we have a complete binary tree.

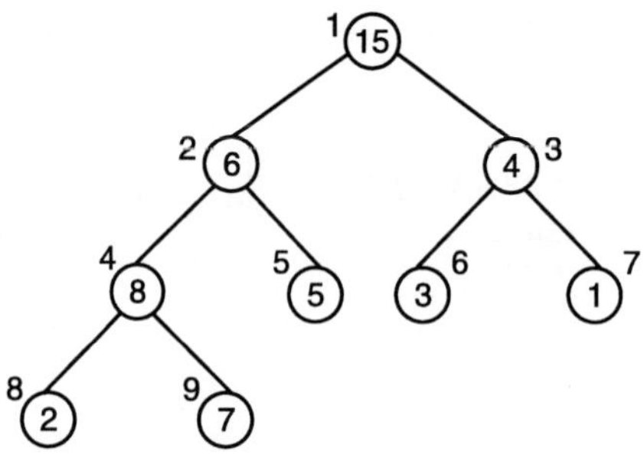

Fig. 13.6.

In this complete binary tree, the subtrees of 6 are not max-heap. So we call MAX-HEAPIFY (A, 2)

$$l = 4$$

$$r = 5 \quad \text{and} \quad i = 2$$

Here heap-size (A) = 9

$$l \le \text{heap-size} \quad \text{and} \quad A[l] = 8 \quad \text{and} \quad A[i] = 6$$

So, $A[l] > A[i]$

then largest = 4

$$r \le \text{heat-size [A]} \quad \text{and} \quad A[r] = 5$$

$$A[r] < A[\text{largest}]$$

Here, largest ≠ i

then A[2] = 8 and A[4] = 6 after exchanging A[i] and A[largest]

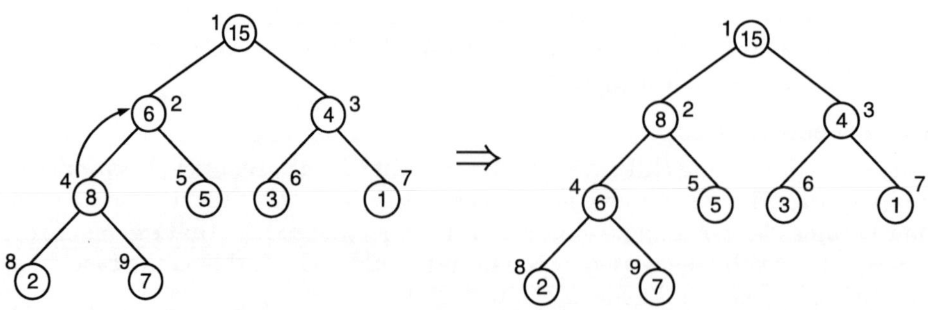

Fig. 13.7.

Now, call MAX-HEAPIFY (A, 4)

$$l = 8$$

$$r = 9 \quad \text{and} \quad A[l] = 2$$

$$A[r] = 7$$

$$A[i] = 6$$

$l \leq$ heap-size [A] and $A[l] < A[i]$ so largest = i

$r \leq$ heap-size [A] and $A[r] > A[\text{largest}]$ so largest = r

largest $\neq$ i then exchange A[largest] $\leftrightarrow$ A[i]

Thus the final tree is

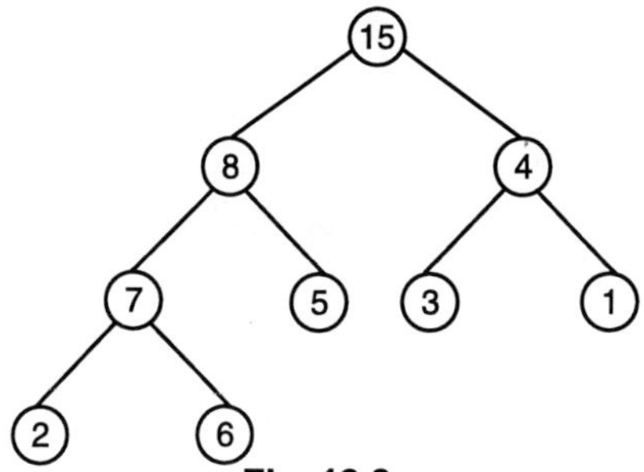

Fig. 13.8.

Building a Heap

We can use the procedure 'Heapify' in a bottom-up fashion to convert an array A[1..n] into a heap.

BUILD-MAX-HEAP (A)

1. heap-size (A) ← length [A]
2. for i ← floor(length [A]/2) down to 1 do
3.   MAX-HEAPIFY (A, i)

Heap-Sort

The heap-sort algorithm starts by using procedure MAX-BUILD-HEAP to build a heap on the input array A[1..n], since the maximum element of the array stored at the root A[1], it can be put into its correct final position by exchanging it with A[n] (the last element in A). If we now discard node n from the heap then the remaining elements can be made into heap. Note that the new element at the root may violate the heap property. All that is needed to restore the heap property.

HEAP-SORT(A)

1. BUILD-MAX-HEAP (A)
2. for i ← length [A] down to 2
3.   do exchange A[1] ↔ A[i]
4.   heap-size [A] ← heap-size [A] – 1
5.   MAX-HEAPIFY (A, 1)

EXAMPLE 13.6. Illustrate the operation of HEAP-SORT on the array

A = (5, 13, 2, 25, 7, 17, 20, 8, 4)

Solution: Originally,

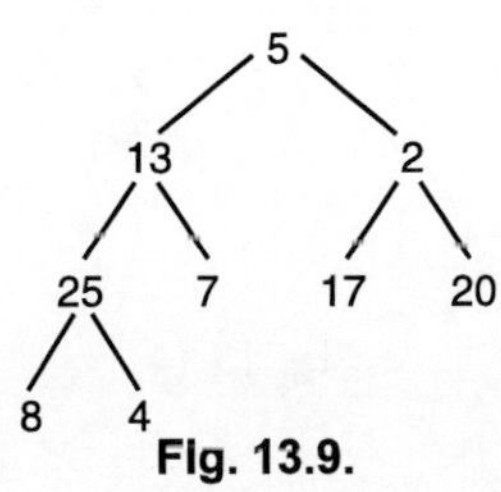

Fig. 13.9.

First we call BUILD-MAX-HEAP

heap-size[A] = 9

So, i = 4 to 1. Call Max-Heapify (A, i)

i.e., first we call MAX-HEAPIFY (A, 4)

A[l] = 8   A[i] = 25   A[r] = 4

A[i] > A[l]

A[i] > A[r]

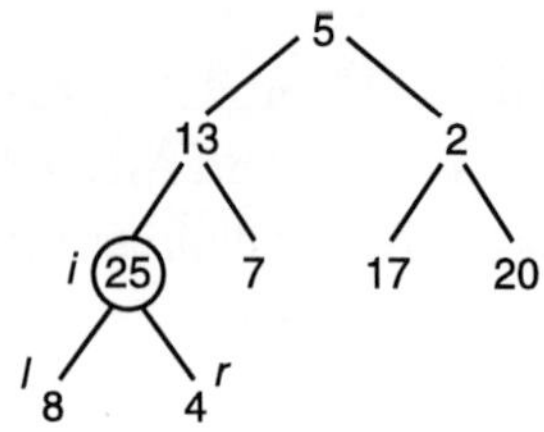

Fig. 13.10.

Now we call MAX-HEAPIFY (A, 3)

A[i] = 2   A[l] = 17   A[r] = 20

A[l] > A[i]

∴ largest = 6

A[r] > A[largest]

20 > 17

∴ largest = 7

largest ≠ i

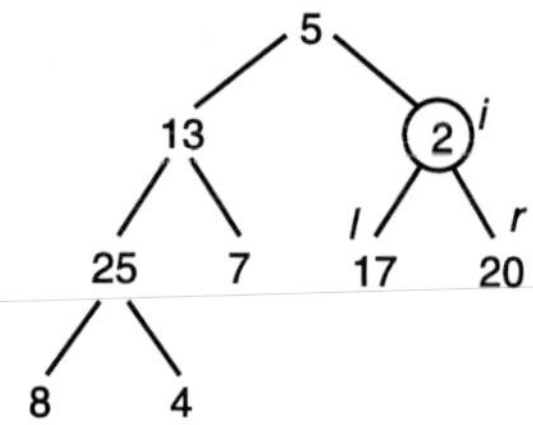

Fig. 13.11.

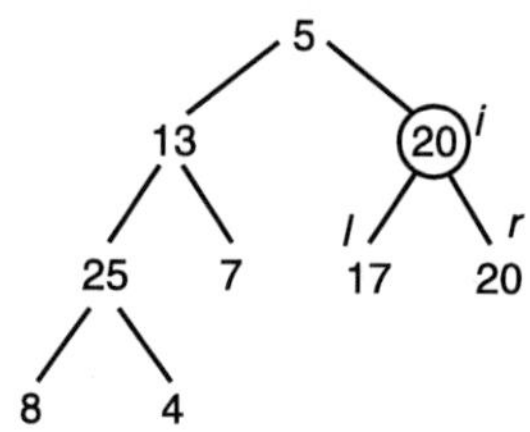

Fig. 13.12.

∴ A[i] ↔ A[largest] i.e., now

A[i] > A[l]

A[i] > A[r]

Now, we call MAX-HEAPIFY (A, 2) i.e.,

A[i] < A[l]

So, largest = 4

A[largest] > A[r]

∴ i ≠ largest, so A[i] ↔ A[largest]

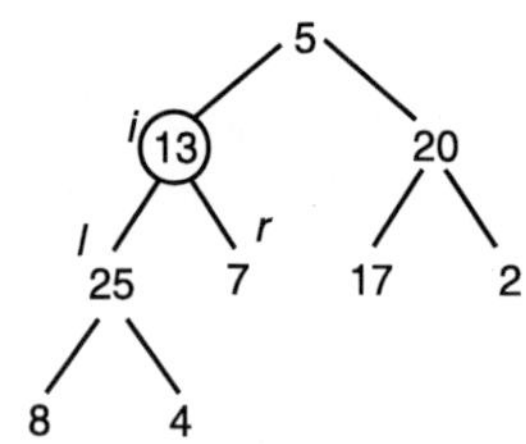

Fig. 13.13.

Now,

A[i] > A[l]

A[i] > A[r]

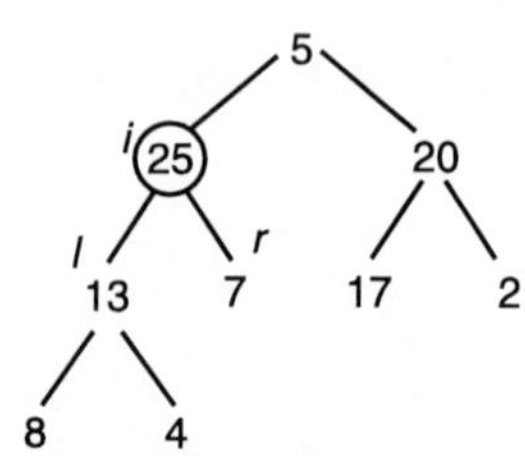

Fig. 13.14.

We call MAX-HEAPIFY (A, 1)

A[i] < A[l]

largest = 2

A[largest] > A[r] and largest ≠ i

A[i] ↔ A[largest]

largest = 2, so i = 2

A[i] < A[l] then largest = 4

and A[largest] > A[r], largest ≠ i

A[i] ↔ A[largest]

Now, A[i] < A[l]

largest = 8 A[largest] > A[r]

largest ≠ i

A[largest] ↔ A[i]

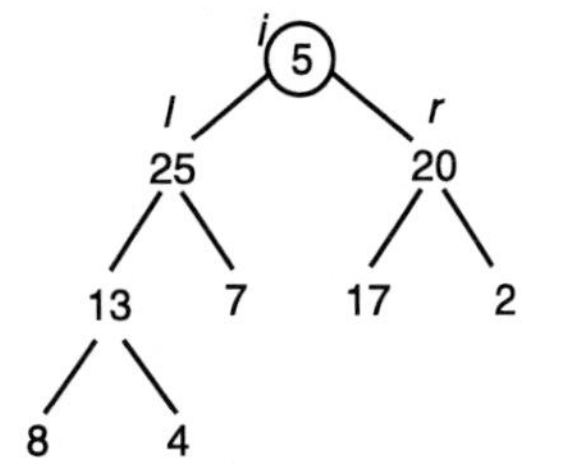

Fig. 13.15.

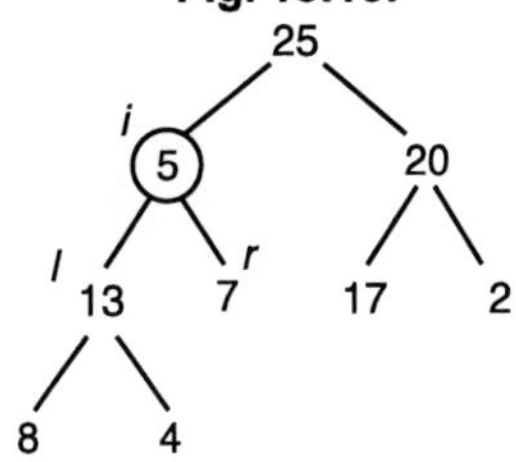

Fig. 13.16.

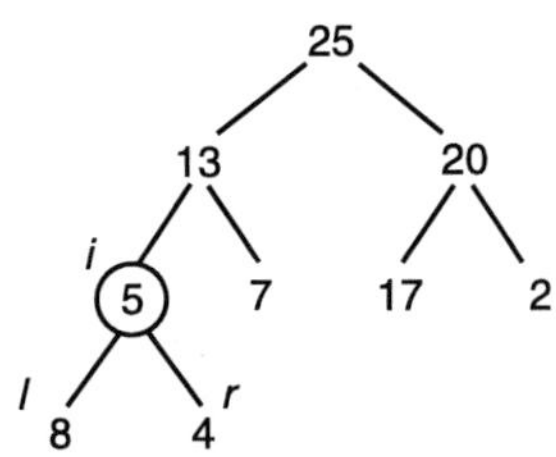

Fig. 13.17.

So, this is the final tree after BUILD-MAX-HEAP.

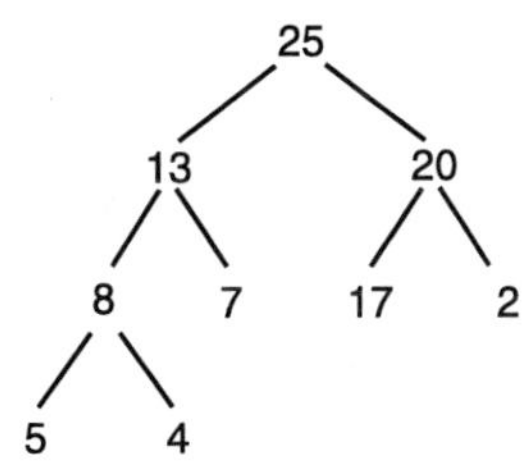

Fig. 13.18.

Now, i = 9 down to 2 exchange A[1] ↔ A[i] and size = size – 1 and call MAX-HEAPIFY (A, 1) each time.

Exchanging A[1] ↔ A[9], we get

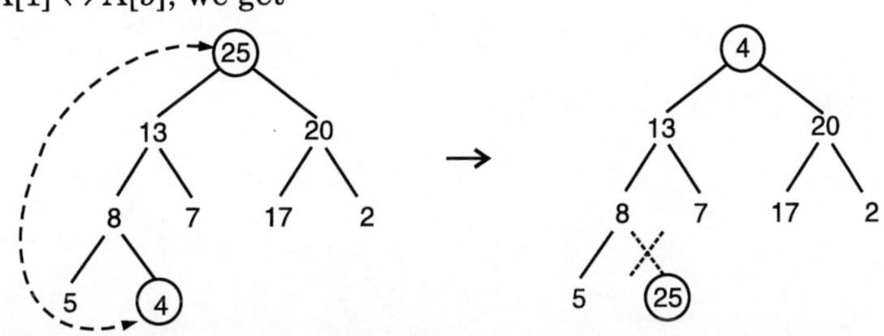

4	13	20	8	7	17	2	5	25

Fig. 13.19.

Now call MAX-HEAPIFY (A, 1) we get

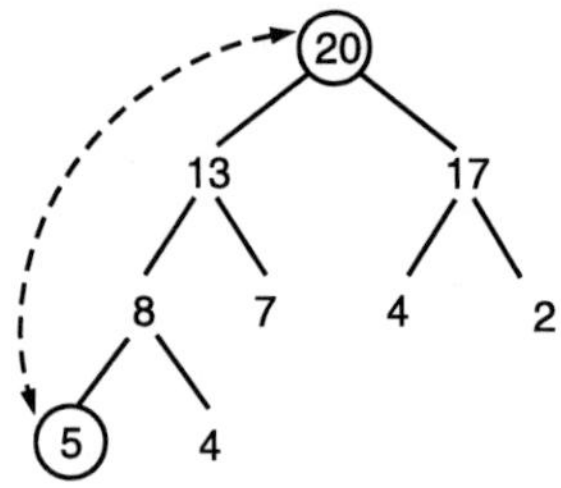

Fig. 13.20.

Now, exchange A[1] and A[8] and size = size – 1 = 8 – 1 = 7.

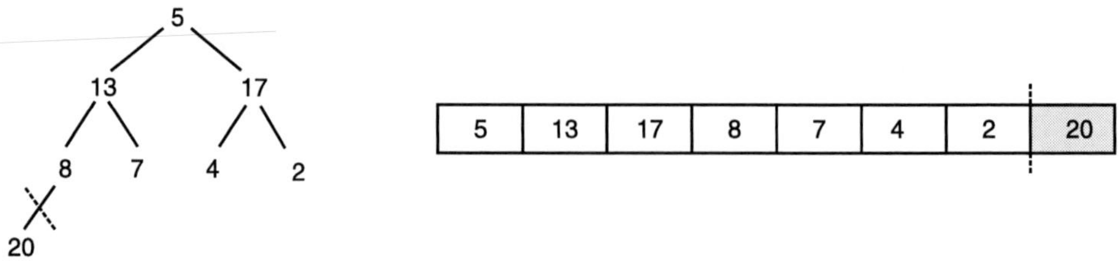

Fig. 13.21.

Again, call MAX-HEAPIFY (A, 1) we get

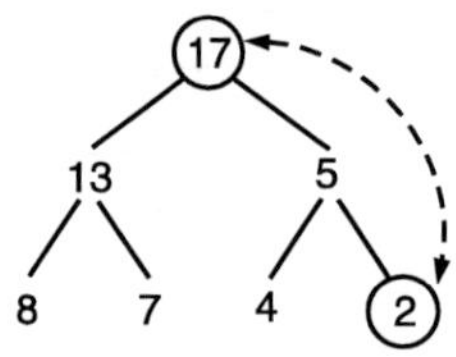

Fig. 13.22.

Exchange A[1] and A[7] and size = size – 1 = 7 – 1 = 6.

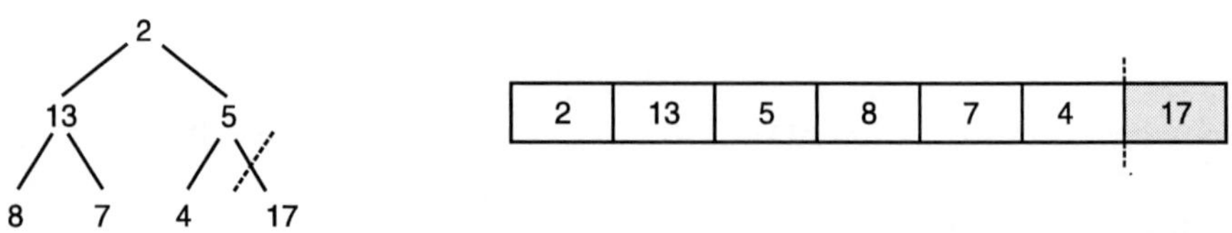

Fig. 13.23.

Again, call MAX-HEAPIFY (A, 1) we get

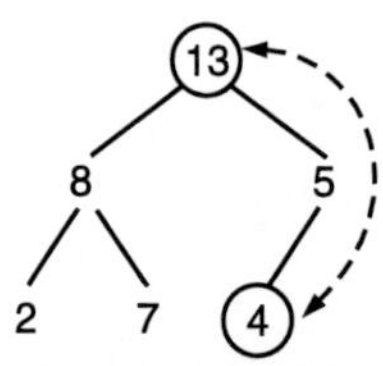

Fig. 13.24.

Exchange A[1] and A[6] and now size = 6 – 1 = 5.

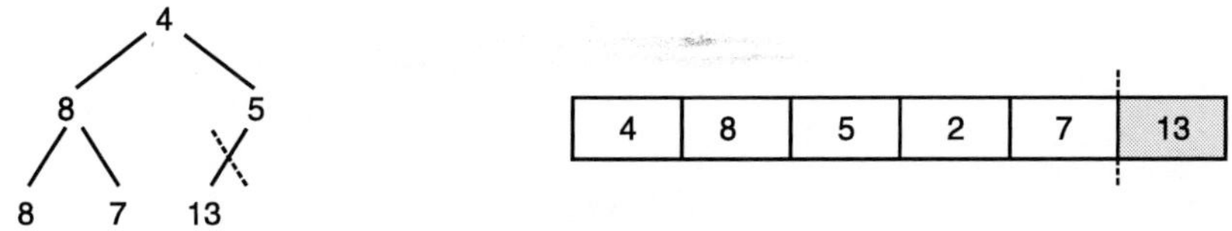

Fig. 13.25.

Again, call MAX-HEAPIFY (A, 1) we get

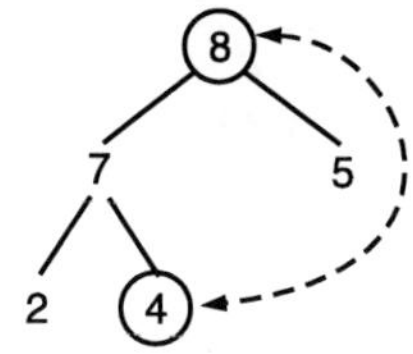

Fig. 13.26.

Exchange A[1] and A[5] and now size = 5 – 1 = 4.

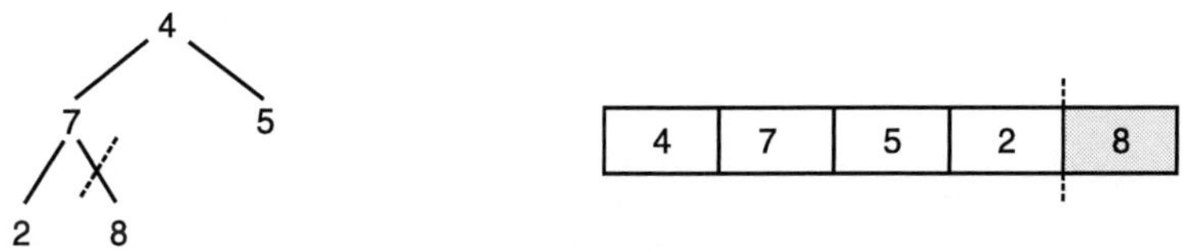

Fig. 13.27.

Again, call MAX-HEAPIFY (A, 1) we get

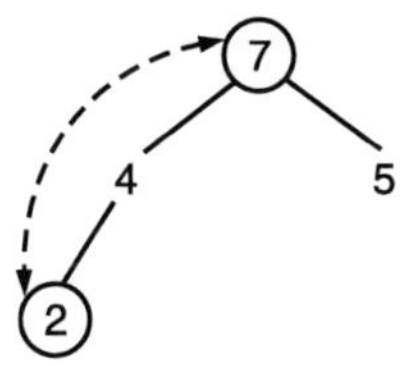

Fig. 13.28.

Exchange A[1] and A[4] and size = 4 – 1 = 3.

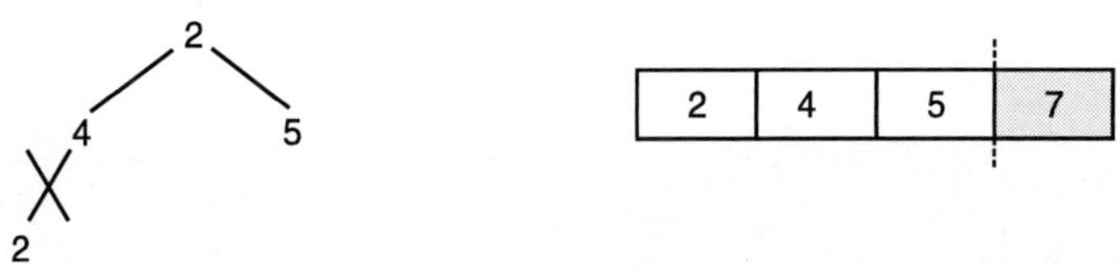

Fig. 13.29.

Call MAX-HEAPIFY (A, 1) we get

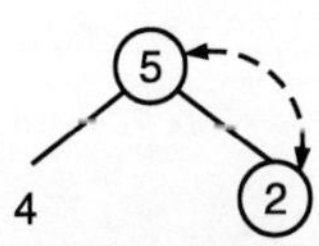

Fig. 13.30.

Exchange A[1] and A[3] and size = 3 – 1 = 2.

Fig. 13.31.

Call MAX-HEAPIFY (A, 1) we get

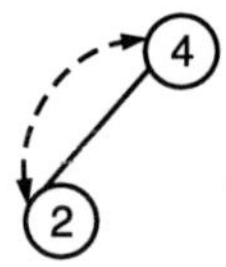

Fig. 13.32.

Exchange A[1] and A[2] and size = 2 – 1 = 1.

Fig. 13.33.

Thus the sorted array

2	4	5	7	8	13	17	20	25

Program /* Implement Heap Sort Using Arrays */

```
#include <stdio.h>
#include <conio.h>
int a[20], n;
void display( )
{
int i;
for(i=0; i<n; i++)
printf("%d", a[i]);
printf("\n");
}
void insert(int num, int loc)
{
int parent;
while(loc>0)
{
```

```
parent=(loc-1)/2;
if (num<=a[parent])
{
a[loc]=num;
return;
}
a[loc]=a[parent];
loc=parent;
}
a[0]=num;
}
void create_heap( )
{
int i;
for(i=0; i<n; i++)
insert(a[i], i);
}
void del_root(int last)
{
int left, right, temp;
i=0;
/*Since every time we have to replace root with last */
/*Exchange last element with the root */
temp = a[i];
a[i] = a[last];
a[last]=temp;
left = 2*i+1; /*left child of root */
right = 2*i+2; /*right child of root */
while (right < last)
{
if (a[i]>=a[left] && a[i] >=a[right])
return;
if (a[right]<=a[left])
{
temp = a[i];
a[i]=a[left];
a[left]=temp;
i=left;
}
else
```

```
{
temp=a[i];
a[i]=a[right];
a[right]=temp;
i=right;
}
left = 2*i+1;
right = 2*i+2;
}
if (left==last-1 && a[i]<a[left])
{
temp=a[i];
a[i]=a[left]
a[left]=temp;
}
}
void heap_sort( )
{
int last;
for(last=n-1; last>0; last--)
del_root(last);
}
void main( )
{
int i;
clrscr( );
printf("\nEnter number of elements :");
scanf("%d", &n);
for(i=0; i<n; i++)
{
printf("\nEnter element %d :", i+1);
scanf("%d", &a[i]);
}
printf("\nEnter list is : \n");
display( );
create_heap( );
printf("\nHeap is : \n");
display( );
heap_sort( );
printf("\nSorted list is : \n");
```

```
display( );
getch( );
}
```

13.11 EXTERNAL SORTING

The internal sorting algorithm applies to arrays that reside in the main memory. When large (voluminous) data have to be sorted they cannot be brought into the main memory at the same time. In this situation, the entire file may reside on some secondary storage device.

At the same time based on the availability of the storage space in the main memory, some portion of the data from the secondary storage device will be brought into the main memory that will be sorted and sent back to the secondary storage device. Then the next portion will be brought into the main memory, sorted, sent to the secondary memory, and so on, till the entire file stored in the secondary storage device is sorted. At the end all, the sorted files are merged to get the sorted data. Here, each portion is called *segment*. The time required to sort the file is more because the time required to transfer data from secondary memory to the main memory itself will take sometime. The most common algorithm used in external sorting is the **merge sort**.

Algorithm

The sequence of steps required to perform the merge sort operation on a file are as follows:

Steps

1. The given file is divided into 2 sub-files P1 and P2.
2. Then these files P1 and P2 are compared one pair of records at a time and merged and written to the output file M1 in the sorted order.
3. Then once again, one item is read from P1 and P2 and these two records are written in sorted order in another file called M2.
4. Thus, the alternate pair of sorted items are stored in files M1 and M2.
5. This process is repeated till the sub-files P1 and P2 come to an end.
6. The following steps have to be repeated twice.
   (*a*) Read one item from M1 and M2 and write them in sorted order in file P1.
   (*b*) Read one item from M1 and M2 and write then in sorted order in file P2.
7. Now the following steps have to performed 4 times.
   (*a*) Read one item from file P1 and P2 and write the sorted records in M1.
   (*b*) Read one item from file P1, P2 and write the sorted records in M2.
8. Stop.

Exercises

1. What do you mean by the term sorting ? List the different types of sorting techniques.
2. How the sorting algorithms can be compared ? Give a chart to show time and space complexity of various sorting methods.
3. Write a C Program to implement bubble sort.
4. Write a Program to sort the set of strings in an ascending order by using the bubble sort method.
5. Why quick sort is called as "quick" ?
6. Write a C Program to implement the quick sort.

7. Write a program to sort a set of strings using insertion sort method.
8. Explain selection sort algorithm.
9. Compare merge sort and insertion sort methods.
10. Differentiate between Heap-sort and Radix-sort.
11. Explain Heap-sort with an example.
12. Distinguish between internal sorting and external sorting.
13. Sort the following using heap sort.
    25 30 59 10 92 85 30
14. Explain external sorting methods.
15. Write a C program to implement merge sort.
16. Write algorithm to sorting a list of integer using heap-sort. Derive the expression for time complexity of heap sort.

Chapter

14 Searching and Hashing Techniques

Chapter Outline

- Introduction
- Linear or Sequential Searching
- Binary Search
- Interpolation Search
- Hash Tables
- Hashing
- Hash Functions
- Collision Resolution Techniques
- Hashing with Open Addressing
- Hashing with Chaining
- Rehashing
- Indexed Search Techniques

14.1 INTRODUCTION

Computer systems are often used to store large amounts of data from which individual records must be retrived according to some search criterion. Let DATA be a collection of data elements in memory and suppose a specific ITEM of information is given. Searching refers to the operation of finding the location LOC of ITEM in DATA or printing some message like ITEM does not appear there. The search is said to be successful if ITEM does appear in DATA and unsuccessful otherwise.

> The process of finding the location of a specific data item or record with a given key value or finding the locations of all records, which satisfy one or more conditions in a list, is called "**Searching**". If the item exists in the given list then search is said to be successful otherwise if the element if not found in the given list then search is said to be unsuccessful.

Searching falls into two categories:

1. **External searching**
2. **Internal searching**

External searching means searching the records using keys where there are many records, which reside in files stored on disks. *Internal searching* means searching the records using keys where there are less number of records residing entirely within the computer's main memory.

There are many different searching algorithms. The algorithm that one chooses generally depends on the way information in DATA is arranged. Following are the three important searching techniques:

- **Linear or Sequential Searching**
- **Binary Searching**
- **Interpolation Search**

In data structures when we use the word **searching**, we actually refer to **internal searching**.

The complexity of searching algorithms is measured in terms of number *f*(*n*) of comparisons required to find ITEM in data where DATA contains *n* elements. The time required for a search operation depends on the complexity of the searching algorithm. Basically we have to consider three cases when we search for a particular element in the list.

1. **Best Case:** The best case is that in whcih the element is found during the first comparison.
2. **Worst Case:** The worst case is that in which the element is found only at the end *i.e.,* in the last comparison.
3. **Average Case:** The average case is that in which the element is found in comparisons more than best case but less than worst case.

14.2 LINEAR OR SEQUENTIAL SEARCHING

Suppose DATA is a linear array with *n* elements. Given no other information about DATA, The most intuitive way to search for a given ITEM in DATA is to compare ITEM with each element of DATA one-by-one. That is, first we test whether DATA (1) = ITEM, and then we test whether DATA (2) =ITEM, and so on. This method, which traverses DATA sequentially to locate ITEM, is called linear search or sequential search.

In linear search, each element of an array is read one-by-one sequentially and it is compared with the desired element. A search will be unsuccessful if all the elements are read and the desired element is not found.

Linear search is the least efficient search technique among the quantity dependent search techniques. This technique should be chosen for searching the records when the records are stored without considering the order or when the storage medium lacks the direct access facility. Some important points are:

- It is the simplest way for finding an element in a list.
- It searches the element sequentially in a list, no matter whether list is sorted or unsorted.
  - In case of sorted list in ascending order, the search is started from 1st element and continued until the desired element is found or the element whose value is greater than the value being searched.
  - In case of sorted list in descending order, the serach is started from 1st element and continued until the desired element is found or the element whose value is smaller than the value being searched.
  - If the list is unsorted searching started from 1st location and continued until the element is found or the end of the list is reached.

Drawbacks

- It is a very slow process.
- It is used only for small amount of data.
- It is a very time consuming method.

Algorithm

Let A be an array of n elements, A (1), A (2), A (3), ... A (n). "data" is the element to be searched. Then this algorithm will find the location "loc" of data in A. Set loc = – 1, if the search is unsuccessful.

1. Input an array A of n element and "data" to be searched and initialise loc = – 1.
2. Initialise i = 0; and repeat through step 3 if (i < n) by incrementing i by one.
3. If (data = A [i])
   (a) loc = i
   (b) GOTO step 4
4. If (loc > 0)
   Print "data is found and searching is successful"
   Else
   Print "data is found and searching is unsuccessful"
5. Exit

14.2.1 *C* Implementation of Linear Search

```
#include<stdio.h>
#include<conio.h>
void main()
{
char ch;
int arr[50],n,i,item;
clrscr();
printf("\nHow many elements you want to enter in the array:");
scanf("%d",&n);
for(i=0;i < n;i++)
{
printf("\nEnter element %d:",i+1);
scanf("%d",&arr[i]);
}
printf("\n\nPress any key to continue.....");
getch();
do
{
clrscr();
printf("\nEnter the element to be searched:");
scanf("%d",&item);
for(i=0;i < n;i++)
{
if (item == arr[i])
{
printf("\n%d found at position %d\n",item,i+1);
break;
}
```

```
}
if (i == n)
printf("\nItem %d not found in array\n",item);
printf("\n\nPress (Y/y) to continue: ");
fflush(stdin);
scanf("%c",&ch);
}while(ch == 'Y' || ch == 'y');
}
```

14.2.2 Complexity/Efficiency of Linear Search

The algorithm is called linear search because it's complexity/efficiency can be expressed as a linear function *i.e.*, the number of comprisons to find a target increases linearly as the size of the data. Linear search provide complexity for finding an element in an array because linear search is a step-by-step process, in which specific element is compared with each element of array.

Best Case

In the best case, the desired element is present in the first position of the array, *i.e.*, only one comparison is made. So $T(n) = O(1)$.

Average Case

Here we asume that ITEM does appear, and that is equally likely to occur at any position in the array. Accordingly the number of comparisons can be any of the number 1, 2, 3, ..., n and each number occurs with the probability p = 1/n. Then

$$
\begin{aligned}
T(n) &= 1 . (1/n) + 2 . (1/n) + 3 . (1/n) ..... + n . (1/n) \\
&= (1 + 2 + 3 + ... + n) . (1/n) \\
&= n . (n + 1)/2. (1/n) \\
&= (n + 1)/2 \\
&= O((n + 1)/2)
\end{aligned}
$$

Worst Case

Clearly the worst case occurs when ITEM is the last element is the array or is not there at all. In either situation we have

$$T(n) = n + 1$$

Accordingly T(n) = O(n + 1) is the worst case complexity of the linear search algorithm.

Model	**Number of Comparisons (for *n* = 100000)**	**Comparisons as a function of *n***
Best case (fewest comparisons)	1 (target is first item)	1
Worst case (most comparisons)	100000 (target is last item)	*n*
Average case (average number of comparisons)	50000 (target is middle item)	*n*/2

14.3 BINARY SEARCH

If we place our items in an array and sort them in either ascending or descending order on the key first, then we can obtain much better performance with an algorithm called **binary search.**

Binary search is an extremely efficient algorithm when it is compared to linear search. In binary search, we first compare the key with the item in the middle position of the array. If there's a match, we can return immediately. If the key is less than the middle key, then the item sought must lie in the lower half of the array; if it's greater then the item sought must lie in the upper half of the array. So we repeat the procedure on the lower (or upper) half of the array.

14.3.1 General Idea About Binary Search Algorithm

1. Find the middle element of the array (*i.e.*, *n*/2 is the middle element if the array or the sub-array contains *n* elements).
2. Compare the middle element with the data to be searched, then there are following three cases.
   (*a*) If it is a desired element, then search is successful.
   (*b*) If it is less than desired data, then search only the first half of the array, *i.e.*, the elements which come to the left side of the middle element.
   (*c*) If it is greater than the desired data, then search only the second half of the array, *i.e.*, the elements which come to the right side of the middle element.

Repeat the same step until an element is found or exhaust the search area.

Algorithm

1. Input an array A of n elements an "data" to be sorted.
2. LB = 0, UB = n; mid = int ((LB + UB)/2)
3. Repeat step 4 and 5 while (LB <= UB) and (A [mid] ! = data)
4. If (data < A [mid])
   UB = mid – 1
   else
   LB = mid + 1
5. Mid = int ((LB + UB)/2)
6. If (A [mid] == data)
   Print "the data is found"
   else
   Print "the data is not found"
7. Exit

14.3.2 C Implementation of Binary Search

```
#include<stdio.h>
#include<conio.h>

void main()
{
char ch;
```

```
int arr[20],start,end,mid,n,i,data;
clrscr();
printf("\nHow many elements you want to enter in the array:");
scanf("%d",&n);
for(i=0; i < n; i++)
{
printf("\nEnter element %d:",i+1);
scanf("%d",&arr[i]);
}
printf("\n\nPress any key to continue...");
getch();
do
{
clrscr();
printf("\nEnter the element to be searched:");
scanf("%d",&data);
start=0;
end=n - 1;
mid=(start + end)/2;
while(data!=arr[mid] && start <=end)
{
if(data > arr[mid])
start=mid+1;
else
end=mid-1;
mid=(start+end)/2;
}
if(data==arr[mid])
printf("\n%dfound at position %d\n",data,mid + 1);
if(start>end)
printf("\n%d not found in array\n",data);
printf("\n\n Press <Y or y> to continue:");
fflush(stdin);
scanf("%c",&ch);
}while(ch == 'Y' || ch == 'y');
}
```

EXAMPLE 14.1. **Suppose an array A [] whose elements are**

5	8	15	25	30	40	55
0	1	2	3	4	5	6

Suppose data = 30

Solution: Following steps are used if we use binary search to search *a* data = 30 from the above array.

Here n = 6

So, LB = 0, UB = 6

$$\text{mid} = \frac{(0+6)}{2} = 3 \text{ i.e., A [mid]} = \text{A [3]} = 25$$

Since, A [3] < data *i.e.,* 25 < 30.

So, reintialise the variable LB as LB = mid + 1

i.e., LB = 3 + 1 = 4

Now, $\text{mid} = \frac{4+6}{2} = 5$

i.e., A[mid] = A [5] = 40

Now, data < A [mid] *i.e.,* 30 < 40

So, reintialise the variable UB = mid – 1 = 5 – 1 = 4

Now, LB = 4 and UB = 4

So, $\text{mid} = \frac{4+4}{2} = \frac{8}{2} = 4$ i.e., A[4] = 30

since (A[mid] = data) *i.e.,* 30 = 30

Thus searching is successful and 30 is found in the given array.

Time Complexity

In binary serach in each comparison the size of the search area is reduced by half. So, the efficiency of the binary serach method at the worst case is $\log_2 n + 1$, *i.e.,* $O(\log_2 n + 1)$ where n is the total number of items that will be used for the binary search.

Drawbacks

The drawbacks of the binary search are:

(*a*) Binary search requires that the items in the array should be sorted.

(*b*) Binary search cannot be used where there are many insertions or deletions.

14.4 INTERPOLATION SEARCH

Another technique for searching an ordered array is called *interpolation search*. This method is even more efficient than binary search, if the elements are uniformly distributed (or sorted) in an array A.

In a binary search, the search space is always divided in two to gaurantee logarithmic time; however, when we search for "Ajay" in the phone book, we don't start in the middle–we start toward the front and work from there. That is the idea of an interpolation search. Instead of cutting the search space by a fixed half, we cut it by an amount that seems most likely to succeed. This amount is determined by interpolation.

Consider an array A of n elements and the elements are uniformly distributed (or the elements are arranged in a sorted array). Initially, as in binary serach, low is set to 0 and high is set to n – 1.

Now we are searching an element key in an array between A [low] and A [high]. The key would be expected to be at mid, which is an approximately position.

mid = low + (high – low) × ((key – A [low]/A[high] – A[low]))

If key is lower than A [mid], reset high to mid – 1; else reset low to mid + 1. Repeat the process until the key has found or low > high.

Algorithm

Suppose *A* be array of sorted elements and key is the elements to be searched and *low* represents the lower bound of the array and *high* represents higher bound of the array.

1. Input a sorted array of n elements and the key to be searched.
2. Initialise low = 0 and high = n – 1
3. Repeat the steps 4 through 7 until if (low < high)
4. Mid = low + (high – low) × ((key – A[low])/(A[high] – A[low]))
5. If (key == A[mid])
   print "key is found"
6. If (key < A[mid])
   high = mid – 1
   else if (key > A[mid])
   low = mid + 1
   else
   print "The key is not in the array and Exit"
7. STOP.

EXAMPLE 14.2. **Let us consider 7 numbers as**

4, 15, 20, 35, 45, 55, 65

Solution: Suppose we are searching 45 from the given array

Here n = 7, key = 45, low = 0, high = n – 1 = 6

$$\text{mid} = 0 + (6 - 0) \times \{(45 - 4)/(65 - 4)\}$$
$$= 0 + 6 \times (41/61) = 4.032$$

Consider only the integer part of the mid i.e., mid = 4.

if (key == A[mid])

i.e., key == A[4]

45 = 45

Thus, key is found.

14.5 HASH TABLES

Hash tables support one of the most efficient types of searching: *hashing*. Fundamentally, a hash table consists of an array in which data is accessed via a special index called a *key*. The primary idea behind a hash table is to establish a mapping between the set of all possible keys and positions in the array using a *hash function*. A hash function accepts a key and returns its *hash coding, hash value*. Keys vary in type, but hash codings

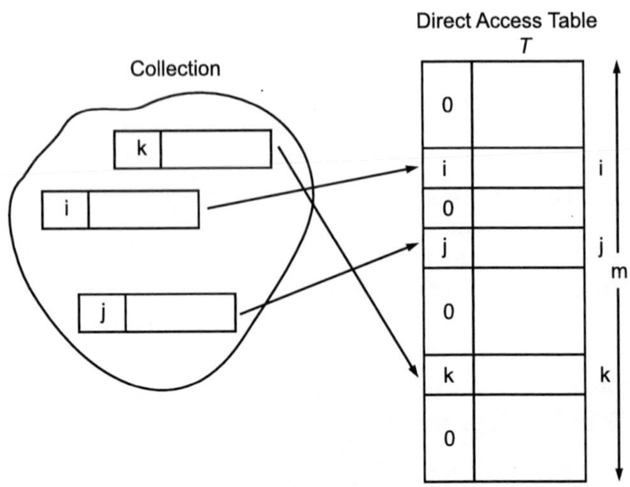

Fig. 14.1.

are always integers. Since both computing a hash value and indexing into an array can be performed in constant time, the beauty of hashing is that we can use it to perform constant time searches. When a hash function can guarantee that no two keys will generate the same hash coding, the resulting hash table is said to be *directly addressed.*

This is ideal, but direct addressing is rarely possible in practice.

Typically, the number of entries in a hash table is small relative to the universe of possible keys. Consequently, most hash functions map some keys to the same position in the table. When two keys map to the same position, they *collide.* A good hash function minimises collisions, but we must still be prepared to deal with them.

14.5.1 Applications of Hash Tables

Some applications of hash tables are:

1. **Database Systems.** Specifically, those that require efficient random access. Generally, database systems try to optimise between two types of access method : sequential and random. Hash tables are an important part of efficient random access because they provide a way to locate data in a constant amount of time.
2. **Symbol Tables.** The tables used by compilers to maintain information about symbols from a program. Compilers access information about symbols frequently. Therefore, it is important that symbol tables be implemented very efficiently.
3. **Tagged Buffers.** A mechanism for storing and retrieving data in a machine-independent manner. Each data member resides at a fixed offset in the buffer. A hash table is stored in the buffer so that the location of each tagged member can be ascertained quickly. One use of a tagged buffer is sending structured data across a network to a machine whose byte ordering and structure alignment may not be same as the original host's. The buffer handles these concerns as the data is stored and extracted member-by-member.
4. **Data Dictionaries.** Data structures that support adding, deleting, and searching for data. Although the operations of a hash table and a data dictionary are similar, other data structures may be used to implement data dictionaries. Using a hash table is particularly efficient.

14.6 HASHING

We have seen different searching techniques where search time is basically dependent on the number of elements. Sequential search, binary search and all the search trees are totally dependent on number of elements and so many key comparisons are also involved. For example,

Unsorted sequential array implementation

- **insert:** add to back of arrays; O (1)
- **find:** search through the keys one at a time, potentially all of the keys; O(n)
- **remove:** find and replace removed node with last node; O(n)

Sorted sequential array implementation

- **insert:** add in sorted order; O(n)
- **find:** binary search $O(\log_2 n)$
- **remove:** find, remove node and shuffle down; O(n)

Linked list (unsorted or sorted) implementation

- **insert:** add to front; O(1) or O(n) for a sorted list

- **find:** search through potentially all the keys, one at a time; O(n) still O(n) for a sorted list
- **remove:** find, remove using pointer alternations; O(n)

AVL tree implementation

- An AVL tree, ordered by key
- **insert:** a standard insert; $O(\log_2 n)$
- **find:** a standard find, $O(\log n)$
- **remove:** a standard remove; $O(\log_2 n)$

Now our need is to serach the element in constant time and less key comparisons should be involved. Suppose all the elements are in array size *N*. Let us take all the keys are unique and in the range 0 to $N - 1$. Now we are storing the records in array based on the key where array index and keys are same. Now we can access the record in constant time and there no key comparisons are involved. Let us take the 5 records where keys are:

9 4 6 7 2

It will be stored in array as:

	[0]	[1]	[2]	[3]	[4]	[5]	[6]	[7]	[8]	[9]
arr			2		4		6	7		9

Here we can see the record which has key value '2' can be directly accessed through array index arr[2].Now the idea that comes in picture is *hashing* where we will convert the key into array index and put the records in array and in the same way for searching the record, convert the key into array index and get the records from array. This can be described as:

For **storing record**:

Key
↓
Generate array index
↓
Store the records on that array index

For **accessing record**:

Key
↓
Generate array index
↓
Get the record from that array index

The generation of array index uses **hash function**, which converts the keys into array index and the array which supports hashing for storing record or searching record called **hash table**.

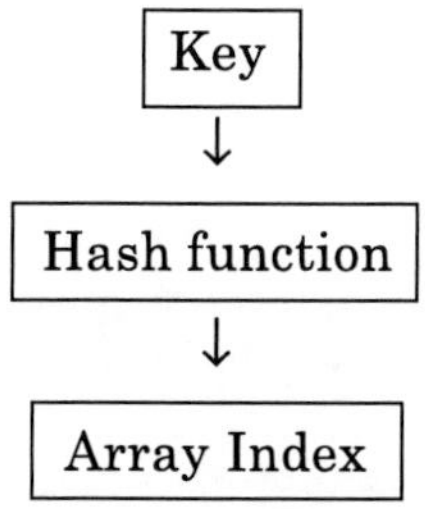

So we can say each key is mapped on a particular array index through hash function. If each key is mapped on a unique hash table address then this situation is ideal situation but there may be possibility that our hash function is generating same hash table address for different keys, this situation is called **collision**. So our hash function should generate unique address but it is not possible, only one thing we can do is to take the good hash function for minimizing collision.

14.7 HASH FUNCTIONS

Hash functions is a function which, when applied to the key, produced an integer which can be used as an address in a hash table. The intent is that elements will be relatively randomly and uniformly distributed. **Perfect Hash Function** is a function which, when applied to all the members of the set of items to be stored in a hash table, produces a unique set of integers within some suitable range. Such function produces no collisions. **Good Hash Function** minimises collissions by spreading the elements uniformly throughout the array.

There is no magic formula for the creation of the hash function. It can be any mathematical transformation that produces a relatively random and unique distribution of values within the address space of the storage.

14.7.1 Characteristics of a Good Hash Function

There are *four* main characteristics of a good hash function:

- The hash value is fully determined by the data being hashed.
- The hash function uses all the input data.
- The hash function "uniformly" distributes the data across the entire set of possible hash values.
- The hash function generates very different hash values for similar strings.

Let's examine why each of these is important:

Rule 1 If something else besides the input data is used to determine the hash, then the hash value is not as dependent upon the input data, thus allowing for a worse distribution of the hash values.

Rule 2 If the hash function doesn't use all the input data, then slight variations to the input data would cause an inappropriate number of similar hash values resulting in too many collisions.

Rule 3 If the hash function does not uniformly distribute the data across the entire set of possible hash values, a large number of collisions will result, cutting down on the efficiency of the hash table.

Rule 4 In real world applications, many data sets contain very similar data elements. We would like these data elements to still be distributable over a hash table.

However, finding a perfect hash function that works for a given data set can be extremely time consuming and very often it is just impossible. Therefore, we must live with collissions and learn how to handle them.

14.7.2 Types of Hash Functions

Some common Hash functions are:

1. Division Method

In the division method for creating hash functions, we map a key k into one of m slots by taking the remainder of k divided by m. That is, the hash function is:

$$h(k) = k \bmod m \quad \text{or} \quad h(k) = k \bmod m + 1$$

For example, if the hash table has size m = 12 and the key is k = 100, then h(k) = 4. Since it requires only a single division operation, hashing by division is quite fast.

When using the division method, we usually avoid certain values of m. For example, m should not be a power of 2, since if $m = 2^p$, then h(k) is just be p lowest-order bits of k. Powers of 10 should be avoided if the application deals with decimal number of keys.

2. Multiplication Method

The **multiplication method** for creating hash functions operates in two steps: First, we multiply the key k by a constant A in the range 0 < A < 1 and extract the fractional part of kA. Then, we multiply this value by m and take the floor of the result. In short, the hash function is:

$$h(k) = \lfloor m(k\,A \bmod 1) \rfloor,$$

where "k A mod 1" means the fractional part of kA, that is, $k\,A - \lfloor kA \rfloor$.

This method is not affected too much by the choice of m (which is the size of the hash table), but the value of A will impact the effectiveness of this method. **'Knuth'** has suggested in his study that the following value of A is likely to work reasonably

$$A \approx \frac{(\sqrt{5}-1)}{2} = .6180339887 \text{ ..... is a good choice.}$$

3. Midsquare Method

In midsquare method we square the key, after getting number we take some digits from the middle of that number as an address. Let us take some 4 digit number as a key:

1337 1273 1391 1026

Now square these keys:

1337	1273	1391	1026
1787569	1620529	1934881	1052676

Now take the 3rd and 4th digit from each number and that will be the hash adress of these keys. Let us assume here table size is 1000. So the hash address for these keys will be 75, 05, 48 and 26.

4. Folding Method

One easiest way to compute the key is break the key into pieces, add them and get the hash address. Let us take some 8 digits key:

82394561, 87139465, 83567271, 85943228

Now chop them in pieces of 3, 2 and 3 digits and add them. So the address will be:

82394561 = 823 + 94 + 561 = 1478

87139465 = 871 + 39 + 465 = 1375

$$83567271 = 835 + 67 + 271 = 1173$$
$$85943228 = 859 + 43 + 228 = 1130$$

Now truncate them upto the digit based on the size of hash table. Suppose the table size is 1000, so the hash address can be from 0 to 999. So we will truncate here the higher digit of number. Now the hash address of keys will be as:

$$H(82394561) = 478$$
$$H(87139465) = 375$$
$$H(83567271) = 173$$
$$H(85943228) = 130$$

14.7.3 Hash Function for Strings

Most of the places strings are used as keys. It can be alphabetic or alphanumeric. Best example is the English word distionary. Every character has some ASCII value that can be used for calculation in generating hash key value and that value can be used with modulus operation for mapping with hash table. Suppose key has alphabetic character and 8 character long then maximum number of combinations is (the total number of words in Englsh for 8 characters are $26^8 =$) 208827064576. But practically English language has not this much of word which are of 8 characters. So taking actual table size is a big problem and most of the table space will be free. Let us take table size is M(97) and key is "suresh". We can add the ASCII value of each character and then we can apply modulus operation on this value as:

$$\text{Suresh} = s + u + r + e + s + h$$
$$= 115 + 117 + 114 + 101 + 115 + 104$$

After modulus operation:

$$H(\text{suresh}) = 666 \% 97 = 84$$

So the key "suresh" can be mapped on the 84th position in the hash table.

Second method we can apply as, multiply the each character ASCII value by 127 and add all the values then do the modulus operation. Let us take the key "suresh" and table sze 997.

$$\text{Suresh} = s \times 127 + u \times 127 + r \times 127 + e \times 127 + s \times 127 + h \times 127$$
$$= 115 \times 127 + 117 \times 127 + 114 \times 127 + 101 \times 127 + 115 \times 127 + 104 \times 127$$
$$= 84582$$

After modulus operation:

$$H(\text{suresh}) = 84582 \% 997$$
$$= 834$$

Here we use the value 127 for multiplication because ASCII character has maximum value 127.

Here the key "suresh" can be mapped on the 834th position in the table.

Here we can see that if the key is long then it will take time in calculation. So we have a need to take some approach for this calculation. We can first add the ASCII values of all characters and then multiplication can take place.

$$H(\text{suresh}) = (115 + 117 + 114 + 101 + 115 + 104) * 127 \% 997$$
$$= 666 * 127 \% 997$$
$$= 84582 \% 997$$
$$= 834$$

14.8 COLLISION RESOLUTION TECHNIQUES

Suppose we want to add a new record *R* with key *k* to our file *F*, but suppose the memory location address *H*(*k*) is already occupied. This situation is called **collision.** That is, a collision occurs when more than one keys map to same hash value in the hash table. The following ways resolve the collisions:

1. ***Collision resolution by open addressing.***
2. ***Collision resolution by separate chaining.***

The performance of these methods depends on **load factor** *i.e.*, ratio $\lambda = n/m$. This is the ratio of the number *n* of keys in *k* to the number *m* of hash addresses.

The efficiency of a hash function with a collision resolution procedure is measured by the average number of **probes** (key comparisions) needed to find the location of the record with a given key *k*. The efficiency depends mainly on the load factor λ and the following conditions of search.

$S(\lambda)$: average number of probes for a successful search.

$U(\lambda)$: average number of probes for an unsuccessful search.

14.9 HASHING WITH OPEN ADDRESSING

In **open addressing**, all elements are stored in the hash table itself. That is, each table entry contains either an element of the dynamic set or NIL. When searching for an element, we systematically examine table slots until the desired element is found or it is clear that the element is not in the table. Thus, in open addressing, the load factor λ can never exceed 1.

The **advantage** of open addressing is that it avoids pointers altogether. Instead of following pointers, we compute the sequence of slots to be examined. The extra memory freed by not storing pointers provides the hash table with a larger number of slots for the same amount of memory, potentially yielding fewer collisions and faster retrieval.

The process of examining the locations in the hash table is called '**Probing**'.

To perform insertion using open addressing, we successively examine, or probe, the hash table until we find an empty slot in which to put the key.

HASH-INSERT (T, k)

```
i ← 0
repeat j ← h(k, i)
     if T[j] = NIL
          then T[j] ← k
          return j
     else i ← i + 1
until i = m
error "hash table overflow"
```

The algorithm for searching for key k probes the same sequence of slots that the insertion algorithm examined when key k was inserted. Therefore, the search can terminate (unsuccessfully) when it finds an empty slot, since k would have been inserted there and not later in its probe sequence. (Note that this argument assumes that keys are not deleted from the hash table). The procedure HASH-SEARCH takes as input a hash table T and a key k, returning j if slot j is found to contain key k, or NIL if key k is not present in table T.

HASH-SEARCH (T, K)

```
i ← 0
repeat j ← h(k, i)
    if T(j) = j
        then return j
            i ← i + 1
until T(j) = NIL or i = m
return NIL
```

Deletion from an open-address hash table is difficult. When we delete a key from slot i, we cannot simply mark that slot as empty by storing NIL in it. Doing so might make it impossible to retrive any key k during whose insertion we had probed slot i and found it occupied. One solution is to mark the slot by storing in it the special value DELETED instead of NIL. We would then modify the procedure HASH-SEARCH so that it keeps on looking when it sees the value DELETED, while HASH-INSERT would treat such a slot as if it were empty so that a new key can be inserted. When we do this, though, the search times are no longer dependent on the load factor λ, and for this reason chaining is mode commonly selected as a collision resolution technique when keys must be deleted.

Three techniques are commonly used to compute the probe sequences required for open adressing: **linear probing, quadratic probing, and double hashing.**

14.9.1 Linear Probing

Given an ordinary hash function h′: U [0, 1, ..., m – 1], the method of **linear probing** uses the hash function.

$$h(k, i) = (h'(k) + i) \bmod m$$

where 'm' is the size of the hash table and h′ (k) = k mod m (basic hash function). For i = 0, 1, ..., m – 1. Given key k, the first slot probed is T [h′ (k)]. We next probe slot T [h′ (k) + 1], and so on upto slot T[m – 1]. Then we wrap around to slots T[0], T[1], ..., until we finally probe slot T[h′ (k) – 1]. Since the initial probe position determines the entire probe sequence, only m distinct probe sequences are used with linear probing.

EXAMPLE 14.3. Consider inserting the key 26, 37, 59, 76, 65, 86 into a hash-table of size m = 11 using linear probing, consider the primary hash function is h′ (k) = k mod m.

Solution: Initial state of the hash table

	0	1	2	3	4	5	6	7	8	9	10
T	/	/	/	/	/	/	/	/	/	/	/

1. Insert 26. We know h(k, i) = [h′ (k) + i] mod m

Now h(26, 0) = [26 mod 11 + 0] mod 11 = (4 + 0) mod 11 = 4 mod 11 = 4

Since T[4] is free, insert key 26 at this place.

2. Insert 37. Now h(37, 0) = [37 mod 11 + 0] mod 11 = [4 + 0] mod 11 = 4

Since T[4] is not free, the next probe sequence is computed as

h(37, 1) = [37 mod 11 + 1] mod 11 = [4 + 1] mod 11 = 5 mod 11 = 5

T[5] is free, insert key 37 at this place.

3. Insert 59. Now h(59, 0) = [59 mod 11 + 0] mod 11 = [4 + 0] mod 11 = 4

T[4] is not free, so the next probe sequence is computed as

$$h[59, 1] = [59 \bmod 11 + 1] \bmod 11 = 5$$

T[5] is also not free, so the next probe sequence is computed.

$$h(59, 2) = (59 \bmod 11 + 2) \bmod 11 = 6 \bmod 11 = 6$$

T[6] is free. Insert key 59 at this place.

4. Insert 76. h(76, 0) = (76 mod 11 + 0) mod 11 = (10 + 0) mod 11 = 10

T[10] is free, insert key at this place.

5. Insert 65. h(65, 0) = (65 mod 11 + 0) mod 11 = (10 + 0) mod 11 = 10

T[10] is occupied, the next probe sequence is computed as

$$h(65, 1) = (65 \bmod 11 + 1) \bmod 11$$
$$= (10 + 1) \bmod 11 = 11 \bmod 11 = 0$$

T[0] is free, insert key 65 at this place.

6. Insert 86. h(86, 0) = (86 mod 11 + 0) mod 11 = 9 mod 11 = 9

T[9] is free, insert key 86 at this place.

Thus,

0	1	2	3	4	5	6	7	8	9	10
65	/	/	/	26	37	59	/	/	86	76

/* Insert keys into an array with linear probing technique of collision resolution technique. */

```
#include<stdio.h>
#include<conio.h>
#define MAX 10

int a[MAX];
void lp(int, int[]);
void lpsr(int key, int a[MAX]);
void display(int a[MAX]);

void main( )
{
int i, key,ch ;
clrscr();
for(i = 0;i<MAX;i++)
     a[i] = '\0';
do{
     printf("\n\n Program for insertion/searching keys with linear probing ");
     printf( "\n ************************************************\n\n");
     printf("\n 1. Insert keys ");
     printf("\n 2. Search key ");
```

```
        printf("\n 3. Display keys ");
        printf("\n 4. Exit ");
        printf("\n Select operation ");
        scanf("%d",&ch);
        switch(ch)
        {
        case 1: do{
            printf("\n Enter key value [type -1 for termination] ");
            scanf("%d",&key);
            if (key != -1)
                lp(key,a);
            }while(key!=-1);
            display(a);
            break;
        case 2:
            printf("\n Enter search key value ");
            scanf("%d",&key);
            lpsr(key,a);
            break;
        case 3:
            display(a);
            break;
        }
        }while(ch!=4);
    }

    /* function lp find a location for key and insert it */
    void lp(int key, int a[MAX])
    {
    int loc;
    loc = key % MAX;
    while (a[loc] != '\0')
        loc = ++ loc % MAX;
    a[loc] = key;
    }

    /* function lpsr find a location for a key */
    void lpsr(int key, int a[MAX])
    {
```

```
int loc;
loc = key % MAX;
while ((a[loc] != key) && (a[loc] != '\0'))
     loc= ++loc%MAX;
if (a[loc] != '\0')
     printf("\n Search successful at index %d",loc);
else
     printf("\n Search unsuccessful ");
}

void display(int a[MAX])
{
int i;
printf("\n List of keys ('0' indicate that the location is empty): \n");
for (i = 0;i<MAX;i++)
     printf(" %d",a[i]);
}
```

One main disadvantage of linear probing is that records tend to '**Cluster**' *i.e.,* appear next to one another, when the load factor is greater than 50%. Such clustering substantially increases the average search time for a record. Two techniques to minimise clustering are as follows:

14.9.2 Quadratic Probing

Suppose a record R with key k has the hash address H(k) = h then instead of searching the locations with addresss h, h + 1, h + 2, ... We linarly search the locations with addresses.

$$h, h + 1, h + 4, h + 9, \ldots, h + i^2, \ldots$$

Quadratic probing uses a hash function of the form

$$\boxed{h(k, i) = (h'(k) + c_1 i + c_2 i^2) \bmod m}$$

where (as in linear probing) h' is an auxiliary hash function, c_1 and $c_2 \neq 0$ are auxiliary constants, and $i = 0, 1, \ldots, m - 1$. The initial position probed is $T(h'(k))$; later positions probed are offset by amounts that depend in a quadratic manner on the probe number i. This method works much better than linear probing, but to make full use of the hash table, the values of c_1, c_2 and m are constrained.

EXAMPLE 14.4. Consider inserting the key 76, 26, 37, 59, 21, 65 into a hash table of size m = 11 using quadratic probing with $c_1 = 1$ and $c_2 = 3$. Further consider that the primary hash function is $h'(k) = k \bmod m$.

Solution: For quadratic probing, we have

$$h(k, i) = [k \bmod m + c_1 i + c_2 i^2] \bmod m$$

0	1	2	3	4	5	6	7	8	9	10
/	/	/	/	/	/	/	/	/	/	/

This is the initial state of the hash table.

Here $c_1 = 1$, $c_2 = 3$

$$h(k, i) = [k \bmod m + i + 3i^2] \bmod m$$

1. Insert 76.

$$h(76, 0) = (76 \bmod 11 + 0 + 3 \times 0) \bmod 11$$
$$= (10 + 0 + 0) \bmod 11 = 10$$

T[10] is free, insert the key 76 at this place.

2. Inset 26.

$$h(26, 0) = (26 \bmod 11 + 0 + 3 \times 0) \bmod 11 = (4 + 0 + 0) \bmod 11 = 4$$

T[4] is free, insert the key 26 at this place.

3. Insert 37.

$$h(37, 0) = (37 \bmod 11 + 0 + 3 \times 0) \bmod 11 = (4 + 0 + 0) \bmod 11 = 4$$

T[4] is not free, so next probe sequence is computed as:

$$h(37, 1) = (37 \bmod 11 + 1 + 3 \times 1^2) \bmod 11$$
$$= (4 + 1 + 3) \bmod 11 = 8 \bmod 11 = 8$$

T[8] is free. Insert the key 37 at this place.

4. Insert 59.

$$h(59, 0) = (59 \bmod 11 + 0 + 3 \times 0) \bmod 11$$
$$= (4 + 0 + 0) \bmod 11 = 4 \bmod 11 = 4$$

T[4] is not free, so next probe sequence is computed as:

$$h(59, 1) = (59 \bmod 11 + 1 + 3 \times 1^2) \bmod 11$$
$$= (4 + 1 + 3) \bmod 11 = 8 \bmod 11 = 8$$

T[8] is also not free, so the next probe sequence is computed as:

$$h(59, 2) = (59 \bmod 11 + 2 + 3 \times 2^2) \bmod 11$$
$$= (4 + 2 + 12) \bmod 11 = 18 \bmod 11 = 7$$

T[7] is free, insert key 59 at this place.

5. Insert 21.

$$h(21, 0) = (21 \bmod 11 + 0 + 3 \times 0) \bmod 11$$
$$= (10 + 0 + 0) \bmod 11 = 10 \bmod 11 = 10$$

T[10] is not free, the next probe sequence is computed as:

$$h(21, 1) = (21 \bmod 11 + 1 + 3 \times 1^2) \bmod 11$$
$$= (10 + 1 + 3) \bmod 11 = 14 \bmod 11 = 3$$

T[3] is free, so insert key 21 at this place.

6. Insert 65.

$$h(65, 0) = (65 \bmod 11 + 0 + 3 \times 0) \bmod 11$$
$$= (10 + 0 + 0) \bmod 11 = 10 \bmod 11 = 10$$

Since, T[10] is not free, the next probe sequence is computed as:

$$h(65, 1) = (65 \bmod 11 + 1 + 3 \times 1^2) \bmod 11$$
$$= (10 + 1 + 3) \bmod 11 = 14 \bmod 11 = 3$$

T[3] is not free, so the next probe sequence is computed as:

$$h(65, 2) = (65 \bmod 11 + 2 + 3 \times 2^2) \bmod 11$$
$$= (10 + 2 + 12) \bmod 11 = 24 \bmod 11 = 2$$

T[2] is free, insert the key 65 at this place.

Thus, after inserting all keys, the hash table is:

0	1	2	3	4	5	6	7	8	9	10
/	/	65	21	26	/	/	59	37	/	76

/* Insert keys into an array with quadratic probing technique of collision resolution technique. */

```
#include<stdio.h>
#include<conio.h>
#define MAX 10

int a[MAX];
void qp(int , int[]);
void qpsr(int key, int a[MAX]);
void display(int a[MAX]);

void main()
{
int i,key,ch ;
clrscr();
for(i=0;i<MAX;i++)
     a[i] = '\0';
do{
     printf("\n\n Program for insertion/searching keys with quardratic probing ");
     printf("\n*************************************************\n\n");
     printf("\n 1. Insert keys");
     printf("\n 2. Search key");
     printf("\n 3. Display keys");
     printf("\n 4. Exit");
     printf("\n Select operation");
     scanf("%d",&ch);
     switch(ch)
     {
     case 1: do{
               printf("\n Enter key value [type -1 for termination]");
               scanf("%d",&key);
               if (key != -1)
                    qp(key,a);
```

```
            }while(key!= -1);
            display(a);
            break;
    case 2: printf("\n Enter search key value ");
            scanf("%d", &key);
            qpsr(key,a);
            break;
    case 3: display(a);
            break;
    }
    }while(ch!=4);
}

void qp(int key, int a[MAX])
{
int loc, i = 1;
loc = key % MAX;
while (a[loc] != '\0')
    {
    loc = (key % MAX + i*i) % MAX;
    i++;
    }
a[loc] = key;
}

/* Find a location for a key */
void qpsr(int key, int a[MAX])
{
int loc;
loc = key % MAX;
while ((a[loc] != key) && (a[loc] !=-1))
    loc = ++ loc % MAX;
if (a[loc] != '\0')
    printf("\n Search successful at index %d",loc);
else
    printf("\n Search unsuccessful ");
}
```

```
void display(int a[MAX])
{
int i;
printf("\n List of keys ('0' indicate that the location is empty): \n");
for (i = 0;i<MAX;i++ )
     printf ("%d",a[i]);
}
```

14.9.3 Double Hashing

Double hashing is one of the best methods available for open addressing because the permutations produced have many of the characteristics of randomly chosen permutations.

Double hashing uses a hash function of the form:

$$h(k, i) = (h_1 (k) + ih_2 (k)) \bmod m,$$

where h_1 and h_2 are auxiliary hash functions and m is the size of the hash table.

$h_1(k) = k \bmod m$ or $h_2 (k) = k \bmod m'$. Here m' is slightly less than m (say $m - 1$ or $m - 2$).

The initial position probed is $T(h_1 (k))$ successive probe positions are offest from previous positions by the amount $h_2(k)$ modulo m. Thus, unlike the case of linear or quadratic probing, the probe seqeunce here depends in two ways upon the key k, since the initial probe position, the offset, or both, may vary.

Double hashing represents an improvement over linear or quadratic probing in that $\theta(n^2)$ probe sequences are used, rather than $\theta(m)$, since each possible $(h_1(k), h_2(k))$ pair yields a distinct probe sequence, and as we vary the key, the initial probe position $h_1(k)$ and the offset $h_2(k)$ may vary imdependently. As a result, the performance of double hashng appears to be very close to the performance of the "ideal" scheme of uniform hashing.

EXAMPLE 14.5. Consider inserting the keys 76, 26, 37, 59, 21, 65 into a hash table of size m = 11 using double hashing. Consider that the auxiliary hash functions are $h_1 (k) = k \bmod 11$ and $h_2(k) = k \bmod 9$.

Solution: Initial state of hash table is:

	0	1	2	3	4	5	6	7	8	9	10
T	/	/	/	/	/	/	/	/	/	/	/

1. Insert 76.

$$h_1(76) = 76 \bmod 11 = 10$$
$$h_2(76) = 76 \bmod 9 = 4$$
$$h(76, 0) = (10 + 0 \times 4) \bmod 11 = 10 \bmod 11 = 10$$

T[10] is free, so insert key 76 at this place.

2. Insert 26.

$$h_1(26) = 26 \bmod 11 = 4$$
$$h_2(26) = 26 \bmod 9 = 8$$

$$h(26, 0) = (4 + 0 \times 8) \bmod 11 = 4 \bmod 11 = 4$$

T[4] is free, so insert key 26 at this place.

3. Insert 37.

$$h_1(37) = 37 \bmod 11 = 4$$
$$h_2(37) = 37 \bmod 9 = 1$$
$$h(37, 0) = (4 + 0 \times 1) \bmod 11 = 4 \bmod 11 = 4$$

T[4] is no free, the next probe sequence is computed as

$$h(37, 1) = (4 + 1 \times 1) \bmod 11 = 5 \bmod 11 = 5$$

T[5] is free, so insert key 37 at this place.

4. Insert 59.

$$h_1(59) = 59 \bmod 11 = 4$$
$$h_2(59) = 59 \bmod 9 = 5$$
$$h(59, 0) = (4 + 0 \times 5) \bmod 11 = 4 \bmod 11 = 4$$

Since, T[4] is not free, the next probe sequence is computed as:

$$h(59, 1) = (4 + 1 \times 5) \bmod 11 = 9 \bmod 11 = 9$$

T[9] is free, so insert key 59 at this place.

5. Insert 21.

$$h_1(21) = 21 \bmod 11 = 10$$
$$h_2(21) = 21 \bmod 9 = 3$$
$$h(21, 0) = (10 + 0 \times 3) \bmod 11 = 10 \bmod 11 = 10$$

T[10] is not free, the next probe sequence is computed as

$$h(21, 1) = (10 + 1 \times 3) \bmod 11 = 13 \bmod 11 = 2$$

T[2] is free, so insert key 21 at this place.

6. Insert 65.

$$h_1(65) = 65 \bmod 11 = 10$$
$$h_2(65) = 65 \bmod 9 = 2$$
$$h(65, 0) = (10 + 0 \times 2) \bmod 11 = 10 \bmod 11 = 10$$

T[10] is not free, the next probe sequence is computed as:

$$h(65, 1) = (10 + 1 \times 2) \bmod 11 = 12 \bmod 11 = 1$$

T[1] is free, so insert key 65 at this place.

Thus, after insertion of all keys the final hash table is:

0	1	2	3	4	5	6	7	8	9	10
/	65	21	/	26	37	/	/	/	59	76

EXAMPLE 14.6. Consider inserting the keys 10, 22, 31, 4, 15, 28, 17, 88 and 59 into a hash table of length m = 11 using open addressing with the primary hash function h'(k) = k mod m illustrate the result of inserting these keys using linear probing, using quadratic probing with $c_1 = 1$ and $c_2 = 3$ and using double hashing with $h_2(k) = 1 + (k \bmod (m - 1))$.

Solution: Usng Linear probing the final state of the hash table would be:

0	22
1	88
2	/
3	/
4	4
5	15
6	28
7	17
8	59
9	31
10	10

Using Quadratic probing, with $c_1 = 1$, $c_2 = 3$, the final state of the hash table would be:

$$h(k, i) = (h'(k) + c_1 * i + c_2 * i^2) \bmod m$$

where $m = 11$ and $h'(k) = k \bmod m$

0	22
1	88
2	/
3	17
4	4
5	/
6	28
7	59
8	15
9	31
10	10

Using double hashing the final state of the hash table would be:

0	22
1	/
2	59
3	17
4	4
5	15
6	28
7	88
8	/
9	31
10	10

/* Insert keys into an array with double hashing technique of collision resolution technique. */

```
#include<stdio.h>
#include<conio.h>
#define MAX 10
int a[MAX];
void dh(int , int[]);
void dhsr(int key, int a[MAX]);
void display(int a[MAX]);
void main()
{
int i, key,ch ;
for(i=0;i<MAX;i ++)
     a[i] = '\0';
do{
     printf("\n\n Program for insertion/searching keys with double hashing ");
     printf("\n**********************************************\n\n");
     printf("\n 1. Insert keys ");
     printf("\n 2. Search key ");
     printf("\n 3. Display keys ");
     printf("\n 4. Exit ");
     printf("\n Select operation ");
     scanf("%d",&ch);
     switch(ch)
     {
     case 1: do{
               printf("\n Enter key value [type -1 for termination] ");
               scanf("%d",&key);
               if (key != -1)
                    dh(key,a);
               }while(key!=-1);
               break;
     case 2:   printf("\n Enter search key value");
               scanf("%d",&key);
               dhsr(key,h);
               break;
     case 3:   display(a);
               break;
               }
```

```
            }while(ch!=4);
}

/* Find the location for a key and insert it */
void dh(int key, int a[MAX])
{
int loc, i = 0;
loc = key % MAX;
while (a[loc] != '\0')
     loc = (loc +++i*(key % (MAX -1))) % MAX;
a[loc] = key;
display(a);
}

/* find the location for a key */
void dhsr(int key, int a[MAX])
{
int loc,i = 0;
loc = key % MAX;
while ((a[loc] != key) && (a[loc] != '\0'))
     loc = (loc +++i*(key % (MAX -1))) % MAX;
if (a[loc] != '\0')
     printf("\n Search successful at index %d",loc);
else
     printf("\n Search unsuccessful ");
}

void display(int a[MAX])
{
int i;
printf("\n List of keys, 0 indicate that the location is empty \n");
for (i = 0;i<MAX;i + + )
     printf(" %d ",a[i]);
}
```

14.10 HASHING WITH CHAINING

This method maintains the chain of elements which have same hash address. We can take the hash table as an array of pointers. Size of hash table can be number of records. Here each pointer will point to one linked list and the elements which have same hash address will be maintained in the linked list. We can maintain the linked list in sorted order and each elements of linked list will contain the whole record with key (see Fig. 14.2).

For **inserting one element,** first we have to get the hash value through hash function which will map in the hash table, then that element will be inserted in the linked list.

Searching a key, is also same, first we will get the hash key value in hash table through hash function, then we will search the element in corresponding linked list.

Deletion a key, contains first search operation then same as delete operation of linked list.

The main advantage of this hashing is saving of memory space and perfect collision resolution. Here hash table contains only pointers which point to the linked list, containing records. So there is no wastage of memory space because of empty hash table and the hash table size can be taken as number of records. Hash table space is also minimised because records are stored in linked list. So memory space can be allocated for keeping records at the time of need. From collision resolution point of view, it gives perfect solution. The elements which have same hash address will be in same linked list. So we should use good hash function which distributes elements on different hash address. Basically linked lists will be small, so searching will be fast.

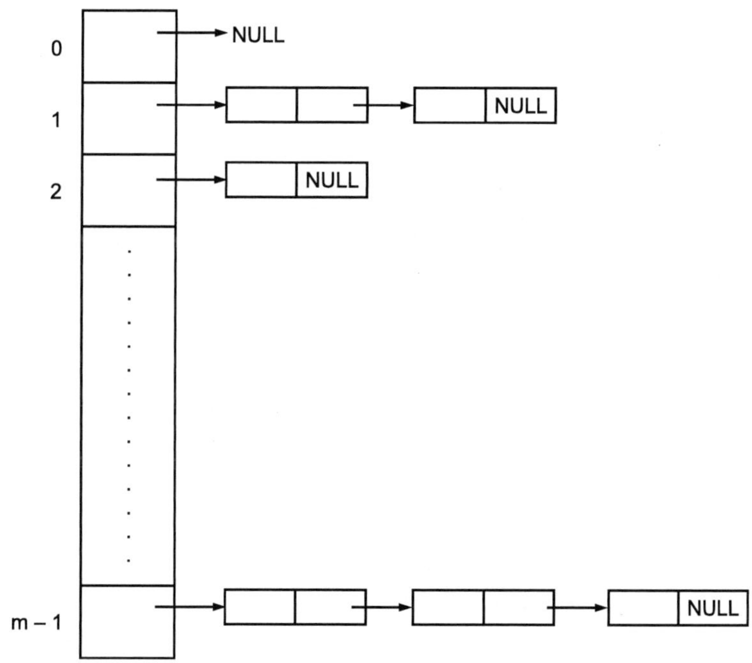

Fig. 14.2.

The main disadvantage is also wastage of memory space. In case where records are very small or contains only key then link part for every element in linked list and pointer space in hash table will be wastage of memory space.

EXAMPLE 14.7. Let us consider the insertions of elements 5, 28, 19, 15, 20, 33, 12, 17, 10 into a chained-hash table. Let us suppose the hash table has 9 slots and the hash function be h(k) = k mod 9.

Solution: The initial state of the chained-hash table

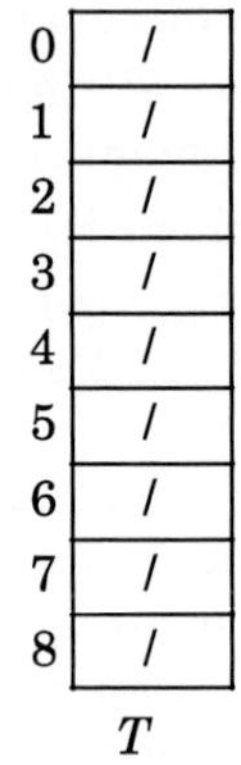

Insert 5:

$$h(5) = 5 \bmod 9 = 5$$

Create a linked-list for *T*(5) and store value 5 in it.

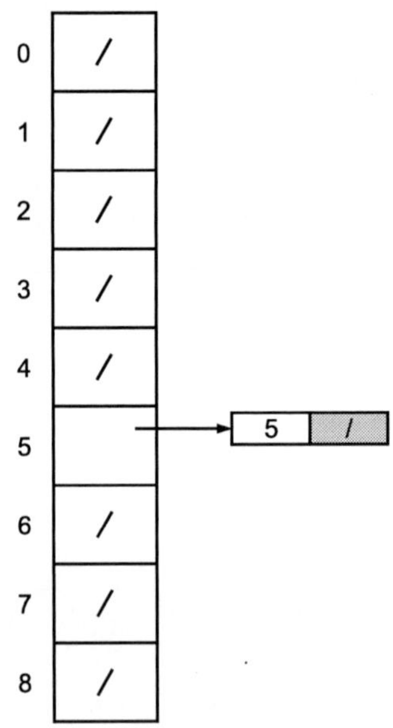

Fig 14.3.

Similarly **insert 28**, h(28) = 28 mod 9 = 1. Create a link-list for T(1) and store value 28 in it.

Now **insert 19** h(19) = 19 mod 9 = 1. Insert value 19 in the slot T(1) in the beginning of the link-list.

Now **insert 15** h(15) = 15 mod 9 = 6. Create a link list for T(6) and store value 15 in it. Similarly, **insert 20**, h(20) = 20 mod 9 = 2 in T(2). **Insert 33**, h(33) = 33 mod 9 = 6, in the beginning of the linked list for T(6). Then,

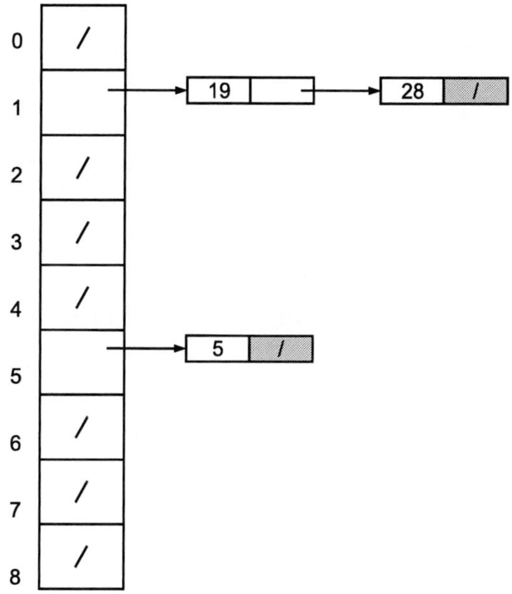

Fig. 14.4.

Insert 12, *i.e.*, h(12) = 12 mod 9 = 3 in T(3)

Insert 17, *i.e.*, h(17) = 17 mod 9 = 8 in T(8)

Insert 10, *i.e.*, h(10) = 10 mod 9 = 1 in T(1)

Thus, the chained-hash-table after inserting key 10 is:

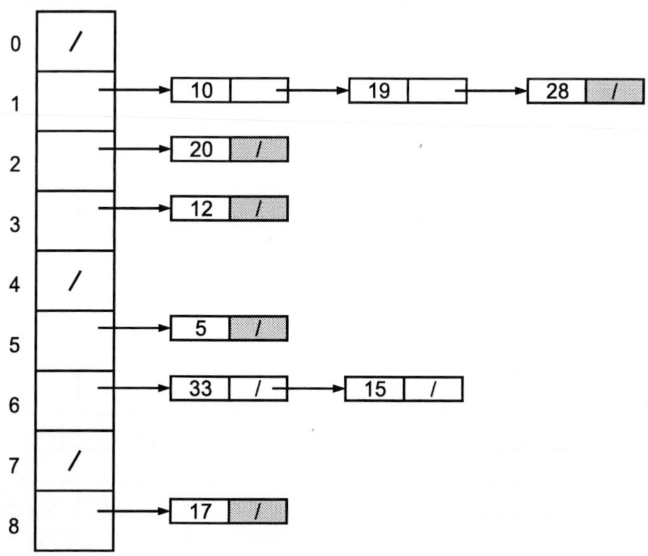

Fig. 14.5.

EXAMPLE 14.8. Consider an open-address hash table with uniform hashing. Give upper bounds on the expected number of probes in an unsuccessful serach and on the expected number of probes in a successful search when the load factor (α) is ¾ and when it is 7/8.

Solution: Given an open address hash table with load factor $\alpha = n/m < 1$, the expected number of probes in an unsuccessful search is at most $1/(1 - \alpha)$, assuming uniform hashing.

$\alpha = \frac{3}{4}$, then the upper bound on the number of probes = $\frac{1}{\left(1-\frac{3}{4}\right)}$ = 4 probes.

$\alpha = \frac{7}{8}$, then the upper bound on the number of probes = $\frac{1}{\left(1-\frac{7}{8}\right)}$ = 8 probes.

Given an open address hash table with load factor (α) = n/m < 1, the expected number of probes in a successful search is at most (1/α) log (1/(1 – α)), assuming uniform, hashing and assuming that each key in the table, is equally likely to be searched for:

$$\alpha = \frac{3}{4}\left(\frac{1}{3/4}\right) \log \frac{1}{1-\frac{3}{4}} = 1.85 \text{ probes}$$

$$\alpha = \frac{7}{8}\left(\frac{1}{0.875}\right) \log \frac{1}{1.875} = 2.38 \text{ probes.}$$

14.11 REHASHING

There are chances of insertion failure when hash table is full. So the solution for this particular case is to create a new hash table with the double size of previous hash table. Here we will use new hash function and we will insert all the elements of the previous hash table. So we will scan the elements of previous hash table one by one and calculate the hash key with new hash function and insert them into new hash table. This technique is called **rehashing**. It ensures the insertion of element in hash table. Let us take a case where hash table size is 13 and we are using linear probing and hash function is H(k) = k mod 13. The elements in the hash table are in this sequences:

7, 18, 48, 10, 36, 25

0	10
1	
2	
3	36
4	25
5	
6	
7	7
8	18
9	
10	43

(*a*)

0	10
1	
2	46
3	36
4	25
5	
6	
7	7
8	18
9	
10	43

(*b*)

Suppose we want to insert the element 46 then the table will be occupied with elements and chances will be for not getting the space for new element. So, we will take new table of size more than double with prime number. The next prime number after double size 22 is 28. So new table will be created of size 23 and all the elements will be inserted with new hash function $H(k) = k \bmod 23$. So the resuilting table will be as:

0	46
1	
2	25
3	
4	
5	
6	
7	7
8	
9	
10	10
11	
12	
13	36
14	
15	
16	
17	
18	18
19	
20	43
21	
22	

This technique is little bit more expensive but it happens very few times. Decision of rehashing can be taken on different conditions like table is occupied more than half, insertion of new element failure or on any given case.

14.12 INDEXED SEARCH TECHNIQUES

This searching technique is useful in searching direct access secondary storage devices. The general strategy of an indexed search is that the key is used to search the index, find the relative record position of the associated data and from there only one access is made to find the data.

14.12.1 Indexed Sequential Search Techniques

The Indexed Sequential Access Method (ISAM) involves considering the disk dependent factors of blocking and track size to build a partial index. ISAM first identifies the region and then within that region, a sequential search is made. Fig. 14.6 shows a one-level indexed sequential file.

The strategy used to conduct an ISAM search is as follows:

(a) Search the main memory directory for a key that is greater than or equal to the target.

(b) Follow the corresponding pointer out to the disk and there sequentially search until we find a match (success) or the key that the directory maintains as the high key within that particular region (failure).

For larger files, it may be advantageous to have more than one level of these directory structures. In this case, a *two level directory structure* may be used.

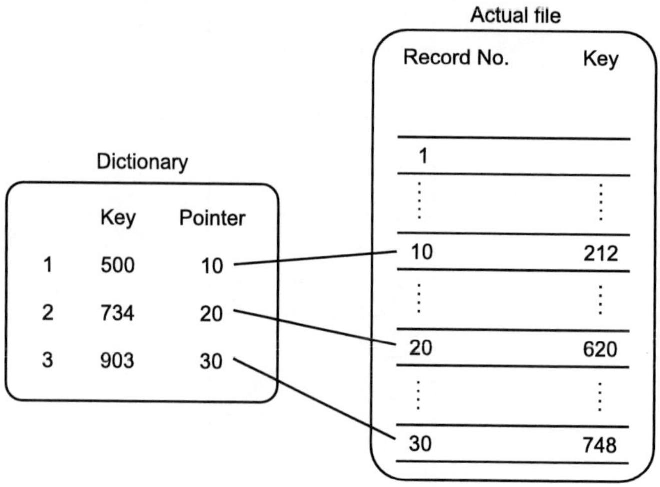

Fig. 14.6. One level indexed sequential file

In a *two-level directory structure* the entire primary directory should be kept in the main memory. The secondary directory should be brought as a single block of the disk file and the data records are stored as some n records per track. The primary directory divides the file into some m regions of records each. The pointer field in the primary directory points a record in the sub-directory instead of the actual file. Here, the search operation is done as follows:

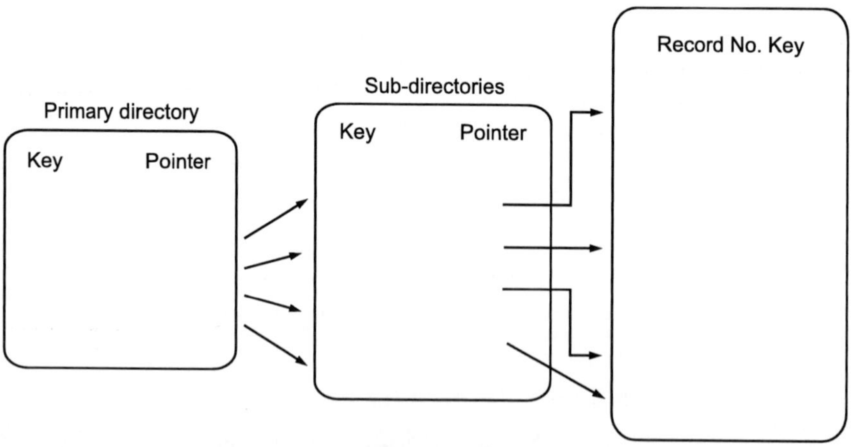

Fig. 14.7. Two level directory structure.

(*a*) The search operation in a primary directory is done in a way to find a key which is greater than or equal to the target. Using the corresponding pointer field, the sub-directory is searched.

(b) In the sub-directory also, the search is made such that the key is greater than or equal to the target. The corresponding pointer field is used to address a particular record in the actual file.

The search efficiency of the indexed sequential file is based on many factors. Some of them are:

1. To what degree the directory structures are able to subdivide the actual file.

2. To what degree the directory structure are able to reside in main memory.
3. The relationship of data records to physical characteristics of the disk such as blocking factors, track size and cylinder size.

This searching method is not suitable for a highly volatile file. It requires that the data records must be stored in a physically increasing or decreasing key order. If there are more records in the overflow area, the search efficiency tends to deteriorate. In that case, ISAM is called **Intrinsically slow Access Method**. To avoid the deterioration problem, the actual file should be reorganised into a new file with no overflow. This reorganisation operaton cannot be done dynamically (*i.e.*, it is done sequentially). It is a time-consuming task. The maintenance problems involved with the ISAM structures lead to the developmet of several dynamic indexing schemes.

Exercises

1. Define the following:
   (*i*) Hash Table
   (*ii*) Rehashing
2. Discuss how one can handle overflow in hashing.
3. Write a C function to retrive an item from a hash table with open addressing and linear probing.
4. Write a short notes on Hash functions.
5. What is hashing ? Explain with example. What is collision ? Explain any one collision resolution technique with example.
6. Compare linear search and Binary search algorithms.
7. What do you understand by hashing ? Name two hashing techniques and explain anyone.
8. Discuss the following:
   (*i*) Hash function
   (*ii*) Hash table implementation.
9. Define hash function. State different types of hash function. Give their algorithm and explain them by suitable diagram.
10. Write an algorithm for binary search and discuss its speed compared with linear serach.
11. What is clustering in a Hash table ? Describe two methods for collision resolution.
12. Explain open addressing technique.

INDEX

INDEX

A

B

C

D

Win Prizes

ATTENTION READERS

Win Prizes

We request for your frank assessment regarding various aspects of the book as given below :

DATA STRUCTURES

By: UDIT AGARWAL

Please Fill up the Sheet in Neat and Clean Writing

(i) What topics of your syllabus, important for your examination point of view, are not covered in this book ?

....

....

....

....

(ii) What are the chapters wherein the treatment of the subject matter is not systematic or organised or updated ?

....

....

....

....

(iii) Have you come across misprints/mistakes/factual inaccuracies in the book? Please specify the chapters and the page numbers.

....

....

....

....

(iv) What other books on the same subject have you found/heard better than the present book? Please specify in terms of price and quality

....

....

....

....

(v) Further suggestions and comments for the improvement of this book.

....

....

....

....

(PS: If need be please attach a separate sheet to write your views)

OTHER DETAILS

(i) Who recommended you this book ? (Please tick (✓) in the box near the option relevant to you.

☐ Teacher ☐ Friends ☐ Bookseller

....

....

....

....

(ii) Name of the recommending teacher, his designation and address :

....

....

....

....

(iii) Name and address of the bookseller you purchased the book from :

....

....

....

....

(iv) Name and address of your institution (Please mention the University or Board as the case may be)

....

....

....

....

(v) Your name and complete postal address :

....

....

....

....

(vi) Course you are enrolled in :

....

....

....

....

The best assessment will be awarded each month: The award will be in the form of our publications as decided by the Editorial Board.

Please mail the filled up coupon at your earliest to :

S.K. KATARIA & SONS®
4885/109, Prakash Mahal, Dr. Subhash Bhargav Lane,
Opposite Delhi Medical Association, Daryaganj, New Delhi-110002 (INDIA)
Phone: +91-11-43551243; Mobile: +91-9871775858, 7678425209
e-mail: katariabook@gmail.com, katariabooks@yahoo.com
Website: www.skkatariaandsons.com